AF361805

The Ins and Outs of miRNAs as Biomarkers

The Ins and Outs of miRNAs as Biomarkers

Editors

Giuseppe Iacomino
Fabio Lauria

MDPI • Basel • Beijing • Wuhan • Barcelona • Belgrade • Manchester • Tokyo • Cluj • Tianjin

Editors
Giuseppe Iacomino
Institute of Food Sciences
National Research Council
Avellino
Italy

Fabio Lauria
Institute of Food Sciences
National Research Council
Avellino
Italy

Editorial Office
MDPI
St. Alban-Anlage 66
4052 Basel, Switzerland

This is a reprint of articles from the Special Issue published online in the open access journal *Genes* (ISSN 2073-4425) (available at: www.mdpi.com/journal/genes/special_issues/miRNA_Biomarkers).

For citation purposes, cite each article independently as indicated on the article page online and as indicated below:

LastName, A.A.; LastName, B.B.; LastName, C.C. Article Title. *Journal Name* **Year**, *Volume Number*, Page Range.

ISBN 978-3-0365-7925-2 (Hbk)
ISBN 978-3-0365-7924-5 (PDF)

Contents

About the Editors

Giuseppe Iacomino

G. Iacomino has more than 30 years of experience in the fields of biochemistry, cell biology, molecular biology, cancer biology, and nutritional biochemistry. His research has focused at the molecular level, on different aspects of nutrition, from chemoprevention to diet-related diseases.

His main expertise is in epigenomics, gene expression, and the role of chromatin modifiers in influencing transcriptional activity, cell cycle, and apoptosis. In this regard, the effects of nutrients with chromatin-modifying abilities are emerging as crucial issues driving the dietary impacts on health. His scientific interests also focus on the role of polyamines in governing nuclear organization: his research group discovered for the first time that polyamines self-assemble in the cell nucleus and generate compounds called nuclear polyamine aggregates (NAPs) that interact with genomic DNA and regulate decisive aspects of DNA physiology, such as conformation, protection and packaging.

His research activity has also focused on microRNA expression and epigenetics, looking for the differential signature of microRNAs in obesity and metabolic diseases.

In addition, his interests include foods and food products that improve consumer health, bioaccessibility, bioprocessing and transport using simulated digestion models.

Fabio Lauria

His team's mission is focused on the understanding of the multifactorial causes of nutrition-related diseases, on the discovery of new biomarkers to study the association between diet and health, and on the dissemination of results.

He is fully involved in large European studies and closely collaborates with outstanding international research groups. His main research activity is focused on the study of the interaction between genetic and nutritional factors in the pathogenesis of diseases associated with nutrition; on the identification of the determinants of eating behaviour in children, adolescents and adults; on the study of the interaction between nutrition and gut microbiota in adults and children with overweight, celiac disease or food allergies in the context of personalized nutritional programs; on the dissemination of knowledge in nutrition field; and on the development of innovative strategies for public health.

Preface to "The Ins and Outs of miRNAs as Biomarkers"

Since the first discovery of a noncoding RNA in C. elegans in 1993, knowledge of miRNAs has continued to the present with thousands of papers clearly defining their crucial role in a wide range of disorders and physiological processes. miRNAs are major components of the cellular epigenetic machinery that act as specific gene silencers through base pairing with a target mRNA. miRNAs have emerged as peacemakers of body homeostasis, playing critical roles in the pathology of a variety of disorders, including cancer. The discovery of circulating miRNAs in plasma and other body fluids has underscored their potential as intercellular signalling molecules and indicators of disease. In addition, data from clinical trials have predicted that anti-miR and miR-mimic compounds constitute a prospective class of drugs for therapeutic applications in next-generation medicine, and numerous biopharmaceutical companies are currently involved in this exciting opportunity.

This Special Issue aims to provide a broad overview of recent knowledge on several open questions and emerging options in the field of miRNA research as early diagnostic tools, providing evidence for the efficacy and implications of this class of small molecules that play a big role in regulating gene expression.

Giuseppe Iacomino and Fabio Lauria
Editors

Editorial

The Landscape of Circulating miRNAs in the Post-Genomic Era

Fabio Lauria and **Giuseppe Iacomino ***

Institute of Food Sciences, National Research Council, Via Roma, 64, 83100 Avellino, Italy; flauria@isa.cnr.it
* Correspondence: piacomino@isa.cnr.it; Tel.: +39-0825299431

Keywords: microRNA; circulating miRNA; gene expression; theranostic biomarkers

Citation: Lauria, F.; Iacomino, G. The Landscape of Circulating miRNAs in the Post-Genomic Era. *Genes* **2021**, *13*, 94. https://doi.org/10.3390/genes13010094

Received: 8 November 2021
Accepted: 15 December 2021
Published: 30 December 2021

Publisher's Note: MDPI stays neutral with regard to jurisdictional claims in published maps and institutional affiliations.

In the past decade, there has been an epochal change in the way that diseases are investigated and diagnosed. The advent of high-throughput technologies has placed us firmly into the post-genomic era. Omics investigation techniques have advanced quickly in the field of personalized medicine, exerting great influence on the progress of epigenetics research [1]. In essence, every step in gene expression flow is finely controlled, and the discovery of small non-coding RNAs (sncRNAs) has added new contributors to the already well-specialised supervisory mechanisms.

Based on their methods of operation, assembly, and structure, sncRNAs can be classified as small interfering RNAs, PIWI-interacting RNAs, endogenous small interfering RNAs, promoter-associated RNAs, small nucleolar RNAs, and microRNAs (miRNAs). These last molecules are sncRNAs (20–24 nucleotides in length), which act as post-transcriptional regulators of gene expression. Up until now, 2599 different human miRNAs have been identified and deposited in the miRTarBase [2]. Since the first discovery of a noncoding RNA in *C. elegans* in 1993 [3], the knowledge of miRNAs has progressed, with thousands of papers clearly defining their crucial role in a wide range of disorders and physiological processes. These molecules typically act as specific gene silencers by base-pairing to the 3' untranslated sequence of a target mRNA, but they have also been reported to bind anywhere along the mRNA sequence. Interestingly, miRNAs work by either suppressing translation or by affecting the stability and degradation of the target mRNA. The nucleotides in positions 2–8 of a mature miRNA have been designated as the "seed sequence" because they are required for base pairing with a target mRNA. Moreover, non-canonical seed-like consensus sequences also mediate a portion of miRNA-target recognition. Sequence identity in the seed region has also been employed to group miRNAs into "families" that share the common ability to recognize clusters of target mRNAs. While a limited number of miRNAs have a typical tissue-specific localization, the majority of them exhibit broad tissue distribution.

A certain miRNA can target a set of transcripts at the same time; moreover, a single mRNA typically includes numerous interaction sites for different miRNAs, thus generating complex regulatory circuits. Even though a specific miRNA commonly exerts limited activity on a particular target, its activity, which influences multiple transcripts in a signalling network, leads to substantial cumulative effects. As a result, Metazoan miRNAs have been denoted as the "sculptors" of the cell transcriptome [4], and in agreement with scientific evidence, endogenous miRNAs are capable of affecting the expression of up to 40–60% of mouse and human genes. However, with millions of hypothetically conceivable miRNA–mRNA interactions, the human miRNAs targetome is far from being determined [5].

Given the ubiquity of nucleases, traditional views suggest that RNAs are not stable molecules in extracellular settings. Conversely, in the recent past, several miRNAs have been shown to retain unexpectedly high concentrations of plasma and other bodily fluids. Circulating miRNAs (c-miRNAs) are not naked molecules and are commonly secreted from cells that are arranged into complexes with proteins, microvesicles, or exosomes.

Interestingly, extracellular miRNAs can be actively secreted and transferred into recipient cells where they may influence the translation of the target gene/s, playing a role in intercellular communication [6]. Despite this, most of the functional significance of c-miRNAs is unknown.

The discovery and validation of biomarkers in prognosis, diagnosis, drug discovery, treatment, and prevention play a key role in the post-genomic era. Therefore, the prognostic and diagnostic potential of c-miRNAs as non-invasive biomarkers has been advocated in the health of both humans and animals [7,8]. In this regard, the identification of about 300 extracellular miRNAs in several biological fluids has emphasized their potential in biomarker development in next-generation medicine [7]. Accordingly, the dysregulation of miRNAs has been established to reflect the status and functions of various tissues and organs, most likely contributing to their anomalies [9]. In this context, many studies support the links between miRNAs and the physiopathology of a variety of processes, including mitochondrial failure, cardiovascular diseases, neurodegenerative disorders, cancer, and other conditions, underlining their relevance for personalized medicine applications. Moreover, the role of miRNAs in maintaining an energy balance and metabolic equilibrium in living organisms, which is achieved by regulating different metabolic pathways, has been established [10]. Likewise, several differentially expressed miRNAs have been characterized in the treatment of infectious diseases [11]. Several studies have also recognized that miRNA expression is altered in response to nutrition and lifestyle variables [12], and numerous microRNA families have been interconnected to different dietary regimes. Accordingly, the recent nutrimiRomics discipline was inspired by the influence of diet on miRNA levels and the sub-sequential effects on gene expression and health-related conditions. Increasing evidence also suggests that dietary miRNAs may survive digestion [13]; nevertheless, the role and capability of food-related miRNAs to modulate cross-species messengers remain puzzling.

Of note, the majority of the 139,968 papers that have been written on miRNAs research that are listed in PubMed (November 2021) show their implications in human diseases, emphasizing that miRNAs are highly selective health-related tools [14], with the bulk of these studies focusing on c-miRNAs as theranostic cancer biomarkers [15–17]. However, the number and variety of the studies reveal the extraordinary complexity of the topic, and, at present, there is still a lack of consensus over specific miRNA profiles that are useful for the early identification of cancer cells in vivo. Most of the working limits are related to approaches that would be useful for c-miRNA recognition, which must be strictly specific and capable of identifying limited amounts of target molecules. Given the conceivable presence of unwanted contaminant inhibitors, these procedures should be strictly specific and capable of identifying limited amounts of target molecules. The inconsistencies among studies are correlated, at least in part, to dissimilarities in the extraction and identification procedures, in the experimental design, and data normalization, with these limitations supporting the need for the development of well-standardized protocols and operational strategies [18,19].

Based on technological and conceptual progress in the field of personalized medicine, new therapeutic opportunities are presented in clinical medicine by molecular drugs that have been inspired by the target sequence. In this context, recent trials have predicted that both anti-miR and miR-mimics compounds can be used as drugs for various therapeutic applications, and numerous biopharmaceutical companies are now involved in this business [20]. However, caution is be advised based on lessons that have been learned in the past by the pioneering trials with antisense oligonucleotides. These synthetic molecules, which act as specific gene silencers by base-pairing to the target RNA, have only offered a limited number of therapies in clinical applications despite expectations and after 30 years of research [21]. Accordingly, the development of miRNA-based therapeutics is in its infancy, and many concerns remain to be resolved, including appropriate delivery systems, stability, bioavailability, etc. [22].

This Special Issue aims to provide a broad overview of the recent knowledge on several open questions and new opportunities in the field of miRNA research and the use of miRNAs as early diagnostic tools, providing evidence of the effectiveness and implications of this class of small molecules that plays a big role in gene expression control.

Author Contributions: Conceptualization, writing—review and editing, F.L. and G.I. All authors have read and agreed to the published version of the manuscript.

Funding: This research received no external funding.

Institutional Review Board Statement: Not applicable.

Informed Consent Statement: Not applicable.

Conflicts of Interest: The authors declare no conflict of interest.

References

1. Cavalieri, V. The Expanding Constellation of Histone Post-Translational Modifications in the Epigenetic Landscape. *Genes* **2021**, *12*, 1596. [CrossRef]
2. Huang, H.Y.; Lin, Y.C.D.; Li, J.; Huang, K.Y.; Shrestha, S.; Hong, H.C.; Huang, H.D. miRTarBase 2020: Updates to the experimentally validated microRNA-target interaction database. *Nucleic Acids Res.* **2020**, *48*, D148–D154. [CrossRef] [PubMed]
3. Lee, R.C.; Feinbaum, R.L.; Ambros, V. The C. elegans heterochronic gene lin-4 encodes small RNAs with antisense complementarity to lin-14. *Cell* **1993**, *75*, 843–854. [CrossRef]
4. Bartel, D.P. Metazoan MicroRNAs. *Cell* **2018**, *173*, 20–51. [CrossRef]
5. Kern, F.; Krammes, L.; Danz, K.; Diener, C.; Kehl, T.; Küchler, O.; Fehlmann, T.; Kahraman, M.; Rheinheimer, S.; Aparicio-Puerta, E.; et al. Validation of human microRNA target pathways enables evaluation of target prediction tools. *Nucleic Acids Res.* **2021**, *49*, 127–144. [CrossRef]
6. Thomou, T.; Mori, M.A.; Dreyfuss, J.M.; Konishi, M.; Sakaguchi, M.; Wolfrum, C.; Rao, T.N.; Winnay, J.N.; Garcia-Martin, R.; Grinspoon, S.K.; et al. Adipose-derived circulating miRNAs regulate gene expression in other tissues. *Nature* **2017**, *542*, 450–455. [CrossRef]
7. Witwer, K.W. Circulating MicroRNA Biomarker Studies: Pitfalls and Potential Solutions. *Clin. Chem.* **2015**, *61*, 56–63. [CrossRef]
8. Miretti, S.; Lecchi, C.; Ceciliani, F.; Baratta, M. MicroRNAs as Biomarkers for Animal Health and Welfare in Livestock. *Front. Veter. Sci.* **2020**, *7*, 578193. [CrossRef] [PubMed]
9. Iacomino, G.; Siani, A. Role of microRNAs in obesity and obesity-related diseases. *Genes Nutr.* **2017**, *12*, 1–16. [CrossRef]
10. Deiuliis, J.A. MicroRNAs as regulators of metabolic disease: Pathophysiologic significance and emerging role as biomarkers and therapeutics. *Int. J. Obes.* **2016**, *40*, 88–101. [CrossRef]
11. Tribolet, L.; Kerr, E.; Cowled, C.; Bean, A.G.D.; Stewart, C.R.; Dearnley, M.; Farr, R.J. MicroRNA Biomarkers for Infectious Diseases: From Basic Research to Biosensing. *Front. Microbiol.* **2020**, *11*, 1197. [CrossRef] [PubMed]
12. Slattery, M.L.; Herrick, J.S.; E Mullany, L.; Stevens, J.R.; Wolff, R.K. Diet and lifestyle factors associated with miRNA expression in colorectal tissue. *Pharm. Pers. Med.* **2016**, *10*, 1–16. [CrossRef]
13. Liang, G.; Zhu, Y.; Sun, B.; Shao, Y.; Jing, A.; Wang, J.; Xiao, Z. Assessing the survival of exogenous plant microRNA in mice. *Food Sci. Nutr.* **2014**, *2*, 380–388. [CrossRef] [PubMed]
14. Keller, A.; Meese, E. Can circulating miRNAs live up to the promise of being minimal invasive biomarkers in clinical settings? *Wiley Interdiscip. Rev. RNA* **2016**, *7*, 148–156. [CrossRef] [PubMed]
15. Calin, G.; Croce, C.M. MicroRNA Signatures in Human Cancers. *Nat. Rev. Cancer* **2006**, *6*, 857–866. [CrossRef] [PubMed]
16. Matsuzaki, J.; Ochiya, T. Circulating microRNAs and extracellular vesicles as potential cancer biomarkers: A systematic review. *Int. J. Clin. Oncol.* **2017**, *22*, 413–420. [CrossRef]
17. Torres, R.; Lang, U.E.; Hejna, M.; Shelton, S.J.; Joseph, N.M.; Shain, A.H.; Yeh, I.; Wei, M.L.; Oldham, M.C.; Bastian, B.C.; et al. MicroRNA Ratios Distinguish Melanomas from Nevi. *J. Investig. Dermatol.* **2020**, *140*, 164–173. [CrossRef]
18. Galvão-Lima, L.J.; Morais, A.H.F.; Valentim, R.A.M.; Barreto, E.J.S.S. miRNAs as biomarkers for early cancer detection and their application in the development of new diagnostic tools. *Biomed. Eng. Online* **2021**, *20*, 1–20. [CrossRef] [PubMed]
19. Seitz, H. Issues in current microRNA target identification methods. *RNA Biol.* **2017**, *14*, 831–834. [CrossRef] [PubMed]
20. Chakraborty, C.; Sharma, A.R.; Sharma, G.; Lee, S.-S. Therapeutic advances of miRNAs: A preclinical and clinical update. *J. Adv. Res.* **2021**, *28*, 127–138. [CrossRef]
21. Dhuri, K.; Bechtold, C.; Quijano, E.; Pham, H.; Gupta, A.; Vikram, A.; Bahal, R. Antisense Oligonucleotides: An Emerging Area in Drug Discovery and Development. *J. Clin. Med.* **2020**, *9*, 2004. [CrossRef] [PubMed]
22. Rupaimoole, R.; Slack, F.J. MicroRNA therapeutics: Towards a new era for the management of cancer and other diseases. *Nat. Rev. Drug Discov.* **2017**, *16*, 203–222. [CrossRef] [PubMed]

Review

miRNAs: The Road from Bench to Bedside

Giuseppe Iacomino

Institute of Food Sciences, National Research Council, Via Roma, 64, 83100 Avellino, Italy; giuseppe.iacomino@isa.cnr.it; Tel.: +39-0825299431

Abstract: miRNAs are small noncoding RNAs that control gene expression at the posttranscriptional level. It has been recognised that miRNA dysregulation reflects the state and function of cells and tissues, contributing to their dysfunction. The identification of hundreds of extracellular miRNAs in biological fluids has underscored their potential in the field of biomarker research. In addition, the therapeutic potential of miRNAs is receiving increasing attention in numerous conditions. On the other hand, many operative problems including stability, delivery systems, and bioavailability, still need to be solved. In this dynamic field, biopharmaceutical companies are increasingly engaged, and ongoing clinical trials point to anti-miR and miR-mimic molecules as an innovative class of molecules for upcoming therapeutic applications. This article aims to provide a comprehensive overview of current knowledge on several pending issues and new opportunities offered by miRNAs in the treatment of diseases and as early diagnostic tools in next-generation medicine.

Keywords: miRNA; circulating miRNA; theranostic biomarkers; RNA-based therapeutic; personalized medicine; translational medicine

Citation: Iacomino, G. miRNAs: The Road from Bench to Bedside. *Genes* **2023**, *14*, 314. https://doi.org/10.3390/genes14020314

Academic Editor: Mark Boldin

Received: 9 January 2023
Revised: 20 January 2023
Accepted: 21 January 2023
Published: 25 January 2023

1. General Concepts

Recent improvements in omics research have accelerated advances in personalized medicine with a driving impact on epigenetic research. The term epigenetics refers to chemical changes in chromatin, inherited during cell division, that do not affect DNA sequence and that include hundreds of posttranslational changes in chromatin chemical composition, such as acetylation, phosphorylation, methylation, and ubiquitination [1,2].

Gene expression is an outstandingly controlled process, and the discovery of small noncoding RNAs (sncRNAs) has provided outstanding contributors to its well-specialized supervisory systems. Most synthesized RNAs are ribosomal ribonucleic acid (rRNA), transfer ribonucleic acid (tRNA), and noncoding RNAs, such as long noncoding RNAs (lncRNAs), circular RNAs (circRNAs), and sncRNAs [3]. sncRNAs are classified as small interfering RNAs, PIWI-interacting RNAs, endogenous small interfering RNAs, promoter-associated RNAs, small nucleolar RNAs, and microRNAs (miRNAs), based on their structure, assembly, and operational modes [4]. The latter have prospects as new therapeutics, with the RNA itself acting as a target or drug.

miRNAs are sncRNAs with a length of 20–24 nucleotides that control gene expression posttranscriptionally [5]. A distinct miRNA species acts by regulating one or multiple transcripts at the same time (up to hundreds), and a single mRNA species often exhibits multiple interaction sites for various miRNAs, thus generating complex regulatory circuits. As a result, even while single miRNAs commonly have a restrained effect on a particular target, their action produces cumulative effects by negatively influencing different transcripts in a signalling network. Accordingly, metazoan miRNAs have been dubbed the sculptors of the cell transcriptome [6].

These short molecules work as gene silencers by way of base-pairing to a target mRNA in the $3'$ untranslated sequence even though they have been observed to bind anywhere alongside the mRNA sequence. Interestingly, miRNAs can act either by influencing the stability and degradation of the target mRNA or by suppressing its translation. Because they are essential for base pairing with a target mRNA, the nucleotides at positions 2–8 of a mature miRNA have been termed a "seed sequence"; however, a fraction of miRNA-target recognition is also mediated through noncanonical seed-like consensus sequences. In addition, seed-region identity has also been used to classify miRNAs into families that have a common capacity to recognize sets of target mRNAs [7].

Currently, 3012 unique human miRNAs have been recognized and stored in the miRTarBase repository, with a huge, estimated capacity of 4,475,477 potential miRNA–target interactions [8]. Furthermore, several issues have been reported regarding the quality of miRNA annotations in publicly available databases, as well as the reproducibility of microRNA studies [9]. Comprehensively, endogenous miRNAs have been reported to regulate most, if not all, protein-coding genes in mammals. However, with hundreds of conceivably miRNA–mRNA interactions, the human miRNA targetome is far from being determined [10].

The miRNA biogenesis pathway is accountable for the handling of a pre-miRNA to a mature miRNA and has been widely reported [11–13]. Of note, miRNAs are themselves sensitive to transcriptional and posttranscriptional control, and emerging approaches for the direct mapping of RNA modifications in sncRNAs have been proposed, with an emphasis on mass spectrometry and nanopore technologies [14]. As an example, the terminal nucleotidyltransferases (rNTrs) can modulate miRNA sequences by adding nontemplated nucleotides to the 3′-end (tailing). Accordingly, miRNA isoforms with diverse 3′-ends (and possibly functions), known as 3′ isomiRs, have been thoroughly detected by NGS [15]. The editing by adenosine deaminase acting on RNA (ADAR) is a relevant regulatory mechanism acting at the posttranscriptional level and is increasingly found in metazoans [16]. The enzyme induces A-to-I RNA editing modification of double-stranded RNA. Editing leads to changes in miRNA processing, as well as the editing of a target mRNA, thus upsetting the mutual complementarity and driving a microRNA–mRNA interactional redirection, distressing diverse cellular processes. Of note, the editing activity of ADAR is increasingly reported to be enhanced in cancer, but its influence on deriving RNA modifications is still unknown [16].

Excitingly, several endogenous cellular RNAs, such as circular RNAs, long noncoding RNAs, transcribed pseudogenes, and mRNAs, act as natural miRNA sponges, able to firmly interact with and disrupt target miRNAs, thus distressing their controlled networks [17].

Although a small number of miRNAs have a tissue-specific distribution, most reveal a broad tissue localization. Given the pervasive presence of nucleases, it is assumed that RNA molecules are prone to rapid degradation in the extracellular environment. Nevertheless, miRNA molecules have been found to retain unexpected stability in plasma and other physiological fluids since they can be packaged from cells in the form of microvesicles and exosomes or protein complexes, such as Argonaute 2 (AGO-2) (taking part in the RNA silencing complex), nucleophosmin-1 (NPM-1) (an RNA-binding protein involved in ribosome nuclear export), or high-density lipoproteins. These forms are able to be actively released into the extracellular space and reach recipient cells, where they can influence the translation of target genes, thus defining an explicit role of miRNAs in cell–cell communication [18]. Accordingly, the discovery of about 300 circulating miRNAs (c-miRNAs) has highlighted miRNAs' potential as intercellular signalling molecules and disease biomarkers [19] since a dysregulated expression of miRNA can unsettle tightly controlled RNA networks in tissues and organs, potentially contributing to abnormalities [20]. Nevertheless, most of the functional significance of c-miRNAs remains unknown to date, and the gap between discovery and function has yet to be bridged.

Many studies have found correlations between altered levels of miRNA and the physiopathology of numerous processes including mitochondrial [21], cardiovascular [22], neurodegenerative [23], immune disease [24], inflammatory [25], rare genetic disorders, and more; most of these studies have focused on miRNAs as biomarkers of cancer [26–30], emphasizing their relevance as personalized theranostic factors. Interestingly, miRNAs are not only prospective markers of the onset and progression of neoplastic disease, but also indicators of responses to cancer therapy, as well as their relationship to drug resistance and the modulation of responses to cancer treatment [31].

By controlling crucial metabolic processes, miRNAs have also been found to play a role in energy balance and the oversight of metabolic pathways in living organisms [20,32–38]. Furthermore, the treatment of infectious diseases has been linked to several miRNAs [39]. Yet, it has been discovered that miRNA expression changes in response to dietary and lifestyle factors [40]; consequently, the recent nutrimiRomics research focuses on the impact of diet on miRNA levels in the human body, as well as their downstream impact on gene expression and subsequent health outcomes. Finally, there is growing evidence that dietary miRNAs may survive digestion [41]. However, the roles and capabilities of food-related miRNAs in modulating interspecies RNA are still an enigma.

2. Technological Advances Offer New Opportunities for miRNAs in Translational Medicine

Since the first report of a noncoding RNA identified in *C. elegans* in 1993 [42], the bulk of the 153,210 papers on miRNA research listed in PubMed (January 2023) reveals their critical impact on human diseases. The substantial translational interest is confirmed by searching Google's patent database; a search for the keywords "microRNA" and "biomarker" yielded 56,021 results (as of January 2023). The quantity and diversity of studies are evocative of the astonishing complexity and limitations of miRNA research that is increasingly projected toward next-generation medicine. According to the unusual levels of miRNAs found in some unhealthy conditions, novel molecular diagnostic strategies are expected, deeply encouraging researchers and producers to focus on miRNAs as relevant noninvasive disease biomarkers. However, so far, there is no agreement on specific miRNAs suitable for early disease detection, even in crucial topic areas such as in vivo cancer research [43]. Current limitations in this discovery field are related to c-miRNA recognition methods, which must be extremely specific and able to recognize small quantities of target molecules, while also taking into account the presence of unwanted contaminants and inhibitors that could interfere with downstream analytical methodologies. Differences in all the critical steps used for miRNA extraction, detection, data normalization, validation, target gene identification, and experimental design are responsible for at least some of the discrepancies between the different studies [44]. Overall, these constraints require the timely development of standard procedures and well-recognized guidelines to accurately decipher the power and versatility of miRNAs in molecular diagnostics [45,46].

Furthermore, the prompt increase of RNA drugs in recent research and clinical development, driven in part by the success of messenger RNA vaccines in the fight against the SARS-CoV-2 pandemic, has spurred the pursuit of mRNA-based drugs for treating other conditions.

Therapeutic agents are classically based on synthetic small molecules, monoclonal antibodies, or large proteins. Traditional drugs may fail to hit intended therapeutic targets due to the inaccessibility of active sites in the target's three-dimensional structure. RNA-based therapies can offer an exceptional opportunity to potentially reach any relevant target with therapeutic purposes. Additionally, techniques based on nucleic acid technologies can avoid the need for laborious synthesis procedures. Yet, the recognition sequence can be quickly revised and adapted to the target. Nevertheless, RNA-based therapies have specificity problems that carry the risk of side effects [47,48]; in addition, RNA-based drugs' susceptibility to degradation may determine poor pharmacodynamics [47,48]. A number of these issues can be mitigated by chemical changes in the structure of the synthetic polymer (RNA or DNA). Of note, RNA-based drugs are usually larger in size than small-molecule therapies and carry electrical charges, which makes their intracellular delivery in their native form difficult.

In this context, the therapeutic potential of miRNA treatments is receiving attention in clinical trials of almost all human diseases [43,49]. Given the ability of miRNAs to target specific mRNAs, inhibitors based on the sequence of overexpressed miRNAs can be used prospectively in the form of anti-miRs to lower elevated levels of miRNAs and, in turn, restore their downregulated transcripts [50]. This opportunity for medically-based interventions is closely linked to the pioneering use of antisense molecules, the first molecular drugs inspired by the target sequence [51]. These synthetic tools have a sequence complementary to a specific mRNA whose levels they can modulate [52]. Anti-miRs are typically based on first-generation antisense single-stranded oligonucleotides (ASOs), or their chemically modified forms as blocked nucleic acids (LNAs), which are opportunely designed to recognize target mRNAs. Yet, anti-miRs with a 2′-O-methoxyethyl substitution are classified as antagomiRs [53]. Furthermore, miRNA mimics are chemically synthesized, double-stranded small RNA molecules which mimic mature endogenous miRNAs (miRNA replacement therapy) after transfection into cells, the action of which is aimed at replacing downregulated or missing miRNA expression [54].

The delivery of virus-mediated miRNA-based therapies has shown great success in animal models, where adenoviruses have been successful in delivering both anti-miR and miRNA mimics. However, despite the high efficiency of virus-based miRNA delivery systems, several concerns continue, including toxicity, immunogenicity, and insert-size constraints [55]. To overcome these difficulties, over the years, nonviral approaches for the production and delivery of miRNA-based drugs have emerged. Substantial improvements in designing, synthesis, binding affinity, stability, and target modulation effects of both miRNA mimics and anti-miRs have been accomplished through chemical changes to the nucleotide backbone since a major challenge for RNA-based therapeutic strategies is the possible degradation of oligonucleotides by RNases in extracellular and endocellular compartments. Accordingly, the oligonucleotide chemistry has been widely altered by modifying the nucleotides, introducing base modifications, modifications on ribose moiety, and modifications to the phosphate group in the sugar−phosphate backbone (Figure 1). As an example, by introducing methylation (2′-O-methoxyethyl), locked nucleic acids (LNAs), a nucleic acid analogue that contains a 2′-O-methoxyethyl, 2′-fluoro and 2′,4′-methylene bridge, peptide nucleic acid (PNA) a synthetic polymer, an analogue to DNA and RNA, in which the sugar–phosphate backbone has been substituted by a unit of N-(2-aminoethyl) glycine, adding phosphorothioate (PS)-like groups (by replacing the nonbridging oxygen in the phosphate group with sulphur) [51], and others [56]. Overall, these changes have been shown to increase the stability of oligonucleotides by making internucleotide linkages more resistant to the degradation of nucleases, and preserving both the ability to activate RNase H (the enzyme responsible for mRNA target cleavage) and their function in suppressing the target gene. In this research field, the diverse anti-miRNA oligonucleotide chemistries were also directly compared, and the in vitro results showed that the combination of 2′-O-methyl and LNA with phosphorothioate ends was about 10 times more active than modification alone, and 2 times as effective as the 2′-O-methyl with LNA changes [57]. Overall, up to now, most efforts to develop therapies based on chemically modified miRNAs have been directed toward the development of anti-miRs using LNA chemistry, whereas currently commercially available miRNA mimics are often modified by methylation [58,59].

Another strategic approach facilitating RNA-based therapeutics in clinical application consists of encapsulating miRNA mimics or anti-miRs in carrying vehicles to confer to them protection from nuclease.

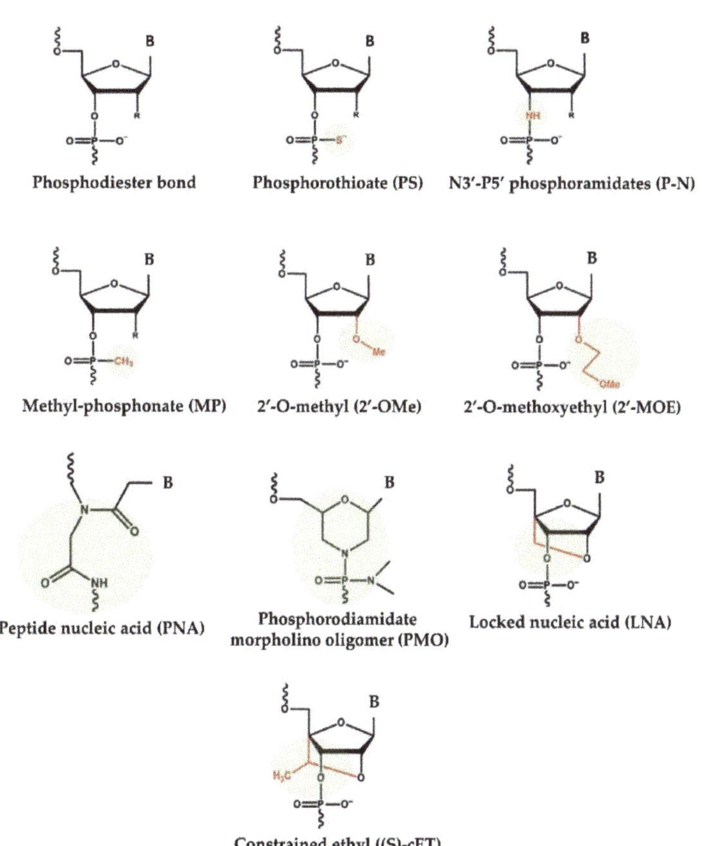

Figure 1. Chemical changes of oligonucleotides used in main clinical trials.

Therapeutic miRNAs include negatively charged polymers that cannot directly cross cell membranes. To get to their intended targets, they need appropriate formulations, including carriers as well as chemical modifications. Accordingly, delivery systems must shield the therapeutic RNAs from serum nucleases, avoid immune-system interference, escape unintended interactions with serum proteins, and prevent renal clearance when given systemically. Previous studies have shown that therapeutic RNAs administered locally or topically can circumvent complications associated with systemic administration, have higher bioavailability, and have a reduced clearance as compared to those administered systemically; nevertheless, this delivery opportunity is restricted to tumours, mucous membranes, eyes, and skin [60]. In this context, the subcutaneous administration of naked mRNA, as compared with mRNA-loaded nanoparticles, leads to a more effective translation of the protein [61].

The nanocarrier encapsulation strategy can both protect and deliver the drug to recipient cells, while biological obstacles, such as immunogenicity and nuclease, are often approached by chemically modifying the nucleotide structure [62]. Various transport methods are increasingly being used to improve bioavailability, including liposomes and biodegradable polymers. In this context, liposomes, long used as immunological adjuvants in vaccination, adequately fulfil this role [63].

Lipid nanoparticles are the most commonly used nonviral delivery technology for nucleic-acid-based drugs and vaccines [64]. They are composed of complexes of anionic nucleic acids and synthetic cationic lipids. The cationic lipids characteristically include amine derivatives, such as primary, secondary, and tertiary amines, in addition to quaternary ammonium, amidinium salts, and various amine combinations [65]. Furthermore, guanidine, imidazole groups (pyridinium, piperazine), and amino acids, including tryptophan, lysine, arginine, and ornithine, have also been employed [66]. A comprehensive list of commonly used cationic lipids in drug formulations was recently reviewed [66]. The advantages of lipid-based transfer systems consist of ease of assembly, biodegradability, the ability to shield nucleic acids entrapped by nucleases, the protection of renal clearance, the ability to promote cellular uptake, and the avoidance of endosomes. Additional carriers with biodegradability, biocompatibility, and low toxicity are employed. Polymeric nanoparticles are among the most-studied carriers and include synthetic and natural cationic polymers, as well as polyethylenimine (PEI), cyclodextrin, chitosan, and poly (lactic-co-glycolic acid) (PLGA) [67]. Furthermore, dendrimers, carbon nanotubes, peptides, gold nanoparticles, silica-derived nanoparticles, iron-oxide-based magnetic nanoparticles, and other nanoparticles are increasingly being studied for medical interventions (Figure 2) [62].

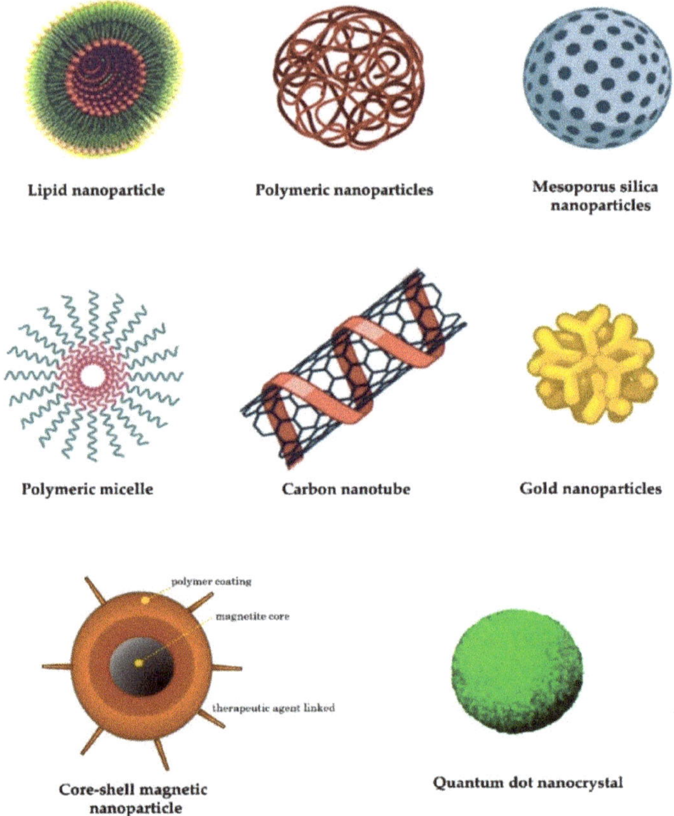

Figure 2. Examples of nanocarriers as drug-delivery systems.

3. RNA Therapeutics

The great expectation in the field of miRNA therapeutics is fully evidenced by the number of studies discovered by querying the keyword "microRNA" in the *Clinical Trials Database* (U.S. National Library of Medicine; http://www.ClinicalTrials.gov, accessed on 20 January 2023, a search of which yielded 1213 results, a result that includes 368 studies in the recruitment or not-yet-recruited phase, 91 ongoing studies, and 414 completed studies.

In this dynamic field, biopharmaceutical companies are increasingly engaged, and ongoing clinical trials point to anti-miR and miR-mimic molecules as an innovative class of drugs for upcoming therapeutic applications in the coming years [68]. As an example, in 2020, the North American market accounted for more than 45% of the comprehensive miRNA market, given the presence of key players with research and development abilities and recognized research infrastructure in terms of genomics, proteomics, oncology, and in silico tools. Interestingly, this market, estimated at USD 225.5 million in the year 2020, is projected to increase at a compound annual growth rate of 13.8% from 2020 to 2027, to reach a revised value of USD 556.1 million [69].

The Lesson of Antisense Technology

The road to development and regulatory approval of emerging miRNA-based therapeutics could be faster than conventional drugs. However, such therapy, as with miRNAs in the biomarker field, is still in the early stages of development, and many aspects, including synthesis, purification, chemical stability, immunogenicity, bioavailability, distribution, metabolism, and body-elimination capacity, need to be further confirmed in preclinical studies [70]. Yet, attention must be paid to the accurate delivery of these novel drugs to target tissues and organs. Furthermore, miRNA-based drug formulations are largely based on the knowledge of antisense drugs previously approved by the U.S. Food and Drug Administration (US FDA) for the treatment of several diseases, some relevant examples of which are given below.

Eteplirsen (Exondys 51®)

It is a drug used to treat some types of Duchenne muscular dystrophy (DMD), caused by specific mutations in the dystrophin gene. The gene deletion determines a reading frameshift causing an early stop codon that prevents translation to functional dystrophin protein by nonsense-mediated RNA decay [71]. Exon skipping may be induced by Eteplirsen, a phosphorodiamidate morpholino oligomer (PMO), that, by selectively binding to exon 51 of the dystrophin pre-mRNA transcript, restores the reading frame phase and allows the assembly of a modified functional protein. The drug only targets specific mutations and can be used to treat about 14% of DMD cases [71].

Fomivirsen (Vitravene)

It was the first antisense drug to be approved by the FDA (in 1998) and by the European Agency for the Evaluation of Medicinal Products (in 1999), and it is used to treat the symptoms of retinitis, the most frequent manifestation of cytomegalovirus (CMV) disease in patients with HIV infection [72]. Fomivirsen is a phosphorothioate oligonucleotide (PS) with potent antiviral properties developed by Ionis Pharmaceuticals and marketed by Novartis CIBA Vision. When injected into a human eye, hybridization of this antisense ODN to CMV mRNA avoids RNA transcription from the immediate-early region-2 gene, thereby limiting viral replication [72]. Furthermore, this agent prevents the adsorption of CMV into host cells via a sequence-independent mechanism.

Givosiran (Givlaari®)

It is a drug used to treat acute hepatic porphyria (AHP) in both adults and adolescents. AHP is an uncommon genetic disorder caused by inborn defects in metabolism, in which the enzyme delta-aminolevulinate synthase 1 (ALAS1), involved in the synthesis of the heme group, is produced in excess. This condition results in a harmful accumulation of porphyrins in the liver and bone marrow that triggers a wide range of highly debilitating symptoms [73]. Givosiran, developed by Alnylam Pharmaceuticals, is a siRNA drug conjugated to GalNAc to accomplish liver-specific distribution, where it leads to a decrease in the ALAS1 enzyme and downstream metabolites [73].

Golodirsen (Vyondys 53™)

It is an antisense PMO oligomer from Sarepta. The drug is used in DMD patients with a definite mutation in the dystrophin gene that can be treated by skipping exon 53. The objective of exon skipping is to allow the muscle tissue to make a shorter form of the modular dystrophin protein. In subjects treated with Golodirsen, levels of dystrophin increased by more than 15 fold [74].

Inotersen (Tegsedi®)

It was originally developed by Ionis Pharmaceuticals in 2018 and marketed globally by Akcea Therapeutics. Inotersen consists of an ASO with PS modifications along the sequence and five 2′-O-MOE nucleotides at the 5′ and 3′ ends, designed to inhibit TTR production [75]. hATTR amyloidosis is an inherited, progressive, and fatal disease for which there are limited treatment options. The mutated protein forms amyloid deposits, which accumulate in tissues and organs throughout the body, including nerves, interfering with their functions. Inotersen reduces transthyretin production, thereby decreasing amyloid formation and alleviating the symptoms of hATTR amyloidosis. Inotersen is effective in treating stage 1 or stage 2 neuropathy in patients with this condition.

Milasen

It is a personalized drug that was explicitly designed for the treatment of a 6-year-old child suffering from neuronal ceroid lipofuscinosis 7 (CLN7) [76]. Neuronal ceroid-lipofuscinosis (NCL) includes a group of genetic, lysosomal storage disorders characterised by progressive intellectual and motor deterioration, vision loss, seizures, and early death [77]. The genomic analysis of the patient recognized the insertion of SVA (SINE-VNTR-Alu) retrotransposon into the MFSD8 gene. The 22-mer ASO contains 2′-O-MOE- and PS-based modifications and targets cryptic splice sites in the MFSD8 pre-mRNA, restoring normal splicing (exons 6–7). Of note, Milasen treatments significantly improved the patient's quality of life.

Nusinersen (Spinraza®)

Spinal muscular atrophy (SMA) represents one of the most common genetic diseases responsible for infantile death due to mutations in the SMN1 gene and the consequent injury of motor neurons. The intronic splicing silencer N1 represents a relevant pharmacological target, and several ASOs are being developed to include exon 7 in the mature mRNA transcript of the SMN2 gene to increase the synthesis of spinal motor neuron proteins. Nusinersen was developed by Biogen and contains both 2′-O-MOE- and PS-based changes [78]. The drug was approved by the FDA in 2016 and shown to be effective in improving motor function and survival in infant- and childhood-onset SMA.

Patisiran (Onpattro®)

It is sold under the brand name Onpattro and was developed by Alnylam. It is the first siRNA-based drug used for the treatment of polyneuropathy in people with hereditary transthyretin (TTR)-mediated amyloidosis, a rare and fatal condition [79]. TTR is a protein involved in the carriage of thyroxin and retinol-binding protein vitamin A complex. Mutations in the TTR gene produce proteins more prone to misfolding and collapse as TTR amyloid fibrils in the extracellular spaces of several organs, finally leading to their dysfunction. In a phase III study, patients treated with Patisiran displayed an 80% reduction of TTR levels in serum.

4. miRNA-Based Therapeutics

The development and commercialization of new diagnostics and therapeutics is a long process. Since the association of miRNAs with human disorders was discovered in 2002, their great potential as next-generation drugs has prompted fields at the interface of biology, chemistry, and medical science to invest heavily in the development and exploitation of miRNA-based therapies. However, despite the considerable number of preclinical studies, the development of both miRNA diagnostic and therapeutic applications is still in its infancy, and only a minor number of miRNA-based therapies have moved on to clinical development. In this scenario, several biotech and pharmaceutical companies have included miRNAs in their line of development to rapidly bring noncoding RNAs from the bench to the patient's bedside, working on two types of drugs: antagomiRs and miRNA mimics. The list of miRNA-based drugs in clinical trials is expanding daily, with several candidates directed against genetic, metabolic, and oncological diseases reaching the clinical-trials stage [56,80].

Miravirsen

It is the first antisense miRNA that passed in clinical trials as a targeted therapy for the treatment of hepatitis C virus (HCV) infection. The drug consists of a 15-mer LNA-PS-modified ASO able to target miR-122, which controls HCV replication in the liver [81]. The long-term efficacy and safety of Miravirsen in patients with chronic HCV infection were evaluated in a phase II study. The drug was supplied to the patient through subcutaneous injection during the study. Treatment with Miravirsen has been established to determine a dose-dependent reduction in viral cargo in chronic HCV subjects. However, severe side effects halted the trial.

RG-012

Regulus Therapeutics has developed an anti-miR-21 for the management of Alport syndrome, an inherited disease caused by mutations in the COL4A3, COL4A4, and COL4A5 genes that result in eye abnormalities, hearing loss, and kidney disease. In this context, glomeruli become less able to function as the disease progresses, resulting in end-stage renal failure. The disease has no approved therapy [82]. miR-21 is a noncoding RNA that negatively controls genes/networks and has been described as being up-regulated in fibrotic kidney disease. Preclinical studies have shown that treatment with an anti-miR-21 significantly attenuates kidney failure by reducing the rate of progression of renal fibrosis. RG-012 has been granted orphan drug status in the United States and Europe.

Cobomarsen (MRG-106)

miR-155 is up-regulated in several lymphoma subtypes, as well as in diffuse large B-cell lymphoma [83]. Cobomarsen is an LNA-based antagomiR targeting miR-155. The molecule developed by miRagen Therapeutics (Viridian Therapeutics Inc) is currently in phase II trials for the treatment of cutaneous T-cell lymphoma, T-cell lymphoma, and leukaemia.

MRG-107

MRG-107 is also being developed by miRagen Therapeutics and, similar to MRG-106, it targets miR-155. The miRNA plays relevant functions in the immune mechanisms and inflammation processes in amyotrophic lateral sclerosis (ALS). In the spinal cords of ALS patients, levels of miR-155 are increased. The inhibition of miR-155 has alleviated symptoms and extended survival in preclinical models of the disease.

MRX34

miR-34a is a natural tumour-suppressor expressed at reduced levels in many tumour types. Clinical studies have reported a negative correlation between reduced miR-34 expression and survival in several tumour types. MRX34 is a liposomal formulation of miR-34a. It is considered to be a first-in-class miRNA mimic for the treatment of a wide range of cancers, including ovarian cancer, colon cancer, cervical cancer, hepatocellular carcinoma, non-small cell lung cancer, and others. The formulation is currently in a phase I clinical trial [84].

RG-101

miR-122 is a liver-specific micro-RNA with relevant functions in this organ's metabolism [20]. This miR is also an essential host factor for HCV. Clinical management with RG-101, an anti-miR-122 ODN conjugated with N-acetylgalactosamine (GalNAc) developed by Regulus Therapeutics, resulted in a significant reduction in viral loads in chronic HCV subjects. The drug is currently in a phase I clinical trial [85].

RGLS4326

Polycystic kidney disease (ADPKD) is an autosomal dominant disease caused by mutations in the PKD1 or PKD2 gene. (Autosomal recessive polycystic kidney disease is less common.) ADPKD represents one of the most frequent monogenetic disorders, and it is the primary genetic cause of end-stage renal disease [86]. Therapeutic opportunities for ADPKD treatment are limited. RGLS4326 is a single-stranded, chemically modified, 9-mer ASO with full complementarity to the seed sequence of miR-17. ASO potentially inhibits the pathologic functions of the miR-17 family in ADPKD [87]. This molecule is in a phase I clinical trial.

MRG-110

It is an antagomiR developed by MiRagen Therapeutics in collaboration with Servier. The drug targets miR-92 to treat ischemic conditions such as heart failure [88]. The phase I clinical trial is currently in the recruitment phase.

Remlarsen (MRG-201)

The drug is intended to mimic the activity of miR-29, a miRNA which downregulates the levels of collagen and other proteins involved in scar formation. Levels of miR-29 family members are typically downregulated in fibrotic diseases [89]. Remlarsen, an LNA RNA mimic, is being investigated to determine if it can limit the formation of fibrous scar tissue when administered by intradermal injection at the site of an excisional wound [90]. The phase II clinical trial is currently underway.

MesomiR

As in other cancer types, miRNA expression in malignant pleural mesothelioma (MPM) exhibits characteristic variations. The MesomiR 1 study is an ongoing phase I study testing the treatment of miR-15/16 (microRNAs implicated as tumour suppressors in MPM) packed in EDV TM nanocells (a 400 nm particle of bacterial origin able to carry a drug cargo) and targeted with EGFR antibodies (TargomiRs) in patients with MPM and refractory lung cancer (NSCLC) [91].

MGN-1374

miRagen Therapeutics is a company developing several miRNA-based drugs, among which ASO MGN-1374 targets the miR-15-family seed region. The 8-mer LNA oligonucleotide is in the preclinical stage for the control of postmyocardial infarction remodelling.

Of note, several other therapeutic miRNAs are under investigation.

5. Conclusions

Advances in understanding the role of miRNAs in pathophysiology, along with optimizing the efficacy and safety of anti-miRs/miRNA-mimic-based strategies, are contributing to the translation of miRNA research into clinical practice. RNA-based therapies have the remarkable potential to theoretically reach any relevant target, offering the opportunity to modulate specific miRNA levels. Currently, several RNA-based drugs have been approved for medical application, and the therapeutic potential of miRNA-based treatments is receiving increasing attention in relation to almost all human diseases, with the list of miRNA-based therapeutics in clinical trials expanding daily. Nevertheless, despite these exciting advancements, miRNA-based drugs have not yet passed phase III clinical trials and received approval from the US FDA for clinical use. In fact, several bottlenecks, such as best target selection, development of more effective modified ASOs, off-target binding, toxicity, immunogenicity, bioavailability, the efficiency of delivery systems to target organs, and vector encapsulation, still limit the possibility of translating miRNA-based approaches into therapeutic realities. Against this backdrop, numerous biotech and pharmaceutical companies are increasingly engaged in including miRNAs in their development pipeline to rapidly bring miRNA-based technology from the bench to the patient's bedside.

Funding: This research received no external funding.

Institutional Review Board Statement: Not applicable.

Informed Consent Statement: Not applicable.

Conflicts of Interest: The authors declare no conflict of interest.

References

1. Iacomino, G.; Medici, M.C.; Napoli, D.; Russo, G.L. Effects of histone deacetylase inhibitors on p55CDC/Cdc20 expression in HT29 cell line. *J. Cell. Biochem.* **2006**, *99*, 1122–1131. [CrossRef]
2. Cavalieri, V. The Expanding Constellation of Histone Post-Translational Modifications in the Epigenetic Landscape. *Genes* **2021**, *12*, 1596. [CrossRef]

3. Nieselt, K.; Herbig, A. Non-Coding RNA, Classification. In *Encyclopedia of Systems Biology*; Dubitzky, W., Wolkenhauer, O., Cho, K.-H., Yokota, H., Eds.; Springer: New York, NY, USA, 2013; pp. 1532–1534. [CrossRef]
4. Hombach, S.; Kretz, M. Non-Coding RNAs: Classification, Biology and Functioning. *Adv. Exp. Med. Biol.* **2016**, *937*, 3–17. [CrossRef]
5. Lauria, F.; Iacomino, G. The Landscape of Circulating miRNAs in the Post-Genomic Era. *Genes* **2022**, *13*, 94. [CrossRef]
6. Bartel, D.P. Metazoan MicroRNAs. *Cell* **2018**, *173*, 20–51. [CrossRef] [PubMed]
7. Bartel, D.P. MicroRNAs: Genomics, biogenesis, mechanism, and function. *Cell* **2004**, *116*, 281–297. [CrossRef]
8. Huang, H.Y.; Lin, Y.C.; Li, J.; Huang, K.Y.; Shrestha, S.; Hong, H.C.; Tang, Y.; Chen, Y.G.; Jin, C.N.; Yu, Y.; et al. miRTarBase 2020: Updates to the experimentally validated microRNA-target interaction database. *Nucleic Acids Res.* **2020**, *48*, D148–D154. [CrossRef] [PubMed]
9. Fromm, B.; Domanska, D.; Hoye, E.; Ovchinnikov, V.; Kang, W.; Aparicio-Puerta, E.; Johansen, M.; Flatmark, K.; Mathelier, A.; Hovig, E.; et al. MirGeneDB 2.0: The metazoan microRNA complement. *Nucleic Acids Res.* **2020**, *48*, D132–D141. [CrossRef]
10. Kern, F.; Krammes, L.; Danz, K.; Diener, C.; Kehl, T.; Kuchler, O.; Fehlmann, T.; Kahraman, M.; Rheinheimer, S.; Aparicio-Puerta, E.; et al. Validation of human microRNA target pathways enables evaluation of target prediction tools. *Nucleic Acids Res.* **2021**, *49*, 127–144. [CrossRef] [PubMed]
11. Ha, M.; Kim, V.N. Regulation of microRNA biogenesis. *Nat. Rev. Mol. Cell Biol.* **2014**, *15*, 509–524. [CrossRef] [PubMed]
12. Lauria, F.; Venezia, A.; Iacomino, G. Circulating MicroRNA (miRNA)s as Biological Markers and Links with Obesity and Obesity-Related Morbid Conditions. In *Biomarkers in Nutrition*; Patel, V.B., Preedy, V.R., Eds.; Springer International Publishing: Cham, Switzerland, 2022; pp. 1–22. [CrossRef]
13. Iacomino, G.; Lauria, F.; Venezia, A.; Iannaccone, N.; Russo, P.; Siani, A. Chapter 6, microRNAs in Obesity and Metabolic Diseases. In *Obesity and Diabetes: Scientific Advances and Best Practice*; Springer Nature: Berlin, Germany, 2020.
14. Decoding noncoding RNAs. *Nat Methods* **2022**, *19*, 1147–1148. [CrossRef]
15. Yang, A.; Bofill-De Ros, X.; Stanton, R.; Shao, T.-J.; Villanueva, P.; Gu, S. TENT2, TUT4, and TUT7 selectively regulate miRNA sequence and abundance. *Nat. Commun.* **2022**, *13*, 5260. [CrossRef] [PubMed]
16. Roberts, J.T.; Patterson, D.G.; King, V.M.; Amin, S.V.; Polska, C.J.; Houserova, D.; Crucello, A.; Barnhill, E.C.; Miller, M.M.; Sherman, T.D.; et al. ADAR Mediated RNA Editing Modulates MicroRNA Targeting in Human Breast Cancer. *Processes* **2018**, *6*, 42. [CrossRef] [PubMed]
17. Alkan, A.H.; Akgül, B. Endogenous miRNA Sponges. In *Methods in Molecular Biology*; Humana: New York, NY, USA, 2022; Volume 2257, pp. 91–104. [CrossRef]
18. Thomou, T.; Mori, M.A.; Dreyfuss, J.M.; Konishi, M.; Sakaguchi, M.; Wolfrum, C.; Rao, T.N.; Winnay, J.N.; Garcia-Martin, R.; Grinspoon, S.K.; et al. Adipose-derived circulating miRNAs regulate gene expression in other tissues. *Nature* **2017**, *542*, 450–455. [CrossRef] [PubMed]
19. Witwer, K.W. Circulating microRNA biomarker studies: Pitfalls and potential solutions. *Clin. Chem.* **2015**, *61*, 56–63. [CrossRef]
20. Iacomino, G.; Siani, A. Role of microRNAs in obesity and obesity-related diseases. *Genes Nutr.* **2017**, *12*, 23. [CrossRef]
21. Catanesi, M.; d'Angelo, M.; Tupone, M.G.; Benedetti, E.; Giordano, A.; Castelli, V.; Cimini, A. MicroRNAs Dysregulation and Mitochondrial Dysfunction in Neurodegenerative Diseases. *Int. J. Mol. Sci.* **2020**, *21*, 5986. [CrossRef]
22. Zhou, S.-S.; Jin, J.-P.; Wang, J.-Q.; Zhang, Z.-G.; Freedman, J.H.; Zheng, Y.; Cai, L. miRNAS in cardiovascular diseases: Potential biomarkers, therapeutic targets and challenges. *Acta Pharmacol. Sin.* **2018**, *39*, 1073–1084. [CrossRef]
23. Swarbrick, S.; Wragg, N.; Ghosh, S.; Stolzing, A. Systematic Review of miRNA as Biomarkers in Alzheimer's Disease. *Mol. Neurobiol.* **2019**, *56*, 6156–6167. [CrossRef]
24. Pauley, K.M.; Cha, S.; Chan, E.K. MicroRNA in autoimmunity and autoimmune diseases. *J. Autoimmun.* **2009**, *32*, 189–194. [CrossRef] [PubMed]
25. Nejad, C.; Stunden, H.J.; Gantier, M.P. A guide to miRNAs in inflammation and innate immune responses. *FEBS. J.* **2018**, *285*, 3695–3716. [CrossRef] [PubMed]
26. Calin, G.A.; Croce, C.M. MicroRNA signatures in human cancers. *Nat. Rev. Cancer* **2006**, *6*, 857–866. [CrossRef]
27. Matsuzaki, J.; Ochiya, T. Circulating microRNAs and extracellular vesicles as potential cancer biomarkers: A systematic review. *Int. J. Clin. Oncol.* **2017**, *22*, 413–420. [CrossRef] [PubMed]
28. Torres, R.; Lang, U.E.; Hejna, M.; Shelton, S.J.; Joseph, N.M.; Shain, A.H.; Yeh, I.; Wei, M.L.; Oldham, M.C.; Bastian, B.C.; et al. MicroRNA ratios distinguish melanomas from nevi. *J. Investig. Dermatol.* **2019**, *140*, 164–173. [CrossRef] [PubMed]
29. Parashar, D.; Singh, A.; Gupta, S.; Sharma, A.; Sharma, M.K.; Roy, K.K.; Chauhan, S.C.; Kashyap, V.K. Emerging Roles and Potential Applications of Non-Coding RNAs in Cervical Cancer. *Genes* **2022**, *13*, 1254. [CrossRef] [PubMed]
30. Vischioni, C.; Bove, F.; De Chiara, M.; Mandreoli, F.; Martoglia, R.; Pisi, V.; Liti, G.; Taccioli, C. miRNAs Copy Number Variations Repertoire as Hallmark Indicator of Cancer Species Predisposition. *Genes* **2022**, *13*, 1046. [CrossRef] [PubMed]
31. Szczepanek, J.; Skorupa, M.; Tretyn, A. MicroRNA as a Potential Therapeutic Molecule in Cancer. *Cells* **2022**, *11*, 1008. [CrossRef]
32. Deiuliis, J.A. MicroRNAs as regulators of metabolic disease: Pathophysiologic significance and emerging role as biomarkers and therapeutics. *Int. J. Obes.* **2016**, *40*, 88–101. [CrossRef]
33. Iacomino, G.; Lauria, F.; Russo, P.; Marena, P.; Venezia, A.; Iannaccone, N.; De Henauw, S.; Foraita, R.; Heidinger-Felso, R.; Hunsberger, M.; et al. Circulating miRNAs are associated with sleep duration in children/adolescents: Results of the I. Family Study. *Exp. Physiol.* **2020**, *105*, 105–347. [CrossRef]

34. Iacomino, G.; Lauria, F.; Russo, P.; Venezia, A.; Iannaccone, N.; Marena, P.; Ahrens, W.; De Henauw, S.; Molnar, D.; Eiben, G.; et al. The association of circulating miR-191 and miR-375 expression levels with markers of insulin resistance in overweight children: An exploratory analysis of the I. Family Study. *Genes Nutr.* **2021**, *16*, 10. [CrossRef] [PubMed]

35. Iacomino, G.; Russo, P.; Marena, P.; Lauria, F.; Venezia, A.; Ahrens, W.; De Henauw, S.; De Luca, P.; Foraita, R.; Gunther, K.; et al. Circulating microRNAs are associated with early childhood obesity: Results of the I.Family Study. *Genes Nutr.* **2019**, *14*, 2. [CrossRef] [PubMed]

36. Iacomino, G.; Russo, P.; Stillitano, I.; Lauria, F.; Marena, P.; Ahrens, W.; De Luca, P.; Siani, A. Circulating microRNAs are deregulated in overweight/obese children: Preliminary results of the I.Family study. *Genes Nutr.* **2016**, *11*, 7. [CrossRef]

37. Lauria, F.; Iacomino, G.; Russo, P.; Venezia, A.; Marena, P.; Ahrens, W.; De Henauw, S.; Eiben, G.; Foraita, R.; Hebestreit, A.; et al. Circulating miRNAs Are Associated with Inflammation Biomarkers in Children with Overweight and Obesity: Results of the I. Family Study. *Genes* **2022**, *13*, 632. [CrossRef]

38. Cappelli, K.; Mecocci, S.; Capomaccio, S.; Beccati, F.; Palumbo, A.R.; Tognoloni, A.; Pepe, M.; Chiaradia, E. Circulating Transcriptional Profile Modulation in Response to Metabolic Unbalance Due to Long-Term Exercise in Equine Athletes: A Pilot Study. *Genes* **2021**, *12*, 1965. [CrossRef] [PubMed]

39. Tribolet, L.; Kerr, E.; Cowled, C.; Bean, A.G.D.; Stewart, C.R.; Dearnley, M.; Farr, R.J. MicroRNA Biomarkers for Infectious Diseases: From Basic Research to Biosensing. *Front. Microbiol.* **2020**, *11*, 1197. [CrossRef] [PubMed]

40. Slattery, M.L.; Herrick, J.S.; Mullany, L.E.; Stevens, J.R.; Wolff, R.K. Diet and lifestyle factors associated with miRNA expression in colorectal tissue. *Pharmgenom. Pers. Med.* **2017**, *10*, 1–16. [CrossRef]

41. Liang, G.; Zhu, Y.; Sun, B.; Shao, Y.; Jing, A.; Wang, J.; Xiao, Z. Assessing the survival of exogenous plant microRNA in mice. *Food Sci. Nutr.* **2014**, *2*, 380–388. [CrossRef] [PubMed]

42. Lee, R.C.; Feinbaum, R.L.; Ambros, V. The C-Elegans Heterochronic Gene Lin-4 Encodes Small Rnas with Antisense Complementarity to Lin-14. *Cell* **1993**, *75*, 843–854. [CrossRef]

43. Diener, C.; Keller, A.; Meese, E. Emerging concepts of miRNA therapeutics: From cells to clinic. *Trends Genet.* **2022**, *38*, 613–626. [CrossRef]

44. Ban, E.; Song, E.J. Considerations and Suggestions for the Reliable Analysis of miRNA in Plasma Using qRT-PCR. *Genes* **2022**, *13*, 328. [CrossRef]

45. Galvao-Lima, L.J.; Morais, A.H.F.; Valentim, R.A.M.; Barreto, E. miRNAs as biomarkers for early cancer detection and their application in the development of new diagnostic tools. *Biomed. Eng. Online* **2021**, *20*, 21. [CrossRef]

46. Seitz, H. Issues in current microRNA target identification methods. *RNA Biol.* **2017**, *14*, 831–834. [CrossRef]

47. Zhu, Y.; Zhu, L.; Wang, X.; Jin, H. RNA-based therapeutics: An overview and prospectus. *Cell Death Dis.* **2022**, *13*, 644. [CrossRef]

48. Jo, S.J.; Chae, S.U.; Lee, C.B.; Bae, S.K. Clinical Pharmacokinetics of Approved RNA Therapeutics. *Int. J. Mol. Sci.* **2023**, *24*, 746. [CrossRef]

49. Saiyed, A.N.; Vasavada, A.R.; Johar, S.R.K. Recent trends in miRNA therapeutics and the application of plant miRNA for prevention and treatment of human diseases. *Futur. J. Pharm. Sci.* **2022**, *8*, 24. [CrossRef] [PubMed]

50. Chakraborty, C.; Sharma, A.R.; Sharma, G.; Doss, C.G.P.; Lee, S.-S. Therapeutic miRNA and siRNA: Moving from Bench to Clinic as Next Generation Medicine. *Mol. Ther. Nucleic Acids* **2017**, *8*, 132–143. [CrossRef] [PubMed]

51. Iacomino, G.; Galderisi, U.; Cipollaro, M.; Di Biase, S.; La Colla, P.; Marongiu, M.E.; De Rienzo, A.; De Falco, S.; Galano, G.; Cascino, A. Phosphorothioated antisense oligonucleotides: Prospects of AIDS therapy. *Life Sci. Adv. Mol. Biol.* **1994**, *13*, 69–74.

52. Hagedorn, P.H.; Hansen, B.R.; Koch, T.; Lindow, M. Managing the sequence-specificity of antisense oligonucleotides in drug discovery. *Nucleic Acids Res.* **2017**, *45*, 2262–2282. [CrossRef]

53. van Rooij, E.; Kauppinen, S. Development of microRNA therapeutics is coming of age. *EMBO Mol Med* **2014**, *6*, 851–864. [CrossRef]

54. Wang, Z. The Guideline of the Design and Validation of MiRNA Mimics. In *MicroRNA and Cancer: Methods and Protocols*; Wu, W., Ed.; Humana Press: Totowa, NJ, USA, 2011; pp. 211–223. [CrossRef]

55. Dasgupta, I.; Chatterjee, A. Recent Advances in miRNA Delivery Systems. *Methods Protoc.* **2021**, *4*, 10. [CrossRef] [PubMed]

56. Dhuri, K.; Bechtold, C.; Quijano, E.; Pham, H.; Gupta, A.; Vikram, A.; Bahal, R. Antisense Oligonucleotides: An Emerging Area in Drug Discovery and Development. *J. Clin. Med.* **2020**, *9*, 2004. [CrossRef] [PubMed]

57. Lennox, K.A.; Behlke, M.A. A Direct Comparison of Anti-microRNA Oligonucleotide Potency. *Pharm. Res.* **2010**, *27*, 1788–1799. [CrossRef]

58. Lima, J.F.; Cerqueira, L.; Figueiredo, C.; Oliveira, C.; Azevedo, N.F. Anti-miRNA oligonucleotides: A comprehensive guide for design. *RNA Biol.* **2018**, *15*, 338–352. [CrossRef]

59. Baumann, V.; Winkler, J. miRNA-based therapies: Strategies and delivery platforms for oligonucleotide and non-oligonucleotide agents. *Future Med. Chem.* **2014**, *6*, 1967–1984. [CrossRef]

60. Broderick, J.A.; Zamore, P.D. MicroRNA therapeutics. *Gene Ther.* **2011**, *18*, 1104–1110. [CrossRef] [PubMed]

61. Sultana, N.; Magadum, A.; Hadas, Y.; Kondrat, J.; Singh, N.; Youssef, E.; Calderon, D.; Chepurko, E.; Dubois, N.; Hajjar, R.J.; et al. Optimizing Cardiac Delivery of Modified mRNA. *Mol. Ther.* **2017**, *25*, 1306–1315. [CrossRef] [PubMed]

62. Sasso, J.M.; Ambrose, B.J.B.; Tenchov, R.; Datta, R.S.; Basel, M.T.; DeLong, R.K.; Zhou, Q.A. The Progress and Promise of RNA Medicine–An Arsenal of Targeted Treatments. *J. Med. Chem.* **2022**, *65*, 6975–7015. [CrossRef]

63. Wang, N.; Chen, M.; Wang, T. Liposomes used as a vaccine adjuvant-delivery system: From basics to clinical immunization. *J. Control Release* **2019**, *303*, 130–150. [CrossRef]
64. Jung, H.N.; Lee, S.Y.; Lee, S.; Youn, H.; Im, H.J. Lipid nanoparticles for delivery of RNA therapeutics: Current status and the role of in vivo imaging. *Theranostics* **2022**, *12*, 7509–7531. [CrossRef] [PubMed]
65. Zak, M.M.; Zangi, L. Lipid Nanoparticles for Organ-Specific mRNA Therapeutic Delivery. *Pharmaceutics* **2021**, *13*, 1675. [CrossRef] [PubMed]
66. Tenchov, R.; Bird, R.; Curtze, A.E.; Zhou, Q. Lipid Nanoparticles–From Liposomes to mRNA Vaccine Delivery, a Landscape of Research Diversity and Advancement. *ACS Nano* **2021**, *15*, 16982–17015. [CrossRef] [PubMed]
67. Zhu, L.; Mahato, R.I. Lipid and polymeric carrier-mediated nucleic acid delivery. *Expert Opin. Drug Deliv.* **2010**, *7*, 1209–1226. [CrossRef] [PubMed]
68. Chakraborty, C.; Sharma, A.R.; Sharma, G.; Lee, S.-S. Therapeutic advances of miRNAs: A preclinical and clinical update. *J. Adv. Res.* **2021**, *28*, 127–138. [CrossRef]
69. *microRNA (miRNA)—Global Market Trajectory & Analytics*; Global Industry Analysts, Inc.: San Jose, CA, USA, 2022; Volume 2022, p. 5519708. Available online: https://www.researchandmarkets.com/reports/5519708/microrna-mirna-global-market-trajectory-and?utm_source=BW&utm_medium=PressRelease&utm_code=g3wfsm&utm_campaign=1742402+-+microRNA+(miRNA)+Global+Market+Report+2022%3a+Increasing+Incidence+of+Infectious+Diseases+and+Chronic+Conditions+to+Drive+Growth&utm_exec=como322prd (accessed on 20 January 2023).
70. Rupaimoole, R.; Slack, F.J. MicroRNA therapeutics: Towards a new era for the management of cancer and other diseases. *Nat. Rev. Drug Discov.* **2017**, *16*, 203–222. [CrossRef] [PubMed]
71. McDonald, C.M.; Shieh, P.B.; Abdel-Hamid, H.Z.; Connolly, A.M.; Ciafaloni, E.; Wagner, K.R.; Goemans, N.; Mercuri, E.; Khan, N.; Koenig, E.; et al. Open-Label Evaluation of Eteplirsen in Patients with Duchenne Muscular Dystrophy Amenable to Exon 51 Skipping: PROMOVI Trial. *J. Neuromuscul. Dis.* **2021**, *8*, 989–1001. [CrossRef]
72. Perry, C.M.; Balfour, J.A. Fomivirsen. *Drugs* **1999**, *57*, 375–380; discussion 381. [CrossRef]
73. Scott, L.J. Givosiran: First Approval. *Drugs* **2020**, *80*, 335–339. [CrossRef]
74. Heo, Y.A. Golodirsen: First Approval. *Drugs* **2020**, *80*, 329–333. [CrossRef] [PubMed]
75. Benson, M.D.; Waddington-Cruz, M.; Berk, J.L.; Polydefkis, M.; Dyck, P.J.; Wang, A.K.; Planté-Bordeneuve, V.; Barroso, F.A.; Merlini, G.; Obici, L.; et al. Inotersen Treatment for Patients with Hereditary Transthyretin Amyloidosis. *N. Engl. J. Med.* **2018**, *379*, 22–31. [CrossRef] [PubMed]
76. Hill, S.F.; Meisler, M.H. Antisense Oligonucleotide Therapy for Neurodevelopmental Disorders. *Dev. Neurosci.* **2021**, *43*, 247–252. [CrossRef]
77. Connolly, K.J.; O'Hare, M.B.; Mohammed, A.; Aitchison, K.M.; Anthoney, N.C.; Taylor, M.J.; Stewart, B.A.; Tuxworth, R.I.; Tear, G. The neuronal ceroid lipofuscinosis protein Cln7 functions in the postsynaptic cell to regulate synapse development. *Sci. Rep.* **2019**, *9*, 15592. [CrossRef] [PubMed]
78. Chiriboga, C.A. Nusinersen for the treatment of spinal muscular atrophy. *Expert Rev. Neurother.* **2017**, *17*, 955–962. [CrossRef] [PubMed]
79. Adams, D.; Gonzalez-Duarte, A.; O'Riordan, W.D.; Yang, C.C.; Ueda, M.; Kristen, A.V.; Tournev, I.; Schmidt, H.H.; Coelho, T.; Berk, J.L.; et al. Patisiran, an RNAi Therapeutic, for Hereditary Transthyretin Amyloidosis. *N. Engl. J. Med.* **2018**, *379*, 11–21. [CrossRef]
80. Smith, E.S.; Whitty, E.; Yoo, B.; Moore, A.; Sempere, L.F.; Medarova, Z. Clinical Applications of Short Non-Coding RNA-Based Therapies in the Era of Precision Medicine. *Cancers* **2022**, *14*, 1588. [CrossRef]
81. Gebert, L.F.; Rebhan, M.A.; Crivelli, S.E.; Denzler, R.; Stoffel, M.; Hall, J. Miravirsen (SPC3649) can inhibit the biogenesis of miR-122. *Nucleic Acids Res.* **2014**, *42*, 609–621. [CrossRef] [PubMed]
82. Gomez, I.G.; MacKenna, D.A.; Johnson, B.G.; Kaimal, V.; Roach, A.M.; Ren, S.; Nakagawa, N.; Xin, C.; Newitt, R.; Pandya, S.; et al. Anti-microRNA-21 oligonucleotides prevent Alport nephropathy progression by stimulating metabolic pathways. *J. Clin. Investig.* **2015**, *125*, 141–156. [CrossRef]
83. Bedewy, A.M.L.; Elmaghraby, S.M.; Shehata, A.A.; Kandil, N.S. Prognostic Value of miRNA-155 Expression in B-Cell Non-Hodgkin Lymphoma. *Turk. J. Hematol.* **2017**, *34*, 207–212. [CrossRef]
84. Beg, M.S.; Brenner, A.J.; Sachdev, J.; Borad, M.; Kang, Y.-K.; Stoudemire, J.; Smith, S.; Bader, A.G.; Kim, S.; Hong, D.S. Phase I study of MRX34, a liposomal miR-34a mimic, administered twice weekly in patients with advanced solid tumors. *Investig. N. Drugs* **2017**, *35*, 180–188. [CrossRef] [PubMed]
85. van der Ree, M.H.; de Vree, J.M.; Stelma, F.; Willemse, S.; van der Valk, M.; Rietdijk, S.; Molenkamp, R.; Schinkel, J.; van Nuenen, A.C.; Beuers, U.; et al. Safety, tolerability, and antiviral effect of RG-101 in patients with chronic hepatitis C: A phase 1B, double-blind, randomised controlled trial. *Lancet* **2017**, *389*, 709–717. [CrossRef] [PubMed]
86. Cordido, A.; Besada-Cerecedo, L.; Garcia-Gonzalez, M.A. The Genetic and Cellular Basis of Autosomal Dominant Polycystic Kidney Disease-A Primer for Clinicians. *Front. Pediatr.* **2017**, *5*, 279. [CrossRef]
87. Bais, T.; Gansevoort, R.T.; Meijer, E. Drugs in Clinical Development to Treat Autosomal Dominant Polycystic Kidney Disease. *Drugs* **2022**, *82*, 1095–1115. [CrossRef]

88. Gallant-Behm, C.L.; Piper, J.; Dickinson, B.A.; Dalby, C.M.; Pestano, L.A.; Jackson, A.L. A synthetic microRNA-92a inhibitor (MRG-110) accelerates angiogenesis and wound healing in diabetic and nondiabetic wounds. *Wound Repair Regen.* **2018**, *26*, 311–323. [CrossRef] [PubMed]

89. Kriegel, A.J.; Liu, Y.; Fang, Y.; Ding, X.; Liang, M. The miR-29 family: Genomics, cell biology, and relevance to renal and cardiovascular injury. *Physiol. Genom.* **2012**, *44*, 237–244. [CrossRef] [PubMed]

90. Gallant-Behm, C.L.; Piper, J.; Lynch, J.M.; Seto, A.G.; Hong, S.J.; Mustoe, T.A.; Maari, C.; Pestano, L.A.; Dalby, C.M.; Jackson, A.L.; et al. A MicroRNA-29 Mimic (Remlarsen) Represses Extracellular Matrix Expression and Fibroplasia in the Skin. *J. Investig. Dermatol.* **2019**, *139*, 1073–1081. [CrossRef]

91. van Zandwijk, N.; McDiarmid, J.; Brahmbhatt, H.; Reid, G. Response to "An innovative mesothelioma treatment based on mir-16 mimic loaded EGFR targeted minicells (TargomiRs)". *Transl. Lung Cancer Res.* **2018**, *7*, S60–S61. [CrossRef] [PubMed]

 *genes*

Article

Circulating miRNAs Are Associated with Inflammation Biomarkers in Children with Overweight and Obesity: Results of the I.Family Study

Fabio Lauria [1,†] , Giuseppe Iacomino [1,*,†] , Paola Russo [1] , Antonella Venezia [1], Pasquale Marena [1], Wolfgang Ahrens [2] , Stefaan De Henauw [3], Gabriele Eiben [4], Ronja Foraita [2], Antje Hebestreit [2] , Yiannis Kourides [5], Dénes Molnár [6] , Luis A. Moreno [7], Toomas Veidebaum [8] and Alfonso Siani [1]
on behalf of the I.Family Consortium

[1] Institute of Food Sciences, National Research Council, 83100 Avellino, Italy; flauria@isa.cnr.it (F.L.); prusso@isa.cnr.it (P.R.); antovenezia@yahoo.it (A.V.); pask.mare@gmail.com (P.M.); asiani@isa.cnr.it (A.S.)
[2] Leibniz Institute for Prevention Research and Epidemiology-BIPS, Achterstraße 30, 28359 Bremen, Germany; ahrens@leibniz-bips.de (W.A.); foraita@leibniz-bips.de (R.F.); hebestr@leibniz-bips.de (A.H.)
[3] Department of Public Health and Primary Care, University Hospital 4K3 C. Heymanslaan, 10, 9000 Ghent, Belgium; stefaan.dehenauw@ugent.be
[4] Sahlgrenska Academy, University of Gothenburg, Medicinaregatan 3, 41390 Göteborg, Sweden; gabriele.eiben@medfak.gu.se
[5] Research and Education Institute of Child Health, 138 Limassol Ave, #205, Strovolos 2015, Cyprus; kourides@cytanet.com.cy
[6] Department of Pediatrics, Medical School, University of Pécs, H-7624 Pecs, Hungary; molnar.denes@pte.hu
[7] University of Zaragoza, Domingo Miral s/n, 50009 Zaragoza, Spain; lmoreno@unizar.es
[8] National Institute for Health Development, Hiiu 42, 11619 Tallinn, Estonia; toomas.veidebaum@tai.ee
* Correspondence: giuseppe.iacomino@isa.cnr.it; Tel.: +39-0825299431
† These authors contributed equally to this work.

Citation: Lauria, F.; Iacomino, G.; Russo, P.; Venezia, A.; Marena, P.; Ahrens, W.; De Henauw, S.; Eiben, G.; Foraita, R.; Hebestreit, A.; et al. Circulating miRNAs Are Associated with Inflammation Biomarkers in Children with Overweight and Obesity: Results of the I.Family Study. *Genes* **2022**, *13*, 632. https://doi.org/10.3390/genes13040632

Academic Editor: Mark Boldin

Received: 22 February 2022
Accepted: 29 March 2022
Published: 1 April 2022

Publisher's Note: MDPI stays neutral with regard to jurisdictional claims in published maps and institutional affiliations.

Abstract: Increasing data suggest that overnutrition-induced obesity may trigger an inflammatory process in adipose tissue and upturn in the innate immune system. Numerous players have been involved in governing the inflammatory response, including epigenetics. Among epigenetic players, miRNAs are emerging as crucial regulators of immune cell development, immune responses, autoimmunity, and inflammation. In this study, we aimed at identifying the involvement of candidate miRNAs in relation to inflammation-associated biomarkers in a subsample of European children with overweight and obesity participating in the I.Family study. The study sample included individuals with increased adiposity since this condition contributes to the early occurrence of chronic low-grade inflammation. We focused on the acute-phase reagent C-reactive protein (CRP) as the primary outcome and selected cytokines as plausible biomarkers of inflammation. We found that chronic low-grade CRP elevation shows a highly significant association with miR-26b-3p and hsa-miR-576-5p in boys. Furthermore, the association of CRP with hsa-miR-10b-5p and hsa-miR-31-5p is highly significant in girls. We also observed major sex-related associations of candidate miRNAs with selected cytokines. Except for IL-6, a significant association of hsa-miR-26b-3p and hsa-miR-576-5p with TNF-α, IL1-Ra, IL-8, and IL-15 levels was found exclusively in boys. The findings of this exploratory study suggest sex differences in the association of circulating miRNAs with inflammatory response biomarkers, and indicate a possible role of miRNAs among the candidate epigenetic mechanisms related to the process of low-grade inflammation in childhood obesity.

Keywords: miRNAs; chronic low-grade inflammation; inflammation-associated biomarkers; overweight and obesity; children/adolescents; sex-related associations

1. Introduction

During the past decade, obesity has become a global epidemic with substantial health, social, and economic implications. According to the World Health Organization, there are nearly 2 billion overweight adults worldwide and approximately 600 million of them are obese [1,2]. Numerous studies have recognized a strong association between an increased body mass index (BMI) and quality and expectancy of life given its interconnected effects with type 2 diabetes, hyperlipidemia, non-alcoholic fatty liver disease, hypertension, heart disease, stroke, arthritis, cancer, depression, asthma, psychological problems, and other non-communicable diseases [3–5].

Despite extensive investigation, most of the genetic variability in obesity remains unresolved and the influence of candidate genes in this context appears limited. Numerous studies have suggested that genes that predispose to obesity are not causal but act in conjunction with a variety of individual, environmental, and lifestyle factors, including obesogenic environments, low levels of education, sedentary habits, and reduced sleep hours, among others [6]. A highly energy-dense diet, as well as low physical activity levels, are considered driving factors [3]. Evidence also indicates that offspring born to obese mothers are prone to a higher BMI and fat accumulation, thus supporting the conceivable transgenerational risk contribution based on epigenetic mechanisms [7].

The basic tissue for energy storage in humans is the white adipose tissue. Aside from its storing function, this tissue is metabolically active. It releases hundreds of different factors, such as hormones including adiponectin and leptin, growth factors including Insulin-Like Growth Factor-1 (IGF-1) and Platelet-Derived Growth Factor (PDGF), and inflammatory mediators including Tumor Necrosis Factor-α (TNF-α), Interleukin-6 (IL-6), and Interleukin-8 (IL-8), all of which contribute to insulin sensitivity and dietary control. Growing data suggest that overnutrition-induced obesity triggers an inflammatory process in adipose tissue [8], with even moderate weight gain linked to inflammatory activation [9] and an upturn in the innate immune system [10]. Moreover, chemokines and their receptors also contribute to obesity development by activating the resident immune surveillance system [11]. This state is firmly interconnected to the local recruitment of macrophages and enhanced immune cell infiltration/proliferation/activation [12]. All these processes collectively point toward adipocyte hypertrophy and impaired adipogenesis [13]. The resulting persistent chronic low-grade inflammation represents a hallmark of obesity that leads to and perpetuates the state of metabolic alterations and insulin resistance in target organs, including the adipose tissue, liver, muscles, and vascular system [14]. Among the inflammatory biomarkers, the acute-phase reactant C-reactive protein (CRP) is considered the major factor associated with overweight and obesity in both adults and children [15,16].

In recent years, numerous biochemical players have been recognized as being involved in the inflammatory response, including microRNAs (miRNAs), small non-coding RNAs expressed in a wide variety of organs and cells and capable of potentially influencing almost all cellular functions. miRNAs are constituents of the epigenetic mechanisms that finely regulate the expression of messenger RNAs [17]. Until now, 2599 different miRNAs have been documented in humans (miRTarBase, release 8.0) [18]. Growing evidence also underlines their relevance as stable, non-invasive, and reliable biomarkers for a variety of pathophysiological conditions [19–21], including the body's energy balance [22]. Many studies have established that altered miRNA profiles are interconnected with obesity [23] and other non-communicable diseases [24,25]. However, nowadays, epigenetics represents a critical but still poorly understood factor among the known molecular mechanisms related to the process of low-grade inflammation in obesity [26]. Among epigenetic players, miRNAs are progressively emerging as key regulators of immune cell development, immune responses, autoimmunity, and inflammation, capable of affecting both pro-and anti-inflammatory responses [26]. Indeed, by influencing definite signaling networks in the immune system, so-called immuno-miRs have been shown to impact both innate and adaptive immune responses in health and disease [27]. Of note, immuno-miRs may exert different functions in different cell types. Therefore, during inflammation, different cells undergo substantial transcriptional activation, thus presenting different sets of targets regulated by a given miRNA.

In this study, we aimed to identify the potential involvement of candidate miRNAs in relation to inflammation-associated biomarkers in a subsample of European children with overweight and obesity (OW/Ob) participating in the I.Family study (www.ifamilystudy.eu, last accessed on 21 February 2022) [28,29]. The primary outcome of this study was the association of miRNA expression with CRP. The association of miRNAs with the selected cytokines constituted a secondary outcome of the study. The study sample included individuals with OW/Ob since this condition is a known risk factor contributing to the early occurrence of chronic low-grade inflammation and interconnected metabolic changes.

2. Materials and Methods

2.1. Participants

The study was conducted on a subgroup of European children/adolescents with OW/Ob belonging to the I.Family study, an EC-funded project aiming at investigating determinants of food choice, lifestyle, and related health outcomes in children and adolescents of eight European countries (Belgium, Cyprus, Estonia, Germany, Hungary, Italy, Spain, and Sweden). A full description of the study designs, selection criteria, and participants' characteristics, from which the subsample data are drawn, has been previously published [30]. A complete explanation of the I.Family study (registration number IS-RCTN62310987) has been earlier reported [29]. The study was conducted according to the criteria set by the Declaration of Helsinki. Approval by the national ethics committees was obtained by each of the participating centers. Anthropometric characteristics, pubertal status, and dietary intake data were collected using standardized procedures; a full description of these methods has been previously published [31,32]. In each country, we first selected 20 children (n = 160) who retained overweight or obesity, i.e., who had a BMI z-score of more than + 1 at baseline and after 2 years at follow-up, respectively, and did not change more than ± 0.1 in BMI z-score per year (defined as overweight/obese) [33]. Out of 160 subjects, the current analysis was performed on a subsample of 79 overweight and obese children/adolescents (Belgium n = 7, Cyprus n = 6, Estonia n = 13, Germany n = 16, Hungary n = 8, Italy n = 12, Spain n = 5, and Sweden n = 12), with a complete dataset for the variables of interest, after the exclusion of 19 participants with hemolyzed samples.

2.2. Biochemical Analysis

The fasting venous blood draws were collected in BD Vacutainer® blood tubes according to standardized operative procedures. A complete description of the sample collection and investigative procedures has been earlier published [34]. Clinical chemistry tests were determined as part of routine laboratory testing, in a central laboratory (University of Bremen, Centre for Biomolecular Interactions Bremen—CBIB). Serum samples stored at $-80\ ^{\circ}$C were used to detect levels of CRP, Interleukin-1 Receptor Antagonist (IL1-Ra), IL-6, IL-8, Interleukin-15 (IL-15), and TNF-α using an electrochemiluminescent multiplex assay (using either single or MULTI-SPOT® Assay Systems, Meso Scale Discovery).

2.3. miRNA Profiling

Taking advantage of the qPCR array technology, we previously reported that an altered circulating miRNA profile is associated with OW/Ob in children and adolescents [30]. In that study, we also identified four circulating miRNAs, hsa-miR-10b-5p (MIMAT0000254), hsa-miR-26b-3p (MIMAT0004500), hsa-miR-31-5p (MIMAT0000089), hsa-miR-576-5p (MIMAT0003241), potentially linked to increased CRP levels in subjects with OW/Ob (unpublished data). Of note, among the four miRNAs characterized, only hsa-miR-10b-5p was confirmed to be associated with OW/Ob. In the current investigation, we aimed to confirm the association of the candidate miRNAs with levels of CRP and the selected inflammatory biomarkers through validation by SYBR green-based real-time quantitative RT-PCR (RT-qPCR) in the new sample of children and adolescents with OW/Ob. Protocols for miRNA extraction and screening from plasma samples have been earlier published [30,32]. Individual plasma samples were first tested for hemoglobin levels and hemolyzed samples were excluded from the analysis [30]. Different assays were performed in triplicate employing the miScript Primer Assays according to the manufacturer's instructions (Qiagen, Germany). miRNA relative levels were determined using the Cel-miR-39 spike-in as the endogenous normalizer [30]. Levels were calculated using the Data Assist v3.1 software package (Life Technologies, Thermo Fisher Scientific, Milan, Italy).

2.4. Statistical Analysis

Statistical analyses were achieved by using IBM SPSS Statistics software (v24.0. Armonk, NY, USA: IBM Corp.). Data collected were calculated as means and 95% confidence intervals (CIs). Associations of miRNA expression with CRP and the different interleukins were assessed using linear regression analyses, adjusting for covariates including country of residence, age, BMI z-score, pubertal status. Since potential sex disparities in CRP levels have been previously reported [35], we considered boys and girls separately in the statistical analysis. To control for the false discovery rate (FDR), the Benjamini–Hochberg (BH) method was adopted. The level of statistical significance was set at $\alpha < 0.05$.

3. Results

3.1. Anthropometric Characteristics and Biochemical Markers of the Study Sample

The anthropometric and metabolic characteristics and levels of inflammatory markers of the 79 participants are reported in Table 1. There were no obvious differences regarding the anthropometric characteristics and tested biochemical markers between males and females.

Table 1. Anthropometric and chemical characteristics of the study sample.

Ow/Ob	Boys (25/11)	Girls (31/12)
Age (years)	12.1 (11.5–12.8)	12.4 (11.9–12.9)
Puberty (% yes)	47.8	52.2
BMI z-score	1.8 (1.6–2.0)	1.7 (1.5–1.9)
CRP (mg/dL)	0.36 (0.08–0.63)	0.45 (0.14–0.77)
TNF-α (pg/mL)	4.3 (2.7–6.0)	3.9 (2.4–5.4)
IL-1Ra (pg/mL)	422.9 (339.7–506.2)	527.8 (398.0–657.6)
IL-6 (pg/mL)	1.2 (−0.1–2.5)	0.8 (0.5–1.0)
IL-8 (pg/mL)	38.5 (−23.1–100.0)	17.6 (1.7–33.6)
IL-15 (pg/mL)	2.6 (2.2–3.0)	3.0 (2.5–3.4)
hsa-miR-10b-5p	2.9 (2.1–3.6)	3.3 (2.7–3.9)
hsa-miR-26b-3p	2.4 (1.6–3.3)	2.7 (0.7–4.6)
hsa-miR-31-5p	1.1 (0.3–1.8)	1.5 (0.8–2.3)
hsa-miR-576-5p	7.3 (5.4–9.1)	6.4 (5.0–7.8)

Data are expressed as mean (CIs) or as frequency (%).

3.2. RT-qPCR Validation in Individual Plasma Samples

After plasma extraction, the single candidate miRNAs were determined in individual assays by RT-qPCR. Differences in miRNA signatures with respect to anthropometric and biochemical variables were investigated. Associations of miRNA expression levels with CRP, the primary outcome of this study, and the selected interleukins were assessed using linear regression analysis stratified by sex. Results reported in Table 2 are adjusted for covariates including country of residence, age, BMI z-score, and pubertal status. CRP values show a significant association with miR-26b-3p and hsa-miR-576-5p exclusively in boys. Moreover, their associations with hsa-miR-10b-5p and hsa-miR-31-5p are highly significant only in girls. In Figure S1 is reported the distribution of selected miRNAs in relation to CRP levels in the girls' and boys' subgroups.

Table 2. Association of miRNA expression levels with CRP.

	Boys (36)	*q*-Value	Girls (43)	*q*-Value
hsa-miR-10b-5p	2.7 (1.7–3.7)	0.399	3.6 (3.0–4.2)	**0.008**
hsa-miR-26b-3p	3.2 (3.0–3.4)	**0.004**	2.1 (−0.9–5.0)	0.553
hsa-miR-31-5p	1.1 (−0.3–2.1)	0.914	1.3 (0.9–1.8)	**0.02**
hsa-miR-576-5p	8.3 (6.8–9.8)	**0.006**	6.8 (5.7–8.0)	0.187

Data are expressed as mean (CIs). Covariates: Country of residence, age, BMI z-score, pubertal status. *q*-values are BH-adjusted *p*-values. Values in bold indicate statistically significant results.

Table 3 reports the results of the associations of candidate miRNAs with the cytokines selected as additional inflammation biomarkers. A significant association of both hsa-miR10b-5p and hsa-miR-26b-3p with TNF-α, IL1-Ra, IL-8, and IL-15 levels was found exclusively in boys. No association of candidate miRNAs with IL-6 has been established. Moreover, none of the candidate miRNAs was associated with cytokine levels in girls.

Table 3. Association of miRNA expression levels with selected cytokines (secondary outcome).

Cytokine	Sex	miR-10b-5p	*q*-Value	miR-26b-3p	*q*-Value	miR-31-5p	*q*-Value	miR-576-5p	*q*-Value
TNF-α	Boys	2.5 (2.0–3.0)	1.000	3.0 (2.6–3.4)	**0.006**	1.3 (0.3–2.2)	0.963	7.7 (6.4–9.0)	**0.005**
	Girls	3.2 (2.4–4.0)	0.635	2.4 (−0.4–5.2)	1.000	1.5 (1.0–2.1)	0.485	7.2 (6.1–8.2)	0.730
IL1-Ra	Boys	2.6 (2.1–3.1)	0.810	3.0 (2.6–3.4)	**0.005**	1.1 (−0.03–2.2)	0.393	7.9 (6.4–9.4)	**0.023**
	Girls	3.3 (2.5–4.1)	0.558	2.3 (−0.8–5.5)	1.000	1.6 (0.9–2.2)	1.000	6.6 (5.3–7.8)	0.735
IL-6	Boys	2.6 (2.0–3.2)	0.743	3.0 (2.1–4.0)	0.461	0.8 (−0.5–2.0)	0.963	7.1 (4.2–10.0)	0.932
	Girls	3.3 (2.5–4.1)	0.949	2.2 (−0.9–5.3)	0.732	1.6 (0.9–2.2)	0.692	6.8 (5.7–8.0)	0.625
IL-8	Boys	2.4 (1.8–3.0)	0.927	2.9 (2.5–3.3)	**0.005**	1.2 (0.2–2.2)	0.410	7.2 (5.6–8.9)	**0.021**
	Girls	3.3 (2.4–4.1)	0.649	2.5 (−0.5–5.4)	1.000	1.6 (1.1–2.1)	1.000	7.3 (6.2–8.4)	0.856
IL-15	Boys	2.6 (2.1–3.1)	0.588	2.9 (2.5–3.3)	**0.005**	1.2 (0.1–2.3)	0.890	7.5 (5.9–9.1)	**0.015**
	Girls	3.3 (2.6–4.0)	0.512	2.3 (0.9–2.3)	0.912	1.6 (0.9–2.2)	0.856	6.8 (5.7–8.0)	0.908

Data are expressed as mean (CIs). Covariates: Country of residence, age, BMI z-score, pubertal status. *q*-values are BH-adjusted *p*-values. Values in bold indicate statistically significant results.

4. Discussion

Inflammation is a physio-pathological process, commonly triggered by injuries and infections, and characterized by a complex flow of dynamically and finely regulated responses. The degree, the dynamics of pro-and anti-inflammatory networks, and the course of an inflammatory reaction may decisively impact the onset, progression, and development of health disorders. Inflammation and its supporting mechanisms are considered closely related to numerous diseases' progression. Various studies have established that specific miRNAs participate in the development of innate and adaptive immunity, acting as crucial players in the fine-tuning of the inflammatory network. In this context, several miRNAs attenuate the response, while others, by depressing specific inhibitors, are capable of intensifying the immune reaction [36], with certain miRNAs essential for mounting the inflammatory response [37]. Notably, chronic inflammation is one of the main factors involved in the process of obesity progression [9,38], and several studies have reported that even limited weight gain is interconnected with the sustained inflammatory process [38].

Previously, we identified specific circulating miRNAs associated with childhood obesity in a subsample of the I.Family study [28,30]. Moreover, we also recognized in that study candidate miRNAs possibly related to CRP levels, among which only hsa-miR-10b-5p was associated with OW/Ob. In the present analysis, we report sex-related associations of these miRNAs with inflammatory biomarkers in children and adolescents with OW/Ob. Of note, the subjects enrolled for the current investigation belong to a different subgroup of the common I.Family population.

Given the different fat distribution and the influence of sex hormones, it is conceivable that the relationship between inflammatory markers and obesity may differ by sex. Accordingly, we stratified the study population by sex. Association analyses were corrected for confounding factors including the country of residence, age, BMI z-score, and pubertal status.

We found a clear indication pointing towards a cross-talk between candidate c-miRNAs and inflammatory biomarkers in the context of raised adiposity. Among the numerous inflammatory biomarkers, we focused on the acute-phase reactant CRP as the primary outcome of the study. CRP has been considered the strongest factor associated with OW/Ob in epidemiological studies [15]. Several reports consider CRP to be a consequence of an obesity condition rather than the cause [15]; conversely, increasing evidence establishes a causal role of CRP elevation in the onset and development of obesity by causing extensive tuning in the innate immune system and energy expenditure system [39].

We also found sex-related associations of the candidate miRNAs with the selected cytokines as plausible inflammation biomarkers. Except for IL-6, a significant association of both hsa-miR-26b-3p and hsa-miR-576-5p with TNF-α, IL1-Ra, IL-8, and IL-15 levels was found exclusively in boys. Moreover, no association of candidate miRNAs with cytokine levels was established in the girls' subgroup.

Among the characterized miRNAs, miR-10b-5p is one of the first identified as abnormal in human cancer and, since its first description, it has been widely studied in this context. Recent papers also suggest that it participates in inflammation control and inflammation-associated diseases by regulating T cells [40,41]. Similarly, miR-26b has been described as a key regulator in carcinogenesis and cancer progression, acting as a tumor suppressor gene in several types of cancer. miR-26b has also been shown to play a role in inflammation as in cytokine secretion [42]. miR-26b also targets the inflammatory factor prostaglandin-endoperoxide synthase 2, which plays relevant roles in inflammatory diseases by inducing the production of prostaglandin E2 [43]. However, abnormal expression of miR-31-5p has been described in various cancers, where this miRNA plays a significant role in tumorigenesis, acting as either an oncogene or tumor suppressor, in a context-dependent manner, although the underlying mechanism remains unclear [44]. Moreover, miR-31 is involved in several inflammation-associated disorders. Interestingly, the molecular role of miR-31-5p activation in early inflammation has been recently defined [45]. Several studies have demonstrated that miR-576-5p acts as a tumor-promoting miRNA in several types of human tumors, highlighting its potential role as a predictor of cancer prognosis [46]. However, in silico studies have shown that miR-576-5p is involved in the regulation of inflammatory, growth, and proliferation signaling pathways.

Our apparently controversial results are in line with recent studies reporting sex influences on the severity and evolution of various inflammatory conditions [47,48]. Numerous studies have confirmed the role of sex hormones in the immune response and recent clinical data have shown significant sex differences in inflammatory markers also in prepubertal children, supporting a genetic contribution [47]. Sex differences occur in both innate and adaptive immune responses and are evolutionarily conserved across species [49]. Overall, there is accumulating evidence that sex is a critical variable that influences innate and adaptive immune responses, resulting in sex-specific outcomes, but the main molecular mechanisms remain elusive [48].

The findings of this exploratory study suggest major differences in the association of circulating miRNAs and inflammatory response biomarkers across sexes, pointing to a conceivable role of miRNAs among the candidate epigenetic mechanisms related to the low-grade inflammation process in childhood obesity, calling for more attention in this largely underexplored area. However, evidence concerning how these molecules may act remains questioned since the experimental design of our cross-sectional analysis, explorative in essence, cannot answer this question.

5. Conclusions

Many regulatory steps are relevant in the transformation and delivery of genetic information to cellular effectors. This network orchestration is finely regulated by both transcriptional and post-transcriptional mechanisms. Sex differences can provide substantial support in defining the course of these supervisory mechanisms [50]. The fascinating emergence of circulating miRNAs as stable and affordable molecules has opened up a promising opportunity for the identification of new crucial players in the onset and progression of inflammation in childhood obesity, as well as their potential application as non-invasive biomarkers. However, it is still uncertain whether the identified miRNAs are drivers of sex-related disparities in obesity-related inflammation or represent epiphenomena. Further molecular-oriented studies are needed to explore the functional relevance of the miRNA species identified.

Supplementary Materials: The following supporting information can be downloaded at: https://www.mdpi.com/article/10.3390/genes13040632/s1, Figure S1. The different panels show the distribution of reported miRNAs in relation to CRP levels in girls and boys.

Author Contributions: G.I., F.L. and A.S. conceived, designed, and oversaw the analyses and drafted the manuscript. A.V. and P.M. conducted the molecular analyses. P.R. contributed to the interpretation of data and provided critical input during the drafting and revision of the manuscript. W.A., S.D.H., G.E., R.F., A.H., Y.K., D.M., L.A.M. and T.V. contributed to the critical revision of the manuscript. All authors have read and agreed to the published version of the manuscript.

Funding: This work was done as part of the I.Family Study (http://www.ifamilystudy.eu/, last accessed on 21 February 2022). We gratefully acknowledge the financial support of the European Community within the Seventh RTD Framework Programme Contract No. 266044.

Institutional Review Board Statement: This study was conducted according to the standards of the Declaration of Helsinki. Approval by the appropriate ethics committees was obtained by each of the eight participating centers carrying out the fieldwork (Ethics Committee of the Ghent University Hospital—Belgium, National Bioethics Committee—Cyprus, Tallinn Medical Research Ethics Committee—Estonia, Ethics Committee of the University of Bremen—Germany, Scientific and Research Ethics Committee of the Medical Research Council of Pécs (TUKEB) and Baranja County Public Health and Medical Officer Service (ANTSZ)—Hungary, Ethics Committee of the Local Health Institute in Avellino (ASL)—Italy, Ethics Committee of Clinical Research of Aragon (CEICA)—Spain, Regional Ethics Committee of the University of Gothenburg—Sweden).

Informed Consent Statement: Participants were not subjected to any study procedure before both the children and their parents gave their oral (children) and written (parents) informed consent for examinations, collection of samples, subsequent analysis, and storage of personal data and collected samples.

Data Availability Statement: All data produced or analyzed during this study are included in this article.

Acknowledgments: The authors would like to thank all children and their families who took part in this study. We recognize that this report would have not been possible without the contributions and efforts of all groups involved in the I.Family study. We gratefully acknowledge the financial support of the European Community within the Seventh RTD Framework Programme Contract No. 266044.

Conflicts of Interest: The authors declare no conflict of interest.

References

1.   WHO. *Library Cataloguing in Publication Data Report of the Commission on Ending Childhood Obesity*; World Health Organization: Geneva, Switzerland, 2016.
2.   Gómez, L.A.; Abden, Z.A.; Hamid, Z.A.; Rmeileh, N.M.A.; Cazares, B.A.; Acuin, C.; Adams, R.J.; Aekplakorn, W.; Afsana, K.; Salinas, C.A.A.; et al. Worldwide trends in body-mass index, underweight, overweight, and obesity from 1975 to 2016: A pooled analysis of 2416 population-based measurement studies in 128.9 million children, adolescents, and adults. *Lancet* **2017**, *390*, 2627–2642. [CrossRef]
3.   Hruby, A.; Hu, F.B. The epidemiology of obesity: A big picture. *Pharmacoeconomics* **2015**, *33*, 673–689. [CrossRef] [PubMed]
4.   Iacomino, G.; Lauria, F.; Venezia, A.; Iannaccone, N.; Russo, P.; Siani, A. microRNAs in Obesity and metabolic diseases. In *Obesity and Diabetes: Scientific Advances and Best Practice*; Faintuch, J., Faintuch, S., Eds.; Springer Nature: Cham, Switzerland, 2020; Chapter 6, pp. 71–95. [CrossRef]
5.   Farag, Y.M.; Gaballa, M.R. Diabesity: An overview of a rising epidemic. *Nephrol. Dial. Transplant. Off. Publ. Eur. Dial. Transpl. Assoc.-Eur. Ren. Assoc.* **2011**, *26*, 28–35. [CrossRef] [PubMed]

6. Huls, A.; Wright, M.N.; Bogl, L.H.; Kaprio, J.; Lissner, L.; Molnar, D.; Moreno, L.A.; De Henauw, S.; Siani, A.; Veidebaum, T.; et al. Polygenic risk for obesity and its interaction with lifestyle and sociodemographic factors in European children and adolescents. *Int. J. Obes.* **2021**, *45*, 1321–1330. [CrossRef]

7. Li, Y.; Pollock, C.A.; Saad, S. Aberrant DNA methylation mediates the transgenerational risk of metabolic and chronic disease due to maternal obesity and overnutrition. *Genes* **2021**, *12*, 1653. [CrossRef] [PubMed]

8. Kammoun, H.L.; Kraakman, M.J.; Febbraio, M.A. Adipose tissue inflammation in glucose metabolism. *Rev. Endocr. Metab. Disord.* **2014**, *15*, 31–44. [CrossRef]

9. Christ, A.; Günther, P.; Lauterbach, M.A.R.; Duewell, P.; Biswas, D.; Pelka, K.; Scholz, C.J.; Oosting, M.; Haendler, K.; Baßler, K.; et al. Western diet triggers NLRP3-dependent innate immune reprogramming. *Cell* **2018**, *172*, 162–175.e14. [CrossRef]

10. Thorburn, A.N.; Macia, L.; Mackay, C.R. Diet, Metabolites, and "Western-Lifestyle" Inflammatory Diseases. *Immunity* **2014**, *40*, 833–842. [CrossRef] [PubMed]

11. Yao, L.; Heuser-Baker, J.; Barlic-Dicen, J. Chemokine receptors on the defensive—The surprising role of CXCR4 in brown adipose tissue. *Recept. Clin. Investig.* **2015**, *2*, e397. [CrossRef]

12. Kraakman, M.J.; Murphy, A.J.; Jandeleit-Dahm, K.; Kammoun, H.L. Macrophage polarization in obesity and type 2 diabetes: Weighing down our understanding of macrophage function? *Front. Immunol.* **2014**, *5*, 470. [CrossRef] [PubMed]

13. Schaffler, A.; Muller-Ladner, U.; Scholmerich, J.; Buchler, C. Role of adipose tissue as an inflammatory organ in human diseases. *Endocr. Rev.* **2006**, *27*, 449–467. [CrossRef] [PubMed]

14. Saltiel, A.R.; Olefsky, J.M. Inflammatory mechanisms linking obesity and metabolic disease. *J. Clin. Investig.* **2017**, *121*, 2111–2117. [CrossRef]

15. Timpson, N.J.; Nordestgaard, B.G.; Harbord, R.M.; Zacho, J.; Frayling, T.M.; Tybjærg-Hansen, A.; Davey Smith, G. C-reactive protein levels and body mass index: Elucidating direction of causation through reciprocal Mendelian randomization. *Int. J. Obes.* **2011**, *35*, 300–308. [CrossRef] [PubMed]

16. Nappo, A.; Iacoviello, L.; Fraterman, A.; Gonzalez-Gil, E.M.; Hadjigeorgiou, C.; Marild, S.; Molnar, D.; Moreno, L.A.; Peplies, J.; Sioen, I.; et al. High-sensitivity C-reactive protein is a predictive factor of adiposity in children: Results of the identification and prevention of dietary- and lifestyle-induced health effects in children and infants (IDEFICS) study. *J. Am. Heart Assoc.* **2013**, *2*, e000101. [CrossRef] [PubMed]

17. Cavalieri, V. The expanding constellation of histone post-translational modifications in the epigenetic landscape. *Genes* **2021**, *12*, 1596. [CrossRef]

18. Huang, H.Y.; Lin, Y.C.; Li, J.; Huang, K.Y.; Shrestha, S.; Hong, H.C.; Tang, Y.; Chen, Y.G.; Jin, C.N.; Yu, Y.; et al. miRTarBase 2020: Updates to the experimentally validated microRNA-target interaction database. *Nucleic Acids Res.* **2020**, *48*, D148–D154. [CrossRef]

19. Kim, Y.K. Extracellular microRNAs as biomarkers in human disease. *Chonnam Med. J.* **2015**, *51*, 51–57. [CrossRef]

20. Condrat, C.E.; Thompson, D.C.; Barbu, M.G.; Bugnar, O.L.; Boboc, A.; Cretoiu, D.; Suciu, N.; Cretoiu, S.M.; Voinea, S.C. miRNAs as biomarkers in disease: Latest findings regarding their role in diagnosis and prognosis. *Cells* **2020**, *9*, 276. [CrossRef]

21. Lauria, F.; Iacomino, G. The landscape of circulating miRNAs in the post-genomic era. *Genes* **2022**, *13*, 94. [CrossRef]

22. Dumortier, O.; Hinault, C.; Van Obberghen, E. MicroRNAs and metabolism crosstalk in energy homeostasis. *Cell Metab.* **2013**, *18*, 312–324. [CrossRef]

23. Iacomino, G.; Siani, A. Role of microRNAs in obesity and obesity-related diseases. *Genes Nutr.* **2017**, *12*, 23. [CrossRef]

24. Kunej, T.; Jevsinek Skok, D.; Zorc, M.; Ogrinc, A.; Michal, J.J.; Kovac, M.; Jiang, Z. Obesity gene atlas in mammals. *J. Genom.* **2013**, *1*, 45–55. [CrossRef] [PubMed]

25. Hartig, S.M.; Hamilton, M.P.; Bader, D.A.; McGuire, S.E. The miRNA interactome in metabolic homeostasis. *Endocrinol. Metab.* **2015**, *26*, 733–745. [CrossRef] [PubMed]

26. Raghuraman, S.; Donkin, I.; Versteyhe, S.; Barres, R.; Simar, D. The emerging role of epigenetics in inflammation and immunometabolism. *Trends Endocrinol. Metab.* **2016**, *27*, 782–795. [CrossRef] [PubMed]

27. Kroesen, B.J.; Teteloshvili, N.; Smigielska-Czepiel, K.; Brouwer, E.; Boots, A.M.; van den Berg, A.; Kluiver, J. Immuno-miRs: Critical regulators of T-cell development, function and ageing. *Immunology* **2015**, *144*, 1–10. [CrossRef]

28. Iacomino, G.; Russo, P.; Stillitano, I.; Lauria, F.; Marena, P.; Ahrens, W.; De Luca, P.; Siani, A. Circulating microRNAs are deregulated in overweight/obese children: Preliminary results of the I.Family study. *Genes Nutr.* **2016**, *11*, 7. [CrossRef] [PubMed]

29. Ahrens, W.; Siani, A.; Adan, R.; De Henauw, S.; Eiben, G.; Gwozdz, W.; Hebestreit, A.; Hunsberger, M.; Kaprio, J.; Krogh, V.; et al. Cohort profile: The transition from childhood to adolescence in European children-how I.Family extends the IDEFICS cohort. *Int. J. Epidemiol.* **2017**, *46*, 1394j–1395j. [CrossRef] [PubMed]

30. Iacomino, G.; Russo, P.; Marena, P.; Lauria, F.; Venezia, A.; Ahrens, W.; De Henauw, S.; De Luca, P.; Foraita, R.; Gunther, K.; et al. Circulating microRNAs are associated with early childhood obesity: Results of the I.Family study. *Genes Nutr.* **2019**, *14*, 2. [CrossRef] [PubMed]

31. Bammann, K.; Lissner, L.; Pigeot, I.; Ahrens, W. (Eds.) *Instruments for Health Surveys in Children and Adolescents*; Springer International Publishing: Berlin/Heidelberg, Germany, 2019.

32. Iacomino, G.; Lauria, F.; Russo, P.; Marena, P.; Venezia, A.; Iannaccone, N.; De Henauw, S.; Foraita, R.; Heidinger-Felso, R.; Hunsberger, M.; et al. Circulating miRNAs are associated with sleep duration in children/adolescents: Results of the I.Family study. *Exp. Physiol.* **2020**, *105*, 347–356. [CrossRef]

33. Cole, T.J.; Lobstein, T. Extended international (IOTF) body mass index cut-offs for thinness, overweight and obesity. *Pediatr. Obes.* **2012**, *7*, 284–294. [CrossRef]
34. Peplies, J.; Fraterman, A.; Scott, R.; Russo, P.; Bammann, K. Quality management for the collection of biological samples in multicentre studies. *Eur. J. Epidemiol.* **2010**, *25*, 607–617. [CrossRef]
35. Shanahan, L.; Copeland, W.E.; Worthman, C.M.; Erkanli, A.; Angold, A.; Costello, E.J. Sex-differentiated changes in C-reactive protein from ages 9 to 21: The contributions of BMI and physical/sexual maturation. *Psychoneuroendocrinology* **2013**, *38*, 2209–2217. [CrossRef] [PubMed]
36. Lindsay, M.A. microRNAs and the immune response. *Trends. Immunol.* **2008**, *29*, 343–351. [CrossRef]
37. Nejad, C.; Stunden, H.J.; Gantier, M.P. A guide to miRNAs in inflammation and innate immune responses. *FEBS J.* **2018**, *285*, 3695–3716. [CrossRef]
38. Piening, B.D.; Zhou, W.; Contrepois, K.; Rost, H.; Gu Urban, G.J.; Mishra, T.; Hanson, B.M.; Bautista, E.J.; Leopold, S.; Yeh, C.Y.; et al. Integrative personal omics profiles during periods of weight gain and loss. *Cell Syst.* **2018**, *6*, 157–170.e8. [CrossRef]
39. Li, Q.; Wang, Q.; Xu, W.; Ma, Y.; Wang, Q.; Eatman, D.; You, S.; Zou, J.; Champion, J.; Zhao, L.; et al. C-reactive protein causes adult-onset obesity through chronic inflammatory mechanism. *Front. Cell Dev. Biol.* **2020**, *8*, 18. [CrossRef]
40. Chen, L.; Al-Mossawi, M.H.; Ridley, A.; Sekine, T.; Hammitzsch, A.; de Wit, J.; Simone, D.; Shi, H.; Penkava, F.; Kurowska-Stolarska, M.; et al. miR-10b-5p is a novel Th17 regulator present in Th17 cells from ankylosing spondylitis. *Ann. Rheum. Dis.* **2017**, *76*, 620–625. [CrossRef]
41. Tu, J.; Han, D.; Fang, Y.; Jiang, H.; Tan, X.; Xu, Z.; Wang, X.; Hong, W.; Li, T.; Wei, W. MicroRNA-10b promotes arthritis development by disrupting CD4$^+$ T cell subtypes. *Mol. Ther.-Nucleic Acids* **2022**, *27*, 733–750. [CrossRef]
42. Sun, J.; Yan, P.; Chen, Y.; Chen, Y.; Yang, J.; Xu, G.; Mao, H.; Qiu, Y. MicroRNA-26b inhibits cell proliferation and cytokine secretion in human RASF cells via the Wnt/GSK-3β/β-catenin pathway. *Diagn. Pathol.* **2015**, *10*, 72. [CrossRef]
43. Ji, Y.; He, Y.; Liu, L.; Zhong, X. MiRNA-26b regulates the expression of cyclooxygenase-2 in desferrioxamine-treated CNE cells. *FEBS Lett.* **2010**, *584*, 961–967. [CrossRef]
44. Yu, T.; Ma, P.; Wu, D.; Shu, Y.; Gao, W. Functions and mechanisms of microRNA-31 in human cancers. *Biomed. Pharmacother.* **2018**, *108*, 1162–1169. [CrossRef] [PubMed]
45. Tian, Y.; Xu, J.; Li, Y.; Zhao, R.; Du, S.; Lv, C.; Wu, W.; Liu, R.; Sheng, X.; Song, Y.; et al. MicroRNA-31 reduces inflammatory signaling and promotes regeneration in colon epithelium, and delivery of mimics in microspheres reduces colitis in mice. *Gastroenterology* **2019**, *156*, 2281–2296.e6. [CrossRef] [PubMed]
46. Zhang, L.; Chen, J.; Wang, L.; Chen, L.; Du, Z.; Zhu, L.; Cui, M.; Zhang, M.; Song, L. Linc-PINT acted as a tumor suppressor by sponging miR-543 and miR-576-5p in esophageal cancer. *J. Cell Biochem.* **2019**, *120*, 19345–19357. [CrossRef] [PubMed]
47. Lefèvre, N.; Corazza, F.; Duchateau, J.; Desir, J.; Casimir, G. Sex differences in inflammatory cytokines and CD99 expression following in vitro lipopolysaccharide stimulation. *Shock* **2012**, *38*, 37–42. [CrossRef] [PubMed]
48. Wegner, A.; Benson, S.; Rebernik, L.; Spreitzer, I.; Jäger, M.; Schedlowski, M.; Elsenbruch, S.; Engler, H. Sex differences in the pro-inflammatory cytokine response to endotoxin unfold in vivo but not ex vivo in healthy humans. *Innate Immun.* **2017**, *23*, 432–439. [CrossRef]
49. Klein, S.L.; Flanagan, K.L. Sex differences in immune responses. *Nat. Rev. Immunol.* **2016**, *16*, 626–638. [CrossRef]
50. Sharma, S.; Eghbali, M. Influence of sex differences on microRNA gene regulation in disease. *Biol. Sex Differ.* **2014**, *5*, 3. [CrossRef]

Article

miRNAs Copy Number Variations Repertoire as Hallmark Indicator of Cancer Species Predisposition

Chiara Vischioni [1,2], Fabio Bove [3], Matteo De Chiara [2], Federica Mandreoli [3], Riccardo Martoglia [3], Valentino Pisi [3], Gianni Liti [2] and Cristian Taccioli [1,*]

1 Department of Animal Medicine, Production and Health, University of Padova, 35020 Legnaro, Italy; chiara.vischioni@phd.unipd.it
2 IRCAN, CNRS, INSERM, Université Côte d'Azur, 06107 Nice, France; matteo.dechiara@unice.fr (M.D.C.); gianni.liti@unice.fr (G.L.)
3 Department of Physics, Informatics and Mathematics, University of Modena and Reggio Emilia, 41125 Modena, Italy; fabio.bove.dr@gmail.com (F.B.); federica.mandreoli@unimo.it (F.M.); riccardo.martoglia@unimo.it (R.M.); valentino.pisi@gmail.com (V.P.)
* Correspondence: cristian.taccioli@unipd.it

Abstract: Aging is one of the hallmarks of multiple human diseases, including cancer. We hypothesized that variations in the number of copies (CNVs) of specific genes may protect some long-living organisms theoretically more susceptible to tumorigenesis from the onset of cancer. Based on the statistical comparison of gene copy numbers within the genomes of both cancer-prone and -resistant species, we identified novel gene targets linked to tumor predisposition, such as CD52, SAT1 and SUMO. Moreover, considering their genome-wide copy number landscape, we discovered that microRNAs (miRNAs) are among the most significant gene families enriched for cancer progression and predisposition. Through bioinformatics analyses, we identified several alterations in miRNAs copy number patterns, involving miR-221, miR-222, miR-21, miR-372, miR-30b, miR-30d and miR-31, among others. Therefore, our analyses provide the first evidence that an altered miRNAs copy number signature can statistically discriminate species more susceptible to cancer from those that are tumor resistant, paving the way for further investigations.

Keywords: DNA copy number variation; miRNAs; comparative study

Citation: Vischioni, C.; Bove, F.; De Chiara, M.; Mandreoli, F.; Martoglia, R.; Pisi, V.; Liti, G.; Taccioli, C. miRNAs Copy Number Variations Repertoire as Hallmark Indicator of Cancer Species Predisposition. *Genes* **2022**, *13*, 1046. https://doi.org/10.3390/genes13061046

Academic Editors: Giuseppe Iacomino and Fabio Lauria

Received: 5 May 2022
Accepted: 4 June 2022
Published: 10 June 2022

Publisher's Note: MDPI stays neutral with regard to jurisdictional claims in published maps and institutional affiliations.

1. Introduction

Aging is considered one of the risk factors of cancer insurgence due to the mutational burden derived from cell division and DNA replication [1]. Therefore, it is probable that, in order to maintain a high longevity rate, those organisms that live longer should theoretically possess a higher risk of cancer occurrence. Nevertheless, considering different species, according to Peto's Paradox theory [2], the body size of an organism and/or its lifespan expectation are not directly correlated with the species percentage of cancer incidence. After more than 40 years of research, the solution to this puzzling paradox is still an open challenge to be solved. For example, despite its small size, the naked mole rat is, to date, the longest-living member of the rodent family, being able to live more than 30 years. Several studies highlighted that, besides the delayed aging, this species also shows the capacity to resist spontaneous and experimentally induced tumorigenesis [3–6]. Conversely, in some other rodents, the cancer-related mortality can reach 90%, coupled with a species maximum life expectancy of 4–5 years [7]. The long-living *Myotis lucifugus* bat species has been recently identified as a prospective organism for comparative cancer research [8]. Given their extended life-span rates [9], it has been suggested that bats develop a very low number of cancer events, as confirmed by different pathological studies performed in different areas of the world [10,11]. The elephant has been pinpointed as another cancer-resistant species [12], with a cancer incidence rate considerably lower compared to the human one, for example (approximately 22%) [13]. In order to maintain a high longevity, some species might have developed intrinsic molecular mechanisms that protect them from cancer onset or development [14]. Interestingly, various authors recently reported that the genome of the African elephant encodes multiple copies of the TP53 gene, also known as the "guardian of the genome stability". This amplification could be at the basis of the elephant's anti-cancer and longevity mechanisms by leading to increased levels of apoptosis in response to DNA damage [12,15]. Indeed, according to Caulin and Maley (2011) [16], the genome of large long-living organisms can reveal an altered number of tumor suppressors and oncogenes (in multiple or reduced copies), which might represent a possible mechanism underlying their capacity of exceeding the threshold of cancer onset, despite their phenotypic predisposition due to body size and longevity [16]. Copy number variations (CNVs) are duplications or deletions of genomic regions which can be associated with phenotypic alterations, including tumorigenic diseases [17]. In particular, a variation in the gene copy numbers can influence the activity of tumor suppressors and oncogenes, leading to the development of cancer [18]. Within this framework, long-living animals have to rely on compensatory mechanisms to suppress and prevent cancer progression, which can be straightened by different molecular and genomic mechanisms such as altered gene copy numbers that increase the number of tumor suppressors paralogues or reduce copies of oncogenes [19,20]. As previously mentioned, mammals have evolved lifespan and cancer incidence rates that vary among species [21], but the mechanisms underlying these differences are still unclear. In order to test the hypothesis that genomic CNVs are related to the cancer incidence rate of a species, we compared the genome-wide copy number landscapes of nine different mammals (five cancer-resistant and four cancer-prone species) and identified the target genes that can significantly discriminate between these two groups. Contrary to what is usually done, we did not use an a priori list of cancer-related genes but included all human genes in our analysis dataset. In this way, we were able to identify miRNAs, usually removed from evolutionary comparative analyses, as the most enriched elements able to discriminate those organisms that are predisposed to cancer from those that are not.

2. Materials and Methods

2.1. Data Collection

According to the hypothesis that positively selected CNVs tend to recur during cancer progression [22,23], but also during the evolution [24], we have recently developed the VarNuCopy database, a unique database that collects the CNVs landscape for multiple organisms, with the aim to compare patterns of copy number changes across the genome of different species [25]. We used a homemade script written in Perl 5.14 and Python 3 in order to download the CNV data from Ensembl comparative genomics resources (http://www.ensembl.org accessed on 1 March 2019) [26], an ideal system to perform and support vertebrate comparative genomic analyses, given the consistency of gene annotation across the genomes of different vertebrate species. We leveraged Ensembl's "gene gain/loss tree" feature, which displays the number of copies of extant homologous genes for each species in a taxonomic tree view [27]. These data are estimated through CAFE (Computational Analysis of gene Family Evolution), a computational tool commonly used to study gene family evolution, which takes into account a priori the species phylogenetic tree [27,28]. The Perl API script provided by the Ensembl website was used to access the genomic databases and used to download all the available CNVs data. We encoded a new homemade Python script to arrange the CNVs data counts in a readable tab-delimited format and used this matrix to perform the subsequent analysis.

2.2. Statistical Comparison

Using a comparative approach, we analyzed the variation landscape of the gene copies among the genomes of nine organisms sub-set in two categories: "cancer resistant" (*Heterocephalus glaber* (Hg), *Nannospalax galili* (Ng), *Dasypus novemcinctus* (Dn),

Loxodonta africana (La) and *Myotis lucifugus* (Ml)) and "cancer prone" (*Mus musculus* (Mm), *Rattus norvegicus* (Rn), *Canis familiaris* (Cf) and *Homo sapiens* (Hs)) species (Supplementary Table S1). We classified as "cancer resistant" those species that, based on the literature review, are known to possess a low cancer incidence rate. Conversely, "cancer prone" organisms are those species for which the percentage of tumors found in a certain number of necropsies is known to be high.

Cancer incidence rate data were collected from different recently published literature [4–6,8,10–12,15,21,29–33]. We performed a statistical comparison between the CNVs of the two different species groups, cancer-prone and -resistant organisms, with the aim to identify new possible gene targets able to discriminate between the two categories. Thus, a statistical unpaired two-group Wilcoxon test was performed using R.3.1.1(R Foundation for Statistical Computing, Vienna, Austria), to compare their entire CNV spectra. To determine whether microRNAs CNVs independently contribute to the variation in cancer incidence percentages among our species, we applied a linear regression model through the PGLS R package [34], in order to check for potential bias due to species phylogeny or population structure (Figure 1D). The phylogenetic tree included in the analysis was derived from VertLife [35] and created and visualized through the Interactive Tree of Life web-tool (Figure 1C) [36]. Data processing and statistical tests were performed with R.3.1.1. Figures were made using the ggplot2 R package, in association with different R Shiny apps such as BoxPlotR, PlotsOfData, ClustVis, and miRTargetLink 2.0 [37–40].

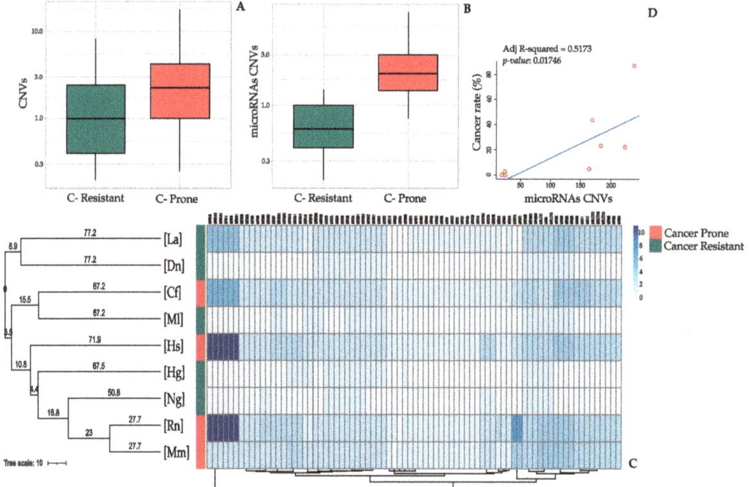

Figure 1. CNV landscape comparisons: (**A**) Boxplot of the distribution of significant gene CNVs in cancer-prone vs. cancer-resistant species. (**B**) Boxplot of the distribution of significant microRNA CNVs in cancer-prone vs. cancer-resistant species. Cancer-resistant species are highlighted in green, cancer-prone species in red. In the boxplots, the *Y*-axis scale has been changed to log one. The boxplots were built considering the average number of copies of each gene in the two different target groups. (**C**) Heatmap representing the microRNA CNV repertoires within the nine analyzed species—(Hg): *Heterocephalus glaber*; (Ng): *Nannospalax galili*; (Dn): *Dasypus novemcinctus*; (La): *Loxodonta Africana*; (Ml): *Myotis lucifugus*; (Mm): *Mus musculus*; (Rn): *Rattus norvegicus*; (Cf): *Canis familiaris*; (Hs): *Homo sapiens*. Hg, Ng, Dn, La and Ml have been previously described as cancer-resistant species. Mm, Rn, Cf and Hs are known to be cancer-prone species. Phylogeny was inferred from VertLife [35], created and visualized through the Interactive Tree of Life web-tool [36]. (**D**) PGLS correlating the cancer incidence rate with the total number of significant microRNAs copies across the nine species included in the analysis. The blue line represents a positive correlation between the two variables (adjusted R^2 = 0.5173; *p*-value = 0.01746).

2.3. Pathways Analysis

To determine if the CNVs are enriched in specific gene families, we used Gene SeT AnaLysis Toolkit, a tool for the interpretation of lists of interesting genes that is commonly used to extract biological insights from targets of interest [41]. The set of significant genes were tested for pathway associations using the hyper-geometric test for over-representation analysis (ORA) [42] (Supplementary Table S4). We selected different pathway enrichment categories (KEGG: https://www.genome.jp/; Wikipathway: https://www.wikipathways.org; Reactome: https://reactome.org/; PANTHER: http://www.pantherdb.org/ accessed on 1 June 2019), considering as over-represented those molecular networks with an FDR significance level lower than 0.05, after a correction with the Benjamini–Hochberg method. In this context, the ORA analysis was the preferred option among the others (e.g., gene set enrichment or network topology-based analysis) in order to obtain biological information underlying the significantly enriched genes, resulting in a reduction in the complexity of the data interpretation [42].

3. Results

A two-group comparison was performed using a Wilcoxon rank sum test, in order to identify an existing distinction in the distribution of the number of gene copies between cancer-prone and cancer-resistant species. A list of the most significant hits (p-value < 0.05), including known tumor suppressors and oncogenes, is reported in Table 1 (see Supplementary Tables S2 and S3 for the extended version). Our analysis, which exclusively considered the variation in number of gene copies among different species, was able to identify those genes involved in biological processes related to cancer development and maintenance.

Table 1. Genomic CNV landscape comparisons. Subset of 25 significant hits resulting from the unpaired 2-group Wilcoxon test (p-value < 0.05). The statistical comparison was made in order to identify those genes able to discriminate between the cancer-prone and -resistant species groups, relying exclusively on the genomic copy number values. Some of these genes are already known to be tumor suppressor and/or oncogenes, whereas the others can play pivotal roles in tumorigenesis events, and, for this reason, can be considered as targets to be further investigated and validated in the context of cancer development.

Gene	p-Value	Known_TS	Known_OG	References
CD52	0.007	NO	NO	[43]
SAT1	0.007	NO	NO	[44]
MIR424	0.009	YES	NO	[45]
MIR372	0.010	NO	YES	[46,47]
DMD	0.014	YES	NO	[48]
EIF5	0.017	NO	NO	[49]
MIR107	0.022	YES	YES	[50,51]
MIR124-1, MIR124-2, MIR124-3	0.022	YES	NO	[52]
SUMO2, SUMO3, SUMO4	0.024	NO	NO	[53,54]
MIR506	0.029	YES	YES	[55]
MIR509-1	0.029	NO	NO	[56]
MIR511	0.029	YES	NO	[57]
MIR514A1, MIR514A3, MIR514B	0.029	NO	NO	[58]
MIR378A	0.030	YES	NO	[59]

Table 1. *Cont.*

Gene	*p*-Value	Known_TS	Known_OG	References
S100A16	0.030	NO	NO	[60]
MBD1, MBD2, MBD3	0.031	NO	YES (MDB1)	[61]
FGFBP1	0.032	NO	NO	[62]
FOXJ1	0.032	NO	NO	[63]
MIR1-1, MIR1-2	0.032	YES	NO	[64]
MIR206	0.032	YES	NO	[65]
MIR340	0.032	YES	NO	[66]
MIR542	0.032	NO	NO	[67]
NUPR1	0.032	YES	NO	[68]
SELENOW	0.032	NO	NO	[69,70]
JUND	0.034	NO	YES	[71]

3.1. Best Candidate Cancer-Related Genes

The distribution of the average number of each gene copies plotted in Figure 1A highlights a difference between the two species categories, which appears even greater if we only refer to the microRNAs CNVs landscape (Figure 1B). Among the most significant genes presenting an altered number of copies, we found CD52 (*p*-value = 0.007), SAT1 (*p*-value = 0.007), DMD (*p*-value = 0.014), EIF5 (*p*-value = 0.017), SUMO2, SUMO3, SUMO4 (*p*-value = 0.024), S100A16 (*p*-value = 0.030), MBD1, MBD2, MBD3 (*p*-value = 0.031), FGFBP1 (*p*-value = 0.032), FOXJ1 (*p*-value = 0.032), NUPR1 (*p*-value = 0.032), SELENOW (*p*-value = 0.032) and JUND (*p*-value = 0.034). Some of these, such as DMD, MDB1, NUPR1 and JUND, have been already well described as tumor suppressors or oncogenes [47,60,67,70], whereas the others do not officially belong to any of these two categories and they have been proposed as key regulators in biological processes such as cell proliferation, migration and cancer development and progression [42,43,48,52,53,59,61,62,68,69]. A Principal Component Analysis (PCA) of the CNV values of the nine species reported in Figure 2A,B showed a clear dichotomy between the cancer-prone and -resistant groups, supporting the hypothesis that an altered CNV landscape is able to discriminate between the two categories. To confirm these results, we performed another unsupervised clustering analysis using Euclidean distance (Figure 2C).

As depicted in the heatmap, each cluster has a distinct set of copy number values, and the main branches representing cancer-prone and -resistant organisms perfectly distinguish the two groups. No additional information (other than copy numbers) was given to the algorithm. In addition, we applied the Euclidean distances, using both the 'complete' and 'ward' methods (criteria that direct how the subclusters are merged) (Supplementary Figures S2–S4)). Remarkably, using this method, the *Loxodonta africana* microRNAs CNV landscape seems to have a different pattern as compared to the other cancer-resistant species (Figure 2C), confirming the elephant as an outlier species of the cancer-resistant group (see Section 4).

3.2. Cancer-Related MicroRNAs Pathways Are among the Most Significantly Enriched Biological Families

Our analysis shows an enrichment of onco-miRNAs amplifications in the cancer-prone species group. In particular, miR-424 (*p*-value = 0.009), miR-372 (*p*-value = 0.010), miR-107 (*p*-value = 0.022), miR-124 (*p*-value = 0.022), miR-506 (*p*-value = 0.029), miR-511 (*p*-value = 0.029), miR-378A (*p*-value = 0.030), miR-1 (*p*-value = 0.032), miR-206 (*p*-value = 0.032) and miR-340 (*p*-value = 0.032) are few examples of the most significant microRNA hits, which possess a suppressor and/or oncogenic role (Figure 1C). Given the high diversity among our species, we used the generalized least squares (PGLS) phylogenetic method [34] in order to assess whether copy number and cancer incidence rates evolved in a dependent manner along the tree, or if their relationship might be the consequence of common ancestry, resulting in similar patterns of miRNAs copy number alteration. Indeed, taking into account the genetic structure of the population, the PGLS comparative method confirmed the association between these traits independently of the shared evolutionary history of the species (Figure 1D and Supplementary File S1).

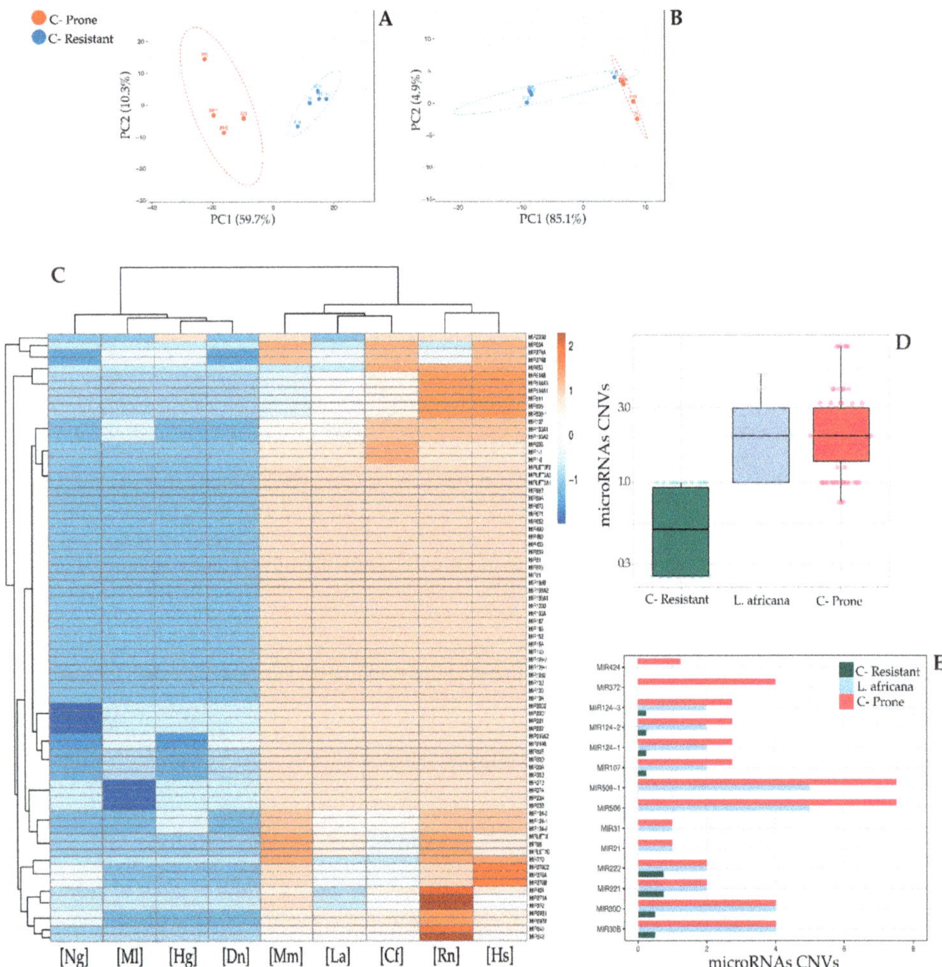

Figure 2. (**A**) PCA based on the CNVs of all the significant genes. (**B**) PCA based on the CNVs of the significant microRNAs subset. Both plots show a dichotomy between cancer-resistant (blue) and cancer-prone species (red). (**C**) Heatmap of the significant microRNAs, clustered with Euclidean distance and complete linkage. (**D,E**) Bar and box plots of the significant microRNAs CNVs in cancer-prone species, cancer-resistant species and *Loxodonta africana*. The microRNAs repertoire of *Loxodonta africana* seems to reflect the cancer-prone miRNAs copy number alteration landscape, rather than the one typical of the cancer-resistant organisms. In the box plots, the *Y*-axis scale was changed to log one. The boxplots are built considering the average number of copies of each gene in the two different target groups.

3.3. ORA Analysis Confirms a Significant Enrichment in the miRNAs Gene Family

We performed an Over-Representation Analysis (ORA) [41] on the complete list of significant genes, in order to identify enriched functional categories potentially related to cancer (Table 2 and Supplementary Figure S1). The most enriched pathways outputted by the ORA analysis were "MicroRNAs in cancer", "miRNAs involved in DNA damage response", "Metastatic brain tumor", "miRNA targets in ECM and membrane receptors", "let-7 inhibition of ES cell reprogramming" and "miRNAs involvement in the immune response in sepsis" [72,73]. These results indicate that the genes more prone to CNVs were those encoding miRNAs involved in cancer initiation, chronic inflammation and immune response. Remarkably, performing the ORA analysis applying the PANTHER algorithm [74], we also found a significant enrichment in the "Cadherin signaling network", which is a well-known molecular pathway described as a key player in cancer [75].

Table 2. Pathway analysis. Gene Over-Representation Analysis (ORA) using KEGG, PANTHER and Wikipathway. The enrichment test used Benjamini–Hochberg's FDR correction (FDR < 0.05). CNV data were previously analyzed by an unpaired 2-group Wilcoxon test (p-value < 0.05). Significant genes altered in their number of copies within the entire genomic landscape were used to perform the ORA analysis, which highlighted a significant enrichment in microRNAs and cancer-related pathways.

	Description	FDR (BH)	Genes
KEGG	MicroRNAs in cancer	0	MIR103A1; MIR103A2; MIR107; MIR124-1; MIR124-2; MIR124-3; MIR1-1; MIR1-2; MIR206; MIR100; MIR10A; MIR10B; MIR129-1; MIR129-2; MIR15A; MIR15B; MIR193B; MIR199A1; MIR199A2; MIR199B; MIR203B; MIR21; MIR223; MIR31; MIR99A; MIRLET7A1; MIRLET7A3; MIRLET7F2; MIR29B1; MIR29B2; MIRLET7G; MIRLET7I; MIR221; MIR222; MIR23A; MIR23B; MIR27A; MIR27B; MIR30C1; MIR30C2; MIR30A; MIR30B; MIR30D; MIR30E.
	Taste transduction	3.16×10^{-10}	TAS2R10; TAS2R13; TAS2R14; TAS2R19; TAS2R20; TAS2R3; TAS2R30; TAS2R31; TAS2R42; TAS2R43; TAS2R45; TAS2R46; TAS2R50; TAS2R7; TAS2R8; TAS2R9
	Progesterone-mediated oocyte maturation	2.43×10^{-4}	SPDYE1; SPDYE11; SPDYE16; SPDYE17; SPDYE2; SPDYE2B; SPDYE3; SPDYE4; SPDYE5; SPDYE6; INS
	Oocyte meiosis	2.73×10^{-4}	PPP3R2; SPDYE1; SPDYE11; SPDYE16; SPDYE17; SPDYE2; SPDYE2B; SPDYE3; SPDYE4; SPDYE5; SPDYE6; INS
PANTHER	Cadherin signaling pathway	4.02×10^{-2}	PCDHB14; PCDHB7; PCDHGB1; PCDHB16; PCDHB6; PCDHGB4; PCDHGA6; PCDHGB6; PCDHGB7
Wikipathway	miRNAs involved in DNA damage response	3.76×10^{-9}	MIR371A; MIR372; MIR542; MIR100; MIR15B; MIRLET7A1; MIR374B; MIR221; MIR222; MIR23A; MIR23B; MIR27A; MIR27B
	Alzheimers Disease	5.31×10^{-5}	MIR124-1; MIR124-2; MIR124-3; MIR10A; MIR129-1; MIR129-2; MIR199B; MIR21; MIR433; MIR671; MIR873; PPP3R2; MIR29B1; MIR30C2; MIR219A2
	Metastatic brain tumor	2.31×10^{-3}	MIRLET7A1; MIRLET7A3; MIRLET7F2; MIR29B1; MIR29B2; MIRLET7G
	miRNA targets in ECM and membrane receptors	2.31×10^{-3}	MIR107; MIR15B; MIR30C1; MIR30C2; MIR30B; MIR30D; MIR30E

Table 2. *Cont.*

Description	FDR (BH)	Genes
MicroRNAs in cardiomyocyte hypertrophy	2.77×10^{-3}	MIR103A1; MIR103A2; MIR140; MIR15B; MIR185; MIR199A1; MIR199A2; MIR23A; MIR27B; MIR30E
Cell Differentiation - Index	1.25×10^{-2}	MIR1-1; MIR206; MIR199A1; MIR199A2; MIR221; MIR222
let-7 inhibition of ES cell reprogramming	1.25×10^{-2}	MIRLET7A1; MIRLET7F2; MIRLET7G; MIRLET7I
miRNAs involvement in the immune response in sepsis	1.43×10^{-2}	MIR187; MIR199A1; MIR199A2; MIR203B; MIR223; MIR29B1; MIRLET7I
Cell Differentiation- Index expanded	2.38×10^{-2}	MIR1-1; MIR206; MIR199A1; MIR199A2; MIR221; MIR222
Role of Osx and miRNAs in tooth development	3.35×10^{-2}	MIRLET7A1; MIRLET7F2; MIR29B1; MIRLET7G; MIRLET7I

4. Discussion

Being theoretically more susceptible to cancer, big and long-living species need additional cancer defense molecular mechanisms. On the other hand, short-living and small-size organisms might not need them because of their lower intrinsic predisposition to cancer due to their short lifespan rate. CNVs can therefore be considered one of the multiple protection ways against tumor insurgence that can explain Peto's Paradox. In fact, we hypothesized that all cancer-resistant organisms implemented a series of molecular mechanisms to counteract their cancer predisposition, which depends on and derives from their own specific evolutionary history. We believe that CNVs that increase the onco-suppressive capacity of specific genes can be an excellent defense against tumor diseases in species at risk. Indeed, some authors have recently suggested that one of the most effective cancer-resistance strategies is represented by an augmentation in the number of copies of tumor suppressor genes [76]. In contrast, a reduced cancer-resistance rate could be caused by a selective loss of the same suppressor genes [77]. For instance, the CD52 gene (higher number of copies in the cancer prone group), a membrane glycoprotein expressed on the surface of mature lymphocytes, monocytes and dendritic cells, was one of the most significant hits of our analysis (*p*-value = 0.007). Recently, Wang and co-authors [43] identified CD52 as a key player in tumor immunity, affecting tumor behavior by regulating the associated tumor microenvironment. With the same significant *p*-value of 0.007, we also identified the SAT1 gene (higher number of copies in the cancer prone group) as one of the possible targets to be further investigated in the context of tumor onset. This gene can regulate and drive brain tumor aggressiveness, promoting molecular pathways that act in response to DNA damage and regulation of the cell cycle [44]. Another significant gene resulting from our analysis was represented by the SUMO protein family members (higher number of copies in the cancer resistant group). During cell cycle progression, many tumor suppressors and oncogenes are regulated via SUMOylation [78], a biological process that, if deregulated, can lead to genome instability and altered cell proliferation. In this context, it is evident that some tumors could be dependent on the functional SUMO pathway, but whether it is required for tumor growth remains to be established. For this reason, SUMO2, SUMO3 and SUMO4 can be potentially exploited in further anti-cancer mechanisms investigations (*p*-value = 0.024 in the present study), in order to shed light on the regulatory mechanisms underlying the activity of SUMO machinery in an oncogenic framework. Among the most significant hits, we also retrieved some genes that are already known to be tumor suppressors or oncogenes (DMD and JUND, respectively). Indeed, mutation or deregulated expression of Duchenne Muscular Dystrophy gene (DMD) is often linked to the development and progression of some major cancer types [48], such as sarcomas, carcinomas, melanomas, lymphomas and brain tumors [79,80], being a well-known tumor suppressor in different types of human cancers. On the other hand, JUND, a member of the AP-1 family that is related to MYC signaling pathway, regulates cell cycle and proliferation and its overexpression is linked to many types of cancers (PCA i.e.,) [71].

Notably, our results show that miRNAs are the most enriched gene family in discriminating between cancer-prone and cancer-resistant species. The specific role of these miRNAs is not yet fully understood, but we speculate that some of them might possess important regulatory functions aimed at defending some species (big size and long lifespan organisms) from cancer, while, at the same time, they are capable of exposing others to tumorigenesis (small-size and short-lifespan mammals). MicroRNAs (miRNAs) are small post-transcriptional molecular regulators that are able to modify gene expression levels, increasing the amount of mRNA degradation or inhibiting protein translation [81]. Since each single miRNA can regulate the expression of dozens of genes, many authors were able to correlate their activity with cell development, homeostasis and disease [82], including cancer [83,84]. Indeed, some tumorigenic events are caused by a malfunction in the heterogeneous regulatory activity of microRNAs inside the eukaryotic cells. Depending on the specific tissue and on the relationship with the immune system, they can behave both as tumor suppressors and as oncogenes [85]. Furthermore, epigenetic factors and species genetic predisposition can drive their double-sided behavior, although some of them are evolutionarily conserved within vertebrate taxonomic families [86]. Several miRNAs have already been described in the literature as oncogenes and tumor suppressors. For example, miR-424 is known to be a human tumor suppressor that can inhibit cell growth enhancing apoptosis or suppressing cell migration [45]. MiR-372, instead, can participate in WNT cancer molecular pathway [46], whereas the overexpression of miR-107, mediating p53 regulation of hypoxic signaling, can suppress tumor angiogenesis and growth in mice [50]. MiR-1 is another example of tumor suppressor microRNA that has been previously found to be significantly down-regulated in squamous carcinoma cells [64]. MiR-30b and miR-30d are considered suppressors in tumors that do not affect immune cells, whereas they have been found to be upregulated in melanoma [87]. In a similar way and for the first time, our analysis revealed several miRNAs candidates that might be involved in a mammalian species cancer predisposition (Figure 1C).

Interestingly, all the miRNAs that we have found show many more copies in the cancer-prone group as compared to the cancer-resistant species, and most of them are well-known oncogenes (miR-221, miR222, and miR-372, etc.). MiR-372, for instance, is not present in cancer-resistant species, whereas it shows multiple copies in those ones belonging to the cancer-prone group. This microRNA plays a pivotal role in the initiation of breast cancer and may activate the WNT and E2F1 pathways during the epithelial–mesenchymal transition process [46,47]. We also found an amplification of miR-221 and miR-222 in the cancer-prone category. Previous literature has extensively described these two RNAs as oncogenes, being deregulated in primary brain tumors and in acute lymphoid leukemia, among other malignancies [88,89]. According to our results, surprisingly, cancer-prone species showed the amplification of miR-15 tumor suppressor, which is known to be able to regulate cancer proliferation initiation by targeting the BCL2 gene [90,91]. Our hypothesis is that this apparent paradox may underlie a defensive role of this microRNA in those species that are, a priori, susceptible to tumor insurgence. On one hand, according to the so-called "gene dosage hypothesis", gains or losses of specific gene copies can have a dramatic impact on the fitness of a species, leading to altered phenotypes due to the change in the expression levels of the affected genes [92]. On the other hand, oncogenes amplification or tumor suppressors deletions are not always detrimental, but can recapitulate tumorigenic events, being drivers or modulators of the disease [93]. As mentioned before, in fact, differences in ecology and evolutionary history are believed to give rise to significant differences between short- and long-living animals [94], and consequently in cancer-prone and -resistant species. In 2020, Tollis and co-authors [20] showed that mammalian lifespan can be correlated to both suppressor gene and oncogene CNVs, a phenomenon that they themselves called "paradoxical". Interestingly, our analysis also leans in the same direction, suggesting that when high copy numbers of oncogenes shorten a lifespan, they must somehow be counterbalanced by higher number of copies of tumor-suppressor genes.

In this framework, the elephant's miRNAs amplification signature resembles that of the organisms of the cancer-prone group (Figure 2D,E). In fact, it showed an alteration in the copy numbers of known oncogenes, such as miR-221 and miR-222, together with miR-30b/d and miR-31. In our opinion, *Loxodonta africana* should be placed in a new category of organisms, which share both oncogenic and cancer-resistant characteristics, being also clustered as an outlier species of the cancer-resistant group (Figure 2B). During their evolution, elephants may have selected certain molecular mechanisms, such as the amplification of TP53 and pseudogenes [12,15], with the aim to defend their cells from the tumorigenic action of a high percentage of onco-miRNAs copy number amplification and high longevity. Consequently, an additional amplification in the number of tumor suppressor microRNAs would have not been sustainable/useful in terms of fitness and/or survival. The hypothesis is that species with bigger sizes and longer lifespans have an expanded number of tumor-suppressor genes (TSGs), which is even higher than the one of their oncogenic counterparts. In this way, the direct elimination of oncogenes, which implies elevated costs in terms of growth and cellular functions maintenance, can be avoided, thus reducing the cancer incidence risk. In support of this, Vazquez and Lynch (2021) [76] reported that, within the Afrotheria order, the tumor-suppressor genes found in an altered number of copies were relatively lower compared to what might be expected. This finding can mirror the trade-off mechanism that natural selection has developed during evolution in order to compensate for the multi-copies effect that can lead to an increased risk of cancer, due to the unbalanced number of copies of the same genes. Indeed, long-living species might possess mechanisms that are capable of maintaining the equilibrium between proliferation and tumor control. Their regulatory networks can create positive feedbacks in which the amplification of tumor suppressor families functions as a buffer against the oncogene co-expansion, or vice-versa [20]. On the other hand, the cancer-prone organisms included in our analysis did not develop these gene defenses because they have a lower lifespan, which does not make them particularly exposed to a severe lack of fitness due to cancer progression (except in the case of *Homo sapiens* that has reached a high lifespan only recently, thanks to the advance of medicine treatments and health care).

5. Limitations and Future Perspectives

Gene duplication is a fundamental process that can lead to the emergence of new phenotypic traits. Analyzing patterns of gene copy number alterations across the genome of large and long-living organisms may reveal new insights about the mechanisms underlying cancer resistance in mammals [12,20,94]. Here, we have developed a simple way to test the hypothesis that CNVs confer protection or increase vulnerability to cancer among species. Using the absolute number of copies of each gene by species, we were able to identify, for the first time, an alteration in miRNA CNVs that are overrepresented and enriched in molecular pathways related to cancer. Further analyses will help to validate these findings by better defining the correlation between miRNAs and their targets. Nowadays, the current challenge is to develop and optimize new experimental design and strategies to be used in human [95] and veterinary biomedical research. Indeed, whenever a potential cancer-suppression mechanism is discovered in a species, there is the real possibility of identifying a new molecular target or therapeutic approach. Therefore, the investigation of genomic alterations, such as CNVs, can direct clinical research towards the discovery of new toolkits able to guide scientists towards the exploration of more focused research topics, such as, for example, specific microRNAs or their targets [96,97].

Supplementary Materials: The following are available online at https://www.mdpi.com/article/10.3390/genes13061046/s1. Table S1: Species description. List of the 9 species included in our analysis; Table S2: Cancer Prone vs. Cancer Resistant: a two-group statistical comparison. List of the significant hits resulting from the unpaired 2-group Wilcoxon test (p-value < 0.05) applied on the genomic CNVs landscape of the selected species; Table S3: Cancer Prone vs Cancer Resistant: a two-group statistical comparison. List of the significant microRNAs resulting from the unpaired 2-group wilcoxon test (*p*-value <0.05) applied on the total genomic CNVs landscape of the selected species; Table S4: Pathway analysis—extended version. Gene Over-Representation Analysis (ORA). The enrichment test used Benjamini-Hochberg's FDR correction (FDR < 0.05). CNVs data were previously analyzed by an unpaired 2-group wilcoxon test (*p*-value < 0.05); Figure S1: MicroRNAs in cancer interaction graph. Figure S2: Heatmap of all the significant genes, clustered with Euclidean distance and ward linkage; Figure S3: Heatmap of all the significant genes, clustered with Euclidean distance and complete linkage; Figure S4: Heatmap of the significant MicroRNAs, clustered with Euclidean distance and ward linkage; File S1: PGLS modelling results: Cancer incidence rate ~ significant miRNAs CNVs.

Author Contributions: Conceptualization, C.V. and C.T.; methodology, C.V. and C.T.; F.B., F.M., R.M. and V.P. established the bioinformatical resource used to generate the data; C.V., F.B., M.D.C., F.M., R.M., V.P. and C.T. performed the analysis and the investigation; data curation, C.V., F.B. and V.P.; data interpretation, C.V. and M.D.C.; writing—original draft preparation, C.V. and C.T.; writing—review and editing, C.V., C.T., M.D.C., F.M., R.M. and G.L.; supervision, C.T., F.M., R.M. and G.L.; funding acquisition, C.T. All authors have read and agreed to the published version of the manuscript.

Funding: This research was supported by the University of Padua through M.A.P.S. Department under the program BIRD213010/21.

Data Availability Statement: All data necessary for confirming the conclusions of the article are present within the article, figures, tables, and its Supplementary Materials.

Acknowledgments: We thank Nicoletta Bianchi for critical reading of our manuscript and for her precious suggestions.

Conflicts of Interest: The authors declare no conflict of interest.

References

1. Serrano, M. Unraveling the Links between Cancer and Aging. *Carcinogenesis* **2016**, *37*, 107. [CrossRef] [PubMed]
2. Peto, R.; Roe, F.J.; Lee, P.N.; Levy, L.; Clack, J. Cancer and Ageing in Mice and Men. *Br. J. Cancer* **1975**, *32*, 411–426. [CrossRef] [PubMed]
3. Buffenstein, R. Negligible Senescence in the Longest Living Rodent, the Naked Mole-Rat: Insights from a Successfully Aging Species. *J. Comp. Physiol. B* **2008**, *178*, 439–445. [CrossRef]
4. Liang, S.; Mele, J.; Wu, Y.; Buffenstein, R.; Hornsby, P.J. Resistance to Experimental Tumorigenesis in Cells of a Long-Lived Mammal, the Naked Mole-Rat (Heterocephalus Glaber): Oncogene Resistance in Naked Mole-Rat Cells. *Aging Cell* **2010**, *9*, 626–635. [CrossRef] [PubMed]
5. Kim, E.B.; Fang, X.; Fushan, A.A.; Huang, Z.; Lobanov, A.V.; Han, L.; Marino, S.M.; Sun, X.; Turanov, A.A.; Yang, P.; et al. Genome Sequencing Reveals Insights into Physiology and Longevity of the Naked Mole Rat. *Nature* **2011**, *479*, 223–227. [CrossRef] [PubMed]
6. Seluanov, A.; Hine, C.; Azpurua, J.; Feigenson, M.; Bozzella, M.; Mao, Z.; Catania, K.C.; Gorbunova, V. Hypersensitivity to Contact Inhibition Provides a Clue to Cancer Resistance of Naked Mole-Rat. *Proc. Natl. Acad. Sci. USA* **2009**, *106*, 19352–19357. [CrossRef]
7. Lipman, R.; Galecki, A.; Burke, D.T.; Miller, R.A. Genetic Loci That Influence Cause of Death in a Heterogeneous Mouse Stock. *J. Gerontol. A Biol. Sci. Med. Sci.* **2004**, *59*, B977–B983. [CrossRef]
8. Boddy, A.M.; Harrison, T.M.; Abegglen, L.M. Comparative Oncology: New Insights into an Ancient Disease. *iScience* **2020**, *23*, 101373. [CrossRef]
9. Wilkinson, G.S.; Adams, D.M. Recurrent Evolution of Extreme Longevity in Bats. *Biol. Lett.* **2019**, *15*, 20180860. [CrossRef]
10. Tollis, M.; Schiffman, J.D.; Boddy, A.M. Evolution of Cancer Suppression as Revealed by Mammalian Comparative Genomics. *Curr. Opin. Genet. Dev.* **2017**, *42*, 40–47. [CrossRef]
11. Wang, L.-F.; Walker, P.J.; Poon, L.L.M. Mass Extinctions, Biodiversity and Mitochondrial Function: Are Bats 'Special' as Reservoirs for Emerging Viruses? *Curr. Opin. Virol.* **2011**, *1*, 649–657. [CrossRef] [PubMed]
12. Abegglen, L.M.; Caulin, A.F.; Chan, A.; Lee, K.; Robinson, R.; Campbell, M.S.; Kiso, W.K.; Schmitt, D.L.; Waddell, P.J.; Bhaskara, S.; et al. Potential Mechanisms for Cancer Resistance in Elephants and Comparative Cellular Response to DNA Damage in Humans. *JAMA* **2015**, *314*, 1850. [CrossRef] [PubMed]
13. Ferlay, J.; Soerjomataram, I.; Dikshit, R.; Eser, S.; Mathers, C.; Rebelo, M.; Parkin, D.M.; Forman, D.; Bray, F. Cancer Incidence and Mortality Worldwide: Sources, Methods and Major Patterns in GLOBOCAN 2012: Globocan 2012. *Int. J. Cancer* **2015**, *136*, E359–E386. [CrossRef]

14. Tian, X.; Seluanov, A.; Gorbunova, V. Molecular Mechanisms Determining Lifespan in Short- and Long-Lived Species. *Trends Endocrinol. Metab.* **2017**, *28*, 722–734. [CrossRef]

15. Sulak, M.; Fong, L.; Mika, K.; Chigurupati, S.; Yon, L.; Mongan, N.P.; Emes, R.D.; Lynch, V.J. TP53 Copy Number Expansion Is Associated with the Evolution of Increased Body Size and an Enhanced DNA Damage Response in Elephants. *eLife* **2016**, *5*, e11994. [CrossRef] [PubMed]

16. Caulin, A.F.; Maley, C.C. Peto's Paradox: Evolution's Prescription for Cancer Prevention. *Trends Ecol. Evol.* **2011**, *26*, 175–182. [CrossRef] [PubMed]

17. Feuk, L.; Carson, A.R.; Scherer, S.W. Structural Variation in the Human Genome. *Nat. Rev. Genet.* **2006**, *7*, 85–97. [CrossRef]

18. Stratton, M.R.; Campbell, P.J.; Futreal, P.A. The Cancer Genome. *Nature* **2009**, *458*, 719–724. [CrossRef]

19. Tollis, M.; Boddy, A.M.; Maley, C.C. Peto's Paradox: How Has Evolution Solved the Problem of Cancer Prevention? *BMC Biol.* **2017**, *15*, 60. [CrossRef]

20. Tollis, M.; Schneider-Utaka, A.K.; Maley, C.C. The Evolution of Human Cancer Gene Duplications across Mammals. *Mol. Biol. Evol.* **2020**, *37*, 2875–2886. [CrossRef]

21. Boddy, A.M.; Abegglen, L.M.; Pessier, A.P.; Aktipis, A.; Schiffman, J.D.; Maley, C.C.; Witte, C. Lifetime Cancer Prevalence and Life History Traits in Mammals. *Evol. Med. Public Health* **2020**, *2020*, 187–195. [CrossRef] [PubMed]

22. Beroukhim, R.; Mermel, C.H.; Porter, D.; Wei, G.; Raychaudhuri, S.; Donovan, J.; Barretina, J.; Boehm, J.S.; Dobson, J.; Urashima, M.; et al. The Landscape of Somatic Copy-Number Alteration across Human Cancers. *Nature* **2010**, *463*, 899–905. [CrossRef] [PubMed]

23. Bignell, G.R.; Greenman, C.D.; Davies, H.; Butler, A.P.; Edkins, S.; Andrews, J.M.; Buck, G.; Chen, L.; Beare, D.; Latimer, C.; et al. Signatures of Mutation and Selection in the Cancer Genome. *Nature* **2010**, *463*, 893–898. [CrossRef] [PubMed]

24. Iskow, R.C.; Gokcumen, O.; Lee, C. Exploring the Role of Copy Number Variants in Human Adaptation. *Trends Genet.* **2012**, *28*, 245–257. [CrossRef]

25. Vischioni, C.; Bove, F.; Mandreoli, F.; Martoglia, R.; Pisi, V.; Taccioli, C. Visual Exploratory Data Analysis for Copy Number Variation Studies in Biomedical Research. *Big Data Res.* **2022**, *27*, 100298. [CrossRef]

26. Howe, K.L.; Achuthan, P.; Allen, J.; Allen, J.; Alvarez-Jarreta, J.; Amode, M.R.; Armean, I.M.; Azov, A.G.; Bennett, R.; Bhai, J.; et al. Ensembl 2021. *Nucleic Acids Res.* **2021**, *49*, D884–D891. [CrossRef]

27. Herrero, J.; Muffato, M.; Beal, K.; Fitzgerald, S.; Gordon, L.; Pignatelli, M.; Vilella, A.J.; Searle, S.M.J.; Amode, R.; Brent, S.; et al. Ensembl Comparative Genomics Resources. *Database* **2016**, *2016*, 17. [CrossRef]

28. De Bie, T.; Cristianini, N.; Demuth, J.P.; Hahn, M.W. CAFE: A Computational Tool for the Study of Gene Family Evolution. *Bioinformatics* **2006**, *22*, 1269–1271. [CrossRef]

29. Seluanov, A.; Gladyshev, V.N.; Vijg, J.; Gorbunova, V. Mechanisms of Cancer Resistance in Long-Lived Mammals. *Nat. Rev. Cancer* **2018**, *18*, 433–441. [CrossRef]

30. Lagunas-Rangel, F.A. Cancer-Free Aging: Insights from Spalax Ehrenbergi Superspecies. *Ageing Res. Rev.* **2018**, *47*, 18–23. [CrossRef]

31. Ma, S.; Gladyshev, V.N. Molecular Signatures of Longevity: Insights from Cross-Species Comparative Studies. *Semin. Cell Dev. Biol.* **2017**, *70*, 190–203. [CrossRef] [PubMed]

32. Gorbunova, V.; Hine, C.; Tian, X.; Ablaeva, J.; Gudkov, A.V.; Nevo, E.; Seluanov, A. Cancer Resistance in the Blind Mole Rat Is Mediated by Concerted Necrotic Cell Death Mechanism. *Proc. Natl. Acad. Sci. USA* **2012**, *109*, 19392–19396. [CrossRef] [PubMed]

33. Shepard, A.; Kissil, J.L. The Use of Non-Traditional Models in the Study of Cancer Resistance—The Case of the Naked Mole Rat. *Oncogene* **2020**, *39*, 5083–5097. [CrossRef] [PubMed]

34. Orme, D. The Caper Package: Comparative Analysis of Phylogenetics and Evolution in R. 36. Available online: https://cran.r-project.org/web/packages/caper/vignettes/caper.pdf (accessed on 4 May 2022).

35. Upham, N.S.; Esselstyn, J.A.; Jetz, W. Inferring the Mammal Tree: Species-Level Sets of Phylogenies for Questions in Ecology, Evolution, and Conservation. *PLOS Biol.* **2019**, *17*, e3000494. [CrossRef] [PubMed]

36. Letunic, I.; Bork, P. Interactive Tree of Life (ITOL) v5: An Online Tool for Phylogenetic Tree Display and Annotation. *Nucleic Acids Res.* **2021**, *49*, W293–W296. [CrossRef]

37. Postma, M.; Goedhart, J. PlotsOfData—A Web App for Visualizing Data Together with Their Summaries. *PLOS Biol.* **2019**, *17*, e3000202. [CrossRef]

38. Spitzer, M.; Wildenhain, J.; Rappsilber, J.; Tyers, M. BoxPlotR: A Web Tool for Generation of Box Plots. *Nat. Methods* **2014**, *11*, 121–122. [CrossRef]

39. Metsalu, T.; Vilo, J. ClustVis: A Web Tool for Visualizing Clustering of Multivariate Data Using Principal Component Analysis and Heatmap. *Nucleic Acids Res.* **2015**, *43*, W566–W570. [CrossRef]

40. Kern, F.; Aparicio-Puerta, E.; Li, Y.; Fehlmann, T.; Kehl, T.; Wagner, V.; Ray, K.; Ludwig, N.; Lenhof, H.-P.; Meese, E.; et al. MiRTargetLink 2.0—Interactive MiRNA Target Gene and Target Pathway Networks. *Nucleic Acids Res.* **2021**, *49*, W409–W416. [CrossRef]

41. Liao, Y.; Wang, J.; Jaehnig, E.J.; Shi, Z.; Zhang, B. WebGestalt 2019: Gene Set Analysis Toolkit with Revamped UIs and APIs. *Nucleic Acids Res.* **2019**, *47*, W199–W205. [CrossRef]

42. Khatri, P.; Sirota, M.; Butte, A.J. Ten Years of Pathway Analysis: Current Approaches and Outstanding Challenges. *PLoS Comput. Biol.* **2012**, *8*, e1002375. [CrossRef] [PubMed]

43. Wang, J.; Zhang, G.; Sui, Y.; Yang, Z.; Chu, Y.; Tang, H.; Guo, B.; Zhang, C.; Wu, C. CD52 Is a Prognostic Biomarker and Associated with Tumor Microenvironment in Breast Cancer. *Front. Genet.* **2020**, *11*, 578002. [CrossRef] [PubMed]

44. Thakur, V.S.; Aguila, B.; Brett-Morris, A.; Creighton, C.J.; Welford, S.M. Spermidine/Spermine N1-Acetyltransferase 1 Is a Gene-Specific Transcriptional Regulator That Drives Brain Tumor Aggressiveness. *Oncogene* **2019**, *38*, 6794–6800. [CrossRef] [PubMed]

45. Xu, J.; Li, Y.; Wang, F.; Wang, X.; Cheng, B.; Ye, F.; Xie, X.; Zhou, C.; Lu, W. Suppressed MiR-424 Expression via Upregulation of Target Gene Chk1 Contributes to the Progression of Cervical Cancer. *Oncogene* **2013**, *32*, 976–987. [CrossRef]

46. Fan, X.; Huang, X.; Li, Z.; Ma, X. MicroRNA-372-3p Promotes the Epithelial-Mesenchymal Transition in Breast Carcinoma by Activating the Wnt Pathway. *J. BUON* **2018**, *23*, 1309–1315.

47. Sun, H.; Gao, D. Propofol Suppresses Growth, Migration and Invasion of A549 Cells by down-Regulation of MiR-372. *BMC Cancer* **2018**, *18*, 1252. [CrossRef]

48. Jones, L.; Naidoo, M.; Machado, L.R.; Anthony, K. The Duchenne Muscular Dystrophy Gene and Cancer. *Cell. Oncol.* **2021**, *44*, 19–32. [CrossRef]

49. Spilka, R.; Ernst, C.; Mehta, A.K.; Haybaeck, J. Eukaryotic Translation Initiation Factors in Cancer Development and Progression. *Cancer Lett.* **2013**, *340*, 9–21. [CrossRef]

50. Yamakuchi, M.; Lotterman, C.D.; Bao, C.; Hruban, R.H.; Karim, B.; Mendell, J.T.; Huso, D.; Lowenstein, C.J. P53-Induced MicroRNA-107 Inhibits HIF-1 and Tumor Angiogenesis. *Proc. Natl. Acad. Sci. USA* **2010**, *107*, 6334–6339. [CrossRef]

51. Turco, C.; Donzelli, S.; Fontemaggi, G. MiR-15/107 MicroRNA Gene Group: Characteristics and Functional Implications in Cancer. *Front. Cell Dev. Biol.* **2020**, *8*, 427. [CrossRef]

52. Wang, P.; Chen, L.; Zhang, J.; Chen, H.; Fan, J.; Wang, K.; Luo, J.; Chen, Z.; Meng, Z.; Liu, L. Methylation-Mediated Silencing of the MiR-124 Genes Facilitates Pancreatic Cancer Progression and Metastasis by Targeting Rac1. *Oncogene* **2014**, *33*, 514–524. [CrossRef] [PubMed]

53. Müller, S.; Ledl, A.; Schmidt, D. SUMO: A Regulator of Gene Expression and Genome Integrity. *Oncogene* **2004**, *23*, 1998–2008. [CrossRef] [PubMed]

54. Schneeweis, C.; Hassan, Z.; Schick, M.; Keller, U.; Schneider, G. The SUMO Pathway in Pancreatic Cancer: Insights and Inhibition. *Br. J. Cancer* **2021**, *124*, 531–538. [CrossRef] [PubMed]

55. Wen, S.-Y.; Lin, Y.; Yu, Y.-Q.; Cao, S.-J.; Zhang, R.; Yang, X.-M.; Li, J.; Zhang, Y.-L.; Wang, Y.-H.; Ma, M.-Z.; et al. MiR-506 Acts as a Tumor Suppressor by Directly Targeting the Hedgehog Pathway Transcription Factor Gli3 in Human Cervical Cancer. *Oncogene* **2015**, *34*, 717–725. [CrossRef]

56. Zhai, Q.; Zhou, L.; Zhao, C.; Wan, J.; Yu, Z.; Guo, X.; Qin, J.; Chen, J.; Lu, R. Identification of MiR-508-3p and MiR-509-3p That Are Associated with Cell Invasion and Migration and Involved in the Apoptosis of Renal Cell Carcinoma. *Biochem. Biophys. Res. Commun.* **2012**, *419*, 621–626. [CrossRef]

57. Squadrito, M.L.; Pucci, F.; Magri, L.; Moi, D.; Gilfillan, G.D.; Ranghetti, A.; Casazza, A.; Mazzone, M.; Lyle, R.; Naldini, L.; et al. MiR-511-3p Modulates Genetic Programs of Tumor-Associated Macrophages. *Cell Rep.* **2012**, *1*, 141–154. [CrossRef]

58. Ren, L.-L.; Yan, T.-T.; Shen, C.-Q.; Tang, J.-Y.; Kong, X.; Wang, Y.-C.; Chen, J.; Liu, Q.; He, J.; Zhong, M.; et al. The Distinct Role of Strand-Specific MiR-514b-3p and MiR-514b-5p in Colorectal Cancer Metastasis. *Cell Death Dis.* **2018**, *9*, 687. [CrossRef]

59. Chen, Q.; Zhou, W.; Han, T.; Du, S.; Li, Z.; Zhang, Z.; Shan, G.; Kong, C. MiR-378 Suppresses Prostate Cancer Cell Growth through Downregulation of MAPK1 in Vitro and in Vivo. *Tumor Biol.* **2016**, *37*, 2095–2103. [CrossRef]

60. Zhu, W.; Xue, Y.; Liang, C.; Zhang, R.; Zhang, Z.; Li, H.; Su, D.; Liang, X.; Zhang, Y.; Huang, Q.; et al. S100A16 Promotes Cell Proliferation and Metastasis via AKT and ERK Cell Signaling Pathways in Human Prostate Cancer. *Tumor Biol.* **2016**, *37*, 12241–12250. [CrossRef]

61. Miremadi, A.; Oestergaard, M.Z.; Pharoah, P.D.P.; Caldas, C. Cancer Genetics of Epigenetic Genes. *Hum. Mol. Genet.* **2007**, *16*, R28–R49. [CrossRef]

62. Zhang, Z.; Liu, M.; Hu, Q.; Xu, W.; Liu, W.; Ye, Z.; Fan, G.; Xu, X.; Yu, X.; Ji, S.; et al. FGFBP1, a Downstream Target of the FBW7/c-Myc Axis, Promotes Cell Proliferation and Migration in Pancreatic Cancer. *Am. J. Cancer Res.* **2019**, *9*, 2650. [PubMed]

63. Xian, S.; Shang, D.; Kong, G.; Tian, Y. FOXJ1 Promotes Bladder Cancer Cell Growth and Regulates Warburg Effect. *Biochem. Biophys. Res. Commun.* **2018**, *495*, 988–994. [CrossRef] [PubMed]

64. Nohata, N.; Sone, Y.; Hanazawa, T.; Fuse, M.; Kikkawa, N.; Yoshino, H.; Chiyomaru, T.; Kawakami, K.; Enokida, H.; Nakagawa, M.; et al. MiR-1 as a Tumor Suppressive MicroRNA Targeting TAGLN2 in Head and Neck Squamous Cell Carcinoma. *Oncotarget* **2011**, *2*, 29–42. [CrossRef] [PubMed]

65. Zhang, L.; Liu, X.; Jin, H.; Guo, X.; Xia, L.; Chen, Z.; Bai, M.; Liu, J.; Shang, X.; Wu, K.; et al. MiR-206 Inhibits Gastric Cancer Proliferation in Part by Repressing CyclinD2. *Cancer Lett.* **2013**, *332*, 94–101. [CrossRef] [PubMed]

66. Wu, Z.; Wu, Q.; Wang, C.; Wang, X.; Huang, J.; Zhao, J.; Mao, S.; Zhang, G.; Xu, X.; Zhang, N. MiR-340 Inhibition of Breast Cancer Cell Migration and Invasion through Targeting of Oncoprotein c-Met. *Cancer* **2011**, *117*, 2842–2852. [CrossRef] [PubMed]

67. Kureel, J.; Dixit, M.; Tyagi, A.M.; Mansoori, M.N.; Srivastava, K.; Raghuvanshi, A.; Maurya, R.; Trivedi, R.; Goel, A.; Singh, D. MiR-542-3p Suppresses Osteoblast Cell Proliferation and Differentiation, Targets BMP-7 Signaling and Inhibits Bone Formation. *Cell Death Dis.* **2014**, *5*, e1050. [CrossRef] [PubMed]

68. Cano, C.E.; Hamidi, T.; Sandi, M.J.; Iovanna, J.L. Nupr1: The Swiss-Knife of Cancer. *J. Cell. Physiol.* **2011**, *226*, 1439–1443. [CrossRef]

69. Peters, K.M.; Carlson, B.A.; Gladyshev, V.N.; Tsuji, P.A. Selenoproteins in Colon Cancer. *Free Radic. Biol. Med.* **2018**, *127*, 14–25. [CrossRef]

70. Yim, S.H.; Clish, C.B.; Gladyshev, V.N. Selenium Deficiency Is Associated with Pro-Longevity Mechanisms. *Cell Rep.* **2019**, *27*, 2785–2797.e3. [CrossRef]

71. Elliott, B.; Millena, A.C.; Matyunina, L.; Zhang, M.; Zou, J.; Wang, G.; Zhang, Q.; Bowen, N.; Eaton, V.; Webb, G.; et al. Essential Role of JunD in Cell Proliferation Is Mediated via MYC Signaling in Prostate Cancer Cells. *Cancer Lett.* **2019**, *448*, 155–167. [CrossRef]

72. Kanehisa, M. Toward Understanding the Origin and Evolution of Cellular Organisms. *Protein Sci.* **2019**, *28*, 1947–1951. [CrossRef]

73. Martens, M.; Ammar, A.; Riutta, A.; Waagmeester, A.; Slenter, D.N.; Hanspers, K.; Miller, R.A.; Digles, D.; Lopes, E.N.; Ehrhart, F.; et al. WikiPathways: Connecting Communities. *Nucleic Acids Res.* **2021**, *49*, D613–D621. [CrossRef] [PubMed]

74. Thomas, P.D. PANTHER: A Library of Protein Families and Subfamilies Indexed by Function. *Genome Res.* **2003**, *13*, 2129–2141. [CrossRef] [PubMed]

75. Kourtidis, A.; Lu, R.; Pence, L.J.; Anastasiadis, P.Z. A Central Role for Cadherin Signaling in Cancer. *Exp. Cell Res.* **2017**, *358*, 78–85. [CrossRef] [PubMed]

76. Vazquez, J.M.; Lynch, V.J. Pervasive Duplication of Tumor Suppressors in Afrotherians during the Evolution of Large Bodies and Reduced Cancer Risk. *eLife* **2021**, *10*, e65041. [CrossRef] [PubMed]

77. Glenfield, C.; Innan, H. Gene Duplication and Gene Fusion Are Important Drivers of Tumourigenesis during Cancer Evolution. *Genes* **2021**, *12*, 1376. [CrossRef]

78. Eifler, K.; Vertegaal, A.C.O. SUMOylation-Mediated Regulation of Cell Cycle Progression and Cancer. *Trends Biochem. Sci.* **2015**, *40*, 779–793. [CrossRef]

79. Gallia, G.L.; Zhang, M.; Ning, Y.; Haffner, M.C.; Batista, D.; Binder, Z.A.; Bishop, J.A.; Hann, C.L.; Hruban, R.H.; Ishii, M.; et al. Genomic Analysis Identifies Frequent Deletions of Dystrophin in Olfactory Neuroblastoma. *Nat. Commun.* **2018**, *9*, 5410. [CrossRef]

80. Ruggieri, S.; De Giorgis, M.; Annese, T.; Tamma, R.; Notarangelo, A.; Marzullo, A.; Senetta, R.; Cassoni, P.; Notarangelo, M.; Ribatti, D.; et al. Dp71 Expression in Human Glioblastoma. *Int. J. Mol. Sci.* **2019**, *20*, 5429. [CrossRef]

81. Schmiedel, J.M.; Klemm, S.L.; Zheng, Y.; Sahay, A.; Blüthgen, N.; Marks, D.S.; van Oudenaarden, A. MicroRNA Control of Protein Expression Noise. *Science* **2015**, *348*, 128–132. [CrossRef]

82. Martinez-Sanchez, A.; Murphy, C. MicroRNA Target Identification—Experimental Approaches. *Biology* **2013**, *2*, 189–205. [CrossRef] [PubMed]

83. Jansson, M.D.; Lund, A.H. MicroRNA and Cancer. *Mol. Oncol.* **2012**, *6*, 590–610. [CrossRef] [PubMed]

84. Iorio, M.V.; Visone, R.; Di Leva, G.; Donati, V.; Petrocca, F.; Casalini, P.; Taccioli, C.; Volinia, S.; Liu, C.-G.; Alder, H.; et al. MicroRNA Signatures in Human Ovarian Cancer. *Cancer Res.* **2007**, *67*, 8699–8707. [CrossRef] [PubMed]

85. Svoronos, A.A.; Engelman, D.M.; Slack, F.J. OncomiR or Tumor Suppressor? The Duplicity of MicroRNAs in Cancer. *Cancer Res.* **2016**, *76*, 3666–3670. [CrossRef]

86. Bartel, D.P. Metazoan MicroRNAs. *Cell* **2018**, *173*, 20–51. [CrossRef] [PubMed]

87. Gaziel-Sovran, A.; Segura, M.F.; Di Micco, R.; Collins, M.K.; Hanniford, D.; Vega-Saenz de Miera, E.; Rakus, J.F.; Dankert, J.F.; Shang, S.; Kerbel, R.S.; et al. MiR-30b/30d Regulation of GalNAc Transferases Enhances Invasion and Immunosuppression during Metastasis. *Cancer Cell* **2011**, *20*, 104–118. [CrossRef]

88. Garofalo, M.; Quintavalle, C.; Romano, G.; Croce, C.M.; Condorelli, G. MiR221/222 in Cancer: Their Role in Tumor Progression and Response to Therapy. *Curr. Mol. Med.* **2012**, *12*, 27–33. [CrossRef]

89. Di Leva, G.; Gasparini, P.; Piovan, C.; Ngankeu, A.; Garofalo, M.; Taccioli, C.; Iorio, M.V.; Li, M.; Volinia, S.; Alder, H.; et al. MicroRNA Cluster 221-222 and Estrogen Receptor α Interactions in Breast Cancer. *JNCI J. Natl. Cancer Inst.* **2010**, *102*, 706–721. [CrossRef]

90. Cimmino, A.; Calin, G.A.; Fabbri, M.; Iorio, M.V.; Ferracin, M.; Shimizu, M.; Wojcik, S.E.; Aqeilan, R.I.; Zupo, S.; Dono, M.; et al. MiR-15 and MiR-16 Induce Apoptosis by Targeting BCL2. *Proc. Natl. Acad. Sci. USA* **2005**, *102*, 13944–13949. [CrossRef]

91. Pekarsky, Y.; Balatti, V.; Croce, C.M. BCL2 and MiR-15/16: From Gene Discovery to Treatment. *Cell Death Differ.* **2018**, *25*, 21–26. [CrossRef]

92. Tang, Y.-C.; Amon, A. Gene Copy-Number Alterations: A Cost-Benefit Analysis. *Cell* **2013**, *152*, 394–405. [CrossRef] [PubMed]

93. Gordon, D.J.; Resio, B.; Pellman, D. Causes and Consequences of Aneuploidy in Cancer. *Nat. Rev. Genet.* **2012**, *13*, 189–203. [CrossRef]

94. Kirkwood, T.B.L. Why and How Are We Living Longer?: Why and How Are We Living Longer? *Exp. Physiol.* **2017**, *102*, 1067–1074. [CrossRef] [PubMed]

95. Holtze, S.; Gorshkova, E.; Braude, S.; Cellerino, A.; Dammann, P.; Hildebrandt, T.B.; Hoeflich, A.; Hoffmann, S.; Koch, P.; Terzibasi Tozzini, E.; et al. Alternative Animal Models of Aging Research. *Front. Mol. Biosci.* **2021**, *8*, 660959. [CrossRef] [PubMed]

96. Hajibabaie, F.; Abedpoor, N.; Assareh, N.; Tabatabaiefar, M.A.; Shariati, L.; Zarrabi, A. The Importance of SNPs at MiRNA Binding Sites as Biomarkers of Gastric and Colorectal Cancers: A Systematic Review. *J. Pers. Med.* **2022**, *12*, 456. [CrossRef] [PubMed]

97. Lange, M.; Begolli, R.; Giakountis, A. Non-Coding Variants in Cancer: Mechanistic Insights and Clinical Potential for Personalized Medicine. *Non-Coding RNA* **2021**, *7*, 47. [CrossRef] [PubMed]

Review

Emerging Roles and Potential Applications of Non-Coding RNAs in Cervical Cancer

Deepak Parashar [1,*], Anupam Singh [2], Saurabh Gupta [2], Aishwarya Sharma [3], Manish K. Sharma [4], Kuldeep K. Roy [5], Subhash C. Chauhan [6,7] and Vivek K. Kashyap [6,7,*]

[1] Department of Obstetrics and Gynecology, Medical College of Wisconsin, Milwaukee, MI 53226, USA
[2] Department of Biotechnology, GLA University, Mathura 281406, Uttar Pradesh, India; anupamsinghsoul@gmail.com (A.S.); saurabh.gupta@gla.ac.in (S.G.)
[3] Sri Siddhartha Medical College and Research Center, Tumkur 572107, Karnataka, India; aishwarya.sharma0909@gmail.com
[4] Department of Biotechnology, IP College, Bulandshahr 203001, Uttar Pradesh, India; mny2096@gmail.com
[5] Department of Pharmaceutical Sciences, School of Health Sciences and Technology, UPES, Dehradun 248007, Uttarakhand, India; kuldeep.roy@ddn.upes.ac.in
[6] Department of Immunology and Microbiology, School of Medicine, University of Texas Rio Grande Valley, McAllen, TX 78504, USA; subhash.chauhan@utrgv.edu
[7] South Texas Center of Excellence in Cancer Research, School of Medicine, University of Texas Rio Grande Valley, McAllen, TX 78504, USA
* Correspondence: dparashar@mcw.edu (D.P.); vivek.kashyap@utrgv.edu (V.K.K.); Tel.: +1-414-439-8089 (D.P.); +1-956-296-1738 (V.K.K.)

Citation: Parashar, D.; Singh, A.; Gupta, S.; Sharma, A.; Sharma, M.K.; Roy, K.K.; Chauhan, S.C.; Kashyap, V.K. Emerging Roles and Potential Applications of Non-Coding RNAs in Cervical Cancer. *Genes* **2022**, *13*, 1254. https://doi.org/10.3390/genes13071254

Academic Editors: Giuseppe Iacomino and Fabio Lauria

Received: 31 May 2022
Accepted: 6 July 2022
Published: 15 July 2022

Publisher's Note: MDPI stays neutral with regard to jurisdictional claims in published maps and institutional affiliations.

Abstract: Cervical cancer (CC) is a preventable disease using proven interventions, specifically prophylactic vaccination, pervasive disease screening, and treatment, but it is still the most frequently diagnosed cancer in women worldwide. Patients with advanced or metastatic CC have a very dismal prognosis and current therapeutic options are very limited. Therefore, understanding the mechanism of metastasis and discovering new therapeutic targets are crucial. New sequencing tools have given a full visualization of the human transcriptome's composition. Non-coding RNAs (NcRNAs) perform various functions in transcriptional, translational, and post-translational processes through their interactions with proteins, RNA, and even DNA. It has been suggested that ncRNAs act as key regulators of a variety of biological processes, with their expression being tightly controlled under physiological settings. In recent years, and notably in the past decade, significant effort has been made to examine the role of ncRNAs in a variety of human diseases, including cancer. Therefore, shedding light on the functions of ncRNA will aid in our better understanding of CC. In this review, we summarize the emerging roles of ncRNAs in progression, metastasis, therapeutics, chemo-resistance, human papillomavirus (HPV) regulation, metabolic reprogramming, diagnosis, and as a prognostic biomarker of CC. We also discussed the role of ncRNA in the tumor microenvironment and tumor immunology, including cancer stem cells (CSCs) in CC. We also address contemporary technologies such as antisense oligonucleotides, CRISPR–Cas9, and exosomes, as well as their potential applications in targeting ncRNAs to manage CC.

Keywords: cervical cancer; non-coding RNAs; diagnosis; prognosis; therapeutics; regulation of gene expression

1. Introduction

Cervical cancer (CC) is the fourth most frequently diagnosed cancer in women with substantial geographical variation in CC morbidity and mortality [1]. CC was accounted to cause approximately 604,000 new cases and 342,000 deaths worldwide in the year 2020 [1]. CC develops in the uterine cervix epithelium, notably at the squamo columnar junction, interface of the ectocervix and endocervix, which is a hotspot for metaplastic activity. Squamous cell carcinomas (SCC) and adenocarcinomas (ADC) are the most frequently diagnosed kinds of CC, accounting for approximately 80–90% and 10–15% of all cervical malignancies [2]. Adenosquamous carcinoma (ADSC) is a rare type of CC [2].

Human papillomavirus (HPV) infections, the most prevalent sexually transmitted infection, is responsible for causing cervical carcinogenesis [3,4]. The viral DNA gets integrated into the host DNA after a long-term high-risk HPV (HR-HPV) infection, and consequently, cervical epithelial cells become malignant, resulting in CC [5,6]. Moreover, precancerous mutations in the cervix lead to the establishment of CC. Additionally, the lag between infection and carcinogenesis is a major factor as to why CC has become a ravaging disease for women. Fortunately, earlier detection, awareness, and effective treatment of CC have been shown to considerably reduce both the morbidity and mortality rate in women. Effective monitoring and vaccination campaigns have resulted in a substantial drop in the CC fatality rate in developed countries over the last four decades [7]. The Papanicolaou (PAP) smear test, visual inspection with acetic acid (VIA), liquid- based cytology (LBC), and HPV testing for HR-HPV strains are some of the current screening approaches utilized for detecting cancer in the early stages [8]. Furthermore, venereal diseases, long-term oral contraception, reproductive factors, and behavioral issues such as smoking, drinking, and obesity have all been identified as CC risk factors [6,9].

Chemotherapy, radiation, and surgery are all available treatments for CC, but none of these improve patient survival rates and can result in serious negative effects. Despite all these advances in the detection and prevention of CC, it remains "a worldwide health crisis", particularly in undeveloped and emerging countries [10,11]. In spite of recent breakthroughs, CC has a poor long-term prognosis due to its resilience and relapsing nature. This necessitates the development of new biomarkers for tracking CC progression, which also serve as putative targets for diagnostic and curative purposes. Expression profiling of several ncRNAs has been shown to be correlated with cancer progression, onset, metastases, and invasion and has emerged as a novel prognostic and diagnostic biomarker in cervical carcinoma [5]. This article provides a comprehensive overview of the function and potential application of ncRNAs in CC.

2. Classification and Biogenesis of ncRNAs

2.1. Classification

According to the literature, ncRNAs can be classified according to their structure, function, biogenesis, localization, and interaction with DNA or protein-coding mRNAs [12,13]. The discovery of the order of activities in the passage of genetic information stored in DNA to working biological processes via proteins has been dubbed the central dogma of molecular genetics by Francis Crick in 1958, and it was a watershed moment in molecular biology [14]. With the emergence of novel technologies and rigorous next-generation sequencing, large international consortiums such as the Functional Annotation of the Mammalian Genome (FANTOM) and the Encyclopedia of DNA Elements (ENCODE) have explained ubiquitous transcription as ~98% of DNA is transcribed into RNA, and only ~2% of that RNA is translated into protein [15–17]. Therefore, in the world of cellular communication, RNA is divided into two distinct types: coding RNAs and ncRNAs. The major chunk of transcribed DNA, i.e., ncRNA, was earlier thought to be evolutionary garbage since it lacked the ability to code for protein, and protein-coding RNA, which is a considerably smaller portion of RNA [18,19].

NcRNAs are basically divided into two domains: structural ncRNAs and regulatory ncRNAs. Structural ncRNAs include transfer RNAs (tRNAs) and ribosomal RNAs (rRNAs). Regulatory ncRNAs are further classified into small (length < 50 nts), medium (length 50–200 nts), and long non-coding RNAs (lncRNAs) (length > 200 nts), based on transcript length (Figure 1) [20–22]. Furthermore, microRNAs (miRNAs), small interfering RNAs (siRNAs), piwi-interacting RNAs (piRNAs), cisRNA, and telomere-specific small RNAs (tel-sRNAs) belong to the category of short non-coding RNA (sncRNAs), having a transcript size ranging between 20–50 nucleotides, and similarly, small nucleolar RNA (snoRNA), prompts, tiRNA, small nuclear RNA (snRNA), and small cytoplasmic RNA (scRNA) can be categorized as medium ncRNAs with a transcript length between 50–200 nucleotides [23,24]. LncRNAs regulate transcripts possessing a size greater than >200 nucleotides. Furthermore, lncRNA could be divided into three main categories. The first category is based on biogenesis of lncRNAs such as intronic, intergenic, sense, antisense, bidirectional, and promoter and enhancer lncRNAs, whereas the second category is based on the mechanism, such as cis-regulatory RNA (cis-RNA), trans-RNA, and competing endogenous RNAs (ceRNAs), long intergenic non-coding RNA (lincRNAs), while the third category is based on structure, such as natural antisense transcripts (NATs), enhancer-derived RNAs (eRNA), and circular RNA (circRNA) [25–27].

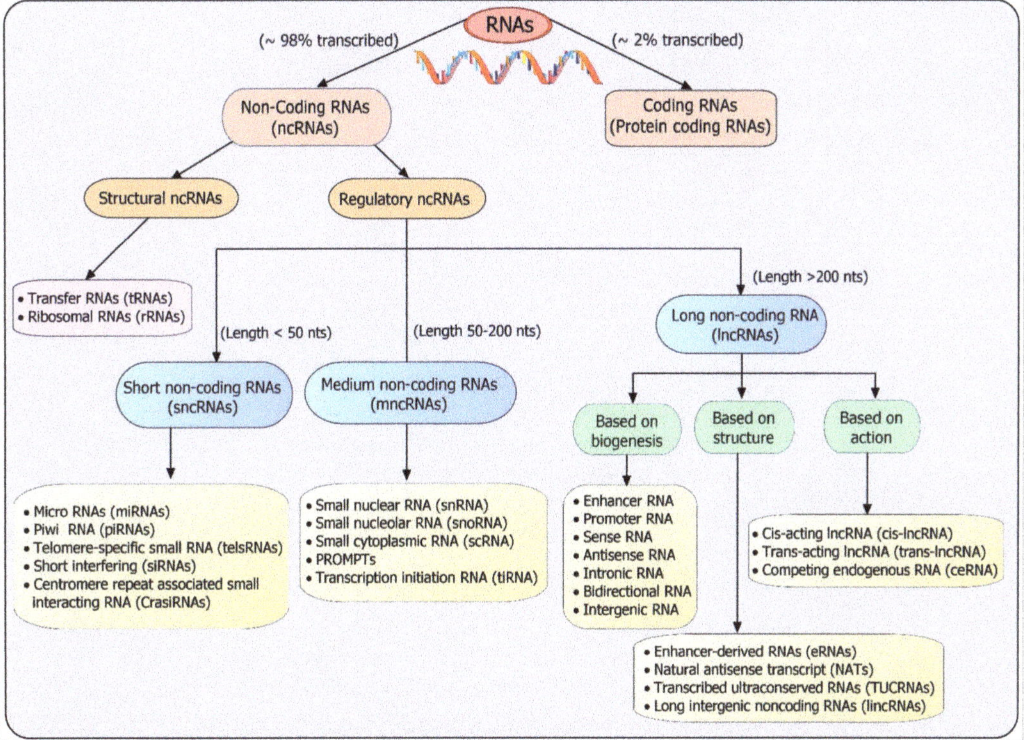

Figure 1. Non-coding RNA (ncRNA) classification. The schematic shows that ncRNAs are divided in to two major categories, such as structural and regulatory ncRNAs. Regulatory ncRNAs are further divided into small (<50 nts length), medium (50–200 nts length), and long non-coding RNAs (>200 nts length), according to their length. LncRNAs are further classified according to their structure, biogenesis, and mechanism of action. See the text for more details.

2.2. Biogenesis of ncRNAs

The biogenesis of ncRNAs is predicated on their characteristics, which are comparable to those of mRNAs. NcRNA play a crucial function in several prospects of human development and diseases [28]. Addressing ncRNA biogenesis is important not only for distinguishing it from the rest of the RNAs but also for assessing its functional relevance [29]. Across the human genome, several genes participate in the generation of various types of ncRNAs [30]. Transcription, nucleosomal maturation, exportation towards the cytoplasm for processing, and production of functional RNA are all quintessential parts of the biogenesis process. RNA polymerase II/III transcribes polycistrons, producing large progenitors (pri-miRNA: hairpin loop structure; 5′capping; 3′polyadenylation) (Figure 2) [31].

After that, it passes through two steps of processing: The microprocessor (DGCR8) identifies and controls the breaking of pri-miRNA via Drosha's, resulting in the emergence of pre-miRNA, which is then translocated from the nucleus to the cytoplasmic region via RAN-GTP and Exportin-5 (XPO5) protein. Furthermore, in the cytoplasm, Dicer acting as RNase III endonuclease chops the progenitor molecule present more towards the terminal end, releasing an RNA duplex that interfaces with Argonaute proteins (AGO-2) present in collaboration with RISC (miRNA-induced silencing complex) [31,32]. However, lncRNA biogenesis proceeds under the influence of the type of cell and phase-specific stimulation governs it [33].

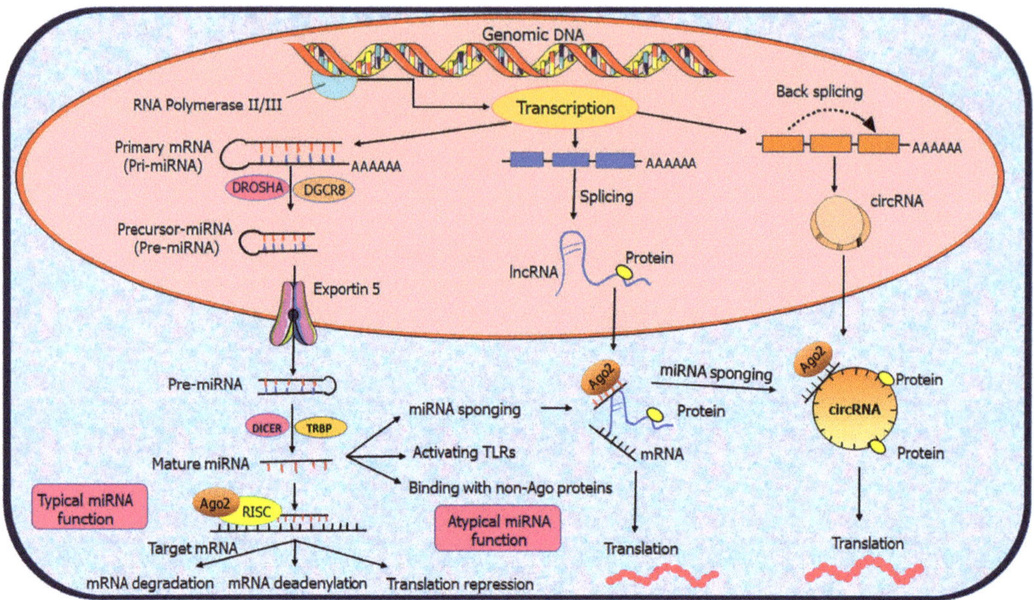

Figure 2. Overview biogenesis and function of ncRNAs in CC cells. MicroRNAs (miRNAs), long non-coding RNAs (lncRNAs), and circular RNAs (circRNAs) are presented along with their fundamental biogenesis and main functional mechanisms. The details are described in the text.

Multiple DNA components in eukaryotic genomes, including enhancers, promoters, and intergenic portions, transcribe distinct kinds of lncRNAs [34]. The principle processes carried out during biogenesis include cleavage by Ribonuclease P (RNase P) to create mature ends, production of snoRNA and small nucleolar ribonucleoprotein (snoRNP) complexes, capping at their ends, and the formation of circular structures [35]. During the synthesis of particular lncRNAs, distinctive sub-nuclear structures known as "paraspeckles" have recently been discovered [36]. Overall, the processes of biosynthetic pathways and regulation of unique ncRNAs are not entirely comprehended. However, we will gain a better understanding of their genesis and applications in the coming years by using a variety of techniques such as ChiRP-Seq (Chromatin Isolation by RNA Purification), RNA structure mapping, crosslinking immunoprecipitation (CLIP), targeted genome engineering with CRISPR–Cas9 and advanced genetic monitoring, ribosome profiling, and phylogenetic lineage tracing [37].

3. Functional Roles and Mechanisms of Action of ncRNAs

3.1. Biological Function of ncRNAs

The biological functions of ncRNAs have been progressively explained, including the regulation of gene expression at the transcriptional and translational levels; instructing DNA synthesis or gene rearrangement; and guarding the genome from foreign nucleic acids [38]. Several recent studies have indicated that ncRNAs are crucial in carcinogenesis by controlling the expression of cancer-associated genes [31,39–41]. Mechanistically, lncRNAs govern gene expression primarily by functioning as transcription factors, controlling chromatin remodeling, or actively contributing to posttranscriptional regulation as ceRNAs [42–44]. MiRNAs, on the other hand, control gene expression at the posttranscriptional level via RNA interference and frequently attach to the $3'$-untranslated region ($3'$UTR) of protein-coding mRNAs and to the ($5'$UTR) or coding sequence [45–48]. Furthermore, through complementary binding with aimed genes, a few tRNA fragments (TRFs) and tRNA-derived stress-induced RNAs (tiRNAs) may contribute to gene regulation and gene silencing, following a mechanism identical to that of miRNA [49]. CircRNAs primarily operate as ceRNAs and control gene expression at three distinct levels, including epigenetic, transcriptional, and posttranscriptional by sponging several miRNAs (Figure 3) [50].

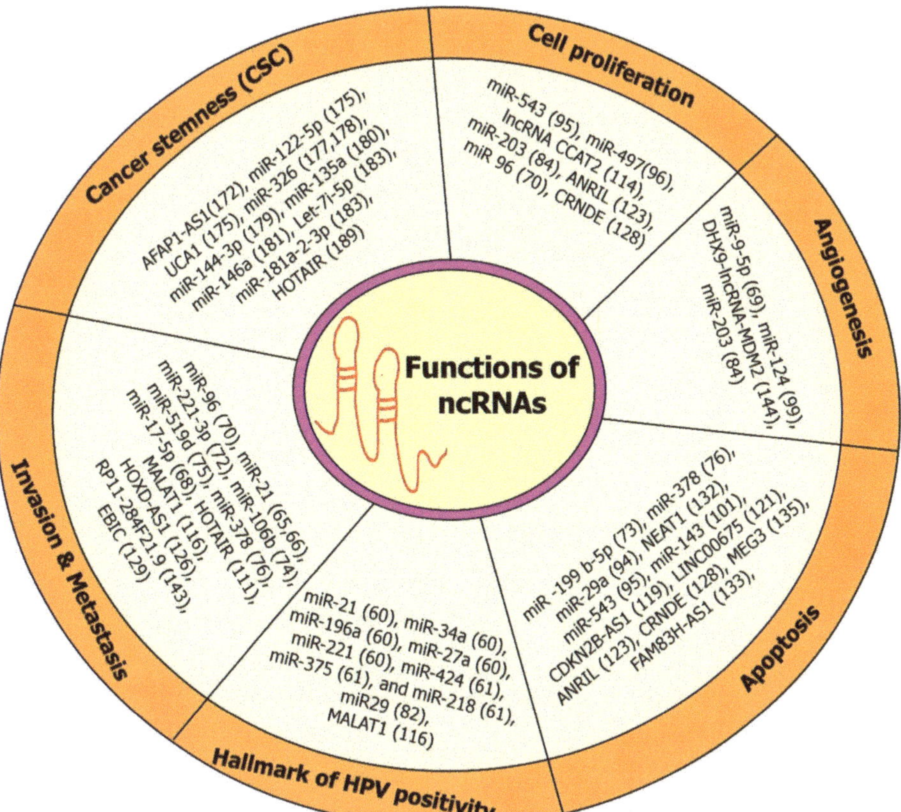

Figure 3. Different biological function of ncRNAs in CC. NcRNAs play a crucial role in the regulation of cell proliferation, cell cycle, cell migration and invasion, epithelial–mesenchymal transition (EMT), and angiogenesis in CC.

3.2. Mechanisms of Action

LncRNAs and miRNAs are structurally similar, and both play crucial roles in the modulation of gene expression. By recruiting multi-subunit chromatin modifying complexes to the DNA molecule (chromatin modulation), some ncRNA regulate multiple biological phenomena such as transcription, nucleosome orientation, chromatin labelling, or histone modifications [19]. All of these modulate gene expression of target genes. Certain miRNAs show interaction with a specific region of the gene promoter. For instance, miR-24-1 acts as an enhancer trigger to stimulate enhancer RNA (eRNA) expression, alters histone modification, and increases the enrichment of p300 and RNA Pol II at the enhancer locus [51]. Some ncRNAs act via splicing regulation and influence disease progression and essential physiological functions by adhering to distinct protein networks that regulate gene expression. Spliceosomes are generated by snRNA and proteins, and they are responsible for the splicing mechanism [19]. For example, SHARP, SAF-A, and LBR are key proteins that are associated with Xist lncRNA for Xist-mediated transcriptional silencing on the X chromosome [52]. Some act via crosstalk in proteomics. Such a mechanism leads to the production of ceRNA and governs translation, transcription, epigenetics, pathological, and physiological processes, exemplified by certain lncRNAs playing a quintessential role during oncogenesis and tumor suppressor cascade and by miR-124, miR-375, and let-7b, which inhibit erbB2/erbB3 to cure breast cancer [53]. A few ncRNAs work by interacting with mRNA (antisense transcription). A mechanism in which the targeted gene is inhibited by the transcription of lncRNA from the opposite template could be a life-changing event in the treatment of hereditary disorders such as Angleman syndrome and others. Furthermore, miRNA–mRNA interactions may suppress mRNA expression. For example, the cell adhesion molecule 1 (CADM1) gene in bone cancer works by sponging miRNA, opening the way for the development of novel therapeutics [54].

4. Expression and Function of ncRNAs in CC

4.1. Dysregulated miRNAs in CC Onset/Progression

Several researchers have investigated the levels of miRNA expression in cervical carcinoma biopsies, exfoliated cervical cells, and cervical mucus, as well as in the serum of women who have been diagnosed with CC. Lui et al. reported the differential expression patterns of six miRNAs (miR-143, miR-143, miR-23b, miR-21, let-7b, and let-7c) which are unique to human CC cell lines [55]. Since then, extensive research has been carried out to characterize the mechanisms that cause miRNA dysregulation as well as profile the expression levels of miRNA in CC and normal cervical epithelial tissues. Indeed, gene knockdown, gene amplifications, or mutations in miRNA loci, coupled with epigenetic silencing such as DNA methylation or dysregulation of miRNA processors (e.g., Drosha) and transcription factors, are all attributed to abnormal miRNA expression patterns observed in cancer, including CC [56]. Muralidhar et al. uncovered 16 dysregulated miRNAs in advanced cervical SCC, including miR-203, miR-31, miR-29a, and miR-21, which have all been attributed to the overexpression of the miRNA processor (Drosha) transcripts and the acquisition of chromosome 5p [57]. Gupta et al. revel that miR-34a or miR-16 may regulate senescence, autophagy, apoptosis, and the functional G1/S checkpoint. Individually, miR-449a may influence senescence and apoptosis and coordinate autophagy in HeLa cells in a synergistic way with miR-16 and/or miR-34a [58]. Multiple studies have previously been conducted to examine the expression profile of miRNAs to find that substantial changes occur during the progression from low to high grade cervical malignancies and to invasive cervical carcinoma, concerning the recognition of unique biomarkers for the determination of cancer stage as well as for resolving diagnosis and prognosis purposes [59]. Gocze et al. conducted miRNA profiling and consequently identified the upregulation of miR-21, miR-34a, miR-196a, miR-27a, and miR-221, which serves as a distinct hallmark of HPV positivity in cervical malignancy samples, regardless of the clinical tumor grade [60]. Tian et al. revealed that the use of single miR-424 and/or miR-375 detection, a miR-424/miR-375/miR-218-based multi-marker panels, is more effective than using cytology in cervical exfoliated cells in gynecological clinics for screening HPV-positive women [61]. MiR-424/miR-375/miR-34a/miR-218 exhibited a statistically significant reduction in expression in high-grade cervical intraepithelial neoplasia (CIN) and abnormal cytology compared to low-grade CIN and normal cytology [61]. Cervical mucus analysis has also been shown to be an effective technique for detecting cervical neoplastic tumors. MiRNAs (miR-20b-5p, miR-126-3p, miR-451a, and miR-144-3p) found in cervical mucus have been shown to be helpful in detecting CC and high-grade intraepithelial lesions [62]. Taken together, the discovery and development of unique tumor biomarkers in cervical exfoliated cells and biological fluids could help with cancer screening and/or reappearance monitoring after treatment. However, further research is needed to determine the clinical implications of miRNA for cancer diagnosis and prognosis.

4.1.1. Oncogenic miRNAs

Evidence from prior studies shows that the overexpression of ncRNAs encourages the growth and development of cervical carcinoma cells and tissues [63,64]. Various studies have demonstrated that the dysregulation of miRNAs significantly contributes to the progression and proliferation of cancerous cells, and they play a crucial role in spreading cancer via advancing cancer growth, development, progression, invasion, angiogenesis, and metastases. miR-21 is found to be overexpressed in aggressive CC tissues, and researchers have shown that miR-21 increases the proliferative index and enhances the migratory and invasion abilities of cervical cells in HeLa cell lineages by considerably repressing the expression of the tumor-suppressive Phosphatase and tensin homolog (PTEN) gene [65,66]. Xu et al. identified that oncogenic miR-21 downregulates PTEN gene expression and increases cell proliferation and migratory and colony forming ability in invasive CC [67]. Overwhelming evidence revealed the involvement of numerous miRNAs in CC progression, as listed in Table 1.

Table 1. Summary of selected oncogenic miRNAs with their functional effects in CC.

MiRNAs	miRNA Expression Profile	Target Gene/Pathway	Biological Function of Oncogenic miRNAs	Ref.
miR-17-5p	Upregulated	TGFBR2	Promotes CC cell metastasis and proliferation	[68]
miR-9-5p	Upregulated	SOCS5.	Promotes angiogenesis, cell proliferation, and invasion	[69]
miR-96	Upregulated	CAV-1	Promotes cell proliferation, migration, and invasion	[70]
miR-21	Upregulated	RASA1	Promotes metastasis and enhances the invasiveness of CC cells	[71]
miR-221-3p	Upregulated	TWIST2/THBS2	Promotes metastases of the lymph nodes in CC	[72]
miR-199 b-5p	Upregulated	KLK10	Promotes cell proliferation, migration, and inhibits apoptosis	[73]
miR-106b	Upregulated	DAB2/TGF-β1	Induces migration of CC cells	[74]
miR-519d	Upregulated	Smad7	Promotes invasiveness and migration abilities of CC cells and prevent cell autophagy	[75]
miR-378	Upregulated	ATG12	Promotes metastases and inhibits apoptosis	[76]
miR-20a	Upregulated	TIMP2 and ATG7	Increases histopathological grade, tumor size, and distant metastases	[77]
miR-106a	Upregulated	TIMP2	Promotes the cell migration and invasion	[78]
miR-150	Upregulated	PDCD4	Promotes cell invasion and migration	[79]
miR-31	Upregulated	BAP1	Promotes cell proliferation and modulates the EMT	[80]
miR-155	Upregulated	LKB1	Promotes CC cell proliferation	[81]

4.1.2. Tumor Suppressor miRNAs

Numerous miRNAs have been identified to be involved in suppressing different malignancies involving CC, indicating that they play a critical function as tumor suppressor miRNAs. For example, HPV-associated miR-29 acts as a tumor suppressor in CC [82]. miR-520d-5p is downregulated in CC and target PTK2 and promotes apoptosis and inhibits CC cell proliferation, migration, and invasion [83]. miR-203 is also downregulated in CC and targets VEGF, which leads to suppressed CC cell proliferation, tumor development, and angiogenesis [84].

Moreover, several studies have demonstrated that circRNA also plays a crucial role in the development and progression of cancer [5,85]. The majority of circRNAs are found in the cytoplasm and typically function as competitive endogenous RNAs (ceRNAs) by sponging miRNAs and enhancing downstream gene expression [86,87]. In a recent study, Ma et al. discovered that circRNA-000284 are upregulated in CC tissues and promotes cell proliferation and invasion in CC cells. Further, the knockdown of circRNA-000284 suppresses the proliferation and migration of CC cells by sponging miR-506 and down-regulates the expression of Snail-2 [88]. Circ 0087429, which is controlled by EIF4A3, may reverse EMT and inhibit CC development through the miR-5003-3p/OGN axis and it is predicted to become a potential target for CC therapy [89]. Recently, Chen et al. found that miR-138 inhibits CC tumor growth by specifically targeting EZH2, showing that DNA methylation at the miR-138 promoter contributes to its downregulation. This study suggests that miR-138 might be used to predict CC metastasis and/or used as a therapeutic target [90]. CircLMO1 overexpression inhibited the growth and metastasis of CC cells both in vitro and in vivo whereas its knockdown increased the proliferation and invasion of CC cells. Mechanistically, CircLMO1 functioned as a competitive endogenous RNA (ceRNA) by sponging miR-4192 to inhibit target gene ACSL4, suggesting that it might be a promising biomarker for the clinical management of CC [91]. In depth research suggests that several miRNAs play a prominent role in CC suppression, many of which are listed in Table 2.

Table 2. Summary of selected tumor suppressor miRNAs with their functional effect in CC.

MiR-RNA	Expression Pattern of miRNAs	Target Gene/Pathway /Molecule	Biological Function of Tumor Suppressor miRNAs	Ref.
miR-520d-5p	Downregulated	PTK2	Promotes apoptosis and inhibits CC cell proliferation, invasion, and migration	[83]
miR-125	Downregulated	VEGF and PI3K/AKT	Inhibits CC cell growth and tumor progression	[92]
miR-23b	Downregulated	AKT/mTOR	Inhibits CC cell multiplication invasion and migration abilities	[93]
miR-29a	Downregulated	DNMT1-SOCS1/NF-κB	Inhibits proliferation, migration, and invasion and promotes CC cell apoptosis	[94]
miR-543	Downregulated	P13K/AKT, p38/MAPK and TRPM7	Inhibits cell proliferation, migration, and invasion; induces cell cycle arrest and boost apoptosis	[95]
miR-497	Downregulated	IGF-1R	Inhibits cell proliferation and arrest cells at S phase of cell cycle	[96]
miR-218	Downregulated	Survivin (BIRC5)	Inhibits clonogenicity, invasion, and migration	[97]
miR-200b	Downregulated	Rho-E	Inhibits migration potential of CC cells and therefore their ability to metastasize	[98]
miR-124	Downregulated	AmotL1	Inhibits angiogenesis, migration, and invasion	[99]

Table 2. *Cont.*

MiR-RNA	Expression Pattern of miRNAs	Target Gene/Pathway /Molecule	Biological Function of Tumor Suppressor miRNAs	Ref.
miR-214	Downregulated	EZH2	Inhibits proliferation of CC cells	[100]
miR-203	Downregulated	VEGFA	Inhibits cell proliferation, tumor development, and angiogenesis	[84]
miR-143	Downregulated	Bcl-2	Inhibits cell proliferation and promoted apoptosis	[101]
miR-101-5p	Downregulated	CXCL6	Inhibits colony formation, invasion, and migration	[102]
miR-132	Downregulated	SMAD2	Inhibits lymph node metastasis	[103]
miR-129-5p	Upregulation	ZIC2	Inhibits tumorigenesis and angiogenesis	[104]
miR-138-5p	Downregulated	SIRT1	Inhibits the tumorigenesis and metastasis	[105]
miR-142-3p	Downregulated	CDC25C	Inhibits cell proliferation	[106]
miR-148b	Downregulated	CASP3	Inhibits cell proliferation and promoted apoptosis	[107]
miR-182	Downregulated	DBMT3a	Induces apoptosis and inhibits cell proliferation	[108]
miR-195	Downregulated	Smad3	Inhibits cell proliferation, migration, and invasion	[109]
miR-196b	Downregulated	VEGF	Inhibits angiogenesis	[110]

4.2. Dysregulated lncRNAs in CC Onset/Progression

LncRNAs possess the potential to bind proteins, mRNAs and miRNAs, and they are found to be engaged in a wide range of biological events as well as cancer formation. Numerous lncRNAs such as HOTAIR, MALAT-1, H19, CCAT2, GAS5, SPRY4-IT1 LET, CCHE1, MEG3, EBIC, and PVT1 are known to perform important roles in cervical tumorigenesis, growth, development, migration, metastases, dissemination, invasion, as well as radio-resistance [111]. NORAD, a long non-coding RNA, could be a key regulator in tumor progression. Huo et al. show that NORAD expression was observed to be significantly upregulated in CC tissues and cell lineages and promotes the development and dissemination of CC by sponging miR-590-3p and targeting SIP1. Aberrant expression of NORAD is attributed to advanced FIGO stage, vascularization lymph node metastases, and poor overall survivability of CC patients. On the other hand, silencing of its expression lowered CC cell division, incursions, and EMT processes [112].

Another lncRNA, CCHE1, was also found to be deregulated in CC and its aberrant expression was linked to a poor prognosis in CC patients, indicating that CCHE1 could be used as a prognostic biomarker [113]. Similarly, lncRNA CCAT2 prominently contributes to CC and it was reported that the knocking down of CCAT2 impeded cervical tumor cell proliferation and caused CC cells to enter the G1 phase of their cycle and stimulated them to undergo autophagy [114]. Taken together, lncRNAs perform various functions and aid in the diagnosis, treatment, and prognosis of CC. However, further studies are required to provide a better understanding.

4.2.1. Oncogenic lncRNAs

Evidence from prior findings reveals that the dysregulation of lncRNA influences the growth and development of CC cells and tissues. Several lncRNAs have been involved in the cancer progression such as HOX transcript antisense intergenic RNA (HOTAIR), H19, and X-inactive specific transcript (XIST), plasmacytoma variant translocation 1 (PVT1), cervical carcinoma high-expressed 1 (CCHE1), and metastasis-associated lung cancer adenocarcinoma transcript 1 (MALAT-1) (Table 3). MALAT-1 has been demonstrated to produce epigenetic modifications and affect gene expression, nuclear organization, and alternative splicing regulation by functioning as a splicing factor decoy [115]. MALAT-1 is expressed exclusively in cervical carcinoma cell lineages and tumor tissues contaminated with HR-HPV [116]. It functions by sponging numerous miRNAs, such as miR-145, and thus encourages the development and progression of cervical carcinoma through the induction of EMT [116].

Table 3. Summary of selected oncogenic LncRNAs with their functional effects in CC.

LncRNA	Expression Pattern lncRNA	Target Gene /Pathways/Molecules	Biological Function of Oncogenic lncRNA	Ref.
HOTAIR	Upregulated	BCL2, miR-143-3p	Promotes CC cell growth	[117]
LINC01535	Upregulated	miR-214/EZH2 feedback loop	Promotes progression and metastasis of CC	[118]
CDKN2B-AS1	Upregulated	miR-181a-5p/TGFβI axis	Promotes tumor cell growth and inhibits apoptosis	[119]
CASC11	Upregulated	Wnt/β-catenin	Promotes cell proliferation	[120]
LINC00675	Upregulated	Wnt/β-catenin	Promotes cancer cell growth, invasiveness, migration, and repressed cell apoptosis	[121]
MALAT-1	Upregulated	HPV16 E6/E7	Promotes cell proliferation, migration, and invasion and modulates EMT expression	[122]
ANRIL	Upregulated	Cyclin D1, CDK4, CDK6, E-cadherin, vimentin, and N-cadherin.	Promotes cell proliferation, migration, and invasion and inhibits apoptosis	[123]
BLACAT1	Upregulated	Cyclin B1, and CDC25C, N-Cadherin, E-Cadherin	Enhances CC cell proliferation and invasion	[124]
PVT1	Upregulated	Smad3, miR-140-5p sponging	Promotes cell proliferation and metastasis	[125]
HOXD-AS1	Upregulated	Ras/ERK,	Enhances cell proliferation, migration, and invasion	[126]
DLX6-AS1	Upregulated	miR-16-5p/ARPP19 axis	Increases cell proliferation and invasion	[127]
CRNDE	Upregulated	PI3K/AKT	Promotes cell proliferation and inhibits apoptosis	[128]
CCAT2	Upregulated	Cell cycle	Promotes cell multiplication and penetration	[114]
EBIC	Upregulated	EZH2, E-cadherin	Promotes metastasis and invasion	[129]
RSU1P2	Upregulated	IGF1R, N-myc, let-7a, EphA4	Promotes tumor development	[130]
SPRY4-IT1	Upregulated	miR-101-3p, ZEB1	Promotes cell proliferation, migration, and invasion and modulates EMT expression	[131]
NEAT1	Upregulated	miR-377/FGFR1 axis	Increases CC cell survival and motility and inhibits apoptosis	[132]
FAM83H-AS1	Upregulated	E6-p300 pathway	Promotes cell proliferation and migration and inhibits apoptosis	[133]
C5orf66-AS1	Upregulated	miR-637/RING1 axis	Promotes progression and proliferation of CC cells	[134]

4.2.2. Tumor Suppressor lncRNAs

In CC, a few lncRNAs act as tumor suppressors (Table 4). Maternally expressed Gene 3 (MEG3) is a well-recognized suppressive lncRNA that increases apoptosis and suppresses the multiplication of CC cells through specifically linking with p-STAT3 and consequently causes its ubiquitination and destruction [135]. In other study, STXBP5-AS1 lncRNA is also downregulated in CC. STXBP5-AS1 decreases the invasion and migration ability of cervical cancer cells via miR-641/PTEN axis [136].

Table 4. Summary of selected tumor suppressor lncRNAs with their functional effects in CC.

LncRNAs	Expression Pattern	Target Genes /Pathways/Molecule	Biological Function of Tumor Suppressor lncRNA	Ref.
MEG3	Downregulated	p-STAT3	Inhibits cell proliferation and increases apoptosis	[135]
GAS5	Downregulated	miR-205, miR-196a	Inhibits growth and metastases	[137]
GAS5-AS1	Downregulated	Increase GAS5 stability by epigenetic modulation	Suppresses growth and metastasis	[138]
STXBP5-AS1	Downregulated	miR-96-5p/PTEN axis	Inhibits cell proliferation and invasiveness of CC cells	[136]
TUSC8	Downregulated	miR-641/PTEN axis	Inhibits migration and invasion	[139]
XLOC_010588	Downregulated	c-Myc	Inhibits proliferation	[140]
LINC00861	Downregulated	PTEN/AKT/mTOR miR-513b-5p	Inhibit the progression of CC cells	[141]
ZNF667-AS1	Downregulated	Sponge miR-93-3p and upregulate PEG3	Inhibits cell proliferation, invasion, and metastasis	[142]
RP11-284F21.9	Downregulated	PPWD1, miR-769-3p	Inhibits cell proliferation, migration, and invasion	[143]
Lnc-CCDST	Downregulated	DHX9-MDM2	Inhibits angiogenesis and invasion	[144]
DGCR5	Downregulated	WNT signaling	Suppresses migration and invasion	[145]

5. Role of ncRNAs (miRNAs and lncRNAs) in the Tumor Microenvironment (TME) of CC Onset/Progression

TME is a complex and dynamic network composed of tumor cells and their surroundings, which includes tumor-linked immune cells, vascular endothelial cells, fibroblasts, pericytes, adipocytes, extracellular matrix (ECM), cytokines, and chemokines [146]. The ECM and various types of stromal cells comprise the TME. Crosstalk between tumor cells and their TME is a crucial event in tumor progression and metastasis [147]. The emerging data suggest that ncRNAs (miRNA and lncRNA) play a significant role in modulating TME as well as tumor progression [148]. However, more research is needed for a better understanding of the physiological and pathological functioning of ncRNAs in the TME.

Matrix metalloproteinases (MMPs) are extracellular proteinases that have an impact on primary tumor invasion and metastasis. Clinical studies in CC indicated that miR-183 decreases CC cell proliferation and metastasis by inhibiting MMP-9 [149]. Similarly, activation of angiogenesis is required for solid tumor development and metastasis in TME [150]. In CC cells, miR-124 has been shown to target the angiomotin-like protein AmotL1 and subsequently decrease clonogenicity and cellular proliferation [99]. Additionally, cervical squamous cancer cells release exosomal miR-221-3p, which has been demonstrated to facilitate angiogenesis via targeting Thrombospondin-2 [151].

Similarly, lncRNAs have been demonstrated to facilitate crosstalk between tumor cells and stromal cells, and the deregulation of their expression in these cells might result in carcinogenesis [152]. The lncRNA MALAT-1 (metastasis-associated lung adenocarcinoma transcript 1), for example, is substantially expressed in patients suffering from non-small-cell lung cancer (NSCLC), and that of exosomal MALAT-1 is linked to the tumor, node, and metastasis (TNM) stage [153]. The lncRNA can also help tumor cells evade immune recognition by promoting the formation of an immunosuppressive microenvironment [154]. Collectively, tumor onset, progression, dissemination, metastasis, and other malignant biological characteristics can all be influenced by information exchange in the TME [155]. Clinically, ncRNA-mediated modulation of the TME and crosstalk between cancer and immune cells has emerged as a promising and appealing diagnostic and therapeutic.

6. Role of ncRNAs (miRNAs and lncRNAs) in the Tumor Immunology of Onset/Progression

The immune system is well acknowledged for its participation in cancer onset and development, and it can have both pro-carcinogenic and anti-carcinogenic effects contingent on the microenvironment [156]. The adaptive immune system provides highly specialized procedures that eliminate pathogens, while the innate immune system is the initial line of defense against foreign pathogens [157]. Importantly, the immune system can kill cancer cells in addition to defending against foreign invaders. In recent decades, researchers and physicians have focused on effectively activating the immune system to better combat cancer, and this treatment is referred to as "immunotherapy". Due to its exceptional and long-lasting efficacy, immunotherapy has been designated as the fourth treatment cornerstone of cancer therapy [158]. However, only a small proportion of patients benefited from immunotherapy. The data suggest that ncRNAs are active participants in several stages of tumor immunity. NcRNAs, which include miRNAs, lncRNAs, and circRNAs, influence a wide range of cellular activities in the development and progression of cancer [159].

A better understanding of the function of ncRNA in the control of cancer immunity will lead to the development of novel treatment targets. Therefore, extensive research is needed to understand the function of ncRNAs in cancer immunity and obtain new insights into cancer diagnostics and immunotherapeutic therapy. Effector cells, such as macrophages, natural killer (NK) cells, and neutrophils, are crucial components of the innate immune response [157]. NcRNAs play a crucial role in the regulation of these effector cells. For example, enforced expression of miR-511-3p, has been shown to suppress tumor formation by downregulating the protumoral gene profile of mannose receptor-1 (MRC1)$^+$ tumor-associated macrophages (TAMs) [160]. Several research studies have been undertaken to determine the role of ncRNAs in macrophage polarization since it is a critical component of many disease states, including cancer [158]. TCONS_00019715, an lncRNAs, play a key role in driving macrophage polarization to the M1 phenotype, which improves tumoricidal capabilities [159]. In addition, miRNA-19a-3p, miR-33, and lncRNA-MM2P impact M2 macrophage polarization [161]. MiR-21 regulates colony-stimulating factor 1 receptor (CSF-1R) for macrophage repolarization [162], whereas a double feedback loop regulated by miRNA-23a/27a/24-2 effectively regulates macrophage polarization and regulates cancer progression [163]. Surprisingly, the roles of macrophages in tumors must be contextualized within the unique microenvironment since macrophages exhibit intensities of cytokines, hence serving as either anti-carcinogenic or pro-carcinogenic [164]. NK cells have anti-cancerous properties, and it was shown that ncRNAs play a significant role in NK cell biology in the domains of growth, inflammation, and tumor monitoring [165]. He et al. revealed that the presence of various miRNAs in circulation, such as miR-122, miR-21, miR-15b, and miR-155, can stimulate NK cells via Toll-like receptor signaling and inhibit tumor formation [166]. Evidence suggests that ncRNAs play a critical role in adaptive immunity and influence tumor development and dissemination. For example, miRNA let-7a expression in colorectal cancer tissue may be negatively correlated to T-cell density and positively associated with colorectal cancer cell death [167]. Hui et al. recently identified and validated six immune-related lncRNAs (AC006126.4, EGFR-AS1, RP4-647J21.1, LINC00925, EMX2OS, and BZRAP1-AS1) of CC and revealed an immune-related risk model for predicting clinical outcomes, indicated the intensity of immune cell infiltration in the TIME, and predicted potential compounds in the immunotherapy treatment for CC [168].

7. Role of ncRNAs (miRNAs and lncRNAs) in Cancer Stem Cells (CSCs) of CC

Despite HPV infection being the most common cause of CC, CSCs also play an important role in the disease's development, metastasis, recurrence, and prognosis [169]. CSCs play a significant role in the recurrence and metastasis of patients with cervical carcinoma [170,171]. Several recent studies have reported that the stemness properties are partly regulated by the interaction of ncRNAs in CC stem cells.

In recent research, Xia et al. found that AFAP1-AS1 suppresses cancer stemness, cell cycle progression, and EMT in CD44v6 (+) CC cells, and that the miR-27b-3p/VEGF-C axis is a direct target of AFAP1-AS1, allowing AFAP1-AS1 to modulate stemness characteristics in CC cells [172].

Another study suggests that urothelial carcinoma-associated 1 (UCA1) is a lncRNA with aberrant expression in a number of malignant tumors [173]. There has been less research on the involvement of UCA1 from CC cell-derived exosomes in CC development. UCA1 overexpression reduces the cytoplasmic levels of free miR-122-5p, reducing miR-122-5ps ability to regulate its target mRNAs [174]. Another study also supports that CaSki- exosomes can influence CC stem cell self-renewal and differentiation, but silencing UCA1 or increased expression of miR-122-5p inhibits CC stem cell self-renewal and differentiation [175].

Transcription factor 4 (TCF-4) is a transcription factor that interacts with β-catenin to activate target gene transcription in response to Wnt activation signaling [176]. The uncontrolled activation of the smoothened (Smo) signal transducer of the oncogenic Hedgehog (Hh) pathway in chronic myeloid leukemia has been linked to the downregulation of miR-326. Restoring miR-326 expression may also aid in the elimination of CD34 + CML stem/progenitor cells [177]. In patients receiving CSC-targeted treatment, CD133 might be used as a specific CC stem cell marker [170]. Zhang et al. reveal that the overexpression of miR-326 significantly decreased TCF-4 protein expression. Furthermore, miR-326 inhibited CaSki cell growth and CSC-like properties in vitro by targeting TCF-4 [178].

Human bone marrow mesenchymal stem cell (hBMSCs)-derived extracellular vesicle (EVs)-loaded miR-144-3p altered the biology of recipient CC cells by curbing cell proliferation, migration, invasion, and clonogenicity while inducing apoptosis, all of which lead to a decreased propensity in the development and progression of CC [179]. Another study reported that miR-135a triggered the development of a CD133+ subpopulation in an HPV-immortalized cervical epithelial cell line. In both in vitro and in vivo studies, miR-135a induced the formation of a subpopulation of cells with CSC characteristics, and the Wnt/β-catenin signaling pathway is required to maintain its tumorigenicity [180]. Dong et al. show that miR-146a downregulation promotes tumorigenesis in CC stem cells via the VEGF/CDC42/PAK1 signaling pathway [181].

CC is often related to HPV infection and the HPV 16 E5 gene has been shown to promote EGFR expression by blocking the degradation of internalized EGFR [13], and HPV 16 E6/E7 has also been demonstrated to increase EGFR levels [182]. The study reports on the link between let-7i-5p, miR-181a-2-3p, and the EGF/PI3K/SOX2 axis, which is essential for the survival of CSCs in CC, Let-7i-5p, miR-181a-2-3p, or SOX2, could be possible treatment targets for cervical CSCs, if more research is carried out on CC tissue samples and in vivo [183].

Several cancers express the homeobox A11 antisense lncRNA (HOXA11-AS), which is near the HOXA11 gene, supporting the concept that it promotes CC progression [184]. NcRNAs dominate homeobox gene cluster intergenic transcripts, which comprise short miRNA and lncRNAs that are antisense to their conventional HOX neighbors. HOX transcription factors promote embryonic development in both humans and mice [185]. In vitro, HOXA11-AS overexpression increased cell proliferation, migration, and tumor invasion, whereas HOXA11-AS knockdown decreased these biologic aggressive characteristics [186]. HOTAIR functions as an oncogenic lncRNA and plays a critical role in regulating stemness properties in various cancers, including CC [187,188]. HOTAIR is significantly elevated in association with the enrichment of CC stem cells, and its knockdown dramatically reduces the expression of stemness markers. The level of HOTAIR was found to be linked to the expression of miR-203, which helps EMT and is controlled by ZEB1 [189].

8. Therapeutic Approaches for Targeting ncRNAs in CC

The use of ncRNAs as a therapeutic target for CC might be very effective. Antisense oligonucleotides (ASOs), the CRISPR–Cas9 system, exosomes, and other methods are currently being used to exploit the therapeutic value of lncRNAs.

ASOs are single-stranded antisense oligonucleotides having a central DNA stretch (>6mers) which can be native or phosphorothioated (chemically modified), and RNA nucleotides at flanking sections of the molecule [190]. Several diseases have been successfully treated with ASOs, which are often employed to change mRNA expression [191,192]. They may be used to inhibit cancerous ncRNAs that are overexpressed in cancer cells. A recent study has shown that LncRNA MALAT-1-specific ASOs suppress cancer cell metastasis in vitro and in vivo [193]. CC therapy based on ASOs needs further investigation at this point. As antisense oligonucleotide technology continues to evolve, research into the clinical use of ASO as a therapy for CC is likely to move quickly.

Therapeutic targeting of coding and non-coding genes is now possible using the CRISPR–Cas9 system [194,195]. There are many ways to employ this system: genome editing using active CRISPR–Cas9 or Cas12a (Cpf1), interfering with gene activity or activation using catalytically dead (d)Cas9 linked to an activating or repressive effector domain, or RNA editing using the Cas13 variant [196–198]. Ex vivo CRISPR–Cas9 genome editing clinical studies are now underway. Several lncRNAs have been shown to either promote (SAF, MALAT-1, HEAL) or inhibit (GAS5, 7 SK, NRON, TAR-gag, lincRNA-p21, NEAT1). [199]. Viral proteins can also modify the biological activities of lncRNAs through direct or indirect binding, hence altering their protein and/or nucleic acid interactomes. For instance, HPV16 E7 has been shown to communicate with HOTAIR, potentially impairing its ability to suppress polycomb-regulated genes [200]. The lncRNA urothelial carcinoma-associated 1 (UCA1) was recently shown to be critical in human heme biosynthesis and erythrocyte development of CD34 + HSCs. In this study, Liu et al. identified that lncRNA UCA1 serves as a scaffold for recruiting PTBP1 to ALAS2 mRNA and stabilizing it through PTBP1 [201]. This demonstrates that lncRNAs may increase the number of therapeutic CRISPR–Cas9 ex vivo editing targets [201]. Delivery concerns, particularly with the big Cas proteins, must be resolved before the CRISPR–Cas9 system can be used in vivo, and immunological responses must be carefully examined. In spite of its numerous benefits, the CRISPR–Cas9 system may have detrimental impacts owing to off-target effects that cannot be ignored and might have major implications [202]. The CRISPR–Cas9 technology has several constraints that need to be solved in order to enhance its therapeutic use.

In addition, exosomes are nanovesicles that facilitate communication between cells. Exosomes are important in the development and progression of cancer [203]. Exosomes include functional components such as proteins, lipids, mRNA molecules, and ncRNAs which act as carriers of extracellular information [204]. With the advancement of scientific technology, we expect that exosomes encapsulating lncRNAs or specialized drugs targeting lncRNAs will be developed for cancer-targeted treatment.

9. Approaches for Systemic Delivery of Therapeutics ncRNAs in CC

Ensuring ncRNA therapeutics reach their intended target organ and cell type, as well as cross cell membranes to accomplish their intracellular activities, is one of the major hurdles in the field. Oligonucleotide delivery is limited by its instability, negative charge, and hydrophilic nature, which hinders diffusion across cell membranes [205]. Lipid and polymer-based vectors, as well as ligand–oligonucleotide conjugate delivery systems, are all being employed as delivery methods. Endosomal escape of the RNA therapy must be made easier to avoid lysosomal degradation because of the variety of endocytosis mechanisms used to pick up these delivery systems.

Lipid nanoparticles (LNPs) are readily manipulated, may be linked to targeting moieties, and have great biodegradability and biocompatibility with low immunogenicity. The elevated lncRNA ceruloplasmin (NRCP) was suppressed in an ovarian cancer mouse model utilizing a phosphocholine-derived 1,2-dioleoyl-sn-glycero-3-phosphocholine (DOPC) nanoliposome carrying siRNA. There was a significant decrease in tumor development and greater sensitivity to cisplatin [206].

Polymer-based carriers differ from lipid-based ones in that they are more versatile in terms of size, molecular composition, and structure. Several polymers, including polyethylene imine (PEI), polylactic-co-glycolic acid (PLGA), poly-amidoamine (PAMAM), and chitosan, have been extensively investigated for the delivery of miRNA mimics or antimiRs, alone or in combination with chemotherapies for improved therapeutics. Several miRNAs such as miR-150, miR-221/miR-222, miR-21, miR-34a, miR-145, and miR-33a have all been delivered systemically or locally through polymers [207–211].

LOcal Drug EluteR (LODERTM) is a new biodegradable polymeric matrix that protects drugs from enzymatic degradation and releases siRNA against G12D-mutated KRAS (siG12D), and a phase 1/2a clinical study of siG12D-LODER was recently completed [212]. Combinatorial therapies can potentially benefit from the development of nanoparticles (NPs) that can deliver multiple therapies at once, and a lot of progress has been made in this area [213].

The oligonucleotide conjugation of diverse entities is being further investigated for their delivery. Ligand conjugation is a frequent clinical strategy for RNA therapeutic delivery to cancer, allowing selective administration through a receptor-mediated mechanism [214]. A recent study showed that the inhibition of the ubiquitin-conjugating enzyme E2 N by miRNA-590-3p reduced the cell growth of CC [215]. Furthermore, in an A549 xenograft model, T7-conjugated Co-ASOs-LNPs (Co-ASOs-LNPs) displayed improved anticancer efficacy, prolonged overall survival time, and tumor targeting activity [216]. Another promising preclinical strategy is conjugation of oligonucleotides to antibodies gates. In this technique, oligonucleotide or drug may be conjugated to antibodies through electrostatic interactions, affinity conjugation using biotin or avidin, direct conjugation, or double-strand hybridization [217,218]. Several alternative approaches, such as using an azide-functionalized linker peptide on the antibody and conjugation to dibenzylcyclooctyne-bearing RNAs or antibodies with a reactive lysine residue paired with β-lactam linker-functionalized RNAs, have also been investigated [219,220].

Numerous advances have been achieved in this sector, which is especially relevant for combinatorial therapies, such as the development of NPs.

By optimizing the targeted delivery of medicines specifically to tumor regions coupled with enhanced efficiency, nanomedicine offers the potential to overcome the limits of traditional therapeutic techniques [221]. The application of NPs as a ncRNA-targeted treatment coupled with immunotherapy seems feasible. However, only a few studies have been carried out to evaluate the use of this delivery system and it will take a while to implement this clinically. Shao et al. successfully developed floral-shaped SiO–PEI NPs which have maximum loads of pDNA/siRNA. These NPs containing a plasmid-expressing miR-let-7c-5p were effective in transferring miR-let-7c-5p to human epithelial cancerous HeLa cells. Furthermore, under relatively low cytotoxic situations, the collaboration of nanotechnology with gene therapy may prevent the onset and progression of cancer. Findings from this study have provided a new anticancer strategy [222]. Similarly, Wang and Liang synthesized a conjugate containing CD59, miRNA-1284, and cisplatin (CDDP), which was subsequently loaded into liposomes (CD/LP-miCDDP). This co-delivery strategy had greater anticancer effects in CC cells, and the apoptosis rate was significantly increased compared to miR-1284 or cisplatin or alone [223]. Similarly, in other cancers, nanomedicines have given good outcomes. For example, to treat lung cancer, Gong et al. efficiently synthesized MALAT-1-targetted ASOs and nucleo-targeted Tat peptide integrated with Au NPs (i.e., ASO-Au-Tat NPS), which might stabilize friable ASOs, improve nuclear uptake, and exhibit excellent biocompatibility. MALAT-1 expression in A549 lung cancerous cells was dramatically reduced after treatment with ASO-Au-Tat NPs [193].

10. Concluding Remarks

Although CC is a curable disease with proven interventions, it remains the most frequently occurring cancer in women globally. Individuals with advanced or metastatic CC have a very poor prognosis and available treatment options are also limited. As a result, it is critical to gain insight into the mechanisms of metastasis and identify new therapeutic targets. The incredible and sophisticated underlying molecular mechanisms that orchestrate life's fundamental concepts are now known to be controlled by a world of highly complex non-coding RNA. The effective implementation of RNA-based therapeutics necessitates a novel multidisciplinary strategy that includes technological advances in molecular biology, immunology, pharmacology, chemistry, and nanotechnology. Considering their active participation in various pathways of cervical tumorigenesis, research on ncRNAs has emerged as a focal point for expanding our knowledge of cancer biology and offering additional research opportunities. An ideal RNA therapeutic should be rigorously assessed for immunogenicity, chemically altered to improve pharmacokinetics and pharmacodynamics, analyzed for biodistribution and potential intracellular escape mechanisms, target specificity and interactions, and be dosed at optimum concentrations to yield desired outcomes. Repeated attempts to divulge the functions of all types of ncRNAs in tumor immunity will lay the foundation for an even better understanding, control, and cancer therapy, as well as make immunotherapy more coherent with an individual's biological properties. Additionally, modifying the tumor microenvironment can provides striking results in the prevention and pathological management of CC, as evidenced by the current report of clinical trials in oncology. Successfully prepared nanomedicines bring a massive shift and ongoing clinical research in oncology demonstrates that nanoscience will shortly provide unique therapeutic approaches for thousands of CC patients globally.

Author Contributions: All authors listed have made a substantial, direct, and intellectual contribution to the work. D.P., A.S (Anupam Singh)., S.G. and V.K.K. conceived the idea and wrote the majority of manuscript. The manuscript was also written and edited by A.S. (Aishwarya Sharma), M.K.S., K.K.R. and S.C.C. All authors have read and agreed to the published version of the manuscript.

Funding: This research received no external funding.

Institutional Review Board Statement: Not applicable.

Informed Consent Statement: Not applicable.

Data Availability Statement: Not applicable.

Acknowledgments: We thank Andrew Massey for the manuscript Proofreading.

Conflicts of Interest: The authors declare no conflict of interest.

References

1. Siegel, R.L.; Miller, K.D.; Fuchs, H.E.; Jemal, A. Cancer statistics, 2022. *CA A Cancer J. Clin.* **2022**, *72*, 7–33. [CrossRef] [PubMed]
2. Nicolás-Párraga, S.; Alemany, L.; de Sanjosé, S.; Bosch, F.X.; Bravo, I.G. Differential HPV16 variant distribution in squamous cell carcinoma, adenocarcinoma and adenosquamous cell carcinoma. *Int. J. Cancer* **2017**, *140*, 2092–2100. [CrossRef] [PubMed]
3. Kashyap, V.K.; Dan, N.; Chauhan, N.; Wang, Q.; Setua, S.; Nagesh, P.K.B.; Malik, S.; Batra, V.; Yallapu, M.M.; Miller, D.D.; et al. VERU-111 suppresses tumor growth and metastatic phenotypes of cervical cancer cells through the activation of p53 signaling pathway. *Cancer Lett.* **2020**, *470*, 64–74. [CrossRef] [PubMed]
4. Kombe Kombe, A.J.; Li, B.; Zahid, A.; Mengist, H.M.; Bounda, G.-A.; Zhou, Y.; Jin, T. Epidemiology and Burden of Human Papillomavirus and Related Diseases, Molecular Pathogenesis, and Vaccine Evaluation. *Front. Public Health* **2021**, *8*, 552028. [CrossRef]
5. Tornesello, M.L.; Faraonio, R.; Buonaguro, L.; Annunziata, C.; Starita, N.; Cerasuolo, A.; Pezzuto, F.; Tornesello, A.L.; Buonaguro, F.M. The Role of microRNAs, Long Non-coding RNAs, and Circular RNAs in Cervical Cancer. *Front. Oncol.* **2020**, *10*, 150. [CrossRef]

6. Karuri, A.R.; Kashyap, V.K.; Yallapu, M.M.; Zafar, N.; Kedia, S.K.; Jaggi, M.; Chauhan, S.C. Disparity in rates of HPV infection and cervical cancer in underserved US populations. *Front. Biosci.* **2017**, *9*, 254–269. [CrossRef]
7. Chan, C.K.; Aimagambetova, G.; Ukybassova, T.; Kongrtay, K.; Azizan, A. Human Papillomavirus Infection and Cervical Cancer: Epidemiology, Screening, and Vaccination-Review of Current Perspectives. *J. Oncol.* **2019**, *2019*, 3257939. [CrossRef]
8. Ajenifuja, K.O.; Gage, J.C.; Adepiti, A.C.; Wentzensen, N.; Eklund, C.; Reilly, M.; Hutchinson, M.; Burk, R.D.; Schiffman, M. A population-based study of visual inspection with acetic acid (VIA) for cervical screening in rural Nigeria. *Int. J. Gynecol. Cancer* **2013**, *23*, 507–512. [CrossRef]
9. Zhang, S.; Xu, H.; Zhang, L.; Qiao, Y. Cervical cancer: Epidemiology, risk factors and screening. *Chin. J. Cancer Res.* **2020**, *32*, 720–728. [CrossRef]
10. Small, W., Jr.; Bacon, M.A.; Bajaj, A.; Chuang, L.T.; Fisher, B.J.; Harkenrider, M.M.; Jhingran, A.; Kitchener, H.C.; Mileshkin, L.R.; Viswanathan, A.N.; et al. Cervical cancer: A global health crisis. *Cancer* **2017**, *123*, 2404–2412. [CrossRef]
11. Agarwal, S.; Saini, S.; Parashar, D.; Verma, A.; Jagadish, N.; Batra, A.; Suri, S.; Bhatnagar, A.; Gupta, A.; Ansari, A.S.; et al. Expression and humoral respo nse of A-kinase anchor protein 4 in cervical cancer. *Int. J. Gynecol. Cancer* **2013**, *23*, 650–658. [CrossRef]
12. St. Laurent, G.; Wahlestedt, C.; Kapranov, P. The Landscape of long noncoding RNA classification. *Trends Genet.* **2015**, *31*, 239–251. [CrossRef]
13. Erhard, F.; Zimmer, R. Classification of ncRNAs using position and size information in deep sequencing data. *Bioinformatics* **2010**, *26*, i426–i432. [CrossRef]
14. Cobb, M. 60 years ago, Francis Crick changed the logic of biology. *PLoS Biol.* **2017**, *15*, e2003243. [CrossRef]
15. Carninci, P.; Kasukawa, T.; Katayama, S.; Gough, J.; Frith, M.C.; Maeda, N.; Oyama, R.; Ravasi, T.; Lenhard, B.; Wells, C.; et al. The transcriptional landscape of the mammalian genome. *Science* **2005**, *309*, 1559–1563. [CrossRef]
16. Hangauer, M.J.; Vaughn, I.W.; McManus, M.T. Pervasive transcription of the human genome produces thousands of previously unidentified long intergenic noncoding RNAs. *PLoS Genet.* **2013**, *9*, e1003569. [CrossRef]
17. Comfort, N. Genetics: We are the 98%. *Nature* **2015**, *520*, 615–616. [CrossRef]
18. Robertson, M.P.; Joyce, G.F. The origins of the RNA world. *Cold Spring Harb. Perspect. Biol.* **2012**, *4*, a003608. [CrossRef]
19. Bhatti, G.K.; Khullar, N.; Sidhu, I.S.; Navik, U.S.; Reddy, A.P.; Reddy, P.H.; Bhatti, J.S. Emerging role of non-coding RNA in health and disease. *Metab. Brain Dis.* **2021**, *36*, 1119–1134. [CrossRef]
20. Ponting, C.P.; Oliver, P.L.; Reik, W. Evolution and Functions of Long Noncoding RNAs. *Cell* **2009**, *136*, 629–641. [CrossRef]
21. Alvarez-Dominguez, J.R.; Knoll, M.; Gromatzky, A.A.; Lodish, H.F. The Super-Enhancer-Derived alncRNA-EC7/Bloodlinc Potentiates Red Blood Cell Development in trans. *Cell Rep.* **2017**, *19*, 2503–2514. [CrossRef]
22. Dahariya, S.; Paddibhatla, I.; Kumar, S.; Raghuwanshi, S.; Pallepati, A.; Gutti, R.K. Long non-coding RNA: Classification, biogenesis and functions in blood cells. *Mol. Immunol.* **2019**, *112*, 82–92. [CrossRef]
23. Nagano, T.; Fraser, P. No-Nonsense Functions for Long Noncoding RNAs. *Cell* **2011**, *145*, 178–181. [CrossRef]
24. O'Day, E.; Lal, A. MicroRNAs and their target gene networks in breast cancer. *Breast Cancer Res.* **2010**, *12*, 201. [CrossRef]
25. Zhang, P.; Wu, W.; Chen, Q.; Chen, M. Non-Coding RNAs and their Integrated Networks. *J. Integr. Bioinform.* **2019**, *16*. [CrossRef] [PubMed]
26. Derrien, T.; Johnson, R.; Bussotti, G.; Tanzer, A.; Djebali, S.; Tilgner, H.; Guernec, G.; Martin, D.; Merkel, A.; Knowles, D.G.; et al. The GENCODE v7 catalog of human long noncoding RNAs: Analysis of their gene structure, evolution, and expression. *Genome Res.* **2012**, *22*, 1775–1789. [CrossRef] [PubMed]
27. Wang, H.; Chung, P.J.; Liu, J.; Jang, I.C.; Kean, M.J.; Xu, J.; Chua, N.H. Genome-wide identification of long noncoding natural antisense transcripts and their responses to light in Arabidopsis. *Genome Res.* **2014**, *24*, 444–453. [CrossRef] [PubMed]
28. Ning, B.; Yu, D.; Yu, A.-M. Advances and challenges in studying noncoding RNA regulation of drug metabolism and development of RNA therapeutics. *Biochem. Pharmacol.* **2019**, *169*, 113638. [CrossRef] [PubMed]
29. Robles, V.; Valcarce, D.G.; Riesco, M.F. Non-coding RNA regulation in reproduction: Their potential use as biomarkers. *Noncoding RNA Res.* **2019**, *4*, 54–62. [CrossRef]
30. Kung, J.T.Y.; Colognori, D.; Lee, J.T. Long noncoding RNAs: Past, present, and future. *Genetics* **2013**, *193*, 651–669. [CrossRef]
31. Statello, L.; Guo, C.-J.; Chen, L.-L.; Huarte, M. Gene regulation by long non-coding RNAs and its biological functions. *Nat. Rev. Mol. Cell Biol.* **2021**, *22*, 96–118. [CrossRef]
32. Beermann, J.; Piccoli, M.T.; Viereck, J.; Thum, T. Non-coding RNAs in Development and Disease: Background, Mechanisms, and Therapeutic Approaches. *Physiol. Rev.* **2016**, *96*, 1297–1325. [CrossRef]
33. Akerman, I.; Tu, Z.; Beucher, A.; Rolando, D.M.Y.; Sauty-Colace, C.; Benazra, M.; Nakic, N.; Yang, J.; Wang, H.; Pasquali, L.; et al. Human Pancreatic β Cell lncRNAs Control Cell-Specific Regulatory Networks. *Cell Metab.* **2017**, *25*, 400–411. [CrossRef]
34. Wu, H.; Yang, L.; Chen, L.L. The Diversity of Long Noncoding RNAs and Their Generation. *Trends Genet.* **2017**, *33*, 540–552. [CrossRef]
35. Ojha, S.; Malla, S.; Lyons, S.M. snoRNPs: Functions in Ribosome Biogenesis. *Biomolecules* **2020**, *10*, 783. [CrossRef]
36. Naganuma, T.; Hirose, T. Paraspeckle formation during the biogenesis of long non-coding RNAs. *RNA Biol.* **2013**, *10*, 456–461. [CrossRef]
37. Salehi, S.; Taheri, M.N.; Azarpira, N.; Zare, A.; Behzad-Behbahani, A. State of the art technologies to explore long non-coding RNAs in cancer. *J. Cell Mol. Med.* **2017**, *21*, 3120–3140. [CrossRef]

38. Cech, T.R.; Steitz, J.A. The noncoding RNA revolution-trashing old rules to forge new ones. *Cell* **2014**, *157*, 77–94. [CrossRef]
39. Grillone, K.; Riillo, C.; Scionti, F.; Rocca, R.; Tradigo, G.; Guzzi, P.H.; Alcaro, S.; Di Martino, M.T.; Tagliaferri, P.; Tassone, P. Non-coding RNAs in cancer: Platforms and strategies for investigating the genomic "dark matter". *J. Exp. Clin. Cancer Res.* **2020**, *39*, 117. [CrossRef]
40. Diamantopoulos, M.A.; Tsiakanikas, P.; Scorilas, A. Non-coding RNAs: The riddle of the transcriptome and their perspectives in cancer. *Ann. Transl. Med.* **2018**, *6*, 3. [CrossRef]
41. Slack, F.J.; Chinnaiyan, A.M. The Role of Non-coding RNAs in Oncology. *Cell* **2019**, *179*, 1033–1055. [CrossRef]
42. Santoleri, D.; Lim, H.W.; Emmett, M.J.; Stoute, J.; Gavin, M.J.; Sostre-Colón, J.; Uehara, K.; Welles, J.E.; Liu, K.F.; Lazar, M.A.; et al. Global-run on sequencing identifies Gm11967 as an Akt-dependent long noncoding RNA involved in insulin sensitivity. *iScience* **2022**, *25*, 104410. [CrossRef]
43. Liu, G.X.; Tan, Y.Z.; He, G.C.; Zhang, Q.L.; Liu, P. EMX2OS plays a prognosis-associated enhancer RNA role in gastric cancer. *Medicine* **2021**, *100*, e27535. [CrossRef]
44. Lee, Y.E.; Lee, J.; Lee, Y.S.; Jang, J.J.; Woo, H.; Choi, H.I.; Chai, Y.G.; Kim, T.K.; Kim, T.; Kim, L.K.; et al. Identification and Functional Characterization of Two Noncoding RNAs Transcribed from Putative Active Enhancers in Hepatocellular Carcinoma. *Mol. Cells* **2021**, *44*, 658–669. [CrossRef]
45. Ambros, V. The functions of animal microRNAs. *Nature* **2004**, *431*, 350–355. [CrossRef]
46. Huang, S.; Wu, S.; Ding, J.; Lin, J.; Wei, L.; Gu, J.; He, X. MicroRNA-181a modulates gene expression of zinc finger family members by directly targeting their coding regions. *Nucleic Acids Res.* **2010**, *38*, 7211–7218. [CrossRef]
47. Ørom, U.A.; Nielsen, F.C.; Lund, A.H. MicroRNA-10a binds the 5′UTR of ribosomal protein mRNAs and enhances their translation. *Mol. Cell* **2008**, *30*, 460–471. [CrossRef]
48. Parashar, D.; Geethadevi, A.; Aure, M.R.; Mishra, J.; George, J.; Chen, C.; Mishra, M.K.; Tahiri, A.; Zhao, W.; Nair, B.; et al. miRNA551b-3p Activates an Oncostatin Signaling Module for the Progression of Triple-Negative Breast Cancer. *Cell Rep.* **2019**, *29*, 4389–4406. [CrossRef]
49. Shen, Y.; Yu, X.; Zhu, L.; Li, T.; Yan, Z.; Guo, J. Transfer RNA-derived fragments and tRNA halves: Biogenesis, biological functions and their roles in diseases. *J. Mol. Med.* **2018**, *96*, 1167–1176. [CrossRef] [PubMed]
50. Fu, L.; Jiang, Z.; Li, T.; Hu, Y.; Guo, J. Circular RNAs in hepatocellular carcinoma: Functions and implications. *Cancer Med.* **2018**, *7*, 3101–3109. [CrossRef] [PubMed]
51. Xiao, M.; Li, J.; Li, W.; Wang, Y.; Wu, F.; Xi, Y.; Zhang, L.; Ding, C.; Luo, H.; Li, Y.; et al. MicroRNAs activate gene transcription epigenetically as an enhancer trigger. *RNA Biol.* **2017**, *14*, 1326–1334. [CrossRef] [PubMed]
52. McHugh, C.A.; Chen, C.K.; Chow, A.; Surka, C.F.; Tran, C.; McDonel, P.; Pandya-Jones, A.; Blanco, M.; Burghard, C.; Moradian, A.; et al. The Xist lncRNA interacts directly with SHARP to silence transcription through HDAC3. *Nature* **2015**, *521*, 232–236. [CrossRef] [PubMed]
53. Tang, J.; Ahmad, A.; Sarkar, F.H. The role of microRNAs in breast cancer migration, invasion and metastasis. *Int. J. Mol. Sci.* **2012**, *13*, 13414–13437. [CrossRef] [PubMed]
54. Palmini, G.; Marini, F.; Brandi, M.L. What Is New in the miRNA World Regarding Osteosarcoma and Chondrosarcoma? *Molecules* **2017**, *22*, 417. [CrossRef]
55. Lui, W.O.; Pourmand, N.; Patterson, B.K.; Fire, A. Patterns of known and novel small RNAs in human cervical cancer. *Cancer Res.* **2007**, *67*, 6031–6043. [CrossRef]
56. Lin, S.; Gregory, R.I. MicroRNA biogenesis pathways in cancer. *Nat. Rev. Cancer* **2015**, *15*, 321–333. [CrossRef]
57. Muralidhar, B.; Goldstein, L.D.; Ng, G.; Winder, D.M.; Palmer, R.D.; Gooding, E.L.; Barbosa-Morais, N.L.; Mukherjee, G.; Thorne, N.P.; Roberts, I.; et al. Global microRNA profiles in cervical squamous cell carcinoma depend on Drosha expression levels. *J. Pathol.* **2007**, *212*, 368–377. [CrossRef]
58. Gupta, S.; Panda, P.K.; Hashimoto, R.F.; Samal, S.K.; Mishra, S.; Verma, S.K.; Mishra, Y.K.; Ahuja, R. Dynamical modeling of miR-34a, miR-449a, and miR-16 reveals numerous DDR signaling pathways regulating senescence, autophagy, and apoptosis in HeLa cells. *Sci. Rep.* **2022**, *12*, 4911. [CrossRef]
59. He, Y.; Lin, J.; Ding, Y.; Liu, G.; Luo, Y.; Huang, M.; Xu, C.; Kim, T.K.; Etheridge, A.; Lin, M.; et al. A systematic study on dysregulated microRNAs in cervical cancer development. *Int. J. Cancer* **2016**, *138*, 1312–1327. [CrossRef]
60. Gocze, K.; Gombos, K.; Juhasz, K.; Kovacs, K.; Kajtar, B.; Benczik, M.; Gocze, P.; Patczai, B.; Arany, I.; Ember, I. Unique microRNA expression profiles in cervical cancer. *Anticancer. Res.* **2013**, *33*, 2561–2567.
61. Tian, Q.; Li, Y.; Wang, F.; Li, Y.; Xu, J.; Shen, Y.; Ye, F.; Wang, X.; Cheng, X.; Chen, Y.; et al. MicroRNA detection in cervical exfoliated cells as a triage for human papillomavirus-positive women. *J. Natl. Cancer Inst.* **2014**, *106*, dju241. [CrossRef]
62. Kawai, S.; Fujii, T.; Kukimoto, I.; Yamada, H.; Yamamoto, N.; Kuroda, M.; Otani, S.; Ichikawa, R.; Nishio, E.; Torii, Y.; et al. Identification of miRNAs in cervical mucus as a novel diagnostic marker for cervical neoplasia. *Sci. Rep.* **2018**, *8*, 7070. [CrossRef]
63. Croce, C.M. Causes and consequences of microRNA dysregulation in cancer. *Nat. Rev. Genet.* **2009**, *10*, 704–714. [CrossRef]
64. Cheng, T.; Huang, S. Roles of Non-Coding RNAs in Cervical Cancer Metastasis. *Front. Oncol.* **2021**, *11*, 646192. [CrossRef]
65. Xu, J.; Zhang, W.; Lv, Q.; Zhu, D. Overexpression of miR-21 promotes the proliferation and migration of cervical cancer cells via the inhibition of PTEN. *Oncol. Rep.* **2015**, *33*, 3108–3116. [CrossRef]
66. Deng, Z.M.; Chen, G.H.; Dai, F.F.; Liu, S.Y.; Yang, D.Y.; Bao, A.Y.; Cheng, Y.X. The clinical value of miRNA-21 in cervical cancer: A comprehensive investigation based on microarray datasets. *PLoS ONE* **2022**, *17*, e0267108. [CrossRef]

67. Peralta-Zaragoza, O.; Deas, J.; Meneses-Acosta, A.; De la O-Gómez, F.; Fernández-Tilapa, G.; Gómez-Cerón, C.; Benítez-Boijseauneau, O.; Burguete-García, A.; Torres-Poveda, K.; Bermúdez-Morales, V.H.; et al. Relevance of miR-21 in regulation of tumor suppressor gene PTEN in human cervical cancer cells. *BMC Cancer* **2016**, *16*, 215. [CrossRef]
68. Cai, N.; Hu, L.; Xie, Y.; Gao, J.H.; Zhai, W.; Wang, L.; Jin, Q.J.; Qin, C.Y.; Qiang, R. MiR-17-5p promotes cervical cancer cell proliferation and metastasis by targeting transforming growth factor-β receptor 2. *Eur. Rev. Med. Pharmacol. Sci.* **2018**, *22*, 1899–1906. [CrossRef]
69. Wei, Y.Q.; Jiao, X.L.; Zhang, S.Y.; Xu, Y.; Li, S.; Kong, B.H. MiR-9-5p could promote angiogenesis and radiosensitivity in cervical cancer by targeting SOCS5. *Eur. Rev. Med. Pharmacol. Sci.* **2019**, *23*, 7314–7326. [CrossRef]
70. Chen, Y.; Liu, C.; Xie, B.; Chen, S.; Zhuang, Y.; Zhang, S. miR-96 exerts an oncogenic role in the progression of cervical cancer by targeting CAV-1. *Mol. Med. Rep.* **2020**, *22*, 543–550. [CrossRef]
71. Zhang, L.; Zhan, X.; Yan, D.; Wang, Z. Circulating MicroRNA-21 Is Involved in Lymph Node Metastasis in Cervical Cancer by Targeting RASA1. *Int. J. Gynecol. Cancer* **2016**, *26*, 810–816. [CrossRef]
72. Wei, W.F.; Zhou, C.F.; Wu, X.G.; He, L.N.; Wu, L.F.; Chen, X.J.; Yan, R.M.; Zhong, M.; Yu, Y.H.; Liang, L.; et al. MicroRNA-221-3p, a TWIST2 target, promotes cervical cancer metastasis by directly targeting THBS2. *Cell Death Dis.* **2017**, *8*, 3220. [CrossRef]
73. Xu, L.J.; Duan, Y.; Wang, P.; Yin, H.Q. MiR-199b-5p promotes tumor growth and metastasis in cervical cancer by down-regulating KLK10. *Biochem. Biophys. Res. Commun.* **2018**, *503*, 556–563. [CrossRef]
74. Cheng, Y.; Guo, Y.; Zhang, Y.; You, K.; Li, Z.; Geng, L. MicroRNA-106b is involved in transforming growth factor β1-induced cell migration by targeting disabled homolog 2 in cervical carcinoma. *J. Exp. Clin. Cancer Res.* **2016**, *35*, 11. [CrossRef]
75. Zhou, J.Y.; Zheng, S.R.; Liu, J.; Shi, R.; Yu, H.L.; Wei, M. MiR-519d facilitates the progression and metastasis of cervical cancer through direct targeting Smad7. *Cancer Cell Int.* **2016**, *16*, 21. [CrossRef]
76. Tan, D.; Zhou, C.; Han, S.; Hou, X.; Kang, S.; Zhang, Y. MicroRNA-378 enhances migration and invasion in cervical cancer by directly targeting autophagy-related protein 12. *Mol. Med. Rep.* **2018**, *17*, 6319–6326. [CrossRef]
77. Zhao, S.; Yao, D.; Chen, J.; Ding, N.; Ren, F. MiR-20a promotes cervical cancer proliferation and metastasis in vitro and in vivo. *PLoS ONE* **2015**, *10*, e0120905. [CrossRef]
78. Li, X.; Zhou, Q.; Tao, L.; Yu, C. MicroRNA-106a promotes cell migration and invasion by targeting tissue inhibitor of matrix metalloproteinase 2 in cervical cancer. *Oncol. Rep.* **2017**, *38*, 1774–1782. [CrossRef]
79. Zhang, Z.; Wang, J.; Li, J.; Wang, X.; Song, W. MicroRNA-150 promotes cell proliferation, migration, and invasion of cervical cancer through targeting PDCD4. *Biomed. Pharmacother.* **2018**, *97*, 511–517. [CrossRef]
80. Wang, N.; Li, Y.; Zhou, J. miR-31 Functions as an Oncomir Which Promotes Epithelial-Mesenchymal Transition via Regulating BAP1 in Cervical Cancer. *Biomed. Res. Int.* **2017**, *2017*, 6361420. [CrossRef]
81. Lao, G.; Liu, P.; Wu, Q.; Zhang, W.; Liu, Y.; Yang, L.; Ma, C. Mir-155 promotes cervical cancer cell proliferation through suppression of its target gene LKB1. *Tumour. Biol.* **2014**, *35*, 11933–11938. [CrossRef] [PubMed]
82. Kwon, J.J.; Factora, T.D.; Dey, S.; Kota, J. A Systematic Review of miR-29 in Cancer. *Mol. Ther. Oncolytics* **2019**, *12*, 173–194. [CrossRef] [PubMed]
83. Zhang, L.; Liu, F.; Fu, Y.; Chen, X.; Zhang, D. MiR-520d-5p functions as a tumor-suppressor gene in cervical cancer through targeting PTK2. *Life Sci.* **2020**, *254*, 117558. [CrossRef] [PubMed]
84. Zhu, X.; Er, K.; Mao, C.; Yan, Q.; Xu, H.; Zhang, Y.; Zhu, J.; Cui, F.; Zhao, W.; Shi, H. miR-203 suppresses tumor growth and angiogenesis by targeting VEGFA in cervical cancer. *Cell Physiol. Biochem.* **2013**, *32*, 64–73. [CrossRef] [PubMed]
85. Fontemaggi, G.; Turco, C.; Esposito, G.; Di Agostino, S. New Molecular Mechanisms and Clinical Impact of circRNAs in Human Cancer. *Cancers* **2021**, *13*, 3154. [CrossRef]
86. Kulcheski, F.R.; Christoff, A.P.; Margis, R. Circular RNAs are miRNA sponges and can be used as a new class of biomarker. *J. Biotechnol.* **2016**, *238*, 42–51. [CrossRef]
87. Zhao, Z.J.; Shen, J. Circular RNA participates in the carcinogenesis and the malignant behavior of cancer. *RNA Biol.* **2017**, *14*, 514–521. [CrossRef]
88. Ma, H.-B.; Yao, Y.-N.; Yu, J.-J.; Chen, X.-X.; Li, H.-F. Extensive profiling of circular RNAs and the potential regulatory role of circRNA-000284 in cell proliferation and invasion of cervical cancer via sponging miR-506. *Am. J. Transl. Res.* **2018**, *10*, 592–604.
89. Yang, M.; Hu, H.; Wu, S.; Ding, J.; Yin, B.; Huang, B.; Li, F.; Guo, X.; Han, L. EIF4A3-regulated circ_0087429 can reverse EMT and inhibit the progression of cervical cancer via miR-5003-3p-dependent upregulation of OGN expression. *J. Exp. Clin. Cancer Res.* **2022**, *41*, 165. [CrossRef]
90. Chen, R.; Gan, Q.; Zhao, S.; Zhang, D.; Wang, S.; Yao, L.; Yuan, M.; Cheng, J. DNA methylation of miR-138 regulates cell proliferation and EMT in cervical cancer by targeting EZH2. *BMC Cancer* **2022**, *22*, 488. [CrossRef]
91. Ou, R.; Lu, S.; Wang, L.; Wang, Y.; Lv, M.; Li, T.; Xu, Y.; Lu, J.; Ge, R.-s. Circular RNA circLMO1 Suppresses Cervical Cancer Growth and Metastasis by Triggering miR-4291/ACSL4-Mediated Ferroptosis. *Front. Oncol.* **2022**, *12*, 858598. [CrossRef]
92. Fu, K.; Zhang, L.; Liu, R.; Shi, Q.; Li, X.; Wang, M. MiR-125 inhibited cervical cancer progression by regulating VEGF and PI3K/AKT signaling pathway. *World J. Surg. Oncol.* **2020**, *18*, 115. [CrossRef]
93. Li, Y.M.; Li, X.J.; Yang, H.L.; Zhang, Y.B.; Li, J.C. MicroRNA-23b suppresses cervical cancer biological progression by directly targeting six1 and affecting epithelial-to-mesenchymal transition and AKT/mTOR signaling pathway. *Eur. Rev. Med. Pharmacol. Sci.* **2019**, *23*, 4688–4697. [CrossRef]

94. Gong, Y.; Wan, J.H.; Zou, W.; Lian, G.Y.; Qin, J.L.; Wang, Q.M. MiR-29a inhibits invasion and metastasis of cervical cancer via modulating methylation of tumor suppressor SOCS1. *Future Oncol.* **2019**, *15*, 1729–1744. [CrossRef]

95. Liu, X.; Gan, L.; Zhang, J. miR-543 inhibites cervical cancer growth and metastasis by targeting TRPM7. *Chem. Biol. Interact.* **2019**, *302*, 83–92. [CrossRef]

96. Luo, M.; Shen, D.; Zhou, X.; Chen, X.; Wang, W. MicroRNA-497 is a potential prognostic marker in human cervical cancer and functions as a tumor suppressor by targeting the insulin-like growth factor 1 receptor. *Surgery* **2013**, *153*, 836–847. [CrossRef]

97. Kogo, R.; How, C.; Chaudary, N.; Bruce, J.; Shi, W.; Hill, R.P.; Zahedi, P.; Yip, K.W.; Liu, F.F. The microRNA-218~Survivin axis regulates migration, invasion, and lymph node metastasis in cervical cancer. *Oncotarget* **2015**, *6*, 1090–1100. [CrossRef]

98. Cheng, Y.X.; Chen, G.T.; Chen, C.; Zhang, Q.F.; Pan, F.; Hu, M.; Li, B.S. MicroRNA-200b inhibits epithelial-mesenchymal transition and migration of cervical cancer cells by directly targeting RhoE. *Mol. Med. Rep.* **2016**, *13*, 3139–3146. [CrossRef]

99. Wan, H.Y.; Li, Q.Q.; Zhang, Y.; Tian, W.; Li, Y.N.; Liu, M.; Li, X.; Tang, H. MiR-124 represses vasculogenic mimicry and cell motility by targeting amotL1 in cervical cancer cells. *Cancer Lett.* **2014**, *355*, 148–158. [CrossRef]

100. Yang, Y.; Liu, Y.; Li, G.; Li, L.; Geng, P.; Song, H. microRNA-214 suppresses the growth of cervical cancer cells by targeting EZH2. *Oncol. Lett.* **2018**, *16*, 5679–5686. [CrossRef]

101. Liu, L.; Yu, X.; Guo, X.; Tian, Z.; Su, M.; Long, Y.; Huang, C.; Zhou, F.; Liu, M.; Wu, X.; et al. miR-143 is downregulated in cervical cancer and promotes apoptosis and inhibits tumor formation by targeting Bcl-2. *Mol. Med. Rep.* **2012**, *5*, 753–760. [CrossRef]

102. Shen, W.; Xie, X.Y.; Liu, M.R.; Wang, L.L. MicroRNA-101-5p inhibits the growth and metastasis of cervical cancer cell by inhibiting CXCL6. *Eur. Rev. Med. Pharmacol. Sci.* **2019**, *23*, 1957–1968. [CrossRef]

103. Zhao, J.L.; Zhang, L.; Guo, X.; Wang, J.H.; Zhou, W.; Liu, M.; Li, X.; Tang, H. miR-212/132 downregulates SMAD2 expression to suppress the G1/S phase transition of the cell cycle and the epithelial to mesenchymal transition in cervical cancer cells. *IUBMB Life* **2015**, *67*, 380–394. [CrossRef]

104. Wang, Y.F.; Yang, H.Y.; Shi, X.Q.; Wang, Y. Upregulation of microRNA-129-5p inhibits cell invasion, migration and tumor angiogenesis by inhibiting ZIC2 via downregulation of the Hedgehog signaling pathway in cervical cancer. *Cancer Biol. Ther.* **2018**, *19*, 1162–1173. [CrossRef]

105. Ou, L.; Wang, D.; Zhang, H.; Yu, Q.; Hua, F. Decreased Expression of miR-138-5p by lncRNA H19 in Cervical Cancer Promotes Tumor Proliferation. *Oncol. Res.* **2018**, *26*, 401–410. [CrossRef]

106. Cao, X.C.; Yu, Y.; Hou, L.K.; Sun, X.H.; Ge, J.; Zhang, B.; Wang, X. miR-142-3p inhibits cancer cell proliferation by targeting CDC25C. *Cell Prolif.* **2016**, *49*, 58–68. [CrossRef]

107. Mou, Z.; Xu, X.; Dong, M.; Xu, J. MicroRNA-148b Acts as a Tumor Suppressor in Cervical Cancer by Inducing G1/S-Phase Cell Cycle Arrest and Apoptosis in a Caspase-3-Dependent Manner. *Med. Sci. Monit.* **2016**, *22*, 2809–2815. [CrossRef]

108. Sun, J.; Ji, J.; Huo, G.; Song, Q.; Zhang, X. miR-182 induces cervical cancer cell apoptosis through inhibiting the expression of DNMT3a. *Int. J. Clin. Exp. Pathol.* **2015**, *8*, 4755–4763. [PubMed]

109. Zhou, Q.; Han, L.R.; Zhou, Y.X.; Li, Y. MiR-195 Suppresses Cervical Cancer Migration and Invasion Through Targeting Smad3. *Int. J. Gynecol. Cancer* **2016**, *26*, 817–824. [CrossRef] [PubMed]

110. How, C.; Hui, A.B.; Alajez, N.M.; Shi, W.; Boutros, P.C.; Clarke, B.A.; Yan, R.; Pintilie, M.; Fyles, A.; Hedley, D.W.; et al. MicroRNA-196b regulates the homeobox B7-vascular endothelial growth factor axis in cervical cancer. *PLoS ONE* **2013**, *8*, e67846. [CrossRef] [PubMed]

111. Aalijahan, H.; Ghorbian, S. Long non-coding RNAs and cervical cancer. *Exp. Mol. Pathol.* **2019**, *106*, 7–16. [CrossRef]

112. Huo, H.; Tian, J.; Wang, R.; Li, Y.; Qu, C.; Wang, N. Long non-coding RNA NORAD upregulate SIP1 expression to promote cell proliferation and invasion in cervical cancer. *Biomed. Pharmacother.* **2018**, *106*, 1454–1460. [CrossRef]

113. Chen, Y.; Wang, C.X.; Sun, X.X.; Wang, C.; Liu, T.F.; Wang, D.J. Long non-coding RNA CCHE1 overexpression predicts a poor prognosis for cervical cancer. *Eur. Rev. Med. Pharmacol. Sci.* **2017**, *21*, 479–483.

114. Wu, L.; Jin, L.; Zhang, W.; Zhang, L. Roles of Long Non-Coding RNA CCAT2 in Cervical Cancer Cell Growth and Apoptosis. *Med. Sci. Monit.* **2016**, *22*, 875–879. [CrossRef]

115. Sun, Y.; Ma, L. New Insights into Long Non-Coding RNA MALAT1 in Cancer and Metastasis. *Cancers* **2019**, *11*, 216. [CrossRef]

116. Lu, H.; He, Y.; Lin, L.; Qi, Z.; Ma, L.; Li, L.; Su, Y. Long non-coding RNA MALAT1 modulates radiosensitivity of HR-HPV+ cervical cancer via sponging miR-145. *Tumour. Biol.* **2016**, *37*, 1683–1691. [CrossRef]

117. Liu, M.; Jia, J.; Wang, X.; Liu, Y.; Wang, C.; Fan, R. Long non-coding RNA HOTAIR promotes cervical cancer progression through regulating BCL2 via targeting miR-143-3p. *Cancer Biol. Ther.* **2018**, *19*, 391–399. [CrossRef]

118. Song, H.; Liu, Y.; Jin, X.; Liu, Y.; Yang, Y.; Li, L.; Wang, X.; Li, G. Long non-coding RNA LINC01535 promotes cervical cancer progression via targeting the miR-214/EZH2 feedback loop. *J. Cell Mol. Med.* **2019**, *23*, 6098–6111. [CrossRef]

119. Zhu, L.; Zhang, Q.; Li, S.; Jiang, S.; Cui, J.; Dang, G. Interference of the long noncoding RNA CDKN2B-AS1 upregulates miR-181a-5p/TGFβI axis to restrain the metastasis and promote apoptosis and senescence of cervical cancer cells. *Cancer Med.* **2019**, *8*, 1721–1730. [CrossRef]

120. Hsu, W.; Liu, L.; Chen, X.; Zhang, Y.; Zhu, W. LncRNA CASC11 promotes the cervical cancer progression by activating Wnt/beta-catenin signaling pathway. *Biol. Res.* **2019**, *52*, 33. [CrossRef]

121. Ma, S.; Deng, X.; Yang, Y.; Zhang, Q.; Zhou, T.; Liu, Z. The lncRNA LINC00675 regulates cell proliferation, migration, and invasion by affecting Wnt/β-catenin signaling in cervical cancer. *Biomed. Pharmacother.* **2018**, *108*, 1686–1693. [CrossRef]

122. Jiang, Y.; Li, Y.; Fang, S.; Jiang, B.; Qin, C.; Xie, P.; Zhou, G.; Li, G. The role of MALAT1 correlates with HPV in cervical cancer. *Oncol. Lett.* **2014**, *7*, 2135–2141. [CrossRef]
123. Zhang, W.Y.; Liu, Y.J.; He, Y.; Chen, P. Down-regulation of long non-coding RNA ANRIL inhibits the proliferation, migration and invasion of cervical cancer cells. *Cancer Biomark* **2018**, *23*, 243–253. [CrossRef]
124. Shan, D.; Shang, Y.; Hu, T. Long noncoding RNA BLACAT1 promotes cell proliferation and invasion in human cervical cancer. *Oncol. Lett.* **2018**, *15*, 3490–3495. [CrossRef]
125. Chang, Q.Q.; Chen, C.Y.; Chen, Z.; Chang, S. LncRNA PVT1 promotes proliferation and invasion through enhancing Smad3 expression by sponging miR-140-5p in cervical cancer. *Radiol. Oncol.* **2019**, *53*, 443–452. [CrossRef]
126. Hu, Y.C.; Wang, A.M.; Lu, J.K.; Cen, R.; Liu, L.L. Long noncoding RNA HOXD-AS1 regulates proliferation of cervical cancer cells by activating Ras/ERK signaling pathway. *Eur. Rev. Med. Pharmacol. Sci.* **2017**, *21*, 5049–5055. [CrossRef]
127. Xie, F.; Xie, G.; Sun, Q. Long Noncoding RNA DLX6-AS1 Promotes the Progression in Cervical Cancer by Targeting miR-16-5p/ARPP19 Axis. *Cancer Biother Radiopharm.* **2020**, *35*, 129–136. [CrossRef]
128. Yang, H.Y.; Huang, C.P.; Cao, M.M.; Wang, Y.F.; Liu, Y. Long non-coding RNA CRNDE may be associated with poor prognosis by promoting proliferation and inhibiting apoptosis of cervical cancer cells through targeting PI3K/AKT. *Neoplasma* **2018**, *65*, 872–880. [CrossRef]
129. Sun, N.X.; Ye, C.; Zhao, Q.; Zhang, Q.; Xu, C.; Wang, S.B.; Jin, Z.J.; Sun, S.H.; Wang, F.; Li, W. Long noncoding RNA-EBIC promotes tumor cell invasion by binding to EZH2 and repressing E-cadherin in cervical cancer. *PLoS ONE* **2014**, *9*, e100340. [CrossRef] [PubMed]
130. Liu, Q.; Guo, X.; Que, S.; Yang, X.; Fan, H.; Liu, M.; Li, X.; Tang, H. LncRNA RSU1P2 contributes to tumorigenesis by acting as a ceRNA against let-7a in cervical cancer cells. *Oncotarget* **2017**, *8*, 43768–43781. [CrossRef] [PubMed]
131. Fan, M.J.; Zou, Y.H.; He, P.J.; Zhang, S.; Sun, X.M.; Li, C.Z. Long non-coding RNA SPRY4-IT1 promotes epithelial-mesenchymal transition of cervical cancer by regulating the miR-101-3p/ZEB1 axis. *Biosci. Rep.* **2019**, *39*, BSR20181339. [CrossRef] [PubMed]
132. Geng, F.; Jia, W.C.; Li, T.; Li, N.; Wei, W. Knockdown of lncRNA NEAT1 suppresses proliferation and migration, and induces apoptosis of cervical cancer cells by regulating the miR-377/FGFR1 axis. *Mol. Med. Rep.* **2022**, *25*, 10. [CrossRef] [PubMed]
133. Barr, J.A.; Hayes, K.E.; Brownmiller, T.; Harold, A.D.; Jagannathan, R.; Lockman, P.R.; Khan, S.; Martinez, I. Long non-coding RNA FAM83H-AS1 is regulated by human papillomavirus 16 E6 independently of p53 in cervical cancer cells. *Sci. Rep.* **2019**, *9*, 3662. [CrossRef] [PubMed]
134. Rui, X.; Xu, Y.; Jiang, X.; Ye, W.; Huang, Y.; Jiang, J. Long non-coding RNA C5orf66-AS1 promotes cell proliferation in cervical cancer by targeting miR-637/RING1 axis. *Cell Death Dis.* **2018**, *9*, 1175. [CrossRef]
135. Zhang, J.; Gao, Y. Long non-coding RNA MEG3 inhibits cervical cancer cell growth by promoting degradation of P-STAT3 protein via ubiquitination. *Cancer Cell Int.* **2019**, *19*, 175. [CrossRef]
136. Shao, S.; Wang, C.; Wang, S.; Zhang, H.; Zhang, Y. LncRNA STXBP5-AS1 suppressed cervical cancer progression via targeting miR-96-5p/PTEN axis. *Biomed. Pharmacother.* **2019**, *117*, 109082. [CrossRef]
137. Yang, W.; Hong, L.; Xu, X.; Wang, Q.; Huang, J.; Jiang, L. LncRNA GAS5 suppresses the tumorigenesis of cervical cancer by downregulating miR-196a and miR-205. *Tumour Biol* **2017**, *39*, 1010428317711315. [CrossRef]
138. Wang, X.; Zhang, J.; Wang, Y. Long noncoding RNA GAS5-AS1 suppresses growth and metastasis of cervical cancer by increasing GAS5 stability. *Am. J. Transl. Res.* **2019**, *11*, 4909–4921.
139. Zhu, Y.; Liu, B.; Zhang, P.; Zhang, J.; Wang, L. LncRNA TUSC8 inhibits the invasion and migration of cervical cancer cells via miR-641/PTEN axis. *Cell Biol. Int.* **2019**, *43*, 781–788. [CrossRef]
140. Liao, L.M.; Sun, X.Y.; Liu, A.W.; Wu, J.B.; Cheng, X.L.; Lin, J.X.; Zheng, M.; Huang, L. Low expression of long noncoding XLOC_010588 indicates a poor prognosis and promotes proliferation through upregulation of c-Myc in cervical cancer. *Gynecol. Oncol.* **2014**, *133*, 616–623. [CrossRef]
141. Liu, H.; Zhang, L.; Ding, X.; Sui, X. LINC00861 inhibits the progression of cervical cancer cells by functioning as a ceRNA for miR-513b-5p and regulating the PTEN/AKT/mTOR signaling pathway. *Mol. Med. Rep.* **2021**, *23*, 24. [CrossRef]
142. Li, Y.J.; Yang, Z.; Wang, Y.Y.; Wang, Y. Long noncoding RNA ZNF667-AS1 reduces tumor invasion and metastasis in cervical cancer by counteracting microRNA-93-3p-dependent PEG3 downregulation. *Mol. Oncol.* **2019**, *13*, 2375–2392. [CrossRef]
143. Han, H.F.; Chen, Q.; Zhao, W.W. Long non-coding RNA RP11-284F21.9 functions as a ceRNA regulating PPWD1 by competitively binding to miR-769-3p in cervical carcinoma. *Biosci. Rep.* **2020**, *40*, BSR20200784. [CrossRef]
144. Ding, X.; Jia, X.; Wang, C.; Xu, J.; Gao, S.-J.; Lu, C. A DHX9-lncRNA-MDM2 interaction regulates cell invasion and angiogenesis of cervical cancer. *Cell Death Differ.* **2019**, *26*, 1750–1765. [CrossRef]
145. Xue, C.; Chen, C.; Gu, X.; Li, L. Progress and assessment of lncRNA DGCR5 in malignant phenotype and immune infiltration of human cancers. *Am. J. Cancer Res.* **2021**, *11*, 1–13.
146. Wang, M.; Zhao, J.; Zhang, L.; Wei, F.; Lian, Y.; Wu, Y.; Gong, Z.; Zhang, S.; Zhou, J.; Cao, K.; et al. Role of tumor microenvironment in tumorigenesis. *J. Cancer* **2017**, *8*, 761–773. [CrossRef]
147. Drak Alsibai, K.; Meseure, D. Tumor microenvironment and noncoding RNAs as co-drivers of epithelial-mesenchymal transition and cancer metastasis. *Dev. Dyn.* **2018**, *247*, 405–431. [CrossRef]
148. Fang, Z.; Xu, J.; Zhang, B.; Wang, W.; Liu, J.; Liang, C.; Hua, J.; Meng, Q.; Yu, X.; Shi, S. The promising role of noncoding RNAs in cancer-associated fibroblasts: An overview of current status and future perspectives. *J. Hematol. Oncol.* **2020**, *13*, 154. [CrossRef]

149. Fan, D.; Wang, Y.; Qi, P.; Chen, Y.; Xu, P.; Yang, X.; Jin, X.; Tian, X. MicroRNA-183 functions as the tumor suppressor via inhibiting cellular invasion and metastasis by targeting MMP-9 in cervical cancer. *Gynecol. Oncol.* **2016**, *141*, 166–174. [CrossRef]
150. Baghban, R.; Roshangar, L.; Jahanban-Esfahlan, R.; Seidi, K.; Ebrahimi-Kalan, A.; Jaymand, M.; Kolahian, S.; Javaheri, T.; Zare, P. Tumor microenvironment complexity and therapeutic implications at a glance. *Cell Commun. Signal.* **2020**, *18*, 59. [CrossRef]
151. Wu, X.G.; Zhou, C.F.; Zhang, Y.M.; Yan, R.M.; Wei, W.F.; Chen, X.J.; Yi, H.Y.; Liang, L.J.; Fan, L.S.; Liang, L.; et al. Cancer-derived exosomal miR-221-3p promotes angiogenesis by targeting THBS2 in cervical squamous cell carcinoma. *Angiogenesis* **2019**, *22*, 397–410. [CrossRef]
152. Chen, D.; Lu, T.; Tan, J.; Li, H.; Wang, Q.; Wei, L. Long Non-coding RNAs as Communicators and Mediators Between the Tumor Microenvironment and Cancer Cells. *Front. Oncol.* **2019**, *9*, 739. [CrossRef]
153. Zhang, R.; Xia, Y.; Wang, Z.; Zheng, J.; Chen, Y.; Li, X.; Wang, Y.; Ming, H. Serum long non coding RNA MALAT-1 protected by exosomes is up-regulated and promotes cell proliferation and migration in non-small cell lung cancer. *Biochem. Biophys. Res. Commun.* **2017**, *490*, 406–414. [CrossRef]
154. Hu, Q.; Ye, Y.; Chan, L.C.; Li, Y.; Liang, K.; Lin, A.; Egranov, S.D.; Zhang, Y.; Xia, W.; Gong, J.; et al. Oncogenic lncRNA downregulates cancer cell antigen presentation and intrinsic tumor suppression. *Nat. Immunol.* **2019**, *20*, 835–851. [CrossRef]
155. Fares, J.; Fares, M.Y.; Khachfe, H.H.; Salhab, H.A.; Fares, Y. Molecular principles of metastasis: A hallmark of cancer revisited. *Signal Transduct. Target. Ther.* **2020**, *5*, 28. [CrossRef]
156. Zhang, L.; Xu, X.; Su, X. Noncoding RNAs in cancer immunity: Functions, regulatory mechanisms, and clinical application. *Mol. Cancer* **2020**, *19*, 48. [CrossRef]
157. Warrington, R.; Watson, W.; Kim, H.L.; Antonetti, F.R. An introduction to immunology and immunopathology. *Allergy Asthma Clin. Immunol.* **2011**, *7* (Suppl. S1), S1. [CrossRef]
158. McCune, J.S. Rapid Advances in Immunotherapy to Treat Cancer. *Clin. Pharmacol. Ther.* **2018**, *103*, 540–544. [CrossRef]
159. Huang, Z.; Luo, Q.; Yao, F.; Qing, C.; Ye, J.; Deng, Y.; Li, J. Identification of Differentially Expressed Long Non-coding RNAs in Polarized Macrophages. *Sci. Rep.* **2016**, *6*, 19705. [CrossRef]
160. Squadrito Mario, L.; Pucci, F.; Magri, L.; Moi, D.; Gilfillan Gregor, D.; Ranghetti, A.; Casazza, A.; Mazzone, M.; Lyle, R.; Naldini, L.; et al. miR-511-3p Modulates Genetic Programs of Tumor-Associated Macrophages. *Cell Rep.* **2012**, *1*, 141–154. [CrossRef] [PubMed]
161. Yang, J.; Zhang, Z.; Chen, C.; Liu, Y.; Si, Q.; Chuang, T.H.; Li, N.; Gomez-Cabrero, A.; Reisfeld, R.A.; Xiang, R.; et al. MicroRNA-19a-3p inhibits breast cancer progression and metastasis by inducing macrophage polarization through down-regulated expression of Fra-1 proto-oncogene. *Oncogene* **2014**, *33*, 3014–3023. [CrossRef] [PubMed]
162. Caescu, C.I.; Guo, X.; Tesfa, L.; Bhagat, T.D.; Verma, A.; Zheng, D.; Stanley, E.R. Colony stimulating factor-1 receptor signaling networks inhibit mouse macrophage inflammatory responses by induction of microRNA-21. *Blood* **2015**, *125*, e1–e13. [CrossRef] [PubMed]
163. Ma, S.; Liu, M.; Xu, Z.; Li, Y.; Guo, H.; Ge, Y.; Liu, Y.; Zheng, D.; Shi, J. A double feedback loop mediated by microRNA-23a/27a/24-2 regulates M1 versus M2 macrophage polarization and thus regulates cancer progression. *Oncotarget* **2016**, *7*, 13502–13519. [CrossRef] [PubMed]
164. Mantovani, A.; Sozzani, S.; Locati, M.; Allavena, P.; Sica, A. Macrophage polarization: Tumor-associated macrophages as a paradigm for polarized M2 mononuclear phagocytes. *Trends Immunol.* **2002**, *23*, 549–555. [CrossRef]
165. Paul, S.; Lal, G. The Molecular Mechanism of Natural Killer Cells Function and Its Importance in Cancer Immunotherapy. *Front. Immunol.* **2017**, *8*, 1124. [CrossRef]
166. He, S.; Chu, J.; Wu, L.-C.; Mao, H.; Peng, Y.; Alvarez-Breckenridge, C.A.; Hughes, T.; Wei, M.; Zhang, J.; Yuan, S.; et al. MicroRNAs activate natural killer cells through Toll-like receptor signaling. *Blood* **2013**, *121*, 4663–4671. [CrossRef]
167. Dou, R.; Nishihara, R.; Cao, Y.; Hamada, T.; Mima, K.; Masuda, A.; Masugi, Y.; Shi, Y.; Gu, M.; Li, W.; et al. MicroRNA let-7, T Cells, and Patient Survival in Colorectal Cancer. *Cancer Immunol. Res.* **2016**, *4*, 927–935. [CrossRef]
168. Yao, H.; Jiang, X.; Fu, H.; Yang, Y.; Jin, Q.; Zhang, W.; Cao, W.; Gao, W.; Wang, S.; Zhu, Y.; et al. Exploration of the Immune-Related Long Noncoding RNA Prognostic Signature and Inflammatory Microenvironment for Cervical Cancer. *Front. Pharmacol.* **2022**, *13*, 870221. [CrossRef]
169. Zuccherato, L.W.; Machado, C.M.T.; Magalhães, W.C.S.; Martins, P.R.; Campos, L.S.; Braga, L.C.; Teixeira-Carvalho, A.; Martins-Filho, O.A.; Franco, T.; Paula, S.O.C.; et al. Cervical Cancer Stem-Like Cell Transcriptome Profiles Predict Response to Chemoradiotherapy. *Front. Oncol.* **2021**, *11*, 639339. [CrossRef]
170. Huang, R.; Rofstad, E.K. Cancer stem cells (CSCs), cervical CSCs and targeted therapies. *Oncotarget* **2017**, *8*, 35351–35367. [CrossRef]
171. Pan, Q.; Li, Q.; Liu, S.; Ning, N.; Zhang, X.; Xu, Y.; Chang, A.E.; Wicha, M.S. Concise Review: Targeting Cancer Stem Cells Using Immunologic Approaches. *Stem. Cells* **2015**, *33*, 2085–2092. [CrossRef]
172. Xia, M.; Duan, L.J.; Lu, B.N.; Pang, Y.Z.; Pang, Z.R. LncRNA AFAP1-AS1/miR-27b-3p/VEGF-C axis modulates stemness characteristics in cervical cancer cells. *Chin. Med. J.* **2021**, *134*, 2091–2101. [CrossRef]
173. Shang, C.; Guo, Y.; Zhang, J.; Huang, B. Silence of long noncoding RNA UCA1 inhibits malignant proliferation and chemotherapy resistance to adriamycin in gastric cancer. *Cancer Chemother. Pharmacol.* **2016**, *77*, 1061–1067. [CrossRef]
174. Zhou, Y.; Meng, X.; Chen, S.; Li, W.; Li, D.; Singer, R.; Gu, W. IMP1 regulates UCA1-mediated cell invasion through facilitating UCA1 decay and decreasing the sponge effect of UCA1 for miR-122-5p. *Breast Cancer Res.* **2018**, *20*, 32. [CrossRef]

175. Gao, Z.; Wang, Q.; Ji, M.; Guo, X.; Li, L.; Su, X. Exosomal lncRNA UCA1 modulates cervical cancer stem cell self-renewal and differentiation through microRNA-122-5p/SOX2 axis. *J. Transl. Med.* **2021**, *19*, 229. [CrossRef]
176. Sareddy, G.R.; Panigrahi, M.; Challa, S.; Mahadevan, A.; Babu, P.P. Activation of Wnt/beta-catenin/Tcf signaling pathway in human astrocytomas. *Neurochem. Int.* **2009**, *55*, 307–317. [CrossRef]
177. Babashah, S.; Sadeghizadeh, M.; Hajifathali, A.; Tavirani, M.R.; Zomorod, M.S.; Ghadiani, M.; Soleimani, M. Targeting of the signal transducer Smo links microRNA-326 to the oncogenic Hedgehog pathway in CD34+ CML stem/progenitor cells. *Int. J. Cancer* **2013**, *133*, 579–589. [CrossRef]
178. Zhang, J.; He, H.; Wang, K.; Xie, Y.; Yang, Z.; Qie, M.; Liao, Z.; Zheng, Z. miR-326 inhibits the cell proliferation and cancer stem cell-like property of cervical cancer in vitro and oncogenesis in vivo via targeting TCF4. *Ann. Transl. Med.* **2020**, *8*, 1638. [CrossRef]
179. Meng, Q.; Zhang, B.; Zhang, Y.; Wang, S.; Zhu, X. Human bone marrow mesenchymal stem cell-derived extracellular vesicles impede the progression of cervical cancer via the miR-144-3p/CEP55 pathway. *J. Cell. Mol. Med.* **2021**, *25*, 1867–1883. [CrossRef]
180. Leung, C.O.N.; Deng, W.; Ye, T.M.; Ngan, H.Y.S.; Tsao, S.W.; Cheung, A.N.Y.; Ziru, N.; Yuen, D.C.K.; Pang, R.T.K.; Yeung, W.S.B. MicroRNA-135a-induced formation of CD133+ subpopulation with cancer stem cell properties in cervical cancer. *Carcinogenesis* **2020**, *41*, 1592–1604. [CrossRef]
181. Dong, Z.; Yu, C.; Rezhiya, K.; Gulijiahan, A.; Wang, X. Downregulation of miR-146a promotes tumorigenesis of cervical cancer stem cells via VEGF/CDC42/PAK1 signaling pathway. *Artif Cells Nanomed. Biotechnol.* **2019**, *47*, 3711–3719. [CrossRef]
182. Akerman, G.S.; Tolleson, W.H.; Brown, K.L.; Zyzak, L.L.; Mourateva, E.; Engin, T.S.; Basaraba, A.; Coker, A.L.; Creek, K.E.; Pirisi, L. Human papillomavirus type 16 E6 and E7 cooperate to increase epidermal growth factor receptor (EGFR) mRNA levels, overcoming mechanisms by which excessive EGFR signaling shortens the life span of normal human keratinocytes. *Cancer Res.* **2001**, *61*, 3837–3843.
183. Chhabra, R. let-7i-5p, miR-181a-2-3p and EGF/PI3K/SOX2 axis coordinate to maintain cancer stem cell population in cervical cancer. *Sci. Rep.* **2018**, *8*, 7840. [CrossRef]
184. Mainguy, G.; Koster, J.; Woltering, J.; Jansen, H.; Durston, A. Extensive polycistronism and antisense transcription in the mammalian Hox clusters. *PLoS ONE* **2007**, *2*, e356. [CrossRef]
185. Krumlauf, R. Hox genes in vertebrate development. *Cell* **1994**, *78*, 191–201. [CrossRef]
186. Kim, H.J.; Eoh, K.J.; Kim, L.K.; Nam, E.J.; Yoon, S.O.; Kim, K.H.; Lee, J.K.; Kim, S.W.; Kim, Y.T. The long noncoding RNA HOXA11 antisense induces tumor progression and stemness maintenance in cervical cancer. *Oncotarget* **2016**, *7*, 83001–83016. [CrossRef]
187. Tang, L.; Zhang, W.; Su, B.; Yu, B. Long noncoding RNA HOTAIR is associated with motility, invasion, and metastatic potential of metastatic melanoma. *Biomed. Res. Int.* **2013**, *2013*, 251098. [CrossRef]
188. Zhou, Y.H.; Cui, Y.H.; Wang, T.; Luo, Y. Long non-coding RNA HOTAIR in cervical cancer: Molecular marker, mechanistic insight, and therapeutic target. *Adv. Clin. Chem.* **2020**, *97*, 117–140. [CrossRef]
189. Zhang, W.; Liu, J.; Wu, Q.; Liu, Y.; Ma, C. HOTAIR Contributes to Stemness Acquisition of Cervical Cancer through Regulating miR-203 Interaction with ZEB1 on Epithelial-Mesenchymal Transition. *J. Oncol.* **2021**, *2021*, 4190764. [CrossRef]
190. Walder, R.Y.; Walder, J.A. Role of RNase H in hybrid-arrested translation by antisense oligonucleotides. *Proc. Natl. Acad. Sci. USA* **1988**, *85*, 5011–5015. [CrossRef]
191. Chiriboga, C.A. Nusinersen for the treatment of spinal muscular atrophy. *Expert. Rev. Neurother.* **2017**, *17*, 955–962. [CrossRef] [PubMed]
192. Jaschinski, F.; Rothhammer, T.; Jachimczak, P.; Seitz, C.; Schneider, A.; Schlingensiepen, K.H. The antisense oligonucleotide trabedersen (AP 12009) for the targeted inhibition of TGF-β2. *Curr. Pharm. Biotechnol.* **2011**, *12*, 2203–2213. [CrossRef] [PubMed]
193. Gong, N.; Teng, X.; Li, J.; Liang, X.-J. Antisense Oligonucleotide-Conjugated Nanostructure-Targeting lncRNA MALAT1 Inhibits Cancer Metastasis. *ACS Appl. Mater. Interfaces* **2019**, *11*, 37–42. [CrossRef] [PubMed]
194. Cong, L.; Ran, F.A.; Cox, D.; Lin, S.; Barretto, R.; Habib, N.; Hsu, P.D.; Wu, X.; Jiang, W.; Marraffini, L.A.; et al. Multiplex genome engineering using CRISPR/Cas systems. *Science* **2013**, *339*, 819–823. [CrossRef]
195. Mali, P.; Yang, L.; Esvelt, K.M.; Aach, J.; Guell, M.; DiCarlo, J.E.; Norville, J.E.; Church, G.M. RNA-guided human genome engineering via Cas9. *Science* **2013**, *339*, 823–826. [CrossRef]
196. Swarts, D.C.; Jinek, M. Cas9 versus Cas12a/Cpf1: Structure-function comparisons and implications for genome editing. *Wiley Interdiscip. Rev. RNA* **2018**, *9*, e1481. [CrossRef]
197. Zhang, Y.; Yin, C.; Zhang, T.; Li, F.; Yang, W.; Kaminski, R.; Fagan, P.R.; Putatunda, R.; Young, W.B.; Khalili, K.; et al. CRISPR/gRNA-directed synergistic activation mediator (SAM) induces specific, persistent and robust reactivation of the HIV-1 latent reservoirs. *Sci. Rep.* **2015**, *5*, 16277. [CrossRef]
198. Cox, D.B.T.; Gootenberg, J.S.; Abudayyeh, O.O.; Franklin, B.; Kellner, M.J.; Joung, J.; Zhang, F. RNA editing with CRISPR-Cas13. *Science* **2017**, *358*, 1019–1027. [CrossRef]
199. Shen, L.; Wu, C.; Zhang, J.; Xu, H.; Liu, X.; Wu, X.; Wang, T.; Mao, L. Roles and potential applications of lncRNAs in HIV infection. *Int. J. Infect. Dis.* **2020**, *92*, 97–104. [CrossRef]
200. Sharma, S.; Mandal, P.; Sadhukhan, T.; Roy Chowdhury, R.; Ranjan Mondal, N.; Chakravarty, B.; Chatterjee, T.; Roy, S.; Sengupta, S. Bridging Links between Long Noncoding RNA HOTAIR and HPV Oncoprotein E7 in Cervical Cancer Pathogenesis. *Sci. Rep.* **2015**, *5*, 11724. [CrossRef]

201. Liu, J.; Li, Y.; Tong, J.; Gao, J.; Guo, Q.; Zhang, L.; Wang, B.; Zhao, H.; Wang, H.; Jiang, E.; et al. Long non-coding RNA-dependent mechanism to regulate heme biosynthesis and erythrocyte development. *Nat. Commun.* **2018**, *9*, 4386. [CrossRef]
202. Yang, J.; Meng, X.; Pan, J.; Jiang, N.; Zhou, C.; Wu, Z.; Gong, Z. CRISPR/Cas9-mediated noncoding RNA editing in human cancers. *RNA Biol.* **2018**, *15*, 35–43. [CrossRef]
203. Gupta, P.; Kadamberi, I.P.; Mittal, S.; Tsaih, S.W.; George, J.; Kumar, S.; Vijayan, D.K.; Geethadevi, A.; Parashar, D.; Topchyan, P.; et al. Tumor Derived Extracellular Vesicles Drive T Cell Exhaustion in Tumor Microenvironment through Sphingosine Mediated Signaling and Impacting Immunotherapy Outcomes in Ovarian Cancer. *Adv. Sci.* **2022**, *9*, e2104452. [CrossRef]
204. Jiang, X.-C.; Gao, J.-Q. Exosomes as novel bio-carriers for gene and drug delivery. *Int. J. Pharm.* **2017**, *521*, 167–175. [CrossRef]
205. Baumann, V.; Winkler, J. miRNA-based therapies: Strategies and delivery platforms for oligonucleotide and non-oligonucleotide agents. *Future Med. Chem.* **2014**, *6*, 1967–1984. [CrossRef]
206. Rupaimoole, R.; Lee, J.; Haemmerle, M.; Ling, H.; Previs, R.A.; Pradeep, S.; Wu, S.Y.; Ivan, C.; Ferracin, M.; Dennison, J.B.; et al. Long Noncoding RNA Ceruloplasmin Promotes Cancer Growth by Altering Glycolysis. *Cell Rep.* **2015**, *13*, 2395–2402. [CrossRef]
207. Arora, S.; Swaminathan, S.K.; Kirtane, A.; Srivastava, S.K.; Bhardwaj, A.; Singh, S.; Panyam, J.; Singh, A.P. Synthesis, characterization, and evaluation of poly (D,L-lactide-co-glycolide)-based nanoformulation of miRNA-150: Potential implications for pancreatic cancer therapy. *Int. J. Nanomed.* **2014**, *9*, 2933–2942. [CrossRef]
208. Zhou, Z.; Kennell, C.; Lee, J.Y.; Leung, Y.K.; Tarapore, P. Calcium phosphate-polymer hybrid nanoparticles for enhanced triple negative breast cancer treatment via co-delivery of paclitaxel and miR-221/222 inhibitors. *Nanomedicine* **2017**, *13*, 403–410. [CrossRef]
209. Gao, S.; Tian, H.; Guo, Y.; Li, Y.; Guo, Z.; Zhu, X.; Chen, X. miRNA oligonucleotide and sponge for miRNA-21 inhibition mediated by PEI-PLL in breast cancer therapy. *Acta Biomater.* **2015**, *25*, 184–193. [CrossRef]
210. Cosco, D.; Cilurzo, F.; Maiuolo, J.; Federico, C.; Di Martino, M.T.; Cristiano, M.C.; Tassone, P.; Fresta, M.; Paolino, D. Delivery of miR-34a by chitosan/PLGA nanoplexes for the anticancer treatment of multiple myeloma. *Sci. Rep.* **2015**, *5*, 17579. [CrossRef]
211. Ibrahim, A.F.; Weirauch, U.; Thomas, M.; Grünweller, A.; Hartmann, R.K.; Aigner, A. MicroRNA replacement therapy for miR-145 and miR-33a is efficacious in a model of colon carcinoma. *Cancer Res.* **2011**, *71*, 5214–5224. [CrossRef]
212. Ramot, Y.; Rotkopf, S.; Gabai, R.M.; Zorde Khvalevsky, E.; Muravnik, S.; Marzoli, G.A.; Domb, A.J.; Shemi, A.; Nyska, A. Preclinical Safety Evaluation in Rats of a Polymeric Matrix Containing an siRNA Drug Used as a Local and Prolonged Delivery System for Pancreatic Cancer Therapy. *Toxicol. Pathol.* **2016**, *44*, 856–865. [CrossRef]
213. Patra, J.K.; Das, G.; Fraceto, L.F.; Campos, E.V.R.; Rodriguez-Torres, M.D.P.; Acosta-Torres, L.S.; Diaz-Torres, L.A.; Grillo, R.; Swamy, M.K.; Sharma, S.; et al. Nano based drug delivery systems: Recent developments and future prospects. *J. Nanobiotechnol.* **2018**, *16*, 71. [CrossRef]
214. Østergaard, M.E.; Yu, J.; Kinberger, G.A.; Wan, W.B.; Migawa, M.T.; Vasquez, G.; Schmidt, K.; Gaus, H.J.; Murray, H.M.; Low, A.; et al. Efficient Synthesis and Biological Evaluation of 5′-GalNAc Conjugated Antisense Oligonucleotides. *Bioconjug Chem.* **2015**, *26*, 1451–1455. [CrossRef]
215. Song, T.-T.; Xu, F.; Wang, W. Inhibiting ubiquitin conjugating enzyme E2 N by microRNA-590-3p reduced cell growth of cervical carcinoma. *Kaohsiung J. Med. Sci.* **2020**, *36*, 501–507. [CrossRef]
216. Cheng, X.; Yu, D.; Cheng, G.; Yung, B.C.; Liu, Y.; Li, H.; Kang, C.; Fang, X.; Tian, S.; Zhou, X.; et al. T7 Peptide-Conjugated Lipid Nanoparticles for Dual Modulation of Bcl-2 and Akt-1 in Lung and Cervical Carcinomas. *Mol. Pharm.* **2018**, *15*, 4722–4732. [CrossRef]
217. Dugal-Tessier, J.; Thirumalairajan, S.; Jain, N. Antibody-Oligonucleotide Conjugates: A Twist to Antibody-Drug Conjugates. *J. Clin. Med.* **2021**, *10*, 838. [CrossRef]
218. Dan, N.; Setua, S.; Kashyap, V.K.; Khan, S.; Jaggi, M.; Yallapu, M.M.; Chauhan, S.C. Antibody-Drug Conjugates for Cancer Therapy: Chemistry to Clinical Implications. *Pharmaceuticals* **2018**, *11*, 32. [CrossRef] [PubMed]
219. Huggins, I.J.; Medina, C.A.; Springer, A.D.; van den Berg, A.; Jadhav, S.; Cui, X.; Dowdy, S.F. Site Selective Antibody-Oligonucleotide Conjugation via Microbial Transglutaminase. *Molecules* **2019**, *24*, 3287. [CrossRef] [PubMed]
220. Nanna, A.R.; Kel'in, A.V.; Theile, C.; Pierson, J.M.; Voo, Z.X.; Garg, A.; Nair, J.K.; Maier, M.A.; Fitzgerald, K.; Rader, C. Generation and validation of structurally defined antibody-siRNA conjugates. *Nucleic. Acids Res.* **2020**, *48*, 5281–5293. [CrossRef] [PubMed]
221. Chaturvedi, V.K.; Singh, A.; Singh, V.K.; Singh, M.P. Cancer Nanotechnology: A New Revolution for Cancer Diagnosis and Therapy. *Curr. Drug Metab.* **2019**, *20*, 416–429. [CrossRef]
222. Shao, L.; Wang, R.; Sun, Y.; Yue, Z.; Sun, H.; Wang, X.; Wang, P.; Sun, G.; Hu, J.; Sun, H.; et al. Delivery of MicroRNA-let-7c-5p by Biodegradable Silica Nanoparticles Suppresses Human Cervical Carcinoma Cell Proliferation and Migration. *J. Biomed. Nanotechnol.* **2020**, *16*, 1600–1611. [CrossRef]
223. Wang, L.; Liang, T.T. CD59 receptor targeted delivery of miRNA-1284 and cisplatin-loaded liposomes for effective therapeutic efficacy against cervical cancer cells. *AMB Express* **2020**, *10*, 54. [CrossRef]

*genes*

Review

Circulating microRNAs as the Potential Diagnostic and Prognostic Biomarkers for Nasopharyngeal Carcinoma

Thuy Ai Huyen Le and Thuan Duc Lao *

Department of Pharmaceutical and Medical Biotechnology, Faculty of Biotechnology, Ho Chi Minh City Open University, Ho Chi Minh City 700000, Vietnam; thuy.lha@ou.edu.vn
* Correspondence: thuan.ld@ou.edu.vn

Abstract: microRNAs are endogenous non-coding miRNAs, 19–25 nucleotides in length, that can be detected in the extracellular environment in stable forms, named circulating miRNAs (CIR-miRNAs). Since the first discovery of CIR-miRNAs, a large number of studies have demonstrated that the abnormal changes in its expression could be used to significantly distinguish nasopharyngeal carcinoma (NPC) from healthy cells. We herein reviewed and highlighted recent advances in the study of CIR-miRNAs in NPC, which pointed out the main components serving as promising and effective biomarkers for NPC diagnosis and prognosis. Furthermore, brief descriptions of its origin and unique characteristics are provided.

Keywords: circulating microRNAs; extracellular miRNA; nasopharyngeal carcinoma; miRNA-based diagnosis

Citation: Le, T.A.H.; Lao, T.D. Circulating microRNAs as the Potential Diagnostic and Prognostic Biomarkers for Nasopharyngeal Carcinoma. *Genes* **2022**, *13*, 1160. https://doi.org/10.3390/ genes13071160

Academic Editors: Giuseppe Iacomino and Fabio Lauria

Received: 9 May 2022
Accepted: 2 June 2022
Published: 27 June 2022

Publisher's Note: MDPI stays neutral with regard to jurisdictional claims in published maps and institutional affiliations.

1. Introduction

Nasopharyngeal carcinoma (NPC), a malignant tumor that arises from the nasopharynx epithelium, is a highly curable disease if detected and diagnosed early. The absence of obvious symptoms in the early stages of NPC has made early diagnosis impossible; therefore, many patients are diagnosed at an advanced stage, leading to decreased survival [1,2]. That is why early diagnosis and screening play significant roles in increasing the success rate of NPC treatment as well as patient survival before it has progressed to an untreatable, ultimately fatal stage.

Biomarkers (biological markers) have gradually been recognized as measurable indicators in the diagnosis and prognosis of human diseases, including NPC [3]. To date, the main clinical practice, screening for NPC, is to detect the presence of Epstein–Barr virus (EBV), one of the etiological factors, via the detection of EBV DNA, EBV-VCA-IgA, EBNA1-IgA, and Rta-IgG [4–6]. The detection rate of NPC varies from 20–100%, leading to the varied efficacy of serology marker examination [7]. Oncoproteins, including EBV latent proteins (LMP-1, LMP-2A, and LMP-2B), and Epstein–Barr virus nuclear antigens (EBNA-1, -2, -3A, -3B, -3C, and –LP) encoded by EBV are also used to implicate the progression of nasopharyngeal tumorigenesis [8–12]. Notably, EBV is widespread in healthy people globally, infecting more than 90% of people [13,14]. The infection of EBV has also been reported in many different pathogeneses of human diseases, such as Burkitt's lymphoma, Hodgkin's disease, non-Hodgkin's lymphoma, breast cancer, gastric cancer, and so on [6,12,15]. This is due mainly to the unsuitability for early diagnosis and screening of NPC based on EBV detection. Therefore, it is urgent that we identify novel and potentially reliable biomarkers that aim towards early clinical diagnosis as well as screening of NPC.

Apart from oncoproteins encoded by EBV, several proteins involved in the regulation of many tumorigenesis processes, including assisting tumor growth and survival, as well as tumor cell immigration and invasion, have been proven to be secreted into the tumor environment [11,16]. Many secreted proteins were reported to be substantially expressed in

NPC cells compared to healthy ones, including heat shock protein 70 (Hsp70), plasminogen activator inhibitor 1 (PAI-1), fibronectin, MAC-2-binding protein (MAC-2 BP), and so on [11,16–20]. Despite the fact that several studies have identified these secreted proteins as possible biomarkers for NPC diagnosis and prognosis, there are still a number of issues that need to be addressed before these proteomic biomarkers may be used in clinical diagnosis and prognosis. Primarily, the concentrations of expressed proteins are dynamic, fluctuating dramatically at times with regard to stress, illness, and/or treatment. Secondly, quantitative discoveries of proteins require advanced high-throughput technologies, which, therefore, lead to many difficulties associated with sample isolation [11,21–23]. Finally, similar to oncoproteins encoded by EBV, there is a lack of further clinical investigations and validations.

Many efforts are now being undertaken to determine prospective biomarkers for the early diagnosis and screening of NPC that are based on the detection of microRNAs (miRNAs), etiological factors of tumorigenesis [2,24]. It is becoming increasingly evident that the aberrant expression of miRNAs, which function as oncogenic miRNAs, known as "oncomirs", and tumor suppressor miRNAs, resulting in numerous cancer-associated phenomena, such as anti-apoptosis, proliferation, migration, metastasis, and so on, are responsible for numerous human pathogeneses, including nasopharyngeal tumorigenesis [12,14,24–27]. Thus far, many researchers have studied to identify the signature of miRNA expression in nasopharyngeal tumorigenesis [27–30]. The study of Zhu et al., which pointed out that miR-106a-5p is markedly up-regulated in NPC specimens, using methods of quantitative real-time PCR, miRNA microarray, and TCGA database analysis findings, is case in point [28]. In their report, they claimed that overexpression of miR-106a-5p was linked to advanced stages, recurrence, and poor clinical outcomes in NPC patients. Furthermore, through its target gene, *BTG3* (*BTG anti-proliferation factor 3*), and activation of autophagy-regulating MAPK signaling, its overexpression inhibits autophagy and promotes the malignant phenotype of nasopharyngeal cancer cells [28]. Their observations may lead to novel insights into the nasopharyngeal carcinoma pathogenesis. Recently, miRNAs existing in body fluids, such as plasma, serum, saliva, and so on, also known as circulating miRNA (CIR-miRNAs) or extracellular miRNA (EC-miRNA), have attracted interest in the investigation of biomarkers for NPC [14,27,31,32]. The discovery of CIR-miRNAs was unanticipated, thus, understanding their significant characteristics and clinical usefulness in cancer diagnosis and screening is urgent. In this article, we focus on summarizing the unique characteristics and usefulness of CIR-miRNAs in NPC diagnosis, screening, and therapy.

2. Brief Introduction of miRNAs and Their Regulation

Since the first miRNA, namely lin-4, in *Caenorhabditis elegans*, was discovered by Victor Ambros and colleagues, the expression patterns of miRNAs have been reported as having fundamental roles in cancer initiation development and progression [14,24–28,30,33]. miRNAs are a family of small non-coding miRNAs, 19–25 nucleotides in length, which play multiple significant roles in a variety of biological processes, including cell proliferation, differentiation, metabolism, stress response, as well as apoptosis, via the regulation of its target genes [24–27]. The majority of post-transcriptional regulation of miRNA target genes is accomplished through the binding of miRNAs to their target sequences at the 3′-untranslated region (3′-UTR), eventually resulting in the degradation and/or repression of mRNA [24–26,34] (Figure 1). Additionally, miRNAs attach with other regions of target genes, including 5′-UTR, as well as promoter and coding regions [24,35]. According to Lewis et al. (2005), more than a third of human genes appear to have been selectively pressured to keep their miRNA seed pairing [36]. The expression of mRNA is down-regulated as a result of the pairing between miRNA and mRNA. This interaction is formed by the binding between a "seed" region located at the 5′ UTR of miRNAs and 3′ UTR of target mRNA conforming to the Watson–Crick rule [24,36]. Four types of sites, including 6 mer-site, 7 mer-A1 site, 7 mer-A8 site, and 8 mer-site, have been identified [24,36,37]. Due to its broad range of target genes, current research has provided much evidence proving that individual miRNA serves as the mediator of biological gene networks related to many biological functions by acting at the post-transcriptional level. According to the data of miRbase (https://www.mirbase.org/, accessed on 30 April 2022 there are more than 4.700 known human miRNAs (within the prefix hsa-) that have been identified.

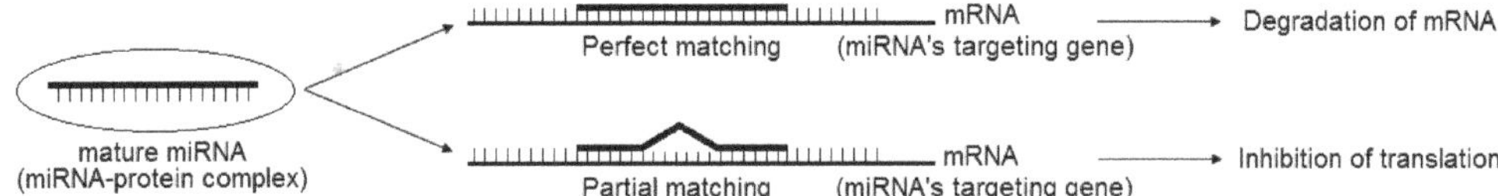

Figure 1. The regulation of gene expression through the binding of miRNA.

The cellular biogenesis of miRNAs involves both the canonical pathway and non-canonical pathway [24,25,38]. Mature miRNA originates from primary miRNA (pri-miRNA) and precursor miRNA (pre-miRNA). The structure of 5′-7 methyl-guanosine capped and 3′ polyadenylated pri-miRNA is transcribed from the intragenic region or intronic region by RNA-polymerase II [24,39,40]. Pri-miRNA is subsequently cleaved into a ~60–70 nucleotide hair-spin structural pre-miRNA by Drosha and Dicer [41,42]. Pre-miRNA is exported into the cytoplasm for further processing by exportin-5 and Ras-related nuclear protein guanosine triphosphate (RAN-GTP) [43]. In the cytoplasm, the stem loop of pre-miRNA is removed by complex of Dicer/transactivation-responsive RNA-binding protein (TRBP) to form a 20–22 nucleotide-long miRNA duplex, which consists of a 5′ phosphorylated strand and 3′overhang, named the mature miRNA guide strand (miRNA) and complementary passenger strand (miRNA*). In the final step, the duplex of miRNA/miRNA* is loaded into an Argonaute protein (Ago protein) to generate an RNA-induced silencing complex (RISC), whereas the other strand (passenger strand) is degraded [24,25,44]. This mature miRNA cooperates with RISC to regulate the expression of target genes through the perfect binding or partial binding of the "seed" region of miRNA to the target mRNA's 3′UTR according to the principle of Watson–Crick complementary base pairing, resulting in the degradation of mRNA and inhibition of translation [45].

3. Circulating miRNAs (CIR-miRNAs), Their Origin and Unique Characteristics

Tumor-associated miRNAs primarily detected in cellular environments can help to more accurately diagnose and monitor human cancer [34,46]. Recently, a handful of CIR-miRNAs, also known as EC-miRNA, have been detected outside of the cellular environment (extracellular environment), including in various bodily fluids [34,47,48]. CIR-miRNAs are not only detected in whole blood and plasma, but also in saliva, tears, urine, breast milk, follicular fluid, semen, and so on [34,47,48]. Since the discovery of CIR-miRNAs, these findings represent a novel approach for human cancer diagnosis and screening due to their being less invasive or non-invasive. For example, Wen et al. identified two miRNA signatures for the highly accurate diagnosis and differential diagnosis of patients with NPC, which represented novel serological biomarkers and potential therapeutic targets for NPC [27]. When studying the expression of serum miRNAs in 74 cases of patients with nasopharyngeal carcinoma and 57 cases of non-cancerous volunteers, they found that miR-17, miR-20a, miR-29c, and miR-223 among CIR-miRNAs were exclusively expressed in the sera of non-cancerous samples compared with that of NPC patients. Additionally, high sensitivity and specificity were recorded by calculating Ct differences, which have been shown to distinguish between NPC cases and controls. Based on their results, the four CIR-miRNAs, miR-17, miR-20a, miR-29c, and miR-223, may provide a novel strategy for NPC diagnosis and screening [14]. The expression patterns of CIR-miRNAs from different types of body fluids reflect the NPC status. Therefore, exploring novel strategies for NPC invasive markers from body fluids is promising, even though challenging, for early diagnosis and screening of NPC.

The mechanism by which CIR-miRNAs enters the extracellular environment is not fully understood. Until now, two major types of CIR-miRNAs, vesicle-associated and non-vesicle-associated forms, have been identified [49]. The non-vesicle-associated type includes CIR-miRNAs that are freely circulating, bound to specific proteins, and that are enclosed in the extracellular environment [50] (Figure 2).

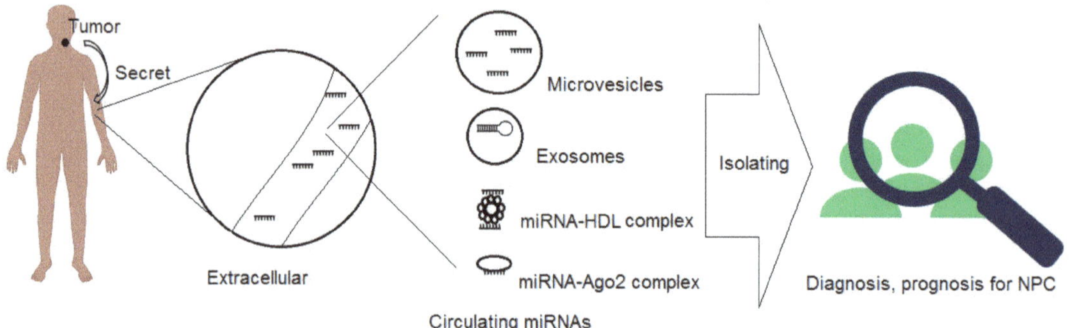

Figure 2. MicroRNAs are exported from cells into circulation by many mechanisms.

Pre-miRNAs and mature miRNAs can be incorporated into small vesicles called exosomes (50–90 nm), which originate from endosomes and are then released from cells by fusing with the plasma membrane, or they can be released by microvesicles (~1 μm) that are released from the cell by blebbing of the plasma membrane [49–51]. Other major studies confirmed that CIR-miRNAs also exist in the non-vesicle-associated form (microparticle-free form). These miRNAs can be incorporated into high-density lipoproteins, or bound to RNA-binding proteins such as Argonaute2 (Ago2), which is the central protein of miRNA-mediated interference, and, together with GW182, was shown to be in charge of extracellular miRNA protection and transport [49,51,52]. It is unclear how these miRNA–protein complexes exit the cell. These miRNAs, exhibiting characteristics from tumor cells, can be produced in two ways: passively, as by-products of dying cells, or actively, in a miRNA-specific manner, via interactions with certain membrane channels or proteins [49,51].

CIR-miRNAs have several unique characteristics [53]. Unlike cellular miRNAs and mRNAs, which are degraded in the extracellular environment in a few seconds, CIR-miRNAs surprisingly persist for lengthy periods of time under adverse environments, such as extreme pH and RNAse digestion [34,54]. Mitchell et al. (2008) found that human plasma and serum contain miRNAs, which present in a remarkably stable form and are resistant to the activities of endogenous RNAse. They also investigated the stability of miRNAs in plasma by incubating plasma at room temperature for up to 24 h and subjecting it to up to eight cycles of freeze–thawing [54]. In the study of Turchinovich et al. (2011), the authors confirmed that circulating mature miRNA is extremely stable in blood plasma and cell culture media. Additionally, CIR-miRNAs can also be maintained in the extracellular environment for at least two months following cell lysis. miRNAs are thought to have originated as by-products of dying cells, and remain stable in the extracellular space for a long time due to the great stability of the Ago2/miRNA complex [55]. These findings suggest that these CIR-miRNAs use certain protective mechanisms to avoid being attacked by RNase in the extracellular environment. These circulating small molecules have unique properties that make them attractive candidates for use as biomarkers in a variety of liquid biopsies, such as saliva, cerebrospinal fluid, ascites, urine, breast milk, and sperm, for cancer diagnosis and prognosis [30,49]. However, mechanisms by which CIR-miRNAs exhibit exceptional stability in the RNase-rich environment of the blood and other bio-fluids are not well understood.

4. CIR-miRNAs as Diagnosis and Prognosis Biomarkers for NPC

Can the expression profile of CIR-miRNAs be identified as an effective biomarker for NPC diagnosis and prognosis? Recently, independent studies have successfully proven that CIR-miRNAs, extracted from whole blood and plasma, as well as serum, can be used as biomarkers for the diagnosis of NPC due to their stability and predictive properties. In the current revision, we focus on the results of recent studies that revealed the function of CIR-miRNAs as diagnosis and prognosis biomarkers for NPC. From the expression of many valuable CIR-miRNAs, including hsa-miR-let-7b-5p, hsa-miR-140-3p, hsa-miR-144-3p, hsa-miR-17-5p, hsa-miR-20a-5p, hsa-miR-20b-5p, hsa-miR-205-5p, hsa-miR-22, hsa-miR-miR-572, hsa-miR-638, hsa-miR-1234, and so on, extracted from whole blood, plasma, and serum, it has been reported that microRNAs possess great diagnostic power in NPC [14,27,56–58] (Table 1).

Table 1. The abnormal expression of CIR-miRNAs in NPC.

Year	CIR-miRNAs candidate	Description	Source	Reference
2020	hsa-miR-let-7b-5p, hsa-miR-140-3p, hsa-miR-144-3p, hsa-miR-17-5p, hsa-miR-20a-5p, hsa-miR-20b-5p, hsa-miR-205-5p	The panel of 7 miRNA, extracted from plasma, performed better in distinguishing NPC patients from healthy controls, the sensitivity and specificity being 0.74 and 0.76, respectively.	Plasma	[56]
2019	hsa-miR-188-5p, hsa-miR-1908, hsa-miR-3196, hsa-miR-3935, hsa-miR-4284, hsa-miR-4433-5p, hsa-miR-4665-3p, hsa-miR-513b	Theses 8 miRNA signatures diagnosed NPC with an accuracy of 97.14%, sensitivity of 96.43%, specificity of 100%, positive predictive value of 100%, and negative predictive value of 87.5% in a group of 84 NPC samples and 21 healthy samples.	Whole blood	[27]
2019	hsa-miR-937-5p, hsa-miR-650, hsa-miR-3612, hsa-miR-4478, hsa-miR-4259, hsa-miR-3714, hsa-miR-4730, hsa-miR-1203, hsa-miR-30b-3p, hsa-miR-1321, hsa-miR-1202, hsa-miR-575	These CIR-miRNAs were significantly down-regulated in saliva of NPC patients compared to healthy controls, detected by miRNA microarray platform with the high accuracy (sensitivity = 100.00%, specificity = 96.00%).	Salivary	[31]
2014	hsa-miR-22, hsa-miR-miR-572, hsa-miR-638, hsa-miR-1234	Different changes were observed in the serum of patients with NPC. The value of prognosis of the TNM staging system was reported. The patients of with high-risk scores had poorer overall survival and low metastasis-free survival than those with the patients with low-risk scores.	Serum	[57]
2014	hsa-miR-548q, hsa-miR-483-5p	miR-548q and miR-483-5p highly expressed in NPC cell lines and 31 plasma samples from NPC patients, compared with 19 non-cancerous controls. Combining these 2 CIR-miRNAs resulted in 67.1% sensitivity and 68.0% specificity.	Plasma	[58]
2014	hsa-miR-483-5p, hsa-miR-103, hsa-miR-29a	Differentially expressed CIR-miRNAs were identified as being effective biomarkers for predicting survival in NPC patients.	Plasma	[26]
2013	hsa-miR-16, miR-21, hsa-miR-24, hsa-miR-155, hsa-and miR-378	Sensitivity and specificity reached 87.7% and 82.0%, respectively, when combining the panel of CIR-miRNAs, hsa-miR-16, miR-21, hsa-miR-24, hsa-miR-155, hsa-and miR-378.	Plasma	[59]
2012	hsa-miR-17, hsa-miR-20a, hsa-miR-29c, hsa-miR-223	miRNAs were differentially expressed in the serum of 20 NPC patients compared with that of 20 non-cancerous controls. Using these 4 CIR-miRNAs, a diagnostic value with sensitivity of 97.3% and specificity of 96.5% was established.	Serum	[14]

miRNA profiles extracted from whole blood in NPC patients were investigated by Wen et al. (2019). In their study, the training group-1, containing 84 NPC samples and 21 healthy samples, and the validation group-1, consisting of 36 NPC samples and 9 healthy samples, were established and enrolled in an NPC diagnostic model and Lasso regression to identify miRNA profiles to diagnose NPC. The profile of eight CIR-miRNAs, including hsa-miR-188-5p, hsa-miR-1908, hsa-miR-3196, hsa-miR-3935, hsa-miR-4284, hsa-miR-4433-5p, hsa-miR-4665-3p, and hsa-miR-513b, were identified as highly accurate in the diagnosis of patients with NPC from healthy ones. The same results of accurate diagnosis were also observed in validation group-1. Additionally, they also reported that high levels of hsa-miR-4790-3p, hsa-miR-188-5p, hsa-miR-5583-5p, and hsa-miR-3615 were frequently observed in NPC samples than in healthy samples. In their study, they compiled a panel of signatures for NPC diagnosis, and their results demonstrated an accuracy of 97.14%, sensitivity of 96.43%, specificity of 100%, positive predictive value of 100%, and negative predictive value of 87.5% [27].

Furthermore, tumor cells have been reported to release miRNAs into the circulation, serum, and plasma; therefore, they have been proposed as potential sources for CIR-miRNAs isolation [54]. Zhang et al. (2020) designed a study with four-stage validation, which consisted of the screening, training, testing, and external validation stages, to identify potential biomarkers for NPC diagnosis by quantitative reverse transcription polymerase chain reaction [56]. In their study, in the validation stage, seven plasma CIR-miRNAs, including hsa-let-7b-5p, hsa-miR-140-3p, hsa-miR-144-3p, hsa-miR-17-5p, hsa-miR-20a-5p, hsa-miR-20b-5p, and hsa-miR-205-5p, were identified to be significantly up-regulated in NPC patients compared with non-cancerous samples. Additionally, among these seven CIR-miRNAs, hsa-let-7b-5p was investigated to be significantly higher in the group of EBV-infected NPC samples, compared with non-EBV-infected NPC samples and healthy samples [56]. In the study of Liu et al. (2013), the combination of five plasma miRNAs, including hsa-miR-16, miR-21, hsa-miR-24, hsa-miR-155, and hsa-miR-378, resulted in values of sensitivity and specificity reaching 87.7% and 82.0%, respectively, for NPC diagnosis. In their study, they also reported that the specificity was reduced when removing hsa-miR-16 in their combination [59]. Collectively, combining panels of CIR-miRNAs can increase the sensitivity and specificity, and thus the accuracy, of CIR-miRNA biomarkers.

miRNA exist not only in blood fluid, such as whole blood, plasma, and serum, but are also detected in human saliva in a stable extracellular form [60]. Wu et al. (2019) was the first to investigate salivary miRNAs as potential indicators for nasopharyngeal cancer. In their study, they highlighted the potential salivary miRNAs as biomarkers for the detection of NPC based on the evaluation of miRNA expression in 22 saliva samples from NPC patients, and 25 healthy controls using microarray miRNA expression. They found that 12 miRNAs were significantly down-regulated in the saliva of NPC patients compared to healthy controls, with high accuracy. Additionally, the regulatory network for differentially expressed miRNAs was also successfully predicted. In their prediction, 11 out of 12 CIR-miRNAs, excepted for hsa-miR-4730, were incorporated into enriched GO pathways. Many target genes, such as platelet-derived growth factor receptor α (PDGFRA), Ras-related C3 botulinum toxin substrate 1 (RAC1), inhibitor of nuclear factor kappa B kinase subunit γ (IKBKG), X-linked inhibitor of apoptosis protein (XIAP), and protein phosphatase, Mg^{2+}/Mn^{2+}-dependent 1D (PPM1D), and so on, were simultaneously regulated by hsa-miR-937-5p, hsa-miR-650, hsa-miR-3612, hsa-miR-4478, hsa-miR-4259, hsa-miR-3714, hsa-miR-1203, hsa-miR-30b-3p, hsa-miR-1321, hsa-miR-1202, and hsa-miR-575 [31]. These findings suggest that differentially expressed saliva miRNAs may play a key role in NPC by targeting their target genes, which are linked to several important pathways.

It is emphasized that CIR-miRNAs profiles could help distinguish nasopharyngeal tumors from other head and neck tumors. In the study of Wen et al. (2019), they successfully identified and validated 16 miRNA signatures to differentiate NPC in comparison with other tumors located in the region of head and neck, and in non-cancerous samples. Additionally, this diagnostic model has an accuracy rate of 100%, specificity of 100%, and sensitivity of 100%. [27]. These results show that the profiles of 16 miRNAs could correctly distinguish NPC patients from head and neck tumor patients and heathy controls. Therefore, it is possible to apply CIR-miRNA non-invasive marker profiles to diagnose NPC.

As prognostic indicators, a stable and circulating presence in non-invasive specimens is also a characteristic aim for prognostic applications. Liu et al. (2014) conducted models to investigate the prognostic value of serum miRNAs in patients with NPC. They identified that four CIR-miRNAs, hsa-miR-22, hsa-miR-miR-572, hsa-miR-638, and hsa-miR-1234, added to the prognostic value to the stage system of TNM. In their results, the analysis of ROC indicated that the model combining the miRNA signature and TNM stage had improved the prognostic value for overall survival (area under ROC (AUROC): 0.69 vs. 0.60, $p = 0.001$; AUROC: 0.69 vs. 0.63, $p = 0.008$) and distant metastasis-free survival (AUROC: 0.71 vs. 0.60, $p < 0.001$; AUROC: 0.71 vs. 0.65, $p = 0.005$) relative to the TNM stage-alone model or miRNA signature-alone model in the training set, which may lead to more personalized therapy [56]. Finding a miRNA profile as an indicator of prognosis, might help to identify individuals who would benefit from more aggressive therapy and, as a result, enhance NPC patient survival. Collectively, these findings are encouraging in terms of using these CIR-miRNAs as a biomarker for the diagnosis and prognosis of NPC.

Referring to miRNA-derived EBV, which is encoded in the host cells, miRNA-derived EBV are transcribed from two regions of the EBV genome, BamH I fragment H rightward open reading frame 1(BHRF1)-cluster and BamH I fragment A rightward transcript (BART)-cluster 1, 2 [61,62]. The first cluster, BHRF1, which produces three miRNA precursors that subsequently encode four mature miRNAs, was discovered by Pfeffer et al. (2004). The second cluster, BART, which produces 22 precursors and encodes 40 mature miRNAs, was also found [60–62]. To date, many studies have revealed that EBV miRNAs play crucial roles in immune evasion, proliferation, apoptosis, invasion, and metastasis [61,63,64]. It is noted that EBV miRNAs, especially circulating EBV miRNA, could serve as biomarkers in NPC [61,62]. Recently, the clinical value of EBV-derived miRNAs for diagnosis and prognosis of EBV-positive nasopharyngeal tumors has been investigated. It has been reported that the plasma levels of BART7-3p and miR-BART13-3p are highly reflective of NPC diagnosis [65]. In their report, quantitative PCR was applied to evaluate the plasma levels of EBV DNA, miR-BART7-3p, and miR-BART13-3p, in 483 treatment-naïve NPC patients, compared to 243 controls. miR-BART7-3p and miR-BART13-3p were detected in 96.1% and 97.7% of NPC cases, whereas only 3.9% and 3.9% were detected in healthy controls, respectively. miR-BART13-3p values of sensitivity and specificity were 97.9% and 96.7%, respectively. The corresponding values for miR-BART7-3p were 96.1% and 96.7%. It is noted that the prognostic performance was assessed by comparing levels to distant metastatic rates during a 55-month follow-up of 245 NPC patients with radiotherapy treatment. After a four-year follow-up, the value of distant metastasis-free survival rates were 89.7% and 61.4% in subjects with detectable miR-BART7-3p and miR-BART13-3p, respectively. Additionally, at diagnosis and following radiation, a combination of plasma levels of miR-BART7-3p and EBV DNA could assist in stratifying individuals based on their risk of poor DMFS [65]. Other study, Wardana et al. (2020) found that circulating miR-BART-7 levels measured in peripheral blood samples can be used as a promising predictor for the clinical outcome and prognosis in NPC patients [66]. Based on these findings, circulating EBV miR-BART7 and miR-BART13 will shed light on the potential serological biomarkers for diagnosis and prognosis of EBV-positive nasopharyngeal malignancies. From their panel of EBV miRNAs, Gao et al. reported that plasma levels of BART2-3P, BART2-5P, BART5-3P, BART5-5P, BART6-3P, BART8-3P, BART9-5P, BART17-5P, BART19-3P, and BART20-3P were significantly increased. Additionally, EBV miRNAs, such as BART8-3P and BART10-3P, could potentially be employed as complementary serological markers if EBV DNA is beyond the lower detection limit or undetectable in plasma samples [67]. Most importantly, EBV miRNA research is still in its early stages, with only a few studies confirming that EBV miRNAs are abundantly expressed in EBV-associated tumors. Better understanding and evaluating, therefore, of the levels of EBV miRNAs might establish novel biomarkers for NPC diagnosis and prognosis. Furthermore, there is a great need for more research aiming to determine the diagnostic and prognostic value of EBV miRNAs in follow-up management of nasopharyngeal patients.

5. Challenges of Circulating miRNAs as Indicators for Diagnosis and Prognosis

Though there are many advantages of using CIR-miRNAs, such as their stability, particular expression profiles in different stages of cancers, and so on, there are still challenges to overcome before clinical application. First, because of their low concentration in the extracellular environment, detecting and evaluating CIR-miRNAs is certainly difficult [49]. Other RNAs in plasma or serum also exist; therefore, from analysis results, it is difficult to distinguish CIR-miRNAs from other RNAs. Second, many methods of detecting and measuring levels of CIR-miRNAs, including qRT-PCR, miRNA microarray, and next-generation sequencing (NGS) have been applied [14,27,31,55–58]. Each method might produce different findings; thus, unifying methodologies and eliminating variance is critical. An additional obstacle is that the understanding of CIR-miRNAs functions, as well as the relationship among CIR-miRNAs and the Epstein–Barr virus (EBV), is still limited. To date, a standard tool for NPC screening has been based on the detection of IgA antibodies against EBV capsid antigen (VCA/IgA) and early antigen (EA/IgA) [27,68]. However, the specificity and sensitivity of combining VCA/IgA and EA/IgA are 50.9% and 95.2%, respectively. Its low sensitivity reflects that EBV infection is involved in other hematological and epithelial malignancies, such as Hodgkin's lymphoma, Burkitt's lymphoma, and other tumors that originate from the oral epithelium, oropharynx, larynx, and so on [69,70]. Therefore, differentiating the CIR-miRNAs profiles between NPC and EBV-related human tumors, including head and neck tumors and esophagus carcinomas, remains a major challenge. Thus, more multicenter prospective trials to verify these diagnostic markers for NPC are still required.

6. Conclusions and Perspectives

The existence of CIR-miRNAs, extracted from human bodily fluid, including whole blood, plasma, serum, salivary, and so on, exhibits the promising and valuable potential of serving as diagnostic and prognostic biomarkers for human cancer, including NPC. CIR-miRNAs are proven to persist for a long time under harsh environmental conditions, meaning they could be effectively applied in clinical environments. However, many obstacles need to be solved urgently. The development of CIR-miRNAs-based diagnostic and prognostic tools for NPC is a lengthy process, hence larger studies, which promote sensitivity and specificity, are necessary.

Author Contributions: All authors contributed to the design and conception of the study. T.D.L.: writing original draft preparation, editing the manuscript. T.A.H.L.: review and editing the manuscript. All authors have read and agreed to the published All authors have read and agreed to the published version of the manuscript.

Funding: This research received funding sponsored by Ho Chi Minh City Open University, Ho Chi Minh City, Vietnam under the grand number E2019.07.3.

Institutional Review Board Statement: Ethical review and approval were waived for this study, as this study is a review article.

Informed Consent Statement: Not applicable.

Data Availability Statement: Not applicable.

Acknowledgments: We wish to express our thanks to the research project sponsored by the Ministry of Education and Training, Vietnam, and Ho Chi Minh City Open University, HCMC, Vietnam.

Conflicts of Interest: The authors declare no conflict of interest.

References

1. Ear, E.N.S.; Irekeola, A.A.; Yean Yean, C. Diagnostic and Prognostic Indications of Nasopharyngeal Carcinoma. *Diagnostics* **2020**, *10*, 611. [CrossRef] [PubMed]
2. Wang, K.H.; Austin, S.A.; Chen, S.H.; Sonne, D.C.; Gurushanthaiah, D. Nasopharyngeal Carcinoma Diagnostic Challenge in a Nonendemic Setting: Our Experience with 101 Patients. *Perm. J.* **2017**, *21*, 16–180. [CrossRef] [PubMed]
3. Hossain, S.F.; Huang, M.; Ono, N.; Morita, A.; Kanaya, S.; Altaf-Ul-Amin, M. Development of a biomarker database toward performing disease classification and finding disease interrelations. *Database* **2021**, *2021*, baab011. [CrossRef]
4. Tan, L.P.; Tan, G.W.; Sivanesan, V.M.; Goh, S.L.; Ng, X.J.; Lim, C.S.; Kim, W.R.; Mohidin, T.B.B.M.; Dali, N.S.M.; Ong, S.H.; et al. Systematic comparison of plasma EBV DNA, anti-EBV antibodies and miRNA levels for early detection and prognosis of nasopharyngeal carcinoma. *Int. J. Cancer* **2020**, *146*, 2336–2347. [CrossRef]
5. Liu, W.; Chen, G.; Gong, X.; Wang, Y.; Zheng, Y.; Liao, X.; Liao, W.; Song, L.; Xu, J.; Zhang, X. The diagnostic value of EBV-DNA and EBV-related antibodies detection for nasopharyngeal carcinoma: A meta-analysis. *Cancer Cell Int.* **2021**, *21*, 164. [CrossRef]
6. Yang, S.; Song, C. A meta-analysis on the EBV DNA and VCA-IgA in diagnosis of Nasopharyngeal Carcinoma. *Pakistan J. Med. Sci.* **2013**, *29*, 885–890. [CrossRef] [PubMed]
7. Chang, C.; Middeldorp, J.; Yu, K.J.; Juwana, H.; Hsu, W.-L.; Lou, P.-J.; Wang, C.-P.; Chen, J.-Y.; Liu, M.-Y.; Pfeiffer, R.M.; et al. Characterization of ELISA detection of broad-spectrum anti-Epstein-Barr virus antibodies associated with nasopharyngeal carcinoma. *J. Med. Virol.* **2013**, *85*, 524–529. [CrossRef] [PubMed]
8. Lao, T.D.; Nguyen, T.A.H.; Ngo, K.D.; Thieu, H.H.; Nguyen, M.T.; Nguyen, D.H.; Le, T.A.H. Molecular Screening of Nasopharyngeal Carcinoma: Detection of LMP-1, LMP-2 Gene Expression in Vietnamese Nasopharyngeal Swab Samples. *Asian Pac. J. Cancer Prev.* **2019**, *20*, 2757–2761. [CrossRef]
9. Lan, Y.-Y.; Hsiao, J.-R.; Chang, K.-C.; Chang, J.S.-M.; Chen, C.-W.; Lai, H.-C.; Wu, S.-Y.; Yeh, T.-H.; Chang, F.-H.; Lin, W.-H.; et al. Epstein-Barr Virus Latent Membrane Protein 2A Promotes Invasion of Nasopharyngeal Carcinoma Cells through ERK/Fra-1-Mediated Induction of Matrix Metalloproteinase 9. *J. Virol.* **2012**, *86*, 6656–6667. [CrossRef]
10. Morris, M.; Laverick, L.; Wei, W.; Davis, A.; O'Neill, S.; Wood, L.; Wright, J.; Dawson, C.; Young, L. The EBV-Encoded Oncoprotein, LMP1, Induces an Epithelial-to-Mesenchymal Transition (EMT) via Its CTAR1 Domain through Integrin-Mediated ERK-MAPK Signalling. *Cancers* **2018**, *10*, 130. [CrossRef]
11. Siak, P.Y.; Khoo, A.S.-B.; Leong, C.O.; Hoh, B.-P.; Cheah, S.-C. Current Status and Future Perspectives about Molecular Biomarkers of Nasopharyngeal Carcinoma. *Cancers* **2021**, *13*, 3490. [CrossRef] [PubMed]
12. Thompson, M.P.; Kurzrock, R. Epstein-Barr Virus and Cancer. *Clin. Cancer Res.* **2004**, *10*, 803–821. [CrossRef] [PubMed]
13. Dowd, J.B.; Palermo, T.; Brite, J.; McDade, T.W.; Aiello, A. Seroprevalence of Epstein-Barr Virus Infection in U.S. Children Ages 6–19, 2003–2010. *PLoS ONE* **2013**, *8*, e64921. [CrossRef] [PubMed]
14. Zeng, X.; Xiang, J.; Wu, M.; Xiong, W.; Tang, H.; Deng, M.; Li, X.; Liao, Q.; Su, B.; Luo, Z.; et al. Circulating miR-17, miR-20a, miR-29c, and miR-223 Combined as Non-Invasive Biomarkers in Nasopharyngeal Carcinoma. *PLoS ONE* **2012**, *7*, e46367. [CrossRef]
15. Sun, K.; Jia, K.; Lv, H.; Wang, S.-Q.; Wu, Y.; Lei, H.; Chen, X. EBV-Positive Gastric Cancer: Current Knowledge and Future Perspectives. *Front. Oncol.* **2020**, *10*, 2096. [CrossRef]
16. Feng, Y.; Xia, W.; He, G.; Ke, R.; Liu, L.; Xie, M.; Tang, A.; Yi, X. Accuracy Evaluation and Comparison of 14 Diagnostic Markers for Nasopharyngeal Carcinoma: A Meta-Analysis. *Front. Oncol.* **2020**, *10*, 1779. [CrossRef]
17. Wang, W.-Y.; Twu, C.-W.; Liu, Y.-C.; Lin, H.-H.; Chen, C.-J.; Lin, J.-C. Fibronectin promotes nasopharyngeal cancer cell motility and proliferation. *Biomed. Pharmacother.* **2019**, *109*, 1772–1784. [CrossRef]
18. Cobanoglu, U.; Mergan, D.; Dülger, A.C.; Celik, S.; Kemik, O.; Sayir, F. Are Serum Mac 2-Binding Protein Levels Elevated in Esophageal Cancer? A Control Study of Esophageal Squamous Cell Carcinoma Patients. *Dis. Markers* **2018**, *2018*, 3610239. [CrossRef]
19. Sang, Y.; Chen, M.; Luo, D.; Zhang, R.-H.; Wang, L.; Li, M.; Luo, R.; Qian, C.-N.; Shao, J.-Y.; Zeng, Y.-X.; et al. TEL2 suppresses metastasis by down-regulating SERPINE1 in nasopharyngeal carcinoma. *Oncotarget* **2015**, *6*, 29240–29253. [CrossRef]
20. Mustikaningtyas, E.; Juniati, S.H.; Romdhoni, A.C. Intracell Heat Shock Protein 70 Expression and Nasopharyngeal Carcinoma Stage. *Indian J. Otolaryngol. Head Neck Surg.* **2019**, *71*, 321–326. [CrossRef]
21. Janvilisri, T. Omics-Based Identification of Biomarkers for Nasopharyngeal Carcinoma. *Dis. Markers* **2015**, *2015*, 762128. [CrossRef] [PubMed]
22. Chandramouli, K.; Qian, P.-Y. Proteomics: Challenges, Techniques and Possibilities to Overcome Biological Sample Complexity. *Hum. Genom. Proteom.* **2009**, *1*. [CrossRef] [PubMed]
23. De Haan, N.; Pučić-Baković, M.; Novokmet, M.; Falck, D.; Lageveen-Kammeijer, G.; Razdorov, G.; Vučković, F.; Trbojević-Akmačić, I.; Gornik, O.; Hanić, M.; et al. Developments and perspectives in high-throughput protein glycomics: Enabling the analysis of thousands of samples. *Glycobiology* **2022**. [CrossRef]
24. Lao, T.D.; Le, T.A.H. MicroRNAs: Biogenesis, Functions and Potential Biomarkers for Early Screening, Prognosis and Therapeutic Molecular Monitoring of Nasopharyngeal Carcinoma. *Processes* **2020**, *8*, 966. [CrossRef]
25. Spence, T.; Bruce, J.; Yip, K.W.; Liu, F.-F. MicroRNAs in nasopharyngeal carcinoma. *Chin. Clin. Oncol.* **2016**, *5*, 17. [CrossRef] [PubMed]

26. Wang, S.; Claret, F.X.; Wu, W. MicroRNAs as therapeutic targets in nasopharyngeal carcinoma. *Front. Oncol.* **2019**, *9*, 756. [CrossRef]
27. Wen, W.; Mai, S.J.; Lin, H.X.; Zhang, M.Y.; Huang, J.L.; Hua, X.; Lin, C.; Long, Z.Q.; Lu, Z.J.; Sun, X.Q.; et al. Identification of two microRNA signatures in whole blood as novel biomarkers for diagnosis of nasopharyngeal carcinoma. *J. Transl. Med.* **2019**, *17*, 186. [CrossRef]
28. Zhu, Q.; Zhang, Q.; Gu, M.; Zhang, K.; Xia, T.; Zhang, S.; Chen, W.; Yin, H.; Yao, H.; Fan, Y.; et al. MIR106A-5p upregulation suppresses autophagy and accelerates malignant phenotype in nasopharyngeal carcinoma. *Autophagy* **2021**, *17*, 1667–1683. [CrossRef]
29. Xia, H.; Ng, S.S.; Jiang, S.; Cheung, W.K.C.; Sze, J.; Bian, X.-W.; Kung, H.; Lin, M.C. miR-200a-mediated downregulation of ZEB2 and CTNNB1 differentially inhibits nasopharyngeal carcinoma cell growth, migration and invasion. *Biochem. Biophys. Res. Commun.* **2010**, *391*, 535–541. [CrossRef]
30. Zhang, L.; Deng, T.; Li, X.; Liu, H.; Zhou, H.; Ma, J.; Wu, M.; Zhou, M.; Shen, S.; Li, X.; et al. microRNA-141 is involved in a nasopharyngeal carcinoma-related genes network. *Carcinogenesis* **2010**, *31*, 559–566. [CrossRef]
31. Wu, L.; Zheng, K.; Yan, C.; Pan, X.; Liu, Y.; Liu, J.; Wang, F.; Guo, W.; He, X.; Li, J.; et al. Genome-wide study of salivary microRNAs as potential noninvasive biomarkers for detection of nasopharyngeal carcinoma. *BMC Cancer* **2019**, *19*, 843. [CrossRef] [PubMed]
32. Xu, X.; Lu, J.; Wang, F.; Liu, X.; Peng, X.; Yu, B.; Zhao, F.; Li, X. Dynamic Changes in Plasma MicroRNAs Have Potential Predictive Values in Monitoring Recurrence and Metastasis of Nasopharyngeal Carcinoma. *Biomed. Res. Int.* **2018**, *2018*, 7329195. [CrossRef] [PubMed]
33. Lee, R.C.; Feinbaum, R.L.; Ambros, V. The C. elegans heterochronic gene lin-4 encodes small RNAs with antisense complementarity to lin-14. *Cell* **1993**, *75*, 843–854. [CrossRef]
34. Sohel, M.H. Extracellular/Circulating MicroRNAs: Release Mechanisms, Functions and Challenges. *Achiev. Life Sci.* **2016**, *10*, 175–186. [CrossRef]
35. Lee, I.; Ajay, S.S.; Yook, J.I.; Kim, H.S.; Hong, S.H.; Kim, N.H.; Dhanasekaran, S.M.; Chinnaiyan, A.M.; Athey, B.D. New class of microRNA targets containing simultaneous 5′-UTR and 3′-UTR interaction sites. *Genome Res.* **2009**, *19*, 1175–1183. [CrossRef]
36. Lewis, B.P.; Burge, C.B.; Bartel, D.P. Conserved Seed Pairing, Often Flanked by Adenosines, Indicates that Thousands of Human Genes are MicroRNA Targets. *Cell* **2005**, *120*, 15–20. [CrossRef]
37. Grimson, A.; Farh, K.K.-H.; Johnston, W.K.; Garrett-Engele, P.; Lim, L.P.; Bartel, D.P. MicroRNA Targeting Specificity in Mammals: Determinants beyond Seed Pairing. *Mol. Cell* **2007**, *27*, 91–105. [CrossRef]
38. Abdelfattah, A.M.; Park, C.; Choi, M.Y. Update on non-canonical microRNAs. *Biomol. Concepts* **2014**, *5*, 275–287. [CrossRef]
39. Lee, Y.; Kim, M.; Han, J.; Yeom, K.-H.; Lee, S.; Baek, S.H.; Kim, V.N. MicroRNA genes are transcribed by RNA polymerase II. *EMBO J.* **2004**, *23*, 4051–4060. [CrossRef]
40. Cai, X.; Hagedorn, C.H.; Cullen, B.R. Human microRNAs are processed from capped, polyadenylated transcripts that can also function as mRNAs. *RNA* **2004**, *10*, 1957–1966. [CrossRef]
41. Sheng, P.; Fields, C.; Aadland, K.; Wei, T.; Kolaczkowski, O.; Gu, T.; Kolaczkowski, B.; Xie, M. Dicer cleaves 5′-extended microRNA precursors originating from RNA polymerase II transcription start sites. *Nucleic Acids Res.* **2018**, *46*, 5737–5752. [CrossRef] [PubMed]
42. Han, J.; Lee, Y.; Yeom, K.-H.; Kim, Y.-K.; Jin, H.; Kim, V.N. The Drosha-DGCR8 complex in primary microRNA processing. *Genes Dev.* **2004**, *18*, 3016–3027. [CrossRef] [PubMed]
43. Wang, X.; Xu, X.; Ma, Z.; Huo, Y.; Xiao, Z.; Li, Y.; Wang, Y. Dynamic mechanisms for pre-miRNA binding and export by Exportin-5. *RNA* **2011**, *17*, 1511–1528. [CrossRef] [PubMed]
44. Ender, C.; Meister, G. Argonaute proteins at a glance. *J. Cell Sci.* **2010**, *123*, 1819–1823. [CrossRef] [PubMed]
45. Bartel, D.P. MicroRNAs: Genomics, biogenesis, mechanism, and function. *Cell* **2004**, *116*, 281–297. [CrossRef]
46. Xu, J.; Qian, X.; Ding, R. MiR-24-3p Attenuates IL-1 β -Induced Chondrocyte Injury Associated With Osteoarthritis by Targeting BCL$_2$L$_{12}$. *J. Orthop. Surg. Res.* **2020**, *16*, 371. [CrossRef]
47. Cortez, M.A.; Bueso-Ramos, C.; Ferdin, J.; Lopez-Berestein, G.; Sood, A.K.; Calin, G.A. MicroRNAs in body fluids—the mix of hormones and biomarkers. *Nat. Rev. Clin. Oncol.* **2011**, *8*, 467–477. [CrossRef]
48. Kosaka, N.; Iguchi, H.; Ochiya, T. Circulating microRNA in body fluid: A new potential biomarker for cancer diagnosis and prognosis. *Cancer Sci.* **2010**, *101*, 2087–2092. [CrossRef]
49. Cui, M.; Wang, H.; Yao, X.; Zhang, D.; Xie, Y.; Cui, R.; Zhang, X. Circulating MicroRNAs in Cancer: Potential and Challenge. *Front. Genet.* **2019**, *10*, 626. [CrossRef]
50. Yáñez-Mó, M.; Siljander, P.R.-M.; Andreu, Z.; Bedina Zavec, A.; Borràs, F.E.; Buzas, E.I.; Buzas, K.; Casal, E.; Cappello, F.; Carvalho, J.; et al. Biological properties of extracellular vesicles and their physiological functions. *J. Extracell. Vesicles* **2015**, *4*, 27066. [CrossRef]
51. Creemers, E.E.; Tijsen, A.J.; Pinto, Y.M. Circulating MicroRNAs. *Circ. Res.* **2012**, *110*, 483–495. [CrossRef] [PubMed]
52. O'Brien, J.; Hayder, H.; Zayed, Y.; Peng, C. Overview of MicroRNA Biogenesis, Mechanisms of Actions, and Circulation. *Front. Endocrinol.* **2018**, *9*, 402. [CrossRef] [PubMed]
53. Ma, R.; Jiang, T.; Kang, X. Circulating microRNAs in cancer: Origin, function and application. *J. Exp. Clin. Cancer Res.* **2012**, *31*, 38. [CrossRef] [PubMed]

54. Mitchell, P.S.; Parkin, R.K.; Kroh, E.M.; Fritz, B.R.; Wyman, S.K.; Pogosova-Agadjanyan, E.L.; Peterson, A.; Noteboom, J.; O'Briant, K.C.; Allen, A.; et al. Circulating microRNAs as stable blood-based markers for cancer detection. *Proc. Natl. Acad. Sci. USA.* **2008**, *105*, 10513–10518. [CrossRef] [PubMed]
55. Turchinovich, A.; Weiz, L.; Langheinz, A.; Burwinkel, B. Characterization of extracellular circulating microRNA. *Nucleic Acids Res.* **2011**, *39*, 7223–7233. [CrossRef] [PubMed]
56. Zhang, H.; Zou, X.; Wu, L.; Zhang, S.; Wang, T.; Liu, P.; Zhu, W.; Zhu, J. Identification of a 7-microRNA signature in plasma as promising biomarker for nasopharyngeal carcinoma detection. *Cancer Med.* **2020**, *9*, 1230–1241. [CrossRef]
57. Liu, N.; Cui, R.-X.; Sun, Y.; Guo, R.; Mao, Y.-P.; Tang, L.-L.; Jiang, W.; Liu, X.; Cheng, Y.-K.; He, Q.-M.; et al. A four-miRNA signature identified from genome-wide serum miRNA profiling predicts survival in patients with nasopharyngeal carcinoma. *Int. J. Cancer* **2014**, *134*, 1359–1368. [CrossRef]
58. Zheng, X.-H.; Cui, C.; Ruan, H.-L.; Xue, W.-Q.; Zhang, S.-D.; Hu, Y.-Z.; Zhou, X.-X.; Jia, W.-H. Plasma microRNA profiles of nasopharyngeal carcinoma patients reveal miR-548q and miR-483-5p as potential biomarkers. *Chin. J. Cancer* **2014**, *33*, 330–338. [CrossRef]
59. Liu, X.; Luo, H.-N.; Tian, W.-D.; Lu, J.; Li, G.; Wang, L.; Zhang, B.; Liang, B.-J.; Peng, X.-H.; Lin, S.-X.; et al. Diagnostic and prognostic value of plasma microRNA deregulation in nasopharyngeal carcinoma. *Cancer Biol. Ther.* **2013**, *14*, 1133–1142. [CrossRef]
60. Xie, Z.; Chen, G.; Zhang, X.; Li, D.; Huang, J.; Yang, C.; Zhang, P.; Qin, Y.; Duan, Y.; Gong, B.; et al. Salivary MicroRNAs as Promising Biomarkers for Detection of Esophageal Cancer. *PLoS ONE* **2013**, *8*, e57502. [CrossRef]
61. Fan, C.; Tang, Y.; Wang, J.; Xiong, F.; Guo, C.; Wang, Y.; Xiang, B.; Zhou, M.; Li, X.; Wu, X.; et al. The emerging role of Epstein-Barr virus encoded microRNAs in nasopharyngeal carcinoma. *J. Cancer* **2018**, *9*, 2852–2864. [CrossRef] [PubMed]
62. Zeng, Z.; Huang, H.; Huang, L.; Sun, M.; Yan, Q.; Song, Y.; Wei, F.; Bo, H.; Gong, Z.; Zeng, Y.; et al. Regulation network and expression profiles of Epstein-Barr virus-encoded microRNAs and their potential target host genes in nasopharyngeal carcinomas. *Sci. China Life Sci.* **2014**, *57*, 315–326. [CrossRef] [PubMed]
63. Židovec Lepej, S.; Matulić, M.; Gršković, P.; Pavlica, M.; Radmanić, L.; Korać, P. miRNAs: EBV Mechanism for Escaping Host's Immune Response and Supporting Tumorigenesis. *Pathogens* **2020**, *9*, 353. [CrossRef] [PubMed]
64. Wang, M.; Yu, F.; Wu, W.; Wang, Y.; Ding, H.; Qian, L. Epstein-Barr virus-encoded microRNAs as regulators in host immune responses. *Int. J. Biol. Sci.* **2018**, *14*, 565–576. [CrossRef]
65. Lu, T.; Guo, Q.; Lin, K.; Chen, H.; Chen, Y.; Xu, Y.; Lin, C.; Su, Y.; Chen, Y.; Chen, M.; et al. Circulating Epstein-Barr virus microRNAs BART7-3p and BART13-3p as novel biomarkers in nasopharyngeal carcinoma. *Cancer Sci.* **2020**, *111*, 1711–1723. [CrossRef]
66. Wardana, T.; Gunawan, L.; Herawati, C.; Oktriani, R.; Anwar, S.; Astuti, I.; Aryandono, T.; Mubarika, S. Circulation EBV Mir-Bart-7 Relating to Clinical Manifestation in Nasopharyngeal Carcinoma. *Asian Pac. J. Cancer Prev.* **2020**, *21*, 2777–2782. [CrossRef]
67. Gao, W.; Wong, T.; Lv, K.; Zhang, M.; TSANG, R.K.; Chan, J.Y. Detection of Epstein–Barr virus (EBV)-encoded microRNAs in plasma of patients with nasopharyngeal carcinoma. *Head Neck* **2018**, *41*, 780–792. [CrossRef]
68. Tay, J.K.; Lim, M.Y.; Kanagalingam, J. Screening in Nasopharyngeal Carcinoma: Current Strategies and Future Directions. *Curr. Otorhinolaryngol. Rep.* **2014**, *2*, 1–7. [CrossRef]
69. Shannon-Lowe, C.; Rickinson, A. The Global Landscape of EBV-Associated Tumors. *Front. Oncol.* **2019**, *9*, 713. [CrossRef]
70. Ayee, R.; Ofori, M.E.; Wright, E.; Quaye, O. Epstein Barr Virus Associated Lymphomas and Epithelia Cancers in Humans. *J. Cancer* **2020**, *11*, 1737–1750. [CrossRef]

 *genes*

Article

Latent Membrane Protein 1 (LMP1) from Epstein–Barr Virus (EBV) Strains M81 and B95.8 Modulate miRNA Expression When Expressed in Immortalized Human Nasopharyngeal Cells

Barbara G. Müller Coan [1], Ethel Cesarman [2], Marcio Luis Acencio [3] and Deilson Elgui de Oliveira [4,5,*]

1   Biosciences Institute of Botucatu, São Paulo State University (UNESP), Botucatu 18618-689, SP, Brazil; barbaramcoan@gmail.com
2   Department of Pathology and Laboratory Medicine, Weill Cornell Medicine, New York, NY 10065, USA; ecesarm@med.cornell.edu
3   Luxembourg Centre for Systems Biomedicine (LCSB), University of Luxembourg, Belvaux, L-4367 Luxembourg, Luxembourg; marcio.acencio@uni.lu
4   Department of Pathology, Medical School, São Paulo State University (UNESP), Botucatu, SP, 18618-687, Brazil
5   ViriCan, Institute for Biotechnology (IBTEC), São Paulo State University (UNESP), Botucatu, SP, 18607-440, Brazil
*   Correspondence: deilson.elgui@unesp.br; Tel.: +55-14-3880-1573

Abstract: The Epstein–Barr virus (EBV) is a ubiquitous γ herpesvirus strongly associated with nasopharyngeal carcinomas, and the viral oncogenicity in part relies on cellular effects of the viral latent membrane protein 1 (LMP1). It was previously described that EBV strains B95.8 and M81 differ in cell tropism and the activation of the lytic cycle. Nonetheless, it is unknown whether LMP1 from these strains have different effects when expressed in nasopharyngeal cells. Thus, herein we evaluated the effects of EBV LMP1 derived from viral strains B95.8 and M81 and expressed in immortalized nasopharyngeal cells NP69^{SV40T} in the regulation of 91 selected cellular miRNAs. We found that cells expressing either LMP1 behave similarly in terms of NF-kB activation and cell migration. Nonetheless, the miRs 100-5p, 192-5p, and 574-3p were expressed at higher levels in cells expressing LMP1 B95.8 compared to M81. Additionally, results generated by in silico pathway enrichment analysis indicated that LMP1 M81 distinctly regulate genes involved in cell cycle (i.e., *RB1*), mRNA processing (i.e., *NUP50*), and mitochondrial biogenesis (i.e., *ATF2*). In conclusion, LMP1 M81 was found to distinctively regulate miRs 100-5p, 192-5p, and 574-3p, and the in silico analysis provided valuable clues to dissect the molecular effects of EBV LMP1 expressed in nasopharyngeal cells.

Keywords: EBV; LMP1; microRNAs; nasopharyngeal cells; expression profiling

Citation: Müller Coan, B.G.; Cesarman, E.; Acencio, M.L.; Elgui de Oliveira, D. Latent Membrane Protein 1 (LMP1) from Epstein–Barr Virus (EBV) Strains M81 and B95.8 Modulate miRNA Expression When Expressed in Immortalized Human Nasopharyngeal Cells. *Genes* **2022**, *13*, 353. https://doi.org/10.3390/genes13020353

Academic Editors: Giuseppe Iacomino and Fabio Lauria

Received: 14 December 2021
Accepted: 15 February 2022
Published: 16 February 2022

Publisher's Note: MDPI stays neutral with regard to jurisdictional claims in published maps and institutional affiliations.

1. Introduction

Cancers are an important cause of human mortality and lethality in adults worldwide, being the second leading cause of global deaths in 2013 [1]. In 2008, 16% of all cancer cases were related to infection, and over 67% of these were viral agents [2–4]. Epstein–Barr virus (EBV) is a human γ herpesvirus associated with many cancers, notably Burkitt lymphoma and undifferentiated nasopharyngeal carcinoma (NPC) [2,5,6]. EBV is ubiquitous and causes lifelong latent infection in over 90% of adults worldwide [2]. The primary infection is usually asymptomatic but can be associated with clinical signs of infectious mononucleosis, mostly in cases of late exposure to EBV [7].

NPC is strongly associated with EBV infection, notably the undifferentiated form—in which EBV is detected within the neoplastic cells in virtually all cases. NPC is an aggressive epithelial cancer, prone to invade adjacent tissue and lymph nodes [8]. The disease has a poor prognosis, and it was responsible for over 86,000 new cases and 50,000 deaths worldwide in 2012, being more prevalent in men [2,9]. The incidence of NPC changes according to geographic localization, and the disease is endemic in southeast Asia, southwest China, and Micronesia. Besides EBV infection, risk factors for NPC include other environmental exposures (e.g., nitrosamides, tobacco, insufficient ventilation of dwellings) and genetic factors, such as polymorphisms in genes encoding leukocyte antigen (HLA) class I and class II molecules, the heat shock 70 kDa protein (HSP 70) or the polymeric immunoglobulin (Ig) A receptor. Furthermore, the NPC incidence may also be influenced by different EBV strains [10–12].

The progression of EBV-associated cancers can be affected by the activity of viral gene products [13,14]. In this regard, the Latent Membrane Protein 1 (LMP1)—one of the major EBV oncoproteins—induces a variety of changes in cell behavior, including proliferative and survival capabilities [15], altered cell adhesion, extracellular matrix (ECM) and vascular remodeling, and an increase in cell motility. Ultimately, these processes may lead to a more aggressive and metastatic carcinoma [16–18].

Different EBV viral strains can distinctly influence the behavior of infected cells. For instance, EBV strains GP202 (isolated from gastric cancer), B95-8, and AKATA (both from lymphomas) were reported to induce cell growth more efficiently than M81 (obtained from NPC), YCCEL1, or SNU719 (both from gastric cancer) [19]. M81 was also reported to behave differently from the viral prototype strain B95-8 (isolated from Burkitt lymphoma), SNU719, GP202, or YCCEL1, showing a higher affinity for epithelial cell infection (consistent with its epithelial cancer origin) and higher capability of lytic cycle induction [19,20]. Notably, chromosomal instability can be achieved by a high expression of the lytic gene BZLF1 and recurring induction of the lytic cycle, leading to increased transformation properties and higher risk for NPC [19].

LMP1 may affect endogenous microRNA (miRNA) expression in both B cells and epithelial cells, causing effects relevant to cancer progression [21–24]. Essentially all phenomena related to cancer progression can be regulated by microRNAs [25–29], and different EBV strains may show unique biological properties. Thus, it is plausible to assume that distinct EBV strains may regulate a unique set of human miRNAs, with possible consequences for EBV-induced carcinogenesis. Based on these premises, in this study, we investigated whether nasopharyngeal cells expressing LMP1 from M81 or B95.8 strains differ in miRNAs expression and examined the possible consequences regarding cell signaling pathways by in silico analysis.

2. Materials and Methods

2.1. Cell Culture

The cell lines HEK293 (RRID:CVCL_0045) and NP69^{SV40T} (hereafter referred to as NP69; RRID:CVCL_F755;) were used in this study. HEK293 is an immortalized human embryonic kidney cell line harboring DNA fragments from type 5 adenovirus [30]. NP69 cells were generated by immortalizing primary nasopharyngeal epithelial cells with the SV40 large T antigen, and they retain many characteristics of normal nasopharyngeal cells (e.g., the profile of keratin expression and responsiveness to TGFβ inhibition). Despite some genetic alterations also found in nasopharyngeal carcinomas, NP69 is non-tumorigenic and highly responsive to the EBV LMP1 expression [31]. Based on that, this cell line was used to assess the expression of human microRNAs and behavioral changes induced by expression of EBV LMP1 derived from strain B95.8 and M81.

Both cell lines had their genetic identity assured by short tandem repeats (STRs) profiling using the GenePrint 10 system (Promega, Madison, WI, USA), and they were verified to be free of mycoplasma contamination by a PCR-based assay. HEK293 cells were maintained in Dulbecco's Modified Eagle Medium (DMEM), supplemented with 10% Fetal Bovine Serum (FBS) and 0.4% gentamicin, while NP69 cells were cultivated with Keratinocyte Serum-Free medium (K-SFM-Life Technologies, Carlsbad, CA, USA) supplemented with 5% FBS, 25 μg/mL of Bovine Pituitary Extract (BPE), 0.2 ng/mL of Epidermal Growth Factor (EGF), and 1% Gentamicin. Both were cultivated at 37 °C in a humid atmosphere with 5% CO_2.

2.2. LMP1 Constructs and Cell Transfections

The open reading frames (ORF) for LMP1 derived from EBV strains B95.8 and M81 were retrieved from recombinant virus constructs kindly provided by Prof. Henri-Jacques Delecluse (German Cancer Research Center, Heidelberg, Germany). Recombinant DNA plasmids were assembled with the F factor-based prokaryotic replicon, pMBO131, with flanking regions for homologous recombination of the viral genome [32,33]. The obtained ORFs were transferred to the pEF1α-IRES-ZsGreen1 backbone vector (Clontech, Mountain View, CA, USA), allowing the detection of LMP1-expressing cells due to the simultaneous expression of a green fluorescent protein (GFP) reporter. The EBV LMP1 ORFs were amplified from the original vectors using primers harboring restriction sites for the enzymes EcoRI and BamHI, used for unidirectional cloning (see Additional File 1: Table S1).

The new LMP1-encoding constructs (Supplementary Figure S1), dubbed pZsG-LMP1-B95.8 and pZsG-LMP1-M81, were validated by Sanger DNA sequencing and the sequences were deposited in GenBank (Accession codes #MN062162 and #MN062163). For transient cellular transfections, 0.7×10^5 cells were seeded in 24-well plates 24 h prior to the assay, carried out with the pZsGreen backbone, pZsG-LMP1B95.8 or pZsG-LMP1M81 constructs, and Lipofectamine 3000 (Thermo Fisher Scientific, Waltham, MA, USA) as the transfection reagent, following the manufacturer's recommendations. Subsequent assays were performed 48 h post-transfection (Supplementary Figure S2).

2.3. Luciferase Assay and Cell Migration In Vitro

Both assays were performed with 0.7×10^5 cells (HEK293 or NP69) seeded into 24-well plates 24 h prior to transfection, and the cells were used for experiments 48 h post-transfection. The luciferase reporter assay was performed with HEK293 cells (given its high susceptibility to liposomal transfection) to validate the EBV LMP1 activity using its effect on NF-κB activation as a proxy. Briefly, HEK293 cells were co-transfected with the pEF1α-IRES-ZsGreen1 backbone vector or the LMP1 B95.8 and M81 constructs, along with pcDNA3.1-NF-κBLuc, for the expression of firefly luciferase under the control of a NF-κB-responsive element) [34], and Promega pGL4.74 (for constitutive expression of *Renilla* luciferase, to normalize results for the transfection efficiency levels). Results were generated in 96-well white plates using the Dual-Luciferase® Reporter Assay System and the GloMax Explorer device (both manufactured by Promega).

The rates of cell migration in vitro were assessed with the scratch wound healing assay. Images were taken at the time of the scratch and 24 h later (48 and 72 h post-transfection) were evaluated using the TScratch software [35] to estimate the percentage of closure of the wounded area in the cell monolayer.

For all assays, the results were obtained from three independent experiments performed with triplicates. The analysis of data was performed with Student's *t*-test, taking a *p*-value ≤ 0.05 as statistically significant.

2.4. MicroRNA Expression Analysis

The analysis of miRNA expression was performed with a custom panel of 91 human miRNAs (see Supplementary Table S2), selected based on literature data showing changes in expression observed in NPC or other human cancers, associated or not with EBV. These experiments were performed with NP69 cells transfected with the pEF1α-IRES-ZsGreen1 backbone vector and the constructs pZsG-LMP1-B95.8 and pZsG-LMP1-M81. After transfection, the cells were subjected to FACS (Supplementary Figure S3 and Table S3) to recover GFP-positive cells, aiming to enrich cells successfully transfected with the vectors for LMP1 expression. Briefly, the transfected cells were cultivated for 48 h post-transfection, then detached and flowed through a 70 μm filter. GFP-positive cells were sorted using a BD FACS Aria III (BD Bioscience, San Jose, CA, USA), pelleted by centrifugation, and cryopreserved in liquid nitrogen. Afterward, the enriched NP69 transfected cells were subjected to miRNAs-enriched RNA extraction with the miRNeasy® Mini Kit, reverse-transcribed using miScript II RT Kit, then validated with the miScript® QC PCR Array system (all Qiagen products), following the manufacturer's instructions. Upon successful quality validation of samples, the qPCR assays were performed using a customized miScript miRNA PCR Array platform (Qiagen, Valencia, CA, USA) with the threshold and baselines determined by the manufacturer's recommendations. Statistics analyses were performed using the online software provided by the GeneGlobe Data Analysis Center (Qiagen's mirScript miRNA PCR Arrays & Assays), using the recommended parameters and three endogenous controls. The experiments were performed with biological triplicates (Supplementary Tables S4 and S5).

2.5. MiRNA's Targets Prediction and Pathway Enrichment Analysis

The miRNAs showing differential expression (*p*-value $\leq$ 0.05) with a fold change of $\pm$ 1.2 were selected based on the following comparation sets: (1) EBV LMP1 B95.8 vs. control (backbone vector); (2) EBV LMP1 M81 vs. control; and (3) EBV LMP1 M81 vs. EBV LMP1 B95.8. The selection of cut-off points was performed as reported previously [36,37]. The differentially expressed miRNAs from sets 1, 2, and 3 were subjected to target prediction analysis using the mirDIP online tool [38,39], and the top 1% target genes were subjected to the pathway-enrichment analysis using the ReactomeFIViz [40] plugin in the Cytoscape software [41]. Further details are available in the Supplementary Material.

3. Results

3.1. Cells Expressing EBV LMP1 from Viral Strains B95.8 and M81 Behaves Similarly in Terms of NF-κB Activation and Cell Migration Rates In Vitro

EBV LMP1 is known to activate NF-κB, and this property can be used as a proxy to validate the functional integrity of LMP1 expressed ectopically. To verify whether the constructs generated to allow the expression of functional LMP1, the activation of NF-kB was measured in HEK293 cells using a luciferase reporter assay with luciferase being expressed in an NF-kB-dependent manner. HEK293 cells expressing EBV LMP1 showed a ninefold increase in NF-κB activation compared to the cells transfected with the empty, backbone vector, and a significant difference was observed considering the activation levels obtained by LMP1 from strains B95.8 and M81 (Figure 1A). Thus, the viral oncoprotein expressed by both LMP1 constructs is functional, and the LMP1 encoded by either of the viral strains activates NF-κB at similar levels.

Furthermore, transiently transfected NP69 cells expressing LMP1, as confirmed by RT-qPCR, showed higher migration rates compared to the control (15% and 26% for M81 or B95.8 variants, respectively, (see Figure 1B)). Additionally, both EBV LMP1-encoding constructs could induce increased cell migration in vitro, but no significant differences were found comparing the LMP1 derived from EBV strains M81 and B95.8.

3.2. EBV LMP1 Modulate miRNA Expression in NP69^{SV40T} Cells

Although it is known that the EBV strains B95.8 and M81 have distinct biological features [20], it remains to be better understood whether their differences also impact the oncogenic properties of the virus. We aimed to investigate this by interrogating whether LMP1 from different viral strains has different effects on miRNA expression, which may also give relevant clues on broad cellular effects of this major EBV oncoprotein. Changes in the levels of selected miRNAs were evaluated in NP69 cells expressing LMP1 derived from viral strains B95.8 or M81, enrichment for GFP-positive cells by FACS (Supplementary Figure S3 and Table S3). The customized qPCR array used allowed the evaluation of 91 selected human miRNAs (Supplementary Table S2), and the raw results were subjected to data normalization prior to the downstream analysis (Supplementary Tables S4 and S5, respectively).

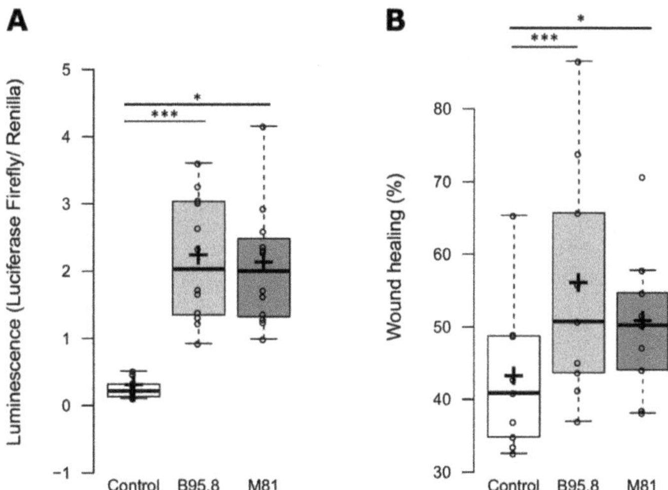

Figure 1. Effects of expression of EBV LMP1 variants M81 or B95.8 in transfected cells in vitro: Luciferase and migration assays with HEK293 (**A**) and NP69SV40T (**B**) cells transfected with pZsGreen (control), pZsG-LMP1-B95.8 or pZsG-LMP1-M81 vectors. (**A**) NF-κB activity assessed via Luciferase reporter assay showing a 9-fold increase in LMP1 expressing HEK293 cells. (**B**) In vitro migration assay showing an increase in migration of 26% and 15% of NP69SV40T cells transfected with LMP1 B95.8 or M81 respectively. The mean values for standard error obtained from at least three independent experiments are shown. p values < 0.05 and < 0.005 are indicated by * and ***, respectively.

Three comparisons were performed: (1) EBV LMP1-B95.8 compared to control (dubbed B95.8 vs. Ctrl); (2) EBV LMP1-M81 compared to control (M81 vs. Ctrl); and (3) EBV LMP1 M81 compared to EBV LMP1 B95.8 (M81 vs. B95.8). We observed the downregulation of 47 miRNAs in B95.8 vs. Ctrl (fold regulation: −1.7 to −2.9; Figure 2A), 2 miRNAs in M81 vs. Ctrl (fold regulation: −1.6 to −2.1; Figure 2B), and upregulation of 3 miRNAs comparing M81 vs. B95.8 (fold regulation varying from 1.7 to 2; Figure 2C). The miR-132-3p was found downregulated in both B95.8 vs. Ctrl and M81 vs. Ctrl comparisons (fold regulation: −2.4 and −2.1, respectively). It is worth noting that the miRNA upregulation in M81 vs. B95.8 indicates that, compared to the control, the EBV LMP1 from B95.8 strain induced a more robust downregulation of miRNAs compared to LMP1 derived from strain M81.

Next, we sought to perform pathway enrichment analysis in silico to extrapolate biological significance for the obtained miRNA expression profiles. For each given comparison (B95.8 vs. Ctrl, M81 vs. Ctrl, and M81 vs. B95.8), we investigated the number of predicted target genes of all differentially expressed miRNAs. We observed 11,045, 3136, and 629 predicted genes for B95.8 vs. Ctrl, M81 vs. Ctrl, and M81 vs. B95.8, respectively. Moreover, 315 (2.8%) targets were shared between all three comparison sets, whereas 2639 (23.5%) were shared only between B95.8 vs. Ctrl and M81 vs. Ctrl (Figure 3A).

The pathway enrichment analysis was used to obtain insights on the most relevant cellular pathways disturbed by the regulation of miRNAs by EBV LMP1. We also evaluated whether a given pathway identified was unique to a given comparison, aiming to identify processes regulated specifically by LMP1 derived from viral strain B95.8 or M81. In the B95.8 vs. Ctrl comparison set, we found pathways for genes involved in cell–cell communication, such as integrins and cadherins (Figures 3B, 4A and 5A), while the M81 vs. Ctrl comparison showed pathways featuring genes involved in the metabolism of RNA (RNA processing), such as *NCBP1, NUP43, NUP58, POM121,* and *RANBP2* (gene IDs in Supplementary Table S8) (Figures 3B, 4A and 5A, and Supplementary Table S7). Furthermore, Death receptors, Integrin and Leptin signaling pathways were found for comparison M81 vs. Ctrl; those three pathways have the gene *SOS1* in common, which encodes a protein that promotes the exchange of Ras-bound GDP by GTP, favoring cell proliferation. Additionally, the M81 vs. B95.8 comparison showed some unique predicted pathways, including involvement in DNA replication through genes *CDC7*, *ORC4*, *MCM10*, and *MCM9* (Figure 5B).

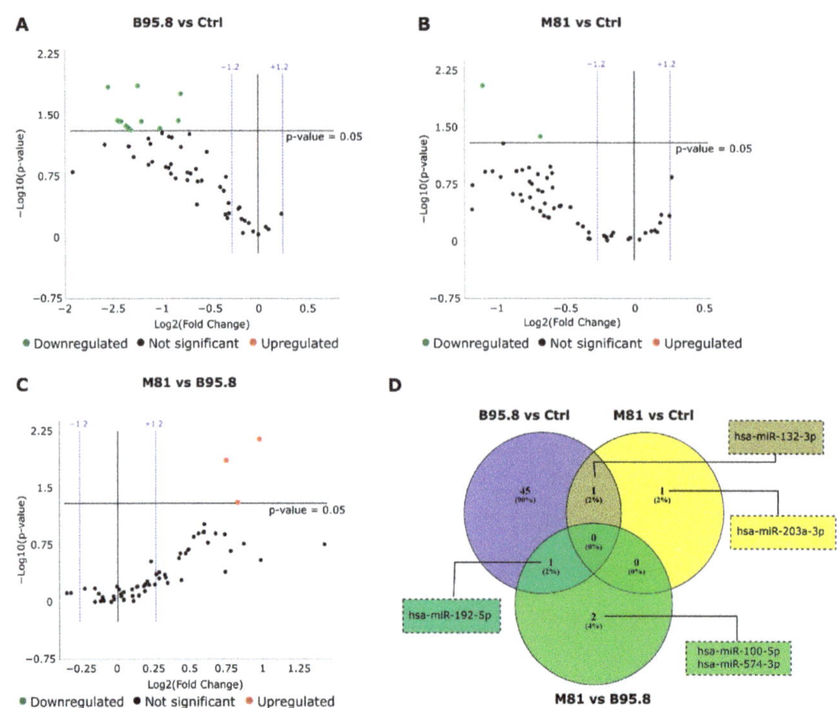

Figure 2. Effects of expression of EBV LMP1 derived from viral strains B95.8 and M81 on miRNA expression in immortalized nasopharyngeal cells NP69^{SV40T}. miRNA RT-qPCR array with 91 selected miRNAs was performed on cells transfected with pZsGreen (control), pZsG-LMP1-B95.8 or pZsG-LMP1-M81 vectors. All results were obtained from three independent experiments. (**A**) A total of 47 miRNAs were downregulated in group LMP1 B95.8 vs. control vector (B95.8 vs. Ctrl), with fold regulation between −1.7 and −2.9. (**B**) 2 miRNAs were downregulated in group LMP1 M81 vs. control (M81 vs. Ctrl) with fold regulation of −1.6 and −2.1. (**C**) A total of 3 upregulated miRNAs were seen in group LMP1 M81 vs. LMP1 B95.8 (M81 vs. B95.8) with fold regulation between 1.7 and 2. (**D**) miRNAs exclusive or commonly altered between groups. MiR-132-3p was found for both B95.8 vs. Ctr or M81 vs. Ctrl and miR-192-5p was altered in groups B95.8 vs. Ctrl and M81 vs. B95.8. Selected miRNAs had *p*-value ≤ 0.05 and fold regulation of ± 1.2.

Both comparison groups involving EBV LMP1 variant M81 (M81 vs. Ctrl/M81 vs. B95.8) were predicted to be uniquely involved in the cell cycle (via *RB1*) and organelle biogenesis and maintenance (via *GABPB1* and *PRKAA2*) (Figures 5A, 5(B5) and Figure S7). Additionally, the LMP1 from the M81 strain was predicted to be involved in mTOR signaling (Figures 3D and 5(D3)), via genes *PRKAA2* and *PPM1A*, for instance. All three comparison sets were predicted to regulate the Wnt signaling through different gene sets, but genes *FZD5* and *CAV1* appear to be regulated in all settings (Figures 4E and 5(B3)). The comparisons M81 vs. Ctrl and M81 vs. B95.8 also were predicted to regulate pathways involved in chromatin organization, extracellular matrix organization, program cell death, vesicle-mediated transport, metabolism of proteins, gene transcription, immune system regulation, and important pathways in cancer, such as MAPK, TGF-β, WNT, VEGF, and IGF1R (Figures 3D, 4E–G and 5).

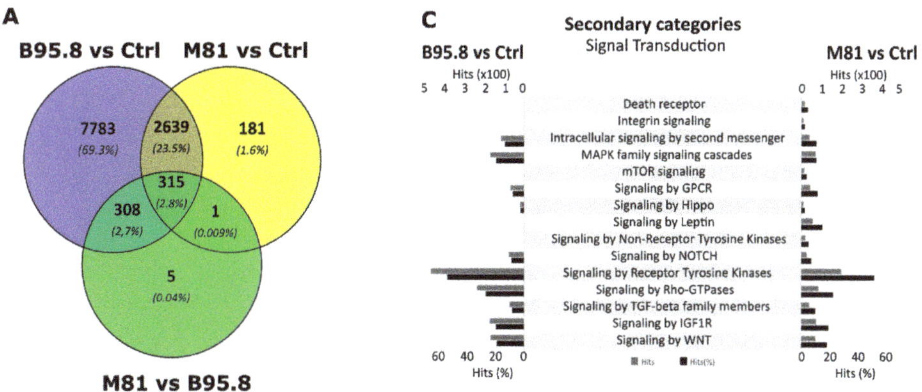

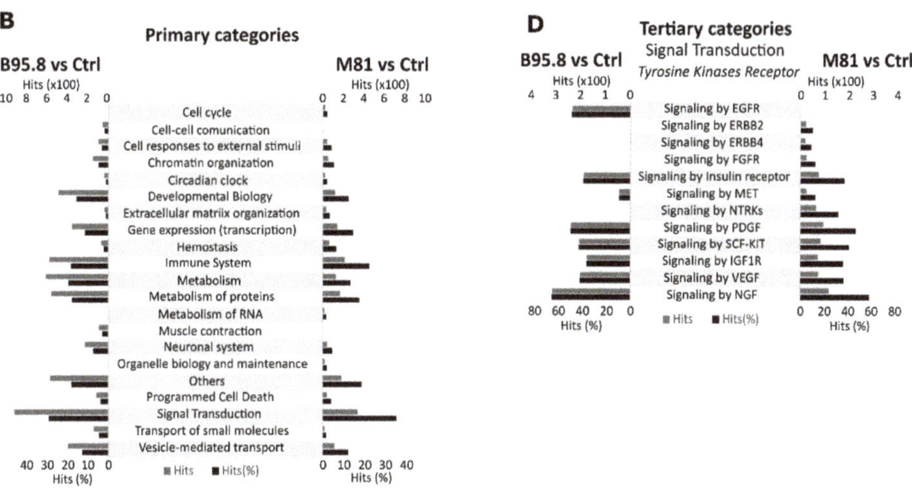

Figure 3. Analysis of target genes from differentially expressed miRNAs in NP69SV40T cells transfected with EBV LMP1 from B95.8 or M81 variants. (**A**) Number and percentage of unique or commonly target genes considering three comparison groups: EBV LMP1 B95.8 vs. Ctrl, EBV LMP1 M81 vs. Ctrl, LMP1 M81 vs. viral LMP1 B95.8. (**B**) Primary categories, (**C**) Secondary categories of "Signal Transduction" and (**D**) Tertiary categories of "Tyrosine kinase receptor signaling" altered by predicted gene targets of miRNA regulated by LMP1 variants B95.8 (**left**) or M81 (**right**). In (**B**–**D**) the light and dark grey indicate, respectively, the absolute number of hits or their percentage considering the respective higher category.

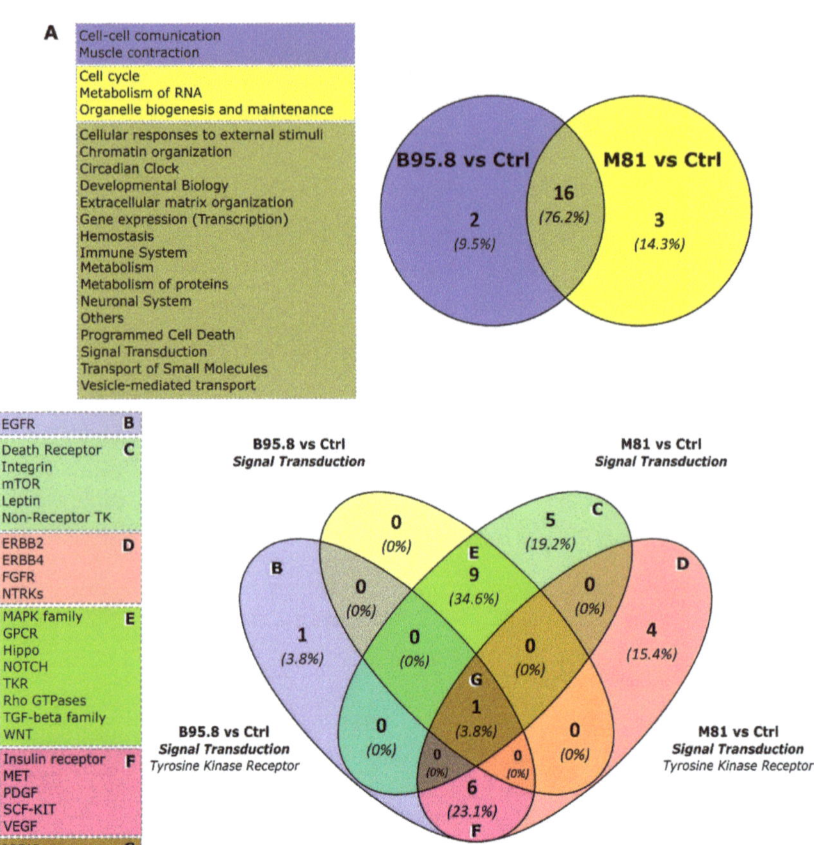

Figure 4. Results of in silico pathway enrichment analysis from predicted target genes of deregulated miRNAs in NP69SV40 cells transfected with LMP1 B95.8, LMP1 M81 or the control vector. (**A**) Unique primary categories for EBV LMP1 B95.8 vs. Ctrl (in blue), LMP1 M81 vs. Ctrl (yellow), and shared categories for by LMPs B95.8 and M81 (overlap). (**B**) Unique "Tyrosine kinase receptor" categories from B95.8 vs. Ctrl comparison. (**C**) Unique "Signal transduction" categories from M81 vs. Ctrl comparison. (**D**) Unique "Tyrosine kinase receptor" categories from M81 vs. Ctrl comparison. (**E**) "Signal transduction" categories shared between B95.8 vs. Ctrl and M81 vs. Ctrl comparisons. (**F**) "Tyrosine kinase receptor" categories shared between B95.8 vs. Ctrl and M81 vs. Ctrl comparisons. (**G**) Category shared between B95.8 vs. Ctrl and M81 vs. Ctrl comparisons, present in "Signal transduction" and Unique "Tyrosine kinase receptor" pathways.

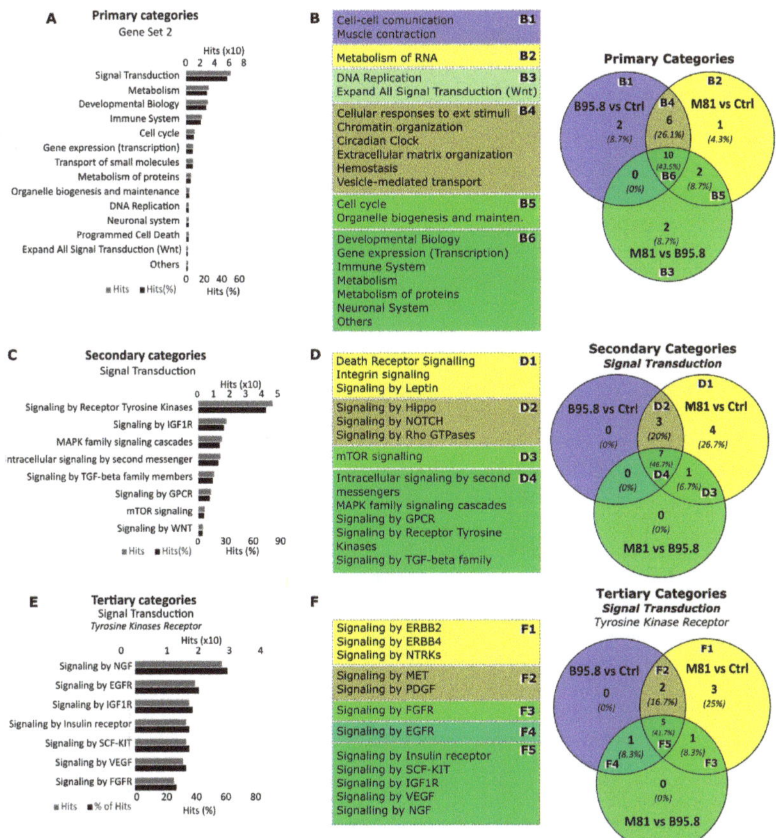

Figure 5. Pathway enrichment analysis of genes targeted by miRNAs deregulated by EBV LMP1 in NP69SV40T cells. Most important primary categories altered by predicted genes comparing LMP1 M81 vs. B95.8. (**B**) Unique and commonly deregulated primary categories between the three comparisons performed. (**C**) "Signal transduction" secondary categories found for LMP1 M81 vs. B95.8. (**D**) Unique and commonly deregulated secondary categories inside "Signal Transduction" for the three comparisons performed. (**E**) Subcategories (tertiary) for "Tyrosine Kinases Receptor" altered by predicted genes comparing LMP1 M81 vs. B95.8. (**F**) Unique and commonly deregulated tertiary categories for "Tyrosine Kinases Receptor" considering the three comparisons performed. In (**A**,**C**,**E**), in light and dark grey indicates, respectively, the absolute number of hits or its percentage considering the respective higher category (Gene set #2, in the case of primary categories depicted in (**A**)).

4. Discussion

There is compelling evidence showing that the viral LMP1 oncoprotein contributes to the progression of EBV-associated cancers [14]. For instance, LMP1 increases migration and invasion of epithelial cells through different mechanisms, such as changes in cell adhesion and motility due to regulation of N-cadherin and integrin-α5 expression, culminating in both individual and collective migration of immortalized nasopharyngeal cells [42]. LMP1 also directly increases the sphingosine kinase 1 (SPHK1) enzyme, which was implicated in a poor prognosis for NPC [43]. SPHK1 activates sphingosine-1-phosphate (S1P), causing increased migration of NPC cells associated with AKT activation [44]. Furthermore, LMP1 represses the Tissue Inhibitor of Metalloproteinase-3 (TIMP-3), leading to extracellular matrix degradation [45], and induces extracellular secretion of HIF1α in exosomes, which ultimately causes epithelial-to-mesenchymal transition (EMT), migration and invasion of EBV-negative nasopharyngeal cells, and NPC [46].

In this study, we found that the expression of LMP1 derived from EBV strains B95.8 and M81 in immortalized nasopharyngeal cells NP69^{SV40T} changes the expression of endogenous miRNAs. Cells expressing LMP1 variant M81 compared to variant B95.8 showed significant upregulation of the human miRNAs 100-5p, 192-3p, and 574-3p (Figure 2C). MiR-100-5p was previously described to behave either as a tumor suppressor or oncomir, in a context-dependent manner. The upregulation of miR-100-5p was implicated on better prognosis in esophageal cancer [47], a decrease in cisplatin resistance in lung cancer [48], and inhibition of tumorigenesis, cell migration, and invasion for human mammary epithelial cells [49]. However, it was also associated with effects expected to favor cancer development and progression, such as resistance against apoptosis in prostate cancer [50] and induction of EMT in human mammary cells [49]. MiR-192-5p seems to behave as a more typical oncomir: its expression stimulates migration, invasion, and proliferation of hepatocellular carcinoma cells in vitro [51], and it was associated with tamoxifen resistance in mammary carcinoma and even higher cancer recurrence and metastasis in both hepatocellular and mammary carcinoma [51,52]. Conversely, miR-574-3p was described with effects mimicking those of tumor suppressor genes. For instance, it was implicated in inhibition of gastric cancer cell proliferation, migration, and invasion [53]. However, its role in NPC still needs elucidation since miRNA effects can be tissue- and context-dependent.

The putative cellular effects can be appreciated by the results obtained by miRNA target prediction and the pathway-enrichment analysis, both performed in silico. The biological processes involved (Figures 4 and 5) include gene expression (e.g., chromatin organization, RNA-pol II transcription, and post-translational protein modification), intracellular signaling pathways (e.g., programmed cell death, tyrosine kinase receptors, MAPK, TGF-β, and Rho GTPases signaling), cellular stress and senescence, modulation of the immune system, and phenomena associated to cell–cell communication (e.g., ECM organization and vesicle transport). This is consistent with previously published data about a range of effects associated with EBV LMP1 expression, including the induction of vasculogenic mimicry in vitro via VEGFA; induction of IGF1 expression and cell proliferation in vitro; MAPK pathway regulation, leading to cell motility and invasion; blocking of TGF1 cell-growth inhibition in vitro; and Wnt pathway regulation, both in mice and in human tumor samples of EBV-positive NPC [54–59].

Some common features were observed when evaluating the effects of LMP1 expression on cellular miRNAs, irrespective of the variant considered (B95.8 vs. Ctrl and M81 vs. Ctrl comparisons). For instance, EBV LMP1 can downregulate miRNAs that are implicated in epigenetic regulation by targeting DNA methyltransferases (DNMTs) genes. It was previously found that EBV-infected cells show downregulation of DNMT1 and upregulation of DNMT3a, admittedly due to LMP1 expression [60,61]. We found that the transcript for DNMT3a is targeted by 10 miRNAs in B95.8 vs. Ctrl, by miR-203a-3p in M81 vs. Ctrl, and by miR-192-5p in M81 vs. B95.8 comparison sets. The downregulation of DNMT3a reduces DNA methylation in specific genomic regions, increasing the expression of *FOXA2* and *HNF4A* (Gene IDs in Supplementary Table S8) [62]. We found that *FOXA2* is targeted by miR-141-3p, while *HNF4A* is targeted by miR-135b-5p and miR-34c-5p; these three human miRNAs were found to be downregulated in cells expressing the EBV LMP1 variant B95.8, compared to control (B95.8 vs. Ctrl). *FOXA2* was previously implicated in cell proliferation, cancer stem cell maintenance, and an increase in relapse in triple-negative breast cancer [63], while *HNF4A* was related to an increase in lymph node and distant metastasis in colon cancer [64,65]. Since all those miRNAs are downregulated in the presence of LMP1, it is expected that the above-described effects will increase, suggesting that LMP1 from both EBV strains B95.8 and M81 regulates the methylation status in cells using a different set of miRNAs.

Even though this study provides relevant clues on common and unique effects features of the EBV LMP1 derived from viral strains M81 and B95.8 in nasopharyngeal cells, some limitations must be carefully considered. First, the model used in this study was based on EBV LMP1 expression in NP69^{SV40T} cells under the control of a CMV promoter, which allows robust ectopic expression of the viral transgene. We initially sought to generate stably transfected cells expressing EBV LMP1, but only transient expression was possible because cells constitutively expressing LMP1 show a high level of cell death after a few weeks. Despite its known antiapoptotic effects, at unconstrained high levels, the EBV LMP1 may actually cause the cell to die by apoptosis [66–68]. To circumvent this issue and minimize the detrimental effects of suboptimal transfection efficiencies, we performed enrichment of LMP1-expressing cells by FACS prior to miRNA expression analysis. Another limitation is that our results based on in silico analysis were not validated experimentally in this study, but other ongoing studies in our laboratory aim to address this. Finally, we observed a much higher number of altered miRNAs when evaluating cells expressing LMP1 B95.8 compared to cells expressing LMP1 M81. As the miRNA panel in this study was defined based on previously published results and considering that there are much more data accumulated for EBV genotype B95.8 compared to M81, we cannot rule out some bias towards the results for miRNAs regulated by B95.8, which is the most studied EBV genotype. Nevertheless, we aimed to reduce any possible bias in this matter by also including in the panel miRNAs reported as altered in cancers in general, not only associated with EBV, and considering both in vitro and in vivo studies.

Despite these limitations, the results obtained consistently showed different profiles of miRNA expression induced by LMP1 derived from viral strains B95.8 and M81 in our model. This allowed us to identify the miRs 100-5p, 192-5p, and 574-3p, as microRNAs with putative roles in the EBV-induced transformation of nasopharyngeal epithelial cells. The LMP1 from EBV strains B95.8 and M81 regulate different miRNA sets, and the data obtained from the in silico analysis suggested putative biological consequences, either some unique for one of the LMP1 variants, but also commonalities, such as changes in cellular pathways involving MAPKs and VEGFA, modulation of the immune system, and apoptosis. Of note, it was previously reported that the EBV strain M81 has a higher capacity to induce the lytic cycle in infected cells [20]. Accordingly, the target gene prediction and pathway enrichment analysis performed in silico in this study indicated that, compared to B95.8, the LMP1 variant M81 had a higher number of genes involved in cell death and survival regulation, suggesting that, to some extent, the M81 biological behavior may be related to its EBV LMP1 variant and the effects of this viral oncoprotein on modulation of cellular miRNAs.

5. Conclusions

This study showed that LMP1 derived from EBV strains B95.8 and M81 can modulate different sets of miRNAs when expressed in NP69 nasopharyngeal cells. The results reported here contribute to a better understanding of how LMP1 from different viral strains may influence the behavior and phenotype of EBV-infected cells, and also indicate novel putative genes and cellular pathways that may play an important role in the pathogenesis of cancers associated with EBV. These differentially expressed miRNAs can also have a role in NPC diagnosis or management since these molecules are known to be found in plasma samples. However, the LMP1 effects on the regulation of endogenous miRNAs are still poorly explored; future studies may focus on how specific miRNAs deregulated by LMP1 affect the cell signaling pathways, which is key to further clarifying the biological and oncogenic properties of this major EBV oncoprotein.

Supplementary Materials: The following supporting information can be downloaded at: https://www.mdpi.com/article/10.3390/genes13020353/s1, Figure S1: Comercial vector pZsGreen after the LMP1 gene insertion (dark green). The vector was assembled through conventional cloning to access differences between LMP1 variants from EBV in selected miRNAs expression after transient transfection in immortalized nasopharyngeal cells (NP69); Figure S2: Analysis of LMP1 expression and GFP positivity in HEK293 or NP69 cells after 48h of transfection with the assembled constructs pZsG-LMP1-B95.8 or pZsG-LMP1-M81. PCR amplification of cDNA from (A) HEK293 cells expressing LMP1 from the construct pZsG-LMP1-B95.8 and pZsG-LMP1-M81 and (B) transfected NP69 cells, indicating the 30bp deletion present in LMP1 variant from M81 strain. expressing LMP1 from the construct pZsG-LMP1-B95.8 and pZsG-LMP1-M81. (C) Example of transfection rates in HEK293 and NP69 cells transfected with pZsGreen vector; Figure S3: FACS of NP69SV40 cells after 48h of transient transfection with pZsGreen (control), pZsG-LMP1-B95.8 or pZsG-LMP1-M81 vectors; Figure S4: Schematic drawing of steps followed after transfection of NP69SV40 cells transfected with pZsGreen (control), pZsG-LMP1-B95.8 or pZsG-LMP1-M81 vectors. and sorted by FACS with pZsGreen vector as control or containing the coding region from LMP1 protein from EBV variants B95.8 or M81. First, 48 h after transfection, GFP positive cells were sorted by FACs, followed by RNA extraction, cDNA production and RT-qPCR array of 91 pre-selected miRNAs. Then, target prediction analysis of differentially expressed miRNAs and genes with high scores were selected for Pathway enrichment analysis; Figure S5: Analysis of target genes from differentially expressed miRNAs in NP69SV40T cells transfected with EBV LMP1 from B95.8 or M81 variants. Example of primary, secondary and tertiary categories organization. Panels C, D and E show results of in silico pathway enrichment analysis of predicted target genes from deregulated miRNAs in NP69SV40 cells transfected with LMP1 B95.8, LMP1 M81 or the control vector; Table S1: Primers sequences, reaction components and cycling utilized for checking LMP1 presence, sequencing and cloning; Table S2: MiRNAs related to NPC, LMP1, EBV, or cancer in general, selected from literature separated by its probable function as tumor suppressor, oncomiR or dual function. Those miRNAs were used in the qPCR miRNA array to analyze their expression in NP69 cells transfected with LMP1 from two distinct EBV strains, B95.8A or M81; Table S3: Results (Cell count and percentage) from FACS of NP69SV40 cells after 48h of transient transfection with pZsGreen (control), pZsG-LMP1-B95.8 or pZsG-LMP1-M81 vectors; Table S4: Raw data (CT) after qPCR array using cDNA from NP69SV40 cells transfected with pZsGreen (control), pZsG-LMP1-B95.8 or pZsG-LMP1-M81 vectors and sorted by FACS; Table S5: Results of RT-qPCR arrays analysis and verification of up or downregulated miRNAs of NP69SV40 cells transfected with pZsGreen (control), pZsG-LMP1-B95.8 or pZsG-LMP1-M81 vectors and sorted by FACS; Table S6: Analysis of pathway enrichment analysis using predicted genes from gene set #2 targeted by differentially expressed miRNAs from NP69SV40 cells transfected with pZsGreen (control), pZsG-LMP1-B95.8 or pZsG-LMP1-M81 vectors and sorted by FACS. Each group has pathways represented in hits (gene in each pathway) and % hits (percentage of hits compared to next higher category); Table S7: Descriptive list of predicted genes found by mirDIP encountered in each primary category after pathway enrichment analysis using differentially expressed miRNAs from NP69SV40 cells transfected with pZsGreen (control), pZsG-LMP1-B95.8 or pZsG-LMP1-M81 vectors and sorted by FACS; Table S8: List of Gene IDs cited on main text. Predicted genes were from differentially expressed miRNAs of transfected immortalized nasopharyngeal cells with LMP1 variant from EBV strain B95.8 or M81.

Author Contributions: Conceptualization, B.G.M.C. and D.E.d.O.; formal analysis, B.G.M.C.; funding acquisition, D.E.d.O.; methodology, B.G.M.C. and D.E.d.O.; resources, B.G.M.C.; supervision, D.E.d.O.; validation, B.G.M.C.; writing—original draft, B.G.M.C.; writing—review and editing, E.C., M.L.A. and D.E.d.O. All authors have read and agreed to the published version of the manuscript.

Funding: This research was funded by FAPESP through BGMC scholarship (Proc. MS 2014/14678-5), and also funded with research grants awarded to DEO by the São Paulo Research Foundation–FAPESP (Proc. AP 2014/17326-9 and AP 2017/23393-9) and the State University of Sao Paulo (UNESP).

Institutional Review Board Statement: Not applicable.

Informed Consent Statement: Not applicable.

Data Availability Statement: The authors confirm that the data supporting the findings of this study are available within the article and its Supplementary Materials.

Acknowledgments: The authors are indebted to Delecluse from the German Cancer Research Center (Heidelberg, Germany) for vectors containing the EBV B95.8 and M81 full genomes (F factor-based prokaryotic replicon, pMBO131), Nancy Raab-Traub for providing NP69 cells via EC laboratory, André Sampaio Pupo for providing HEK293 cells, and Rafael Coan for the Python script that counts the gene targets as unique hits.

Conflicts of Interest: The authors declare no conflict of interest.

Ethics Committee: This research was approved by the Brazilian Committee of Research Ethics (CEP) from the Brazilian government (https://plataformabrasil.saude.gov.br (accessed on 1 July 2016)), registered under #1.440.863.

References

1. Global Burden of Disease Cancer Collaboration. The Global Burden of Cancer 2013. *JAMA Oncol.* **2015**, *1*, 505–527. [CrossRef] [PubMed]
2. Oh, J.-K.; Weiderpass, E. Infection and Cancer: Global Distribution and Burden of Diseases. *Ann. Glob. Health* **2014**, *80*, 384–392. [CrossRef] [PubMed]
3. Schottenfeld, D.; Beebe-Dimmer, J. The Cancer Burden Attributable to Biologic Agents. *Ann. Epidemiol.* **2015**, *25*, 183–187. [CrossRef] [PubMed]
4. Stewart, B.W.; Wild, C.P. World Cancer Report 2014. 2014. Available online: http://www.iahm.org/journal/vol_15/num_3/text/vol15n3p40.pdf (accessed on 20 May 2020).
5. De Martel, C.; Ferlay, J.; Franceschi, S.; Vignat, J.; Bray, F.; Forman, D.; Plummer, M. Global Burden of Cancers Attributable to Infections in 2008: A Review and Synthetic Analysis. *Lancet Oncol.* **2012**, *13*, 607–615. [CrossRef]
6. Vineis, P.; Wild, C.P. Global Cancer Patterns: Causes and Prevention. *Lancet* **2014**, *383*, 549–557. [CrossRef]
7. Dunmire, S.K.; Hogquist, K.A.; Balfour, H.H. Infectious Mononucleosis. In *Epstein Barr Virus Volume 1*; Münz, C., Ed.; Springer International Publishing: Cham, Switzerland, 2015; Volume 390, pp. 211–240. ISBN 978-3-319-22821-1.
8. Elgui de Oliveira, D. DNA Viruses in Human Cancer: An Integrated Overview on Fundamental Mechanisms of Viral Carcinogenesis. *Cancer Lett.* **2007**, *247*, 182–196. [CrossRef]
9. Tang, L.-L.; Chen, W.-Q.; Xue, W.-Q.; He, Y.-Q.; Zheng, R.-S.; Zeng, Y.-X.; Jia, W.-H. Global Trends in Incidence and Mortality of Nasopharyngeal Carcinoma. *Cancer Lett.* **2016**, *374*, 22–30. [CrossRef]
10. Colditz, G.A.; Wolin, K.Y.; Gehlert, S. Applying What We Know to Accelerate Cancer Prevention. *Sci. Transl. Med.* **2012**, *4*, 127rv4. [CrossRef]
11. Brousset, P.; Keryer, C.; Ooca, T.; Corbex, M. EBV-Associated Nasopharyngeal Carcinomas: From Epidemiology to Virus-Targeting Strategies. *Trends Microbiol.* **2004**, *12*, 353–356. [CrossRef]
12. The Enigmatic Epidemiology of Nasopharyngeal Carcinoma | Cancer Epidemiology, Biomarkers & Prevention. Available online: http://cebp.aacrjournals.org/content/15/10/1765.long (accessed on 5 July 2018).
13. Tsao, S.-W.; Tsang, C.M.; To, K.-F.; Lo, K.-W. The Role of Epstein–Barr Virus in Epithelial Malignancies: Role of EBV in Epithelial Malignancies. *J. Pathol.* **2015**, *235*, 323–333. [CrossRef]
14. De Oliveira, D.E.; Müller-Coan, B.G.; Pagano, J.S. Viral Carcinogenesis Beyond Malignant Transformation: EBV in the Progression of Human Cancers. *Trends Microbiol.* **2016**, *24*, 649–664. [CrossRef] [PubMed]
15. Brinkmann, M.M. Regulation of Intracellular Signalling by the Terminal Membrane Proteins of Members of the Gammaherpesvirinae. *J. Gen. Virol.* **2006**, *87*, 1047–1074. [CrossRef] [PubMed]
16. Lin, X.; Tang, M.; Tao, Y.; Li, L.; Liu, S.; Guo, L.; Li, Z.; Ma, X.; Xu, J.; Cao, Y. Epstein–Barr Virus-Encoded LMP1 Triggers Regulation of the ERK-Mediated Op18/Stathmin Signaling Pathway in Association with Cell Cycle. *Cancer Sci.* **2012**, *103*, 993–999. [CrossRef]
17. Masoud, G.N.; Li, W. HIF-1α Pathway: Role, Regulation and Intervention for Cancer Therapy. *Acta Pharm. Sin. B* **2015**, *5*, 378–389. [CrossRef] [PubMed]
18. Wakisaka, N.; Kondo, S.; Yoshizaki, T.; Murono, S.; Furukawa, M.; Pagano, J.S. Epstein–Barr Virus Latent Membrane Protein 1 Induces Synthesis of Hypoxia-Inducible Factor 1. *Mol. Cell. Biol.* **2004**, *24*, 5223–5234. [CrossRef] [PubMed]
19. Tsai, M.-H.; Lin, X.; Shumilov, A.; Bernhardt, K.; Feederle, R.; Poirey, R.; Kopp-Schneider, A.; Pereira, B.; Almeida, R.; Delecluse, H.-J. The Biological Properties of Different Epstein–Barr Virus Strains Explain Their Association with Various Types of Cancers. *Oncotarget* **2016**, *8*, 10238–10254. [CrossRef]
20. Tsai, M.-H.; Raykova, A.; Klinke, O.; Bernhardt, K.; Gärtner, K.; Leung, C.S.; Geletneky, K.; Sertel, S.; Münz, C.; Feederle, R.; et al. Spontaneous Lytic Replication and Epitheliotropism Define an Epstein–Barr Virus Strain Found in Carcinomas. *Cell Rep.* **2013**, *5*, 458–470. [CrossRef]
21. Abba, M.; Patil, N.; Allgayer, H. MicroRNAs in the Regulation of MMPs and Metastasis. *Cancers* **2014**, *6*, 625–645. [CrossRef]
22. Acunzo, M.; Romano, G.; Wernicke, D.; Croce, C.M. MicroRNA and Cancer—A Brief Overview. *Adv. Biol. Regul.* **2015**, *57*, 1–9. [CrossRef]
23. Abba, M.L.; Patil, N.; Leupold, J.H.; Allgayer, H. MicroRNA Regulation of Epithelial to Mesenchymal Transition. *J. Clin. Med.* **2016**, *5*, 8. [CrossRef]
24. Chen, H.-C.; Chen, G.-H.; Chen, Y.-H.; Liao, W.-L.; Liu, C.-Y.; Chang, K.-P.; Chang, Y.-S.; Chen, S.-J. MicroRNA Deregulation and Pathway Alterations in Nasopharyngeal Carcinoma. *Br. J. Cancer* **2009**, *100*, 1002–1011. [CrossRef] [PubMed]
25. Yang, G.-D.; Huang, T.-J.; Peng, L.-X.; Yang, C.-F.; Liu, R.-Y.; Huang, H.-B.; Chu, Q.-Q.; Yang, H.-J.; Huang, J.-L.; Zhu, Z.-Y.; et al. Epstein–Barr Virus_Encoded LMP1 Upregulates MicroRNA-21 to Promote the Resistance of Nasopharyngeal Carcinoma Cells to Cisplatin-Induced Apoptosis by Suppressing PDCD4 and Fas-L. *PLoS ONE* **2013**, *8*, e78355. [CrossRef] [PubMed]

26. Uhlmann, S.; Zhang, J.D.; Schwäger, A.; Mannsperger, H.; Riazalhosseini, Y.; Burmester, S.; Ward, A.; Korf, U.; Wiemann, S.; Sahin, O. MiR-200bc/429 Cluster Targets PLCgamma1 and Differentially Regulates Proliferation and EGF-Driven Invasion than MiR-200a/141 in Breast Cancer. *Oncogene* **2010**, *29*, 4297–4306. [CrossRef] [PubMed]

27. Gu, X.; Li, J.-Y.; Guo, J.; Li, P.-S.; Zhang, W.-H. Influence of MiR-451 on Drug Resistances of Paclitaxel-Resistant Breast Cancer Cell Line. *Med. Sci. Monit. Int. Med. J. Exp. Clin. Res.* **2015**, *21*, 3291–3297. [CrossRef]

28. Raza, U.; Zhang, J.D.; Sahin, O. MicroRNAs: Master Regulators of Drug Resistance, Stemness, and Metastasis. *J. Mol. Med. Berl. Ger.* **2014**, *92*, 321–336. [CrossRef]

29. Calin, G.A.; Dumitru, C.D.; Shimizu, M.; Bichi, R.; Zupo, S.; Noch, E.; Aldler, H.; Rattan, S.; Keating, M.; Rai, K.; et al. Frequent Deletions and Down-Regulation of Micro- RNA Genes MiR15 and MiR16 at 13q14 in Chronic Lymphocytic Leukemia. *Proc. Natl. Acad. Sci. USA* **2002**, *99*, 15524–15529. [CrossRef]

30. Graham, F.L.; Smiley, J.; Russell, W.C.; Nairn, R. Characteristics of a Human Cell Line Transformed by DNA from Human Adenovirus Type 5. *J. Gen. Virol.* **1977**, *36*, 59–72. [CrossRef]

31. Tsao, S.W.; Wang, X.; Liu, Y.; Cheung, Y.C.; Feng, H.; Zheng, Z.; Wong, N.; Yuen, P.W.; Lo, A.K.; Wong, Y.C.; et al. Establishment of Two Immortalized Nasopharyngeal Epithelial Cell Lines Using SV40 Large T and HPV16E6/E7 Viral Oncogenes. *Biochim. Biophys. Acta BBA-Mol. Cell Res.* **2002**, *1590*, 150–158. [CrossRef]

32. Delecluse, H.-J.; Hilsendegen, T.; Pich, D.; Zeidler, R.; Hammerschmidt, W. Propagation and Recovery of Intact, Infectious Epstein–Barr Virus from Prokaryotic to Human Cells. *Proc. Natl. Acad. Sci. USA* **1998**, *95*, 8245–8250. [CrossRef]

33. Feederle, R.; Bartlett, E.J.; Delecluse, H.-J. Epstein–Barr Virus Genetics: Talking about the BAC Generation. *Herpesviridae* **2010**, *1*, 6. [CrossRef]

34. Keller, S.A.; Hernandez-Hopkins, D.; Vider, J.; Ponomarev, V.; Hyjek, E.; Schattner, E.J.; Cesarman, E. NF-KB Is Essential for the Progression of KSHV- and EBV-Infected Lymphomas in Vivo. *Blood* **2006**, *107*, 3295–3302. [CrossRef] [PubMed]

35. Gebäck, T.; Shultz, M.M.P.; Koumoutsakos, P.; Detmar, M. TScratch: A Novel and Simple Software Tool for Automated Analysis of Monolayer Wound Healing Assays. *BioTechniques* **2009**, *46*, 265–274. [CrossRef]

36. Marí-Alexandre, J.; Barceló-Molina, M.; Belmonte-López, E.; García-Oms, J.; Estellés, A.; Braza-Boïls, A.; Gilabert-Estellés, J. Micro-RNA Profile and Proteins in Peritoneal Fluid from Women with Endometriosis: Their Relationship with Sterility. *Fertil. Steril.* **2018**, *109*, 675–684. [CrossRef] [PubMed]

37. Ferrante, S.C.; Nadler, E.P.; Pillai, D.K.; Hubal, M.J.; Wang, Z.; Wang, J.M.; Gordish-Dressman, H.; Koeck, E.; Sevilla, S.; Wiles, A.A.; et al. Adipocyte-Derived Exosomal MiRNAs: A Novel Mechanism for Obesity-Related Disease. *Pediatr. Res.* **2015**, *77*, 447–454. [CrossRef] [PubMed]

38. Tokar, T.; Pastrello, C.; Rossos, A.E.M.; Abovsky, M.; Hauschild, A.-C.; Tsay, M.; Lu, R.; Jurisica, I. MirDIP 4.1—Integrative Database of Human MicroRNA Target Predictions. *Nucleic Acids Res.* **2018**, *46*, D360–D370. [CrossRef] [PubMed]

39. Shirdel, E.A.; Xie, W.; Mak, T.W.; Jurisica, I. NAViGaTing the Micronome—Using Multiple MicroRNA Prediction Databases to Identify Signalling Pathway-Associated MicroRNAs. *PLoS ONE* **2011**, *6*, e17429. [CrossRef]

40. Croft, D.; Mundo, A.F.; Haw, R.; Milacic, M.; Weiser, J.; Wu, G.; Caudy, M.; Garapati, P.; Gillespie, M.; Kamdar, M.R.; et al. The Reactome Pathway Knowledgebase. *Nucleic Acids Res.* **2014**, *42*, D472–D477. [CrossRef]

41. Shannon, P.; Markiel, A.; Ozier, O.; Baliga, N.S.; Wang, J.T.; Ramage, D.; Amin, N.; Schwikowski, B.; Ideker, T. Cytoscape: A Software Environment for Integrated Models of Biomolecular Interaction Networks. *Genome Res.* **2003**, *13*, 2498–2504. [CrossRef]

42. Wasil, L.R.; Shair, K.H.Y. Epstein–Barr Virus LMP1 Induces Focal Adhesions and Epithelial Cell Migration through Effects on Integrin-A5 and N-Cadherin. *Oncogenesis* **2015**, *4*, e171. [CrossRef]

43. Li, W.; Tian, Z.; Qin, H.; Li, N.; Zhou, X.; Li, J.; Ni, B.; Ruan, Z. High Expression of Sphingosine Kinase 1 Is Associated with Poor Prognosis in Nasopharyngeal Carcinoma. *Biochem. Biophys. Res. Commun.* **2015**, *460*, 341–347. [CrossRef]

44. Lee, H.M.; Lo, K.-W.; Wei, W.; Tsao, S.W.; Chung, G.T.Y.; Ibrahim, M.H.; Dawson, C.W.; Murray, P.G.; Paterson, I.C.; Yap, L.F. Oncogenic S1P Signalling in EBV-Associated Nasopharyngeal Carcinoma Activates AKT and Promotes Cell Migration through S1P Receptor 3. *J. Pathol.* **2017**, *242*, 62–72. [CrossRef] [PubMed]

45. Chang, S.-H.; Chang, H.-C.; Hung, W.-C. Transcriptional Repression of Tissue Inhibitor of Metalloproteinase-3 by Epstein–Barr Virus Latent Membrane Protein 1 Enhances Invasiveness of Nasopharyngeal Carcinoma Cells. *Oral Oncol.* **2008**, *44*, 891–897. [CrossRef] [PubMed]

46. Aga, M.; Bentz, G.L.; Raffa, S.; Torrisi, M.R.; Kondo, S.; Wakisaka, N.; Yoshizaki, T.; Shackelford, J. Exosomal HIF1α Supports Invasive Potential of Nasopharyngeal Carcinoma-Associated LMP1-Positive Exosomes. *Oncogene* **2014**, *33*, 4613–4622. [CrossRef] [PubMed]

47. Zhang, H.-C.; Tang, K.-F. Clinical Value of Integrated-Signature MiRNAs in Esophageal Cancer. *Cancer Med.* **2017**, *6*, 1893–1903. [CrossRef]

48. Qin, X.; Yu, S.; Zhou, L.; Shi, M.; Hu, Y.; Xu, X.; Shen, B.; Liu, S.; Yan, D.; Feng, J. Cisplatin-Resistant Lung Cancer Cell–Derived Exosomes Increase Cisplatin Resistance of Recipient Cells in Exosomal MiR-100–5p-Dependent Manner. *Int. J. Nanomed.* **2017**, *12*, 3721–3733. [CrossRef]

49. Chen, D.; Sun, Y.; Yuan, Y.; Han, Z.; Zhang, P.; Zhang, J.; You, M.J.; Teruya-Feldstein, J.; Wang, M.; Gupta, S.; et al. MiR-100 Induces Epithelial-Mesenchymal Transition but Suppresses Tumorigenesis, Migration and Invasion. *PLoS Genet.* **2014**, *10*, e1004177. [CrossRef]

50. Nabavi, N.; Saidy, N.R.N.; Venalainen, E.; Haegert, A.; Parolia, A.; Xue, H.; Wang, Y.; Wu, R.; Dong, X.; Collins, C.; et al. MiR-100-5p Inhibition Induces Apoptosis in Dormant Prostate Cancer Cells and Prevents the Emergence of Castration-Resistant Prostate Cancer. *Sci. Rep.* **2017**, *7*, 4079. [CrossRef]
51. Yan-Chun, L.; Hong-Mei, Y.; Zhi-Hong, C.; Qing, H.; Yan-Hong, Z.; Ji-Fang, W. MicroRNA-192-5p Promote the Proliferation and Metastasis of Hepatocellular Carcinoma Cell by Targeting SEMA3A. *Appl. Immunohistochem. Mol. Morphol.* **2015**, *25*, 251–260. [CrossRef]
52. Kim, Y.S.; Park, S.J.; Lee, Y.S.; Kong, H.K.; Park, J.H. MiRNAs Involved in LY6K and Estrogen Receptor α Contibute to Tamoxifen-Suceptibility in Breast Cancer. *Oncotarget* **2016**, *5*, 42261–42273.
53. Su, Y.; Ni, Z.; Wang, G.; Cui, J.; Wei, C.; Wang, J.; Yang, Q.; Xu, Y.; Li, F. Aberrant Expression of MicroRNAs in Gastric Cancer and Biological Significance of MiR-574-3p. *Int. Immunopharmacol.* **2012**, *13*, 468–475. [CrossRef]
54. Xu, S.; Bai, J.; Zhuan, Z.; Li, B.; Zhang, Z.; Wu, X.; Luo, X.; Yang, L. EBV-LMP1 Is Involved in Vasculogenic Mimicry Formation via VEGFA/VEGFR1 Signaling in Nasopharyngeal Carcinoma. *Oncol. Rep.* **2018**, *40*, 377–384. [CrossRef] [PubMed]
55. Tworkoski, K.; Raab-Traub, N. LMP1 Promotes Expression of Insulin-Like Growth Factor 1 (IGF1) To Selectively Activate IGF1 Receptor and Drive Cell Proliferation. *J. Virol.* **2014**, *89*, 2590–2602. [CrossRef] [PubMed]
56. Dawson, C.W.; Laverick, L.; Morris, M.A.; Tramoutanis, G.; Young, L.S. Epstein–Barr Virus-Encoded LMP1 Regulates Epithelial Cell Motility and Invasion via the ERK-MAPK Pathway. *J. Virol.* **2008**, *82*, 3654–3664. [CrossRef] [PubMed]
57. Fukuda, M.; Kurosaki, W.; Yanagihara, K.; Kuratsune, H.; Sairenji, T. A Mechanism in Epstein–Barr Virus Oncogenesis: Inhibition of Transforming Growth Factor-B1-Mediated Induction of MAPK/P21 by LMP1. *Virology* **2002**, *302*, 310–320. [CrossRef] [PubMed]
58. QingLing, Z.; LiNa, Y.; Liu, L.; Shuang, W.; YuFang, Y.; Yi, D.; Divakaran, J.; Xin, L.; YanQing, D. LMP1 Antagonizes WNT/β-Catenin Signalling through Inhibition of WTX and Promotes Nasopharyngeal Dysplasia but Not Tumourigenesis in LMP1B95−8 Transgenic Mice. *J. Pathol.* **2011**, *223*, 574–583. [CrossRef]
59. Zeng, Z.-Y.; Zhou, Y.-H.; Zhang, W.-L.; Xiong, W.; Fan, S.-Q.; Li, X.-L.; Luo, X.-M.; Wu, M.-H.; Yang, Y.-X.; Huang, C.; et al. Gene Expression Profiling of Nasopharyngeal Carcinoma Reveals the Abnormally Regulated Wnt Signaling Pathway. *Hum. Pathol.* **2007**, *38*, 120–133. [CrossRef]
60. Leonard, S.; Wei, W.; Anderton, J.; Vockerodt, M.; Rowe, M.; Murray, P.G.; Woodman, C.B. Epigenetic and Transcriptional Changes Which Follow Epstein–Barr Virus Infection of Germinal Center B Cells and Their Relevance to the Pathogenesis of Hodgkin's Lymphoma. *J. Virol.* **2011**, *85*, 9568–9577. [CrossRef]
61. Seo, S.Y.; Kim, E.-O.; Jang, K.L. Epstein–Barr Virus Latent Membrane Protein 1 Suppresses the Growth-Inhibitory Effect of Retinoic Acid by Inhibiting Retinoic Acid Receptor-B2 Expression via DNA Methylation. *Cancer Lett.* **2008**, *270*, 66–76. [CrossRef]
62. Liao, J.; Karnik, R.; Gu, H.; Ziller, M.J.; Clement, K.; Tsankov, A.M.; Akopian, V.; Gifford, C.A.; Donaghey, J.; Galonska, C.; et al. Targeted Disruption of DNMT1, DNMT3A and DNMT3B in Human Embryonic Stem Cells. *Nat. Genet.* **2015**, *47*, 469–478. [CrossRef]
63. Perez-Balaguer, A.; Ortiz-Martínez, F.; García-Martínez, A.; Pomares-Navarro, C.; Lerma, E.; Peiró, G. FOXA2 MRNA Expression Is Associated with Relapse in Patients with Triple-Negative/Basal-like Breast Carcinoma. *Breast Cancer Res. Treat.* **2015**, *153*, 465–474. [CrossRef]
64. Yao, H.S.; Wang, J.; Zhang, X.P.; Wang, L.Z.; Wang, Y.; Li, X.X.; Jin, K.Z.; Hu, Z.Q.; Wang, W.J. Hepatocyte Nuclear Factor 4α Suppresses the Aggravation of Colon Carcinoma. *Mol. Carcinog.* **2016**, *55*, 458–472. [CrossRef] [PubMed]
65. Zhao, J.; Adams, A.; Roberts, B.; O'Neil, M.; Vittal, A.; Schmitt, T.; Kumer, S.; Cox, J.; Li, Z.; Weinman, S.A.; et al. Protein Arginine Methyl Transferase 1− and Jumonji C Domain-Containing Protein 6–Dependent Arginine Methylation Regulate Hepatocyte Nuclear Factor 4 α Expression and Hepatocyte Proliferation in Mice. *Hepatology* **2018**, *67*, 1109–1126. [CrossRef] [PubMed]
66. Clorennec, C.L.; Ouk, T.-S.; Youlyouz-Marfak, I.; Panteix, S.; Martin, C.-C.; Rastelli, J.; Adriaenssens, E.; Zimber-Strobl, U.; Coll, J.; Feuillard, J.; et al. Molecular Basis of Cytotoxicity of Epstein–Barr Virus (EBV) Latent Membrane Protein 1 (LMP1) in EBV Latency III B Cells: LMP1 Induces Type II Ligand-Independent Autoactivation of CD95/Fas with Caspase 8-Mediated Apoptosis. *J. Virol.* **2008**, *82*, 6721–6733. [CrossRef] [PubMed]
67. Clorennec, C.L.; Youlyouz-Marfak, I.; Adriaenssens, E.; Coll, J.; Bornkamm, G.W.; Feuillard, J. EBV Latency III Immortalization Program Sensitizes B Cells to Induction of CD95-Mediated Apoptosis via LMP1: Role of NF-KB, STAT1, and P53. *Blood* **2006**, *107*, 2070–2078. [CrossRef] [PubMed]
68. Zhang, X.; Uthaisang, W.; Hu, L.; Ernberg, I.T.; Fadeel, B. Epstein–Barr Virus-Encoded Latent Membrane Protein 1 Promotes Stress-Induced Apoptosis Upstream of Caspase-2-Dependent Mitochondrial Perturbation. *Int. J. Cancer* **2005**, *113*, 397–405. [CrossRef]

Article

Circulating miRNA-192 and miR-29a as Disease Progression Biomarkers in Hepatitis C Patients with a Prevalence of HCV Genotype 3

Amin Ullah [1,2], Irshad Ur Rehman [3], Katharina Ommer [4], Nadeem Ahmed [5], Margarete Odenthal [2,*], Xiaojie Yu [2], Jamshaid Ahmad [3], Tariq Nadeem [5], Qurban Ali [6,*] and Bashir Ahmad [3]

[1] Department of Health and Biological Sciences, Abasyn University, Peshawar 25000, Pakistan
[2] Institute for Pathology, University of Cologne, 50923 Cologne, Germany
[3] Center of Biotechnology and Microbiology, University of Peshawar, Peshawar 25000, Pakistan; bashirdr2015@yahoo.com (B.A.)
[4] Institute of Transfusion Medicine, University of Cologne, 50923 Cologne, Germany
[5] Center of Excellence in Molecular Biology, University of the Punjab, Lahore 54000, Pakistan
[6] Department of Plant Breeding and Genetics, University of the Punjab, Lahore 54000, Pakistan
* Correspondence: m.odenthal@uni-koeln.de (M.O.); saim1692@gmail.com (Q.A.)

Abstract: MicroRNAs miR-29a and miR-192 are involved in inflammatory and fibrotic processes of chronic liver disease, and circulating miR-29a is suggested to diagnose fibrosis progression due to hepatitis C virus (HCV) infection. This study aimed to evaluate the expression profile of circulating miR-192 and 29a in a patient cohort with a high frequency of HCV genotype-3. A total of 222 HCV blood samples were collected and serum were separated. Patients were classified into mild, moderate, and severe liver injury based on their Child–Turcotte–Pugh CTP score. RNA was isolated from the serum and used for quantitative real-time PCR. The HCV genotype-3 (62%) was the predominant HCV genotype. In HCV patients, the serum miR-192 and miR-29a levels were significantly upregulated in comparison to healthy controls ($p = 0.0017$ and $p = 0.0001$, respectively). The progression rate of miR-192 and 29a in the patient group with mild was highly upregulated compared to patients with moderate and severe hepatitis infection. The ROC curve of miR-192 and miR-29a of moderate liver disease had a significant diagnostic performance compared to the other HCV-infected groups. The increase in miR-29a and miR-192 serum levels was even slightly higher in patients with HCV genotype-3 than in non-genotype-3 patients. In conclusion, serum miR-192 and miR-29a levels significantly increased during the progression of chronic HCV infection. The marked upregulation in patients with HCV genotype-3 suggests them as potential biomarkers for hepatic disease, independently of the HCV genotype.

Keywords: hepatitis C virus; liver injury; microRNA-192; microRNA-29a; HCV genotype-3

Citation: Ullah, A.; Rehman, I.U.; Ommer, K.; Ahmed, N.; Odenthal, M.; Yu, X.; Ahmad, J.; Nadeem, T.; Ali, Q.; Ahmad, B. Circulating miRNA-192 and miR-29a as Disease Progression Biomarkers in Hepatitis C Patients with a Prevalence of HCV Genotype 3. *Genes* **2023**, *14*, 1056. https://doi.org/10.3390/genes14051056

Academic Editors: Giuseppe Iacomino and Fabio Lauria

Received: 8 February 2023
Revised: 2 May 2023
Accepted: 3 May 2023
Published: 8 May 2023

1. Introduction

Worldwide, the hepatitis C virus (HCV) is a principal causative agent of human liver infection. It is estimated that 180 million people are infected with HCV globally each year. Chronic HCV infection causes liver failure and hepatocellular carcinoma (HCC), and in North America, it is the primary indication for liver transplantation at present [1]. HCV consists of 9600 nucleotides, with a single-stranded RNA genome containing a single large open reading frame (ORF) [2,3]. Genetic variations are classified into six main genotypes, which include several subtypes. Genotypes 1, 2, and 3 are the most common genotypes, accounting for more than 80% of HCV infections worldwide. In the case of North America and Europe, genotype-1 is predominant, and in Southeast Asia, genotype-3 is the most prevalent, including subtypes 3a, 3b, 3c, 3d, 3e, and 3f [4,5]. HCV infections result in chronic liver inflammatory changes, including liver fibrosis and cirrhosis [3]. Metabolic changes often accompany liver fibrosis, and 70% of HCV infections lead to steatosis. Interestingly, hepatic fat accumulation occurs more frequently and is much more pronounced in patients infected with genotype-3 [6,7].

Most screening studies for the non-invasive marker, indicating the progression of hepatitis-C-caused chronic inflammation, were performed on cohorts with a prevalence of genotype-1, and differences depending on the genotypes are not known. Circulating microRNAs (miRNAs) released from the injured liver are regarded as suggestive indicators of inflammation-mediated disease severity and fibrosis [8,9].

MicroRNAs are small, non-coding moieties of approximately 19–22 nucleotide-long RNA molecules. They are involved in gene expression, targeting mRNA and attenuating translation. MicroRNAs are found in organisms ranging from unicellular to high-multicellular organisms, including mammals and plants. In the case of mammals, miRNAs regulate 60% of the protein-coding genes [10–12]. During maturation, mature miRNAs are generated by successive cleavages of Pol-II-transcribed primary RNA precursors by Drosha and Dicer [10–12]. The RNA-induced silencing complex (RISC) enables microRNAs to modulate gene expression via translational repression and mRNA degradation, with the latter mechanism mediating target inhibition more effectively [2]. Due to their ability to target most mRNAs in the human genome, microRNAs play crucial roles in biological processes, including development, immunity, cancer, and pathogen infections. In addition, miRNA expression is highly tissue-specific, and various human diseases have been linked to altered miRNA expression patterns [13].

Their dysregulation is linked with human pathologies, including cardiovascular diseases, neurological disorders, and chronic liver disease and development of hepatocellular carcinoma (HCC) [8,9,14].

Importantly miRNAs are released from cells by active secretion or passive release uponcell damage [15]. The potential application of extracellular miRNAs released into the bloodstream as non-invasive biomarkers is intensively studied. Various studies have examined the efficacy of microRNAs in diagnosing HCC. The miR-155 and miR-146a have been identified as dysregulated in patients with HCV-associated hepatocellular carcinoma [16–19]. Additionally, a combination of four miRNAs (miR-21, miR-122, miR-192, and miR-223) was highly accurate in diagnosing chronic hepatitis [19,20]. MicroRNAs have also been investigated as a potential diagnostic tool for HCV infection, despite the limitations of current diagnostic methods, which include a high rate of false negatives during the early stages of infection [21]. Thus, some studies have demonstrated that patterns of microRNA expression can distinguish between individuals with chronic HCV infection and healthy controls, as well as between patients with varying stages of liver disease. MiR-122, a liver-specific miRNA, regulates HCV replication, and its downregulation increases viral load and disease progression [19]. MiR-205, downregulated in most liver diseases, could be a diagnostic and treatment molecule. MiR-29a, miR-146a, and miR-155 also modulate HCV replication and host immune response [22]. The microRNA-29a (miR-29a) is a member of miR-29 family, whose serum levels are increased upon HCV mediated liver injury [23]. Importantly, miR-29 targets transcripts encoding ECM proteins, such as fibrillins, β1 integrin, laminins, collagens, and elastin [24–26]. Studies on western patient cohorts with a high prevalence of HCV genotype-1 infections also showed a change in circulating miR-29a serum levels, whereas hepatic tissue levels were reduced [23,26].

Another interesting fact is the close link between miR-192 expression and TGF-β1 signaling during renal fibrosis [27]. Furthermore, miR-192 is a hepatic-enriched miRNA recognized to be upregulated in the serum of HCV patients. Thus, miR-192 is suggested to be involved in processes mediating HCV-infection-associated liver disease and to serve as a potential target against viral pathogenesis [28]. Interestingly, miR-192, in combination with other miRNAs, is used as a novel biomarker for steatosis associated liver injury [28,29].

MicroRNA-192 and miR-29a have been proposed as biomarkers of liver injury in patients with chronic HCV infection. As previous reports have included patients infected with HCV genotype 3 to a limited extent, we focused our study on the circulating profiles of miR-192 and miR-29a in HCV patients with a high prevalence of HCV genotype 3 infection.

2. Materials and Methods

2.1. Sample Collection

The patient's samples and data, containing demographic, clinical characteristics and estimated infection time, were collected by Hayatabad Medical Complex (HMC) and Khyber Teaching Hospital (KTH) Peshawar, Pakistan. The demographic characteristics are shown in Supplementary Table S1A–C.

The HCV genotyping was performed as previously designated by Ullah et al. [2]. The Ethical Committee approved the study, Center of Biotechnology & Microbiology (COBAM), University of Peshawar, Pakistan, and written informed consent was obtained from all the recruited patients. That the experimental samples, including the collection samples material, were confirmed to comply with relevant institutional, national, and international guidelines and legislation with appropriate permissions from authorities of the Department of Health and Biological Sciences, Abasyn University, Peshawar, Pakistan and Center of Biotechnology and Microbiology, University of Peshawar, Peshawar, Pakistan for the collection of sample specimens.

We included patients who showed PCR positivity for the 5′ untranslated region (UTR) of the HCV RNA genome. Patients who were co-infected with other virus types such as HAV or HBV or HDV were excluded. Furthermore, patients were excluded when their viral load was below a titre 500 IU/mL and if the RNA quality in serum was shown to be insufficient.

2.2. RNA Isolation

HCV patients' blood samples were centrifuged, and the serum was taken. RNA was extracted from 200 µL serum using the Qiagen Kit (Qiagen, Hilden, Germany. Cat No./ID: 217204) according to the manufacturer's protocol and as described earlier. Before the RNA extraction, SV40-miRNA (5-UGAGGGCUGAA AUGAGCCUU-3) (Qiagen, Hilden, Germany, Cat No./ID: 331535) was spiked in (2 fmol/200 µL serum) for the later normalization of the miRNA-192 and miR-29a levels [19,30,31].

2.3. Synthesis of cDNA and RT-PCR

The complementary DNA (cDNA)was prepared using the miScript-Reverse Transcription Kit (Qiagen kit. Cat No./ID: 218160) as described earlier For real-time PCR, the miRNA-SYBR Green PCR Kit and SV-40 primers, miRNA-192, and miR-29a from Qiagen (Hilden, Germany) were used. Thermo-cycling conditions of the RT-PCR were as follows: initial denaturation 95 °C/3 min, following cycles of PCR template denaturation 94 °C/30 s, annealing 55 °C/45 s and extension 70 °C/45 s. All the steps were performed in triplicate and in agreement with the supplier's guidelines. A dilution series was generated, and miRNA levels were quantified using the standard curves. Spike-in SV40-miRNA quantification was used to normalize miR-29a and miR-192 extracellular miRNAs levels [19,31,32].

2.4. Data Analysis

The data analysis was conducted using GraphPad Prism 5 and IBM SPSS software 25.0. One-way ANOVA test and *t*-test were used for the statistical significance, indicated as *p*-value. The non-parametric counterpart of the ANOVA, which was obtained using Kruskall–Wallis test, was also employed in cases of deviation from the normality assumption. Bartlett's test was used for equal variance and a strong significant correlation between variables. ROC and AUC analyses were conducted to determine diagnostic performance and the significance of microRNA expression. The best cut-offs were selected by maximizing the Youden index; that is, the sum of specificity and maximized sensitivity. Pearson's correlation was performed between the two variables. Values of less than 0.05 were considered statistically significant (p).

3. Results

3.1. Characteristics of the Patient Cohort

The HCV patients (n = 222) were grouped into mild, moderate, and severe based on the Child–Turcotte–Pugh (CTP) score. Liver disease progression was classified into mild, moderate, and severe, and the score system was as follows: 5–6 was referred to as Child Class A (mild), 7–9 as Child Class B (moderate) and 10–15 as Child Class C (severe) [9]. The patients were admitted to Peshawar's tertiary hospitals from April 2016 to October 2018. The patients' medical data and demographic summary are presented in Supplementary Table S1A,B. The mean age of the patients was 49.03 ± 12.65 years. Importantly, there were significant differences ($p < 0.0001$) in the alanine transaminase enzyme (ALT) and α-fetoprotein (AFP) levels of patients with progressive chronic HCV liver disease. The controls (n = 52) were healthy blood donors, and they are described in Supplementary Table S1C. Notably, in our study, HCV genotype-3 occurred at a very high frequency (61.7%) compared to the other genotypes.

3.2. Serum miR-192 and miR-29a Profile

In this study, the expression profiles of the circulating microRNAs miR-192 and miR-29a were quantified by quantitative PCR. The levels of patients with mild, moderate, and severe HCV hepatitis were compared to those in the control group of healthy blood donors. The parametric and non-parametric analyses of serum miR-192 and 29a levels both revealed that miRNAs were significantly increased compared to the controls, as shown in Figure 1. Importantly, the serum miR-29a and miR-192 levels of each patient group developing mild and moderate stages of liver injury were significantly elevated (Supplementary Figures S1 and S2).

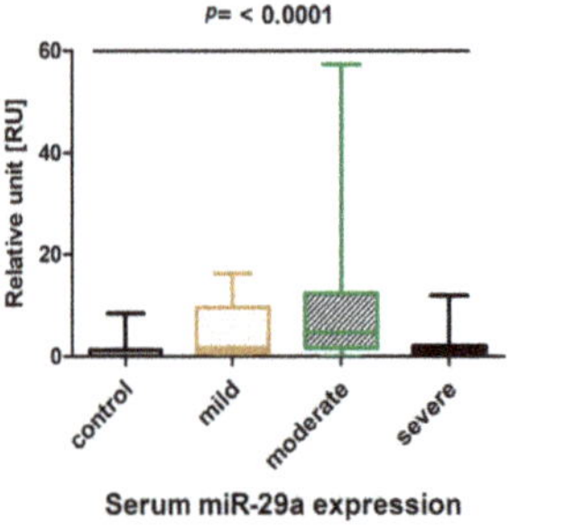

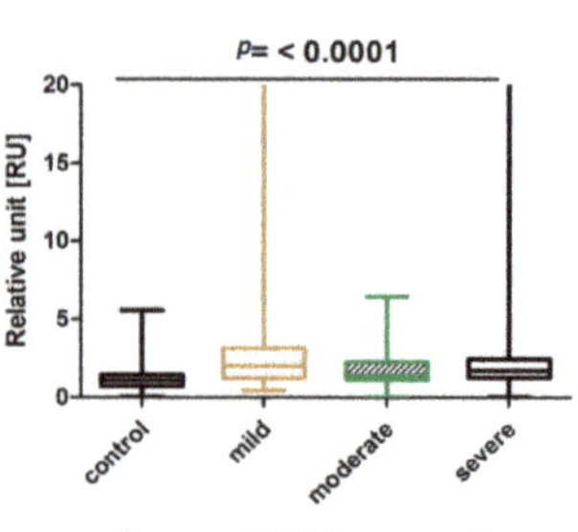

Figure 1. Serum expression profiles of microRNA-192 and 29a in HCV patients. Serum miRNA-192 and miR-29a in patients with mild, moderate or severe liver injury. The Whisker box plots show a significant difference between the levels of miR-192 and miR-29a in the groups of patients with different grades of liver damage.

The receiver-operating characteristic (ROC) curve under the area of (AUC) curve analysis was performed to measure the potential of individual microRNA to discriminate the HCV group with progressive liver injury. The AUC values of miR-192 and 29a showed a significant difference between the HCV-infected cohorts with moderate, mild, and severe liver injury versus the control group (Supplementary Figures S3 and S4, and Supplementary Tables S2 and S3). Figure 2 shows the ROC curve of the HCV-infected cohorts compared to the control, demonstrating that miR-192 and 29a have great potential to estimate the progression of HCV-mediated hepatic disease. The miR-192 ROC curve demonstrated that the miR-192 levels are a more sensitive indicator for mild liver injury than for moderate and severe liver damage. Similarly, the miR-29a ROC curve of the moderate group showed a particularly high sensitivity in HCV patients with mild liver injury. Overall, the analysis revealed that microRNA-29a has a more significant accuracy in diagnosis as compared to microRNA-192 in HCV patients. Therefore, the sensitivity rate of the ROC under the AUC analysis of miR-192 and 29a was considered to be significant, and their diagnostic and precision strength was substantial in HCV patients (Figure 2). According to the Youden index for microRNA 192 HCV patients, the optimal cut-off value was 1.730, the corresponding specificity of 0.65 and sensitivity of 0.8393 were reported in mild HCV patients. For moderate patients, the optimal cutoff value was 1.110, corresponding specificity 0.51 and sensitivity was 0.7778. The optimal cutoff value was 1.395, and the corresponding specificity of 0.67 and sensitivity of 0.7321 were reported in severe liver patients. Similarly, in the microRNA-29a HCV patients, the Youden index of the optimal cutoff value was 0.6250, and the corresponding specificity (0.67) and sensitivity (1.00) were noted in mild patients. For the moderate HCV group, the optimal cutoff value was 1.660, the corresponding specificity was 0.8372 and the sensitivity was 0.7907. The optimal cutoff value was 0.2400, with the corresponding specificity of 0.58 and sensitivity of 1.00, in severe HCV patients.

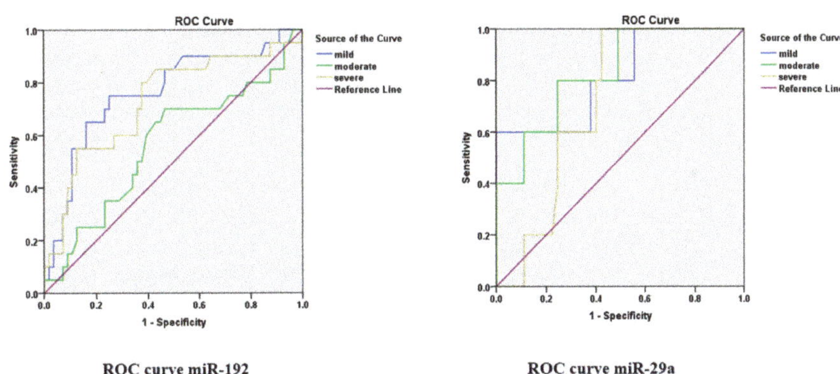

Figure 2. The ROC curve under the AUC for miR-192 and miR-29a in the HCV group patients.

3.3. Correlation of miR-192 and 29a with ALT

Next, we compared the miR-29a and the miR-192 levels with the ALT values of patients in the moderate, mild and severe groups. The Pearson and Spearman correlation were both not significant between the miR-29a or miR-192 levels in the control or HCV-infected groups or with ALT values.

3.4. Expression of miR-192 and miR-29a in HCV Genotypes 3 and Non-Genotypes 3

Since, in the studied cohort, we observed a high frequency of patients with genotype-3, we next investigated if there was a difference between miR-29a and miR-192 levels in patients with HCV genotype-3 infections versus patients with non-genotype-3 infections. Micro-RNA-29a levels in mild, moderate and severe patients were compared with the control cohort using Student's *t*-test. Compared to the control group, the miR-29a levels were significantly elevated in both patient groups with genotype-3 and non-genotype-3 HCV infections (Figure 3). The Bartlett's test in the one-way ANOVA confirmed the genotype-independent increase in miR-29a, showing an equal variance and a strong correlation between miR-29a levels in patients infected with HCV genotype-3 and non-genotype-3 (Figure 3).

Moreover, miR-192 was also significantly increased in patients with genotype-3 and non-genotype-3 versus healthy controls. In Figure 3, Bartlett's test showed that miR-192 was significantly upregulated ($p = 0.0173$) in comparison to healthy controls.

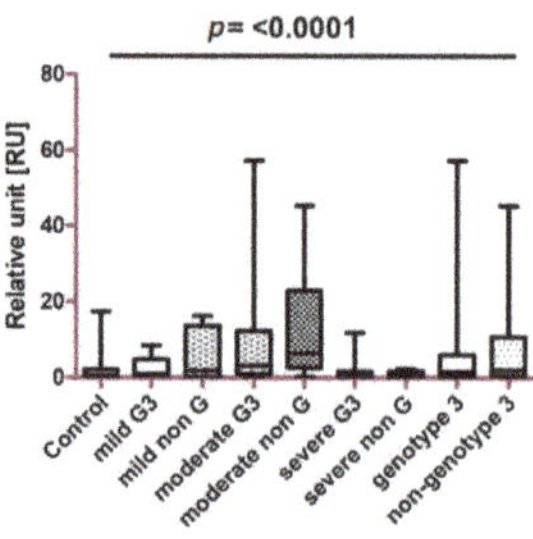

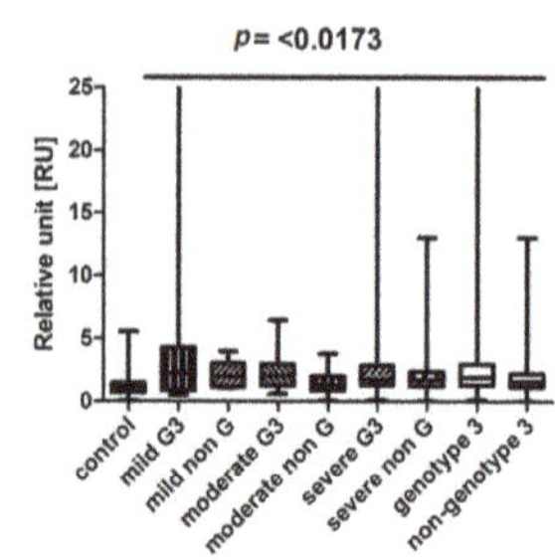

Figure 3. MicroRNA-29a and miR-192 profiles in the HCV genotype 3 and non-genotype 3. The Student test, comparing miR-29 levels of patients with mild, moderate and severe disease, either HCV genotype-3- or non-genotype—infected, with the control group, showed a highly significant increase ($p > 0.001$). The significant increase in the miR-192 levels was 0.0173, and in the patient groups with mild, moderate and severe infection, we observed p values of 0.004, 0.015, 0.0002 and 0.013, respectively, by Student t-test.

3.5. ROC Analysis in miR-192 and miR-29a in the HCV Genotypes 3 and Non-Genotypes 3

The ROC analysis showed no significant difference between the comparisons of miR-29a levels in the mild genotype-3 and mild non-genotype-3 groups compared with the control group. However, in comparison to the control group, the ROC curves showed a significant difference between the HCV genotype-3 and non-genotype-3 groups with moderate fibrosis ($p = 0.0018$ and $p < 0.0001$, respectively). ROC curves for the miR-29a levels of patients with genotype-3 or non-genotype-3 in comparison to healthy controls showed significance ($p = 0.030$ or $p = 0.0007$, respectively) (Figure 4 and Table 1). No significant difference was observed in the miR-29a ROC curves for genotype-3 and non-genotype-3.

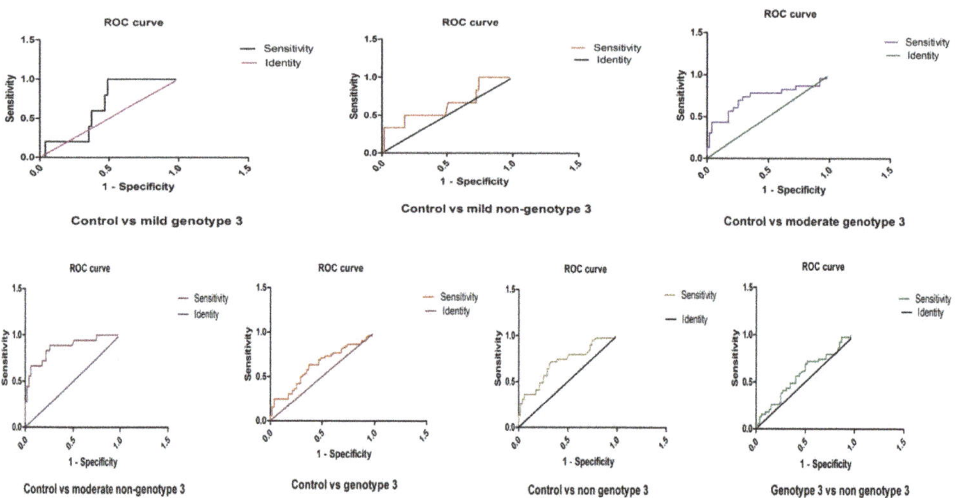

Figure 4. The ROC curve analysis of miR-29a in different groups of HCV patients.

In agreement with the data of miR-29a expression in HCV patients with genotype-3 versus non-genotype-3 infections, ROC analysis indicated that miR-29a might serve as an indicator for the HCV-associated progression of fibrosis independently of the HCV genotypes (Figure 4 and Table 1).

Table 1. MicroRNA-29a performance in HCV patients.

S. No.	Area under the ROC Curve	95% Confidence Interval	Std. Error	*p* Value
Control–Mild genotype 3	0.6549	0.4763 to 0.8335	0.09108	0.2564
Control–Mild non-genotype 3	0.6356	0.3783 to 0.8929	0.1312	0.2806
Control–Moderate genotype 3	0.7276	0.5866 to 0.8686	0.07193	<0.001829
Control–Moderate non-genotype 3	0.8687	0.7674 to 0.9700	0.05168	<0.0001
Control–Severe genotype 3	0.5176	0.3789 to 0.6562	0.07073	0.8071
Control–Severe non-genotype 3	0.5438	0.3880 to 0.6996	0.07949	0.6083
Control vs. Genotype 3	0.6237	0.5152 to 0.7321	0.05532	<0.03055
Control vs. Non-genotype 3	0.7079	0.5987 to 0.8171	0.05569	<0.0007660
Genotype 3 vs. Non-genotype 3	0.5816	0.4625 to 0.7007	0.06076	0.1845

The ROC analysis under the AUC demonstrates that miR-192 is significant in genotype-3 (p = 0.0001) and non-genotype-3 (p = 0.0003) in the control group. No significant expression was observed in the miR-192 and miR-29a ROC curve between HCV genotype-3 and non-genotype-3, as shown in Figure 5 and Table 2. Similarly, no significant difference was observed in the miR-192 ROC curve of moderate non-genotype-3 with the control group. Still, the other groups of miR-192 disclosed significant upregulation with healthy controls (Figure 6 and Supplementary Table S2). Interestingly, the ROC data showed no difference in the potential of miR-192 to indicate infections with HCV genotype-3 versus non-genotype-3 (Figure 5 and Table 2).

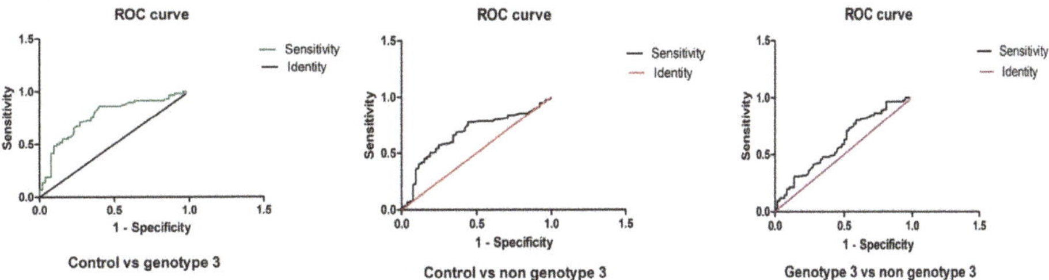

Figure 5. The ROC curve analysis of miR-192 in genotype 3, non-genotype 3 of HCV patients compared with the control group.

Table 2. MicroRNA-192 performance in HCV patients.

S. No.	Area under the ROC Curve	95% Confidence Interval	Std. Error	*p* Value
Control–Mild genotype 3	0.7200	0.4736 to 0.9663	0.1256	<0.04667
Control–Mild non-genotype 3	0.7981	0.6735 to 0.9227	0.06357	<0.002033
Control–Moderate genotype 3	0.7800	0.6703 to 0.8898	0.05596	<0.0001
Control–Moderate non-genotype 3	0.5971	0.4759 to 0.7184	0.06186	0.1142
Control–Severe genotype 3	0.7618	0.6461 to 0.8776	0.05905	<0.0001772
Control–Severe non-genotype 3	0.6945	0.5969 to 0.7920	0.04976	0.0002866
Control vs. Genotype 3	0.7636	0.6728 to 0.8544	0.04631	<0.0001
Control vs. Non-genotype 3	0.6718	0.5852 to 0.7583	0.04416	0.0003750
Genotype 3 vs. Non-genotype 3	0.6102	0.5206 to 0.6997	0.04568	0.01785

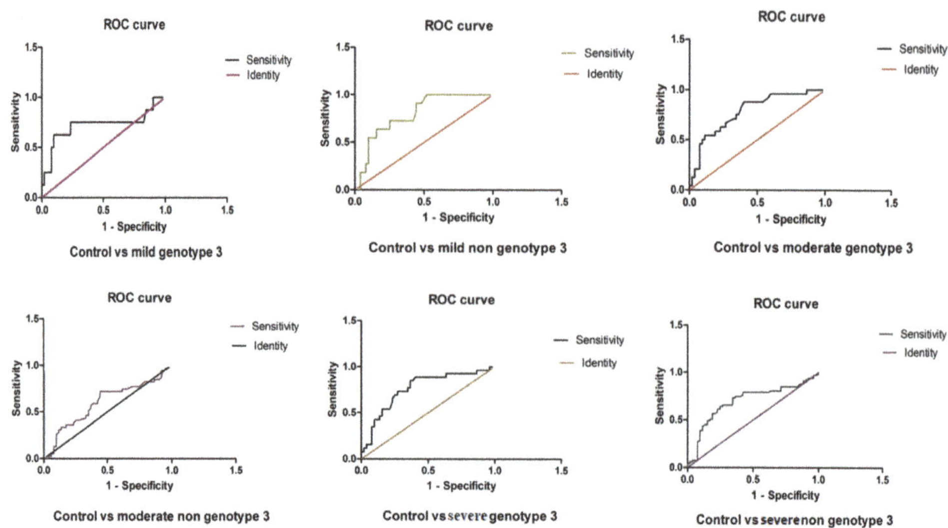

Figure 6. The ROC curve analysis of miR-192 with different genotypes of HCV patients.

4. Discussion

The present study focused on circulating miR-192 and 29a profiles in patients with hepatitis C disease and the comparative analysis at different stages of liver progression/injury. Various studies revealed that miR-192-5p and the miR-29 family are involved in human diseases, such as cancers, liver, kidney, nervous, heart and breast. Significantly, miR-192 and 29a levels are abundant in the urine and serum, and the exosomal stages in the circulation can help to diagnose and predict different diseases, such as hepatic disease, mostly those involving hepatitis infections [23,32]. MicroRNA-29a is known to act as antifibrotic miRNA, and its hepatic loss and secretion during liver disease are shown to be associated with liver fibrosis [23,24,26]. Extracellular miR-192 is induced after metabolic liver disease and HCV-mediated hepatitis and participates in inflammatory processes [32]. Many causative agents damage the liver, but microRNA-192-5p plays a role as a biomarker for diagnosis during hepatic injury. Thus, Roy et al. studied miR-192-5p in the serum of mice and human liver injury. The authors conclude that this miRNA was significantly upregulated in hepatic disease, but, in contrast, downregulated in the liver tissue [33].

In our study, samples of 222 patients with chronic hepatitis C infection were used to compare miR-192 and miR-29a serum levels with those of healthy controls. Both miRNAs, mir-29a and miR-192, were elevated during the liver disease progression of HCV patients. This was in agreement with other reports by Ezaz et al. [29] and Liu et al. [28], which have shown that miR-192-5p and the family 29a are significantly decreased in liver tissue but elevated in serum samples of patients with liver fibrosis [28,29]. The ROC analysis and area under the curve (AUC) were used to assess the diagnostic performance and accuracy of the miR-192 and 29a. Here, we noted that these microRNAs show a significant ROC performance and AUC curve in HCV patients. In the comparative analysis, the mild and moderate groups significantly outperformed the miR-192 and miR-29a, respectively. Overall, comparing these microRNAs showed a significant sensitivity rate in diagnosing hepatic disease. The ROC and AUC revealed that these microRNAs might be used as a biomarker to indicate the prognosis and severity of the liver injury. Similarly, the study of Motawi [34] analyzing the ROC and AUC in chronic HCV patients was closely related to our results. Several approaches and techniques have been used to identify and quantify microRNAs, resulting in discrepancies between reported results. One study found that miR-192 was elevated in HCV-infected patients with cirrhosis relative to those without cirrhosis and that it may boost HCV replication by targeting host genes involved in lipid metabolism. Another study demonstrated that miR-29a was downregulated in HCV-infected cirrhotic individuals and could impede HCV replication by targeting the 3′ UTR of the viral genome. A different study demonstrated that miR-29a could promote HCV replication by targeting host genes involved in lipid metabolism [35,36]. In light of these contradictory findings, the role of miRNA 192 and miRNA 29a in HCV replication remains unknown, and further research is required to identify their diagnostic value as biomarkers for HCV infection.

Most investigations indicate that miR-29a may function as an antiviral host factor by inhibiting HCV replication, and may have therapeutic promise for treating HCV infection. In addition, the overexpression of miR-192 and miR-29a in HCV patients may be attributable to their role in regulating fibrosis and inflammation, which are prominent characteristics of chronic liver disease [35,36]. MiR-192 and miR-29a may be possible non-invasive biomarkers for diagnosing fibrosis progression after chronic liver disease caused by HCV infection, as our investigation supports [36,37]. The causes of the variations in miR-192 and miR-29a levels between liver tissue and serum samples are unknown. It has been argued that sample preparation and analysis procedure variations may contribute to these disparities. Consequently, more studies are needed to establish the causes of these inconsistencies and to develop standardized methodologies for assessing miRNA concentrations in various sample types.

Regardless of the HCV genotype, our study provides more data supporting the potential of miR-192 and miR-29a as biomarkers for hepatic disease. These biomarkers may be useful for identifying the advancement of fibrosis in patients with chronic liver disease due to HCV infection [37,38]. Further research is necessary to assess the therapeutic efficacy of these biomarkers and to develop standardized methods for detecting miRNA levels in diverse sample types. Recent research has studied the diagnostic potential of miR-192 and miR-29a as biomarkers for HCV patients, although the findings are limited. One such limitation is the pilot study's relatively small sample size, which necessitates higher sample sizes to confirm the diagnostic usefulness of these miRNAs.

Various challenges must be overcome to fully comprehend the diagnostic biomarker potential of microRNAs [8,9]. The lack of standardization in miRNA isolation and quantification, inconsistencies in sample preparation, the presence of contaminants and inhibitors, a limited understanding of miRNA biology, and the absence of large-scale validation studies to determine the reliability and robustness of miRNA-based diagnostic tests are among these impediments.

Future studies could emphasize the diagnostic and prognostic capabilities of miR-192 and miR-29a in expanded and more diverse patient populations. Investigating miRNA expression in various HCV genotypes and disease phases could also shed light on the potential use of miRNAs as biomarkers for hepatic disease. In addition, understanding the mechanisms underlying the overexpression of miR-192 and miR-29a throughout chronic HCV infection could reveal prospective therapeutic targets for treating HCV-related liver disorders. By conquering these challenges and pursuing future research routes, we can improve the translation of miRNA research into clinical practice, resulting in patient outcomes.

Overall, our study demonstrates that themiR-29a and miR-192 s are significantly upregulated in patients with HCV genotype-3 and non-genotypes-3 in a similar pattern. Thus, both miRNAs could be an indicator of hepatic disease progression independently of the HCV genotype.

5. Conclusions

The presented study reveals that serum microRNA-192 and miR-29a are significantly increased in patients with chronic hepatic disease infected by HCV genotype 3 and non-genotype 3. Notably, these miRNAs appear to be a sensitive predictor of liver disease and may be better-suited than other regular biochemical assays to monitoring HCV-related liver injury.

Supplementary Materials: The following supporting information can be downloaded at: https://www.mdpi.com/article/10.3390/genes14051056/s1, Table S1: (A): The demographic and clinical features of the enrolled patients; (B): LFTs, Genotypes, Risk factors of chronic HCV Patients; (C): List of healthy controls (n = 52), used in our study; Table S2: MicroRNA-192 performance in HCV patients; Table S3: MicroRNA-29a diagnostic performance in HCV patients; Figure S1. microRNA-192; Figure S2. microRNA-29a; Figure S3. ROC-192; Figure S4. ROC-29a.

Author Contributions: A.U. conducted research under the supervision of B.A. and M.O., X.Y., J.A., K.O. and I.U.R. helped in the data analysis. T.N. and Q.A. helped in the interpretation of results. N.A. and Q.A. made final revisions of the manuscript. All authors have read and agreed to the published version of the manuscript.

Funding: This research was supported by the R&D and Fortune funding of the Medical Faculty of the University of Cologne, Germany. In addition, we are very grateful to the Higher Education Commission of Pakistan that the first author Amin Ullah has got an IRSIP fellowship to visit the Institute for Pathology, University of Cologne, Germany.

Institutional Review Board Statement: This article does not contain any studies with human participants or animals performed by any of the authors. It has been confirmed that the experimental samples of plants, including the collection of plant material, complied with relevant institutional, national, and international guidelines and legislation with appropriate permissions from authorities Department of Health and Biological Sciences, Abasyn University, Peshawar, Pakistan and Center of Biotechnology and Microbiology, University of Peshawar, Peshawar, Pakistan for the collection of specimens.

Informed Consent Statement: Not applicable.

Data Availability Statement: All data generated or analyzed during this study are included in the manuscript and its supplementary files.

Acknowledgments: We are very thankful to the tertiary care hospitals of Peshawar, Pakistan for helping us in collecting the samples and the data. Furthermore, we appreciate the technical support of Ulrike Koitzsch and Hannah Eischeidt-Scholz (Institute for Pathology; University Hospital Cologne, Germany), who assisted in RNA isolation and qPCR quantification.

Conflicts of Interest: The authors declare no conflict of interest.

References

1. AbdElrazek, S.E.; Bilasy, A.E.; AbdElhalim, A.E. Prior to the oral therapy, what do we know about HCV-4 in Egypt: A randomized survey of prevalence and risks using data mining computed analysis. *Medicine* **2014**, *93*, e204. [CrossRef]
2. Lanini, S.; Ustianowski, A.; Pisapia, R.; Zumla, A.; Ippolito, G. Viral hepatitis: Etiology, epidemiology, transmission, diagnostics, treatment, and prevention. *Infect. Dis. Clin.* **2019**, *33*, 1045–1062. [CrossRef] [PubMed]
3. Ullah, A.; Rehman, I.U.; Ahmad, J.; Gohar, M.; Ahmad, S.; Ahmad, B. Hepatitis-C Virus and Cirrhosis: An Overview from Khyber Pakhtunkhwa Province of Pakistan. *Viral Immunol.* **2020**, *33*, 396–403. [CrossRef] [PubMed]
4. Idrees, M.; Riazuddin, S. Frequency distribution of hepatitis C virus genotypes in different geographical regions of Pakistan and their possible routes of transmission. *BMC Infect. Dis.* **2008**, *8*, 69. [CrossRef]
5. Butt, S.; Idrees, M.; Akbar, H.; Rehman, I.U.; Awan, Z.; Afzal, S.; Hussain, A.; Shahid, M.; Manzoor, S.; Rafique, S. The changing epidemiology pattern and frequency distribution of hepatitis C virus in Pakistan. *Infect. Genet. Evol.* **2010**, *10*, 595–600. [CrossRef]
6. Gonzalez, H.C.; Duarte-Rojo, A. Virologic cure of hepatitis C: Impact on hepatic fibrosis and patient outcomes. *Curr. Gastroenterol. Rep.* **2016**, *18*, 32. [CrossRef] [PubMed]
7. Adinolfi, L.E.; Gambardella, M.; Andreana, A.; Tripodi, M.F.; Utili, R.; Ruggiero, G. Steatosis accelerates the progression of liver damage of chronic hepatitisC patients and correlates with specific HCV genotype and visceral obesity. *Hepatology* **2001**, *33*, 1358. [CrossRef]
8. Loosen, S.H.; Schueller, F.; Trautwein, C.; Roy, S.; Roderburg, C. Role of circulating microRNAs in liver diseases. *World J. Hepatol.* **2017**, *9*, 586. [CrossRef]
9. Schueller, F.; Roy, S.; Vucur, M.; Trautwein, C.; Luedde, T.; Roderburg, C. The Role of miRNAs in the Pathophysiology of Liver Diseases and Toxicity. *Int. J. Mol. Sci.* **2018**, *19*, 261. [CrossRef]
10. Dexheimer, P.J.; Cochella, L. MicroRNAs: From mechanism to organism. *Front. Cell Dev. Biol.* **2020**, *8*, 409. [CrossRef]
11. Liang, H.; Gong, F.; Zhang, S.; Zhang, C.Y.; Zen, K.; Chen, X. The origin, function, and diagnostic potential of extracellular microRNAs in human body fluids. *Wiley Interdiscip. Rev. RNA* **2014**, *5*, 285–300. [CrossRef]
12. Fabian, M.R.; Sonenberg, N.; Filipowicz, W. Regulation of mRNA translation and stability by microRNAs. *Annu. Rev. Biochem.* **2010**, *79*, 351–379. [CrossRef] [PubMed]

13. Bueno, M.J.; de Castro, I.P.; Malumbres, M. Control of cell proliferation pathways by microRNAs. *Cell Cycle* **2008**, *7*, 3143–3148. [CrossRef] [PubMed]
14. Lee, C.H.; Kim, J.H.; Lee, S.W. The role of microRNAs in hepatitis C virus replication and related liver diseases. *J. Microbiol.* **2014**, *52*, 445–451. [CrossRef]
15. Yu, X.; Odenthal, M.; Fries, J.W. Exosomes as miRNA Carriers: Formation-Function-Future. *Int. J. Mol. Sci.* **2016**, *17*, 2028. [CrossRef] [PubMed]
16. Ruiz-Manriquez, L.M.; Carrasco-Morales, O.; Osorio-Perez, S.M.; Estrada-Meza, C.; Pathak, S.; Banerjee, A.; Bandyopadhyay, A.; Duttaroy, A.K.; Paul, S. MicroRNA-mediated regulation of key signaling pathways in hepatocellular carcinoma: A mechanistic insight. *Front. Genet.* **2022**, *13*, 910733. [CrossRef] [PubMed]
17. Morishita, A.; Oura, K.; Tadokoro, T.; Fujita, K.; Tani, J.; Masaki, T. MicroRNAs in the pathogenesis of hepatocellular carcinoma: A review. *Cancers* **2021**, *13*, 514. [CrossRef]
18. Cabral, B.; Hoffmann, L.; Bottaro, T.; Costa, P.; Ramos, A.; Coelho, H.; Villela-Nogueira, C.; Ürményi, T.; Faffe, D.; Silva, R. Circulating microRNAs associated with liver fibrosis in chronic hepatitis C patients. *Biochem. Biophys. Rep.* **2020**, *24*, 100814. [CrossRef]
19. Trebicka, J.; Anadol, E.; Elfimova, N.; Strack, I.; Roggendorf, M.; Viazov, S.; Wedemeyer, I.; Drebber, U.; Rockstroh, J.; Sauerbruch, T.; et al. Hepatic and serum levels of miR-122 after chronic HCV-induced fibrosis. *J. Hepatol.* **2013**, *58*, 234. [CrossRef]
20. Xu, J.; Wu, C.; Che, X.; Wang, L.; Yu, D.; Zhang, T.; Huang, L.; Li, H.; Tan, W.; Wang, C.; et al. Circulating microRNAs, miR-21, miR-122, and miR-223, in patients with hepatocellular carcinoma or chronic hepatitis. *Mol. Carcinog.* **2011**, *50*, 136–142. [CrossRef]
21. Kumar, A.; Rajput, M.K.; Paliwal, D.; Yadav, A.; Chhabra, R.; Singh, S. Genotyping & diagnostic methods for hepatitis C virus: A need of low-resource countries. *Indian J. Med. Res.* **2018**, *147*, 445.
22. Cabrera, M.; Kolli, M.; Jaggi, M.; Chauhan, S.C.; Yallapu, M.M. miRNA-205: A future therapeutic molecule for liver diseases. *Future Drug Discov.* **2023**, *4*, FDD78. [CrossRef] [PubMed]
23. Yu, X.; Elfimova, N.; Müller, M.; Bachurski, D.; Koitzsch, U.; Drebber, U.; Mahabir, E.; Hansen, H.P.; Friedman, S.L.; Klein, S.; et al. Autophagy-Related Activation of Hepatic Stellate Cells Reduces Cellular miR-29a by Promoting Its Vesicular Secretion. *Cell. Mol. Gastroenterol. Hepatol.* **2022**, *13*, 1701–1716. [CrossRef]
24. Noetel, A.; Kwiecinski, M.; Elfimova, N.; Huang, J.; Odenthal, M. microRNA are central players in anti-and profibrotic gene regulation during liver fibrosis. *Front. Physiol.* **2012**, *3*, 49. [CrossRef] [PubMed]
25. Van Rooij, E.; Sutherland, L.B.; Thatcher, J.E.; DiMaio, J.M.; Naseem, R.H.; Marshall, W.S.; Olson, E.N. Dysregulation of microRNAs after myocardial infarction reveals a role of miR-29 in cardiac fibrosis. *Proc. Natl. Acad. Sci. USA* **2008**, *105*, 13027–13032. [CrossRef] [PubMed]
26. Roderburg, C.; Urban, G.W.; Bettermann, K.; Vucur, M.; Zimmermann, H.; Schmidt, S.; Luedde, T. Micro-RNA profiling reveals a role for miR-29 in human and murine liver fibrosis. *Hepatology* **2011**, *53*, 209–218. [CrossRef]
27. Kim, J.H.; Lee, C.H.; Lee, S.W. Hepatitis C virus infection stimulates transforming growth factor-β1 expression through up-regulating miR-192. *J. Microbiol.* **2016**, *54*, 520–526. [CrossRef]
28. Liu, X.-L.; Pan, Q.; Cao, H.-X.; Xin, F.-Z.; Zhao, Z.-H.; Yang, R.-X.; Zeng, J.; Zhou, H.; Fan, J.-G. Lipotoxic hepatocyte-derived exosomal microRNA 192-5p activates macrophages through rictor/Akt/forkhead box transcription factor O1 signaling in nonalcoholic fatty liver disease. *Hepatology* **2020**, *72*, 454–469. [CrossRef]
29. Ezaz, G.; Trivedi, H.D.; Connelly, M.A.; Filozof, C.; Howard, K.; Parrish, M.; Kim, M.; Herman, M.A.; Nasser, I.; Afdhal, N.H.; et al. Differential associations of circulating MicroRNAs with pathogenic factors in NAFLD. *Hepatol. Commun.* **2020**, *4*, 670–680. [CrossRef]
30. Ullah, A.; Yu, X.; Odenthal, M.; Meemboor, S.; Ahmad, B.; Rehman, I.U.; Ahmad, J.; Ali, Q.; Nadeem, T. Circulating microRNA-122 in HCV cirrhotic patients with high frequency of genotype 3. *PLoS ONE* **2022**, *17*, e0268526. [CrossRef]
31. Anadol, E.; Schierwagen, R.; Elfimova, N.; Tack, K.; Schwarze-Zander, C.; Eischeid, H.; Noetel, A.; Boesecke, C.; Jansen, C.; Dold, L.; et al. Circulating microRNAs as a marker for liver injury in human immunodeficiency virus patients. *Hepatology* **2015**, *61*, 46–55. [CrossRef] [PubMed]
32. Ren, F.-J.; Yao, Y.; Cai, X.-Y.; Fang, G.-Y. Emerging Role of MiR-192-5p in Human Diseases. *Front. Pharmacol.* **2021**, *12*, 614068. [CrossRef] [PubMed]
33. Roy, S.; Benz, F.; Alder, J.; Bantel, H.; Janssen, J.; Vucur, M.; Gautheron, J.; Schneider, A.; Schüller, F.; Loosen, S.; et al. Downregulation of miR-192-5p protects from oxidative stress-induced acute liver injury. *Clin. Sci.* **2016**, *130*, 1197–1207. [CrossRef]
34. Motawi, T.K.; Shaker, O.G.; El-Maraghy, S.A.; Senousy, M.A. Serum interferon-related microRNAs as biomarkers to predict the response to interferon therapy in chronic hepatitis C genotype 4. *PLoS ONE* **2015**, *10*, e0120794. [CrossRef]
35. Mahdy, M.M.; El-Ekiaby, N.M.; Hashish, R.M.; Salah, R.A.; Hanafi, R.; Azzazy, H.M.E.-S.; Abdelaziz, A.I. miR-29a promotes lipid droplet and triglyceride formation in HCV infection by inducing expression of SREBP-1c and CaV1. *J. Clin. Transl. Hepatol.* **2016**, *4*, 293. [PubMed]
36. Bandyopadhyay, S.; Friedman, R.C.; Marquez, R.T.; Keck, K.; Kong, B.; Icardi, M.S.; Brown, K.E.; Burge, C.B.; Schmidt, W.N.; Wang, Y.; et al. Hepatitis c virus infection and hepatic stellate cell activation downregulate mir-29: Mir-29 overexpression reduces hepatitis c viral abundance in culture. *J. Infect. Dis.* **2011**, *203*, 1753–1762. [CrossRef] [PubMed]

37. Kim, T.H.; Lee, Y.; Lee, Y.-S.; Gim, J.-A.; Ko, E.; Yim, S.Y.; Jung, Y.K.; Kang, S.; Kim, M.Y.; Kim, H.; et al. Circulating mirna is a useful diagnostic biomarker for nonalcoholic steatohepatitis in nonalcoholic fatty liver disease. *Sci. Rep.* **2021**, *11*, 14639. [CrossRef]
38. Iacob, D.G.; Rosca, A.; Ruta, S. Circulating micrornas as non-invasive biomarkers for hepatitis b virus liver fibrosis. *World J. Gastroenterol.* **2020**, *26*, 1113–1127. [CrossRef]

genes

MDPI

Article

MiR-612, miR-637, and miR-874 can Regulate VEGFA Expression in Hepatocellular Carcinoma Cell Lines

Márcia Maria U. Castanhole-Nunes [1], Nathalia M. Tunissiolli [1], André R. C. P. Oliveira [1], Marlon F. Mattos [1], Ana Lívia S. Galbiatti-Dias [1], Rosa S. Kawasaki-Oyama [1], Erika C. Pavarino [1], Renato F. da Silva [2] and Eny M. Goloni-Bertollo [1,*]

[1] Research Unit of Genetics and Molecular Biology (UPGEM), Department of Molecular Biology, Faculty of Medicine of Sao Jose do Rio Preto (FAMERP), Sao Jose do Rio Preto 15090-000, SP, Brazil; mcastanhole@gmail.com (M.M.U.C.-N.); nathaliatunisiolli@yahoo.com (N.M.T.); rodrigueiro@gmail.com (A.R.C.P.O.); marlon.fraga.mattos@outlook.com (M.F.M.); analiviagalbiattidias@gmail.com (A.L.S.G.-D.); rosa.oyama@famerp.br (R.S.K.-O.); erika@famerp.br (E.C.P.)
[2] Department of Surgery and Liver Transplantation, Hospital de Base/FUNFARME/FAMERP-Foundation Faculty of Medicine of Sao Jose do Rio Preto, Sao Jose do Rio Preto 15090-000, SP, Brazil; renatosilva@famerp.br
* Correspondence: eny.goloni@famerp.br; Tel.: +55-1732-015-908

Abstract: MicroRNAs (miRNAs) are short non-coding RNA molecules acting as important posttranscriptional gene and protein expression regulators in cancer. The study goal was to examine VEGFA (vascular endothelial growth factor A) expression in hepatocellular carcinoma (HCC) cell lines upon transfection miR-612, miR-637, or miR-874. **Methods:** MiR-612 mimics, miR-637 mimics, or miR-874 inhibitors were transfected using Lipofectamine RNAiMax in both HCC cell lines, HepG2 and HuH-7. Real-time PCR, Western blotting, and ELISA methods were used to evaluate VEGFA regulation by the miRNAs. **Results:** Gene and protein expression levels of VEGFA were down-expressed in both cell lines, HepG2 and HuH-7, transfected with miR-612 or miR-637. Transfection with miR-874 inhibitor showed an increase in VEGFA gene expression in HepG2 and HuH-7 cell lines; however, no regulation was observed on VEGFA protein expression by miR-874 inhibition. Correlation analysis between miRNAs and VEGFA protein expression showed that miR-637 and miR-874 expression present inversely correlated to VEGFA protein expression. **Conclusions:** VEGFA was down-regulated in response to hsa-miR-612 or hsa-miR-637 overexpression; however, the modulation of VEGFA by miR-874 was observed only at the gene expression and thus, needs further investigation.

Keywords: microRNAs; gene expression; liver neoplasms; angiogenesis; cancer biomarkers; transfection

Citation: Castanhole-Nunes, M.M.U.; Tunissiolli, N.M.; Oliveira, A.R.C.P.; Mattos, M.F.; Galbiatti-Dias, A.L.S.; Kawasaki-Oyama, R.S.; Pavarino, E.C.; da Silva, R.F.; Goloni-Bertollo, E.M. MiR-612, miR-637, and miR-874 can Regulate VEGFA Expression in Hepatocellular Carcinoma Cell Lines. *Genes* **2022**, *13*, 282. https://doi.org/10.3390/genes13020282

Academic Editors: Giuseppe Iacomino and Fabio Lauria

Received: 23 December 2021
Accepted: 28 January 2022
Published: 30 January 2022

Publisher's Note: MDPI stays neutral with regard to jurisdictional claims in published maps and institutional affiliations.

1. Introduction

Hepatocellular carcinoma (HCC) is the most common primary liver malignancy and is a leading cause of cancer-related deaths worldwide [1]. The main risk factors include chronic infection with the hepatitis B or hepatitis C virus, consumption of aflatoxin-contaminated foods, heavy alcohol intake, obesity, smoking, and diabetes [2]. HCC is more prevalent in males than in females (ratio 1:2.4), with high occurrence rates in eastern and southern Asia. As estimated in 2012 by GLOBOCAN, the ratio of incidence and mortality of HCC is 0.95, indicating a poor prognosis [3].

MicroRNAs (miRNAs) are short non-coding RNA molecules acting as important posttranscriptional gene and protein expression regulators [4–7]. Some miRNAs can regulate the vascular endothelial growth factor A (VEGFA) gene, which acts in blood vessel growth in some cancer types, including hepatocellular carcinoma [8–12]. Studies in cancer, including HCC, have shown that miRNAs have an essential role in angiogenesis, tumorigenesis, and metastasis [13–15]. A review highlighted that miRNAs are meaningful for regulating particular endothelial processes downstream of VEGF, thus representing therapeutic targets involved in the VEGF ligand-receptor interaction or VEGFR kinase activity [9].

VEGF is a growth factor that activates receptor tyrosine kinases, initiating the RAS-RAF-MEK (Map kinase)-ERK (extracellular signals regulated kinase) MAPK (Mitogen-activated protein kinase) signaling cascade. This cascade induces key transcription factors as well as the epithelial-mesenchymal transition that results in cell motility and invasion [16]. VEGFA products in cancer have been studied with their interaction in different signaling pathways, such as STAT3 (activator of transcription 3), KRAS (Kirsten rat sarcoma virus), and MAPK, mediated by protein kinase B (PI3K) and regulated by ERK, among other factors. These interactions were observed in different types of cancer, such as liver and lung. In some cases, the activation of these pathways is related to tumor aggressiveness and, therefore, regulation through miRNAs [16,17]. Evidence has suggested that the regulation of miRNAs in cancer can identify biomarkers for the diagnosis and treatment of cancer [7,12]. Because the inhibition of VEGFA signaling interferes in the angiogenesis process, and miRNAs may provide a potential anti-angiogenesis therapy for cancer treatment, we evaluated the expression of VEGFA in HCC cell lines upon treatment with miR-612 and miR-637 mimics and miR-874 inhibitors.

2. Materials and Methods

2.1. Cell Lines

Hepatoma cell line HepG2, derived from a liver HCC of a 15-year-old Caucasian male [18], and HuH-7 cell line from HCC taken from a liver tumor of a 57-year-old Japanese male [19] were used in the study. Both cell lines were cultured in DMEM (Dulbecco's modified Eagle's medium) (Cultilab, Campinas, São Paulo, BR) supplemented with 10% fetal bovine serum (Cultilab, Campinas, São Paulo, BR), 100 U/mL sodium penicillin, 100 mg/mL streptomycin (Cultilab, Campinas, São Paulo, BR), and 1% L-glutamine (Cultilab, Campinas, São Paulo, BR) at 37 °C in a 5% CO_2 atmosphere.

2.2. MiRNA Prediction

Seventeen miRNAs predicted by the miRNAs databases, miRBASE (http://www.mirbase.org/ (accessed on 20 March 2017)), TargetScan (http://www.targetscan.org/vert_71 (accessed on 20 March 2017)), and DIANA-TarBase v7.0 databases (http://diana.imis.athena-innovation.gr/DianaTools (accessed on 16 July 2017)) in our previous study were selected [20]. A previous study analyzing relative miRNA expression in 40 samples (HCC and non-tumor tissue) showed that 9 out of 17 predicted miRNAs were differentially expressed in tumor tissues compared to the control. The relative VEGFA gene expression showed a relationship to the miRNA expression [20]. In the present study, we used the mimics of low expressed miR-612, miR-637, and an inhibitor of highly expressed miR-874 to study the regulation of the VEGFA target gene.

2.3. Transfection of miRNAs in HepG2 and HuH-7 Cell Lines

Transfection assays of the mirVana™ inhibitor for hsa-miR-874 (MH12355, Thermo Scientific, Waltham, Massachusetts, EUA), mirVana™ hsa-miR-612 mimics (MC11461, Thermo Scientific, Waltham, Massachusetts, EUA), and mirVana™ hsa-miR-637 mimics (MC11545, Thermo Scientific, Waltham, Massachusetts, EUA) were conducted using Lipofectamine RNAiMax (Invitrogen, Waltham, Massachusetts, EUA), following the manufacturer's instructions. To determine the best concentration values of the mirVana™ inhibitor, mirVana™ mimics and Lipofectamine RNAiMax were performed concentration tests. Subsequently, the cells were cultured for 48 h in 100 µL of Opti-MEM serum-free medium (Invitrogen, Waltham, Massachusetts, EUA), 1 µL of Lipofectamine RNAiMax (Invitrogen, Waltham, Massachusetts, EUA), and 10 mM of miR-874 inhibitor, miR-612, or miR-637 mimics. Three independent experiments were performed. After this, the RNA was extracted to verify the efficiency of transfection using the respective positive and negative control genes and miRNAs by qPCR (Quantitative real-time PCR). The positive controls for inhibitor assay were the HMGA2 gene and let-7c miRNA. Similarly, the TWF1 gene and miR-1 were positive controls for mimic assays.

2.4. RNA and miRNA Extraction

After transfection, RNA and miRNA were extracted from HepG2 and HuH-7 cell lines using Trizol reagent (Invitrogen, Waltham, Massachusetts, EUA). Complementary DNA (cDNA) was synthesized using the High Capacity cDNA reverse transcription Kit (Thermo Fisher Scientific, Waltham, Massachusetts, EUA). The cDNAs of miRNAs were synthesized using Taqman MicroRNA Reverse Transcription Kit (Applied Biosystems, Waltham, Massachusetts, EUA).

2.5. Quantitative Real-Time PCR for Expression of VEGFA Gene and miRNAs

Expression analysis of the VEGFA (Hs00900055_m1) gene and miR-612 (001579), miR-637 (003307), and miR-874 (00268) was performed by qPCR using specific TaqMan probes (Thermo Fisher Scientific, Waltham, Massachusetts, EUA) on the CFX 96 Real-Time System (Bio-Rad, Hercules, California, EUA). All reactions were performed in duplicate and included a contamination control. The genes GAPDH (Hs03929097_g1) and RPLPO (4333761F) were used as reference genes for the normalization of VEGFA expression data. The genes U6 (001973) and RNU48 (001006) were used for normalization of miR-612, miR-637, and miR-874 expression data (Thermo Fisher Scientific, Waltham, Massachusetts, EUA). Relative quantification (RQ) of genes and miRNA expression levels in HCC lineages were calculated using the $2^{-\Delta\Delta Ct}$ method concerning the negative control [21].

2.6. Extraction and Quantification of Protein

The proteins were extracted using the RIPA Buffer (Sigma-Aldrich, San Luis, Missouri, EUA), and quantified using the Pierce ^{TM} BCA Protein Assay kit (Thermo Fisher Scientific, Waltham, Massachusetts, EUA). Quantification of VEGFA protein in the transfection assays with miR-612, miR-637, miR-874, and negative control was performed using the VEGFA Duo Set ELISA Kit (R&D Systems, Minneapolis, Minnesota, EUA) and Western blotting (WB) method utilizing Anti-VEGFA antibody (ab1316, Abcam, Cambridge, United Kingdom), following the manufacturer's instruction.

2.7. Statistical Analysis

The statistical analysis was performed using GraphPad Prism software version 6. Continuous data distribution was evaluated using D'Agostino and Pearson's normality test. Student's *t* test, Wilcoxon Signed rank test, and Mann–Whitney test were used to evaluate the VEGFA gene and protein expression data. The correlation between the expression of miRNAs and VEGFA proteins was analyzed by Spearman's correlation or Pearson's correlation. Values of $p < 0.05$ were considered significant.

3. Results

3.1. Bioinformatics-TCGA Analysis

The analysis of The Cancer Genome Atlas Program (TCGA) database through the UALCAN [22] website showed that there is a positive correlation between KRAS (Figure 1a), AKT1 (Figure 1b), and STAT3 (Figure 1c) genes in Liver Hepatocellular carcinoma (LIHC) and VEGFA expression.

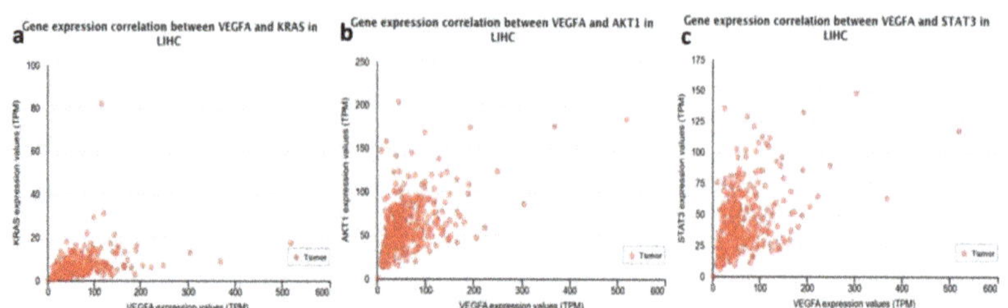

Figure 1. Pearson positive correlation of VEGFA expression with (**a**) KRAS (R = 0.36), (**b**) AKT1 (R = 0.45), and (**c**) STAT3 (R = 0.4) in TCGA analysis in LIHC [22].

In addition, miR-612 (Figure 2a) and miR-874-3p (Figure 2b) expression analyzes were performed in Hepatocellular carcinoma (HCC), comparing normal tissues with primary tumors, allowing visualization of the expression profile in primary tumor tissue samples.

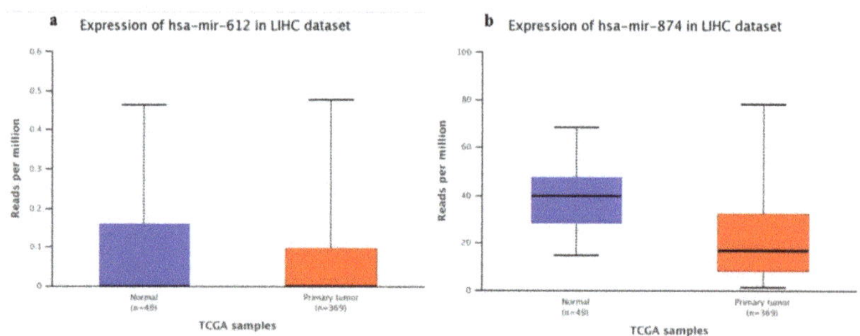

Figure 2. TCGA expression data of (**a**) hsa-miR-612 (p = 0.52) and (**b**) miR-874-3p (p = 0.20) in LIHC, comparing normal tissue and primary tumor [22].

Regarding miR-637, it was not possible to obtain this information as there are no data in the TCGA for this microRNA in HCC.

3.2. Transfection Efficiency Test

The transfection efficiency was calculated by the relative expression levels of TWF1 in cells treated with mirVana ™ miRNA Mimic miR-1 positive control (Applied Biosystems, Waltham, Massachusetts, EUA) in comparison to the negative control, to be approximately 75% in the HuH-7 cell line and 65% in the HepG2 cell line. Transfection efficiency test for the inhibition assay using positive controls, mirVana™ miRNA Inhibitor let-7c (Applied Biosystems, Waltham, Massachusetts, EUA) for HMGA2 gene expression showed a 31% decrease in HepG2 cells and a 57% increase in HuH-7 cells. This divergence between the cell lines may be related to differences in the gene regulation by miRNA and warrants further evaluation.

3.3. MiR-874, miR-612, and miR-637 Expression in Transfected HCC Cells

The expression levels of all three miRNAs were found to be up-regulated as compared to the negative controls in both HCC cell lines. Following are the statistical values obtained for each miRNA in three independent experiments: miR-874 (HuH-7: p = 0.1250; median = 10.52; HepG2: p = 0.1250; median = 5663), miR-612 (HuH-7: p = 0.0313; median = 2.053; HepG2: p = 0.0313; median = 14.03), and miR-637 (HuH-7: p = 0.313; median = 6.115; HepG2: p = 0.313; mean = 12.34).

3.4. VEGFA Expression in Transfected HCC Cell Lines

We performed an assay to evaluate the influence of miR-612 and miR-637 on VEGFA gene expression and observed significant difference in the expression of this gene upon transfection with miR-612 and miR-637 mimics. Expression of VEGFA was lower in miR-637-transfected cells than in miR-612-transfected cells. In the HepG2 cell line, the VEGFA gene expression showed a 51% reduction upon transfection with miR-612 (RQ mean = 0.49, $p < 0.0001$), and a 48% reduction upon transfection with miR-637 (RQ mean = 0.52, $p = 0.0004$). In the HuH-7 cell line, the expression of VEGFA decreased by 4% in cells treated with miR-612 ($p < 0.0001$), and by 73% with miR-637 ($p < 0.0001$). All values were obtained by comparing them with the negative control.

Cells transfected with miR-874 inhibitor showed a 93% increase in VEGFA expression (RQ = 1.93, $p = 0.0005$) in the HepG2 cell line compared to the negative control, and equal or low expression in the HuH-7 cell line when compared to the control (RQ = 1.00, $p < 0.0001$) (Figure 3).

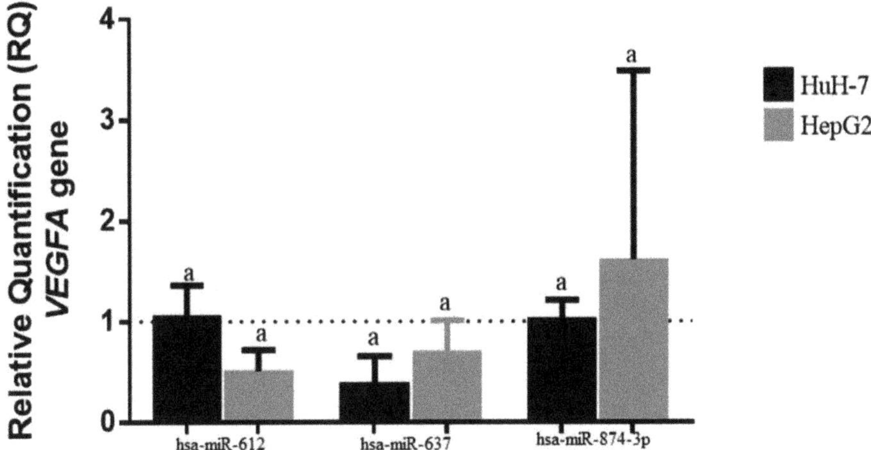

Figure 3. Relative quantification of VEGFA gene, transfected to miR-612, miR-637, and miR-874 on HepG2 and HuH-7 concerning the respective negative control (RQ = 1.00). a—Corresponding to significant *p* values.

3.5. Protein Expression of VEGFA in Transfected HCC Cell Lines

VEGFA protein quantification was performed by ELISA (Figure 4) and WB (Figure 5). WB results showed a reduced expression in miR-612 (HuH-7 cell line) and miR-637 (HuH-7 and HepG2 cell line). However, it was not possible to evaluate the transfection using miR-874-3p. ELISA methods showed the quantification after transfection with the miR-612 (HuH-7: $p = 0.0106$, mean= 26.90; HepG2: $p > 0.9999$, mean = 16.81) and miR-637 (HuH-7: $p = 0.0168$, mean = 31.54; HepG2: $p = 0.0051$, median = 23.30) mimics, and the miR-874 inhibitor (HuH-7: $p = 0.6367$, mean = 39.66; HepG2: $p = 0.6222$, median = 15.30). The protein expression was analyzed by comparing to the same target protein evaluated in transfection performed in the mimic negative controls (HuH-7: mean = 45.24; HepG2: median = 20.68) and inhibitor negative controls (HuH-7: mean = 43.04; HepG2: median = 17.73) (Figure 4).

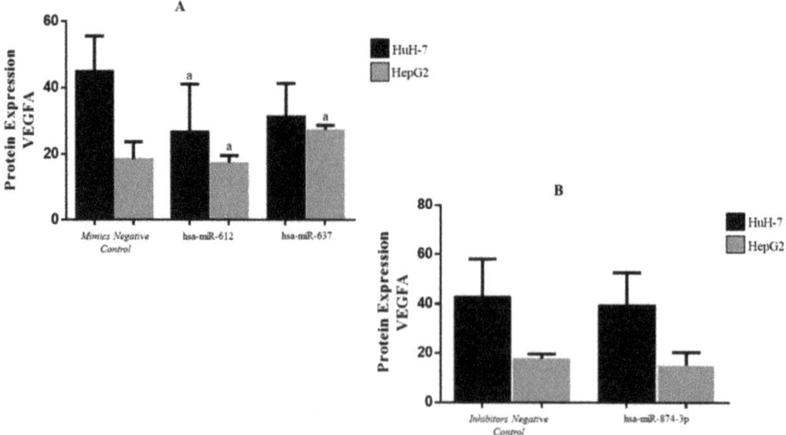

Figure 4. VEGFA protein expression compared to the respective negative control, transfected to miR-612, miR-637 (**A**), and miR-874 (**B**) on HepG2 and HuH-7 cell lines (ELISA method). a—Corresponding to significant *p* values.

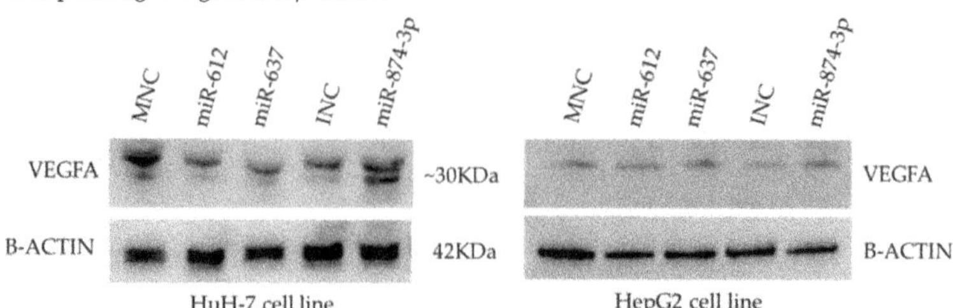

Figure 5. Western blotting analysis of VEGFA treated with the mimics negative control (MNC), hsa-miR-612, hsa-miR-637, inhibitor negative control (INC), and hsa-miR-874-3p, in the HuH-7 and HepG2 cell lines.

The correlation between the gene and protein expression was not significant. However, the miRNA and protein expression correlation analysis showed that expression levels of miR-637 (R = -0.883 *p* = 0.0110) and miR-874 (R = -0.084 *p* = 0.0010) have an inverse correlation with VEGFA protein expression.

4. Discussion

In the present study, we utilized mimics for under-expressed miRNAs and inhibitors for overexpressed miRNA. In our preliminary study in HCC and non-tumor tissue samples (n = 40), we observed higher expression of VEGFA in tumor tissues than in non-tumor samples. Additionally, nine miRNA targets of VEGFA with relative differential expression were identified [20].

Our functional analysis showed that the VEGFA gene was down-regulated by hsa-miR-612 and hsa-miR-637; however, such effect was not observed upon inhibition of hsa-miR-874. VEGF is responsible for new blood vessel growth, and miRNA-mediated decreased expression of VEGF can contribute to angiogenesis inhibition [9].

VEGF is a growth factor that activates receptor tyrosine kinases, initiating the RAS-RAF-MEK (Map kinase)–ERK MAPK signaling cascade. This cascade induces key transcription factors as well as the epithelial-mesenchymal transition that results in cell motility and invasion [16]. The activated RAS-RAF-MEK pathway may be associated with metastasis and aggressive tumors in HCC. Interestingly, miR-612 levels are inversely correlated with HCC tumor size and stage, microvascular invasion, and intrahepatic metastasis, as well as the protein levels of AKT2 (AKT Serine/Threonine Kinase 2) and EMT biomarkers [23]. Tao et al. (2013) [23] showed that transfection of miR-612 in HCC cell lines resulted in increased E-cadherin, decreased vimentin, and EMT suppression, supporting the involvement of this miRNA in EMT through different regulators. Additionally, the authors suggested that reduced expression levels of miR-612 may promote EMT and metastasis in HCC [23].

In the present study, the VEGFA gene was shown to be down-regulated in both cell lines after transfection by miR-637. Zhang et al. (2011) [24] found that miR-637 might be a useful tool for therapy against HCC as it inhibits the activation of signal transducers, such as STAT3 [25]. Under hypoxic conditions, STAT3 is activated and promotes VEGF expression and angiogenesis [25], contributing to cell proliferation and tumorigenesis [26]. Thus, miRNA-mediated regulation of factors involved in hypoxia may contribute to the reduction of tumor progression [25].

MiR-874 Inhibition did not affect on the Huh-7 cell line in our study, once the expression gene in cell lines transfected was equal to the negative control. Probably, the concentrations of inhibitor used for transfection were not sufficient to observe the effect on VEGF gene regulation in this cell line. On the other hand, VEGF gene expression increased significantly in response to miR-874 inhibition in the HepG2 cell line. Zhang et al. (2015) [27] showed a negative correlation between miR-874 and the STAT3 gene in the gastric tumor, suggesting that the down-regulation of miR-874 can contribute to tumor angiogenesis through the STAT3/VEGFA pathway. The analysis of TCGA [22] data showed the VEGFA interaction positive correlation to KRAS, AKT1, and STAT3 genes (Figure 1). Furthermore, it was possible to observe the miR-612 and miR-874 down-expression in tumor samples compared to normal tissue [22], justifying that treatments with mimetic miRNAs would help to reduce the expression of genes involved in tumor development (Figure 2). No data on miR-637 expression in HCC tumor samples were observed, which is new data for the cancer database.

However, our protein expression data showed a reduction in VEGFA levels in both cell lines and a significant negative correlation between miR-874 and VEGFA protein in the HuH-7 cell line. This result may be attributed to the overexpression of miR-874, even after its inhibition. This possibly represents the characteristic of miRNAs posttranscriptional processing, where the transcribed gene is degraded by miR-874 overexpression before its translation [28]. Moreover, other factors may influence the VEGFA protein expression [29].

Because overexpression of miR-612 and miR-637 resulted in down-regulation of the VEGFA gene and protein in HepG2 and HuH-7 cell lines, we can conclude that these miRNAs can regulate VEGFA, while further studies are needed to better understand the miR-874 involvement in HCC. The present study provides information regarding miRNA-mediated regulation of VEGFA and suggests possible molecular mechanisms of liver carcinogenesis.

5. Conclusions

In conclusion, VEGFA is down-regulated in response to hsa-miR-612 and hsa-miR-637 overexpression; however, the modulation of VEGFA by miR-874 observed only at the gene expression level requires further investigation. The new regulatory data with miR-637 in liver cancer have an importance for characterization in the databases, being a potential biomarker for VEGFA.

Author Contributions: M.M.U.C.-N. performed the experiments and analyses, including data generation, computational analysis, methodological application, and creation of the original draft, as well as review and editing of the manuscript; N.M.T. and R.S.K.-O. participated in the practical activities and analysis; M.M.U.C.-N., A.L.S.G.-D., A.R.C.P.O., and M.F.M. contributed with analysis tools and statistics. R.F.d.S. contributed to data analysis and HCC information; M.M.U.C.-N., E.C.P., and E.M.G.-B. participated in conceptualizing the project, supervising the work, and reviewing and editing the manuscript. All authors have read and agreed to the published version of the manuscript.

Funding: We greatly appreciate the grant of the São Paulo Research Foundation (FAPESP, Process number #2015/04403-8), Coordenação de Aperfeiçoamento de Nível Superior-Brazil (CAPES-Finance code 001), and Conselho Nacional de Desenvolvimento Científico e Tecnológico (CNPq productivity—310987/2018-0) for the financial support.

Institutional Review Board Statement: This study was dispensed by local Research Ethics Committee.

Informed Consent Statement: Not applicable.

Data Availability Statement: The data can be found at the Research Unit in Genetics and Molecular Biology (UPGEM), at the Faculty of Medicine of São José do Rio Preto (FAMERP).

Acknowledgments: Carlos H V Nascimento-Filho, Ludimila Leite Marzochi, and Vitoria Scavacini Possebon for technical support. The authors are grateful to Lilian Castiglioni for her support in the statistical analysis. The Faculty of Medicine of São José do Rio Preto, FAMERP and Medical School Foundation, FUNFARME for their institutional support, and UPGEM-Genetics and Molecular Biology Research Unit and GETF-Study Group of Liver Tumors.

Conflicts of Interest: The authors declare no conflict of interest.

References

1. Centers for Disease Control and Prevention. 2021. Available online: https://www.cdc.gov/cancer/liver/ (accessed on 22 December 2021).
2. Bray, F.; Ferlay, J.; Soerjomataram, I.; Siegel, R.L.; Torre, L.A.; Jemal, A. Global cancer statistics 2018: Globocan estimates of incidence and mortality worldwide for 36 cancers in 185 countries. *CA Cancer J. Clin.* **2018**, *68*, 394–424. [CrossRef] [PubMed]
3. Ferlay, J.; Soerjomataram, I.; Dikshit, R.; Eser, S.; Mathers, C.; Rebelo, M.; Parkin, D.M.; Forman, D.; Bray, F. Cancer Incidence and Mortality Worldwide: Sources, methods and major patterns in globocan 2012. *Int. J. Cancer* **2015**, *136*, E359–E386. [CrossRef]
4. Ambros, V. The functions of animal microRNAs. *Nature* **2004**, *431*, 350–355. [CrossRef]
5. Bartel, D.P. MicroRNAs: Genomics, Biogenesis, Mechanism, and Function. *Cell* **2004**, *116*, 281–297. [CrossRef]
6. Bartel, D.P. MicroRNAs: Target Recognition and Regulatory Functions. *Cell* **2009**, *136*, 215–233. [CrossRef] [PubMed]
7. Wong, C.M.; Tsang, F.H.; Ng, I.O. Non-coding RNAs in hepatocellular carcinoma: Molecular functions and pathological im-plications. *Nat. Rev. Gastroenterol. Hepatol.* **2018**, *15*, 137–151. [CrossRef]
8. Kappas, N.C.; Zeng, G.; Chappell, J.C.; Kearney, J.B.; Hazarika, S.; Kallianos, K.G.; Patterson, C.; Annex, B.H.; Bautch, V.L. The VEGF receptor Flt-1 spatially modulates Flk-1 signaling and blood vessel branching. *J. Cell Biol.* **2008**, *181*, 847–858. [CrossRef]
9. Dang, Y.; Luo, D.; Rong, M.; Chen, G. Underexpression of miR-34a in Hepatocellular Carcinoma and Its Contribution towards Enhancement of Proliferating Inhibitory Effects of Agents Targeting c-MET. *PLoS ONE* **2013**, *8*, e61054. [CrossRef]
10. Yu, Z.-Y.; Bai, Y.-N.; Luo, L.-X.; Wu, H.; Zeng, Y. Expression of microRNA-150 targeting vascular endothelial growth factor-A is downregulated under hypoxia during liver regeneration. *Mol. Med. Rep.* **2013**, *8*, 287–293. [CrossRef]
11. Liu, L.; Bi, N.; Wu, L.; Ding, X.; Men, Y.; Zhou, W.; Li, L.; Zhang, W.; Shi, S.; Song, Y.; et al. MicroRNA-29c functions as a tumor suppressor by targeting VEGFA in lung adenocarcinoma. *Mol. Cancer* **2017**, *16*, 50. [CrossRef]
12. Tunissiolli, N.M.; Castanhole-Nunes, M.; Biselli-Chicote, P.M.; Pavarino, É.C.; Da Silva, R.F.; Da Silva, R.D.; Goloni-Bertollo, E.M. Hepatocellular Carcinoma: A Comprehensive Review of Biomarkers, Clinical Aspects, and Therapy. *Asian Pac. J. Cancer Prev.* **2017**, *18*, 863–872.
13. Thurnherr, T.; Mah, W.-C.; Lei, Z.; Jin, Y.; Rozen, S.G.; Lee, C.G. Differentially Expressed miRNAs in Hepatocellular Carcinoma Target Genes in the Genetic Information Processing and Metabolism Pathways. *Sci. Rep.* **2016**, *6*, 20065. [CrossRef] [PubMed]
14. Xu, R.-H.; Wei, W.; Krawczyk, M.; Wang, W.; Luo, H.; Flagg, K.; Yi, S.; Shi, W.; Quan, Q.; Li, K.; et al. Circulating tumour DNA methylation markers for diagnosis and prognosis of hepatocellular carcinoma. *Nat. Mater.* **2017**, *16*, 1155–1161. [CrossRef] [PubMed]
15. Yao, R.; Zou, H.; Liao, W. Prospect of Circular RNA in Hepatocellular Carcinoma: A Novel Potential Biomarker and Thera-peutic Target. *Front. Oncol.* **2018**, *8*, 332. [CrossRef] [PubMed]
16. Lamouille, S.; Xu, J.; Derynck, R. Molecular mechanisms of epithelial–mesenchymal transition. *Nat. Rev. Mol. Cell Biol.* **2014**, *15*, 178–196. [CrossRef]
17. Ceci, C.; Atzori, M.G.; Lacal, P.M.; Graziani, G. Role of VEGFs/VEGFR-1 Signaling and its Inhibition in Modulating Tumor In-vasion: Experimental Evidence in Different Metastatic Cancer Models. *Int. J. Mol. Sci.* **2020**, *21*, 1388. [CrossRef]
18. Knowles, B.B.; Aden, D.P. Human Hepatoma Derived Cell Line, Process for Preparation Thereof, and Uses Therefor. U.S. Patent 4393133, 12 July 1983.
19. Nakabayashi, H.; Taketa, K.; Miyano, K.; Yamane, T.; Sato, J. Growth of human hepatoma cells lines with differentiated func-tions in chemically defined medium. *Cancer Res.* **1982**, *42*, 3858–3863. [PubMed]
20. De Oliveira, A.R.; Castanhole-Nunes, M.M.; Biselli-Chicote, P.M.; Pavarino, É.C.; Da Silva, R.D.; Da Silva, R.F.; Goloni-Bertollo, E.M. Differential expression of angiogenesis-related miRNAs and VEGFA in cirrhosis and hepatocellular carcinoma. *Arch. Med. Sci.* **2020**, *16*, 1150–1157. [CrossRef]

21. Pfaffl, M.W. A new mathematical model for relative quantification in real-time RT-PCR. *Nucleic Acids Res.* **2001**, *29*, e45. [CrossRef]

22. Chandrashekar, D.S.; Bashel, B.; Balasubramanya, S.A.H.; Creighton, C.J.; Ponce-Rodriguez, I.; Chakravarthi, B.V.; Varambally, S. UALCAN: A portal for facilitating tumor subgroup gene expression and survival analyses. *Neoplasia* **2017**, *19*, 649–658. [CrossRef]

23. Tao, Z.-H.; Wan, J.-L.; Zeng, L.-Y.; Xie, L.; Sun, H.-C.; Qin, L.-X.; Wang, L.; Zhou, J.; Ren, Z.-G.; Li, Y.-X.; et al. miR-612 suppresses the invasive-metastatic cascade in hepatocellular carcinoma. *J. Exp. Med.* **2013**, *210*, 789–803. [CrossRef] [PubMed]

24. Zhang, J.F.; He, M.L.; Fu, W.M.; Wang, H.; Chen, L.Z.; Zhu, X.; Chen, Y.; Xie, D.; Lai, P.; Chen, G.; et al. Pri-mate-specific microRNA-637 inhibits tumorigenesis in hepatocellular carcinoma by disrupting signal transducer and ac-tivator of transcription 3 signaling. *Hepatology* **2011**, *54*, 2137–2148. [CrossRef]

25. Gao, P.; Niu, N.; Wei, T.; Tozawa, H.; Chen, X.; Zhang, C.; Zhang, J.; Wada, Y.; Kapron, C.M.; Liu, J. The roles of signal transducer and activator of transcription factor 3 in tumor angiogenesis. *Oncotarget* **2017**, *8*, 69139–69161. [CrossRef] [PubMed]

26. Sakurai, T.; Yada, N.; Hagiwara, S.; Arizumi, T.; Minaga, K.; Kamata, K.; Takenaka, M.; Minami, Y.; Watanabe, T.; Nishida, N.; et al. Gankyrin induces STAT3 activation in tumor microenvironment and sorafenib resistance in hepatocellular car-cinoma. *Cancer Sci.* **2017**, *108*, 1996–2003. [CrossRef] [PubMed]

27. Zhang, X.; Tang, J.; Zhi, X.; Xie, K.; Wang, W.; Li, Z.; Zhu, Y.; Yang, L.; Xu, H.; Xu, Z. miR-874 functions as a tumor suppressor by inhibiting angiogenesis through STAT3/VEGF-A pathway in gastric cancer. *Oncotarget* **2015**, *6*, 1605–1617. [CrossRef]

28. Barbosa, A.S.; Lin, C.J. Gene silencing with RNA interference: A novel tool for the study of physiology and pathophysiology of adrenal cortex. *Arq. Bras. De Endocrinol. Metabol.* **2004**, *48*, 612–619. [CrossRef]

29. Beeghly-Fadiel, A.; Shu, X.O.; Lu, W.; Long, J.; Cai, Q.; Xiang, Y.B.; Zheng, Y.; Zhao, Z.; Gu, K.; Gao, Y.T.; et al. Genetic variation in VEGF family genes and breast cancer risk: A report from the Shanghai Breast Cancer Genetics Study. *Cancer Epidemiol. Biomark. Prev.* **2011**, *20*, 33–41. [CrossRef]

Article

A Possible Cause for the Differential Expression of a Subset of miRNAs in Mesenchymal Stem Cells Derived from Myometrium and Leiomyoma

Mariangela Di Vincenzo [1,†], Concetta De Quattro [2,†], Marzia Rossato [2], Raffaella Lazzarini [1], Giovanni Delli Carpini [3], Andrea Ciavattini [3,*] and Monia Orciani [1]

1   Department of Clinical and Molecular Sciences-Histology, Università Politecnica delle Marche, 60126 Ancona, Italy; m.divincenzo@pm.univpm.it (M.D.V.); r.lazzarini@univpm.it (R.L.); m.orciani@univpm.it (M.O.)
2   Department of Biotechnology, University of Verona, 37134 Verona, Italy; concetta.dequattro@univr.it (C.D.Q.); marzia.rossato@univr.it (M.R.)
3   Clinic of Obstetrics and Gynecology, Department of Clinical Sciences, Università Politecnica delle Marche, 60126 Ancona, Italy; g.dellicarpini@univpm.it
*   Correspondence: a.ciavattini@univpm.it
†   These authors contributed equally to this work.

Abstract: The aetiology of leiomyoma is debated; however, dysregulated progenitor cells or miRNAs appear to be involved. Previous profiling analysis of miRNA in healthy myometrium- (M-MSCs) and leiomyoma- (L-MSCs) derived mesenchymal stem cells (MSCs) identified 15 miRNAs differentially expressed between M-MSCs and L-MSCs. Here, we try to elucidate whether these differentially regulated 15 miRNAs arise as a conversion of M-MSCs along the differentiation process or whether they may originate from divergent cell commitment. To trace the origin of the dysregulation, a comparison was made of the expression of miRNAs previously identified as differentially regulated in M-MSCs and L-MSCs with that detected in MSCs from amniotic fluid (considered as a substitute for embryonic cells). The results do not allow for a foregone conclusion: the miRNAs converging to the adherens junction pathway showed a gradual change along the differentiation process, and the miRNAs which coincided with the other three pathways (ECM-receptor interaction, TGFβ and cell cycle) showed a complex, not linear, regulation and, therefore, a trend along the hypothetical differentiation process was not deduced. However, the role of miRNAs appears to be predominant in the onset of leiomyoma and may follow two different mechanisms (early commitment; exacerbation); furthermore, miRNAs can support the observed (epigenetic) predisposition.

Keywords: miRNAs; progenitor cells; amniotic fluid; leiomyoma; myometrium; divergent cell commitment; linear dysregulation

Citation: Di Vincenzo, M.; De Quattro, C.; Rossato, M.; Lazzarini, R.; Delli Carpini, G.; Ciavattini, A.; Orciani, M. A Possible Cause for the Differential Expression of a Subset of miRNAs in Mesenchymal Stem Cells Derived from Myometrium and Leiomyoma. *Genes* **2022**, *13*, 1106. https://doi.org/10.3390/genes13071106

Academic Editors: Björn Voß, Giuseppe Iacomino and Fabio Lauria

Received: 14 April 2022
Accepted: 17 June 2022
Published: 21 June 2022

Publisher's Note: MDPI stays neutral with regard to jurisdictional claims in published maps and institutional affiliations.

1. Introduction

Uterine fibroids, also called leiomyomas, are the most common benign gynecologic tumors affecting women during their reproductive age, with an incidence directly related to age [1]. Epigenetic mechanisms, gene mutations, chronic inflammation, disrupted controls in progenitor cells, and dysregulation of miRNAs have been all evaluated as potential causes; however, their aetiology has not been fully clarified, [2,3].

Previously [4,5], we demonstrated that mesenchymal stem cells (MSCs) isolated from leiomyomas (L-MSCs) and normal myometrium (M-MSCs) can diversely sustain acute and chronic inflammation promoting a microenvironment suitable for leiomyoma onset; additionally, out of 2646 miRNAs, only 15 miRNAs displayed significantly altered expression between leiomyoma and normal myometrium, supporting the hypothesis that leiomyoma derives from the disruption of specific cellular mechanisms in progenitor cells.

A population of MSCs occurs in almost all human tissues [6]. According to the criteria outlined by Dominici [7], MSCs must be plastic adherent, positive for CD73, CD90, and CD105 and negative for HLA-DR, CD14, CD19, CD34, and CD45 and be able to differentiate towards oste-, chondro- and adipogenic lineages. By fulfilling these criteria, mesenchymal cells derived from adult tissues exhibit tissue-specific features that become even more characteristic during differentiation [8]. Among mesenchymal stem cells, amniotic fluid MSCs (AF-MSCs) are of particular interest, as they express both adult and embryonic cell markers and are therefore considered an intermediate stage between embryonic and adult cells [9]. In a theoretical line of increasing differentiation, the first should be the embryonic cells, followed by AF-MSCs and, finally, by MSCs derived from adult tissues, such as M- and L-MSCs.

miRNAs are switchers able to modulate cell fate by turning on/off specific gene targets, and their aberrant expression can proportionally influence these critical processes leading to various pathological outcomes [10,11]. We recently demonstrated that 15 miRNAs are differentially expressed between MSCs from leiomyoma and healthy myometrium, but we do not know the mechanisms underlying these differences nor their origin. In particular, when do these alterations occur? Do they arise directly from embryonic cells, i.e., differential miRNA expression leads early to different MSCs in normal myometrium and fibroids (as in hypothesis A, Figure 1A)? Alternatively, do they arise later during the last steps of the differentiation process, i.e., embryonic cells produce normal M-MSCs and these, under the control of dysregulated miRNAs, acquire pathological features which give rise to L-MSC (as in hypothesis B, Figure 1B)? Since the use of embryonic cells to answer these research questions is forbidden, AF-MSCs were alternatively used in this study as a substitute for embryonic cells, as it maintains a lower degree of differentiation than MSCs derived from adult tissues and therefore can be used to follow the onset of these alterations during the differentiation process. AF-MSCs are considered as the starting point in the differentiation process and the expression of the 15 dysregulated miRNAs will be compared to those observed in M- and L-MSCs.

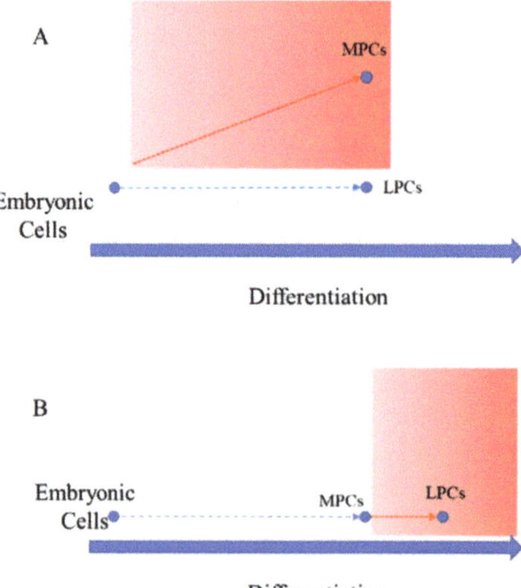

Figure 1. Two different hypotheses for the origin of miRNAs differentially regulated between M-MSCs and L-MSCs. (**A**) M-MSCs and L-MSCs are the results of divergent differentiation from embryonic stem cells; (**B**) embryonic stem cells physiologically differentiate in M-MSCs, and further pathological differentiation produces L-MSCs. The red areas indicate the spectrum of the disease. The red arrows show the pathological differentiation.

2. Materials and Methods

2.1. Tissue Collection

Amniotic fluid (AF) samples ($n = 9$), 3 mL, were obtained by amniocentesis after the 16th week of pregnancy for routine prenatal diagnosis. Gestational age (GA) was determined by ultrasonic measurements of the biparietal diameter and length of the fetus's femur. AF was collected by ultrasound-guided transabdominal puncture. The indications were advanced maternal age (34–36) and the cytogenetic analyses revealed normal karyotypes.

Healthy and fibrotic myometrium samples ($n = 12$), 3 mm^2 in size, were obtained from women of childbearing age (range, 30–35 years) undergoing myomectomy for symptomatic leiomyomas.

All patients provided their written informed consent to participate in the study, which was approved by the institutional ethics committees (2016-0360OR) and conducted in accordance with the Declaration of Helsinki.

Cells were isolated, cultured, and characterized as previously described [4,5,12,13].

Briefly, the solid samples were first mechanically minced into small pieces whereas AF samples were directly centrifuged. The pellets were resuspended and transferred into 6-well plates with MSC growth medium (MSCGM; Lonza-Bioscence, Basel, Switzerland) suitable for the growth of undifferentiated stem cells. The samples were incubated at 37 °C and 5% carbon dioxide. Morphology was assessed by phase-contrast microscopy (Leica DM IL; Leica Microsystems GmbH, Wetzlar, Germany). Vitality and proliferation rate were examined using an automated cell counter (Countess; Invitrogen, Milan, Italy). After sample collection, the cells were characterized by testing the minimal criteria identified by Dominici [7] for mesenchymal definition as previously described [14–16].

For immunophenotyping, 2.5×10^5 cells at the 3rd passage were stained for 45 min with fluorescein isothiocyanate (FITC)-conjugated antibodies (Becton-Dickinson, NJ, USA) against HLA-DR, CD14, CD19, CD34, CD45, CD73, CD90, and CD105. Furthermore, CD9 expression was evaluated in MSCs and fibroblasts (FBs) obtained from the same tissues since it is considered a distinctive marker between MSCs and FBs [17]. Differentiation into osteocytes, chondrocytes and adipocytes was assessed using STEMPRO® Osteogenesis, Chondrogenesis, and Adipogenesis Kits (GIBCO, Invitrogen), respectively. Cells cultured in DMEM/F-12 with 10%FBS were used as negative controls.

Osteogenic differentiation was assessed by Alizarin Red and Alkaline phosphatase (ALP) stainings; adipogenic differentiation was tested by Oil Red staining. For chondrogenesis, cells were grown in a pellet culture system, and the sections were exposed to a Safranin-O solution. Fibroblasts obtained from the same tissues were used to compare the differentiation efficiency.

2.2. miRNA Profiling

Total RNA was extracted in triplicate using Norgen Total RNA Kit (Norgen, Biotek Corporation, Thorold, ON, Canada) from a pool of mixed cells obtained from the twelve cultures of M-MSCs or of L-MSCs and from the nine cultures of AF-MSCs in triplicate. RNA purity and amounts were measured using a NanoDrop Spectrophotometer (NanoDrop Technologies, INK, Wilmington, DE, USA), whereas RNA integrity (RINA $\geq$ 8.0) was assessed using an RNA 6000 Nano Kit (Agilent Technologies, CA, USA). miRNA-sequencing libraries were generated with the QiAseq miRNA kit (QIAGEN, Hilden, Germania), assessed by capillary electrophoretic analysis with the Agilent 4200 Tape station and sequenced using 1×75 bp-reads on an Illumina NextSeq500 generating about 8 million fragments per sample.

2.3. Bioinformatics Analysis

Starting from raw FASTQ files, the quality of reads obtained from each sample was assessed using FastQC software (v.0.10.1) [18] , adapters were trimmed and reads with length < 18 bp or >27 bp were filtered out with cutadapt (1.16). Filtered reads were aligned to human hairpin microRNAs available in the miRBase [19] using SHRiMP2 (v2.2.3. http://compbio.cs.toronto.edu/shrimp/, accessed on 27 April 2020) with the "mirna" mode activated. The number of Unique Molecular Identifiers (UMI) mapped reads were counted for each mature miRNA. Differential miRNA expression analysis between L-MSCs, M-MSCs, and AF-MSCs was performed with DESeq2 (v 1.16.1, https://support.bioconductor.org/, accessed on 27 April 2020). The expression of microRNAs was normalized as CPM (counts per million reads mapped) and subsequently the 15 miRNAs previously identified as differentially expressed between M-MSCs and L-MSCs were clustered by an unsupervised hierarchical clustering using Spearman rank correlation and the average linkage method.

2.4. miRNAs Targets Analysis

To identify molecular pathways potentially altered by the expression of single or multiple miRNA, Diana mir-PathSoftware was used [20]. This web-based application performs enrichment analysis of numerous miRNA target genes comparing each set of miRNA targets to all known KEGG (Kyoto Encyclopedia of Genes and Genomes, Kyoto, Japan) pathways.

The input dataset enrichment in each KEGG pathway is represented by the negative natural logarithm of the P value ($-\ln$P).

In addition, to partially compensate for the lack of functional analysis, the mirTarBase [21] software was used to find validated targets of selected miRNAs and the STRING (Search Tool for the Retrieval of Interacting Genes/Proteins) database [22] to construct a protein−protein interaction network; proteins have been included in the analysis only if directly targeted by at least one miRNA in each KEGG pathways.

3. Results

M-MSCs, L-MSCs, and AF-MSCs, respectively isolated from healthy myometrial, fibroid tissue, and amniotic fluids, met the three criteria established by the International Society for Cellular Therapy (ISCT) for the identification of human mesenchymal stem cells [7]. Cells were adherent to plastic under standard culture conditions, strongly positive for CD105, CD73, and CD90 and negative for CD45, CD34, CD14, CD19, and HLA-DR and were able to differentiate towards mesenchymal lineages in vitro (Figure 2A–D). FBs isolated from the same tissues were induced to differentiate but their ability, as shown in the inserts, was notably inferior to MSCs. Finally, CD9 expression was evaluated by flow cytometry and the percentage of positive cells was 14% and 36% for MSCs and FBs, respectively, confirming the undifferentiated nature of isolated cells (Figure 2E).

RNA obtained from the three cellular pools was used for miRNA profiling by RNA-sequencing [5]. Principal Component Analysis shown in Figure 3 is based on more than 2600 miRNAs expressed in samples and reveals that M-MSCs and L-MSCs exhibit a more similar miRNA expression profile than AF-MSCs, highlighting that MSCs from amniotic fluid are more distant, and so are dissimilar to those of adult tissues.

Subsequently, we recovered the expression of the 15 miRNAs previously found dysregulated in M-MSCs with respect to L-MSCs [5], hsa-miR-10a-3p; hsa-miR-10a-5p; hsa-miR-122-5p; hsa-miR-135b-5p; hsa-miR-146a-5p; hsa-miR-146b-5p; hsa-miR-200a-3p; hsa-miR-335-3p; hsa-miR-335-5p; hsa-miR-576-3p; hsa-miR-595; hsa-miR-658; hsa-miR-924; hsa-miR-1973; and hsa-miR-4284. Clustering analysis sub-grouped the 15 miRNAs into four main clusters based on their level of expression in the cell type analyzed (Figure 4A).

The clustered miRNAs were analyzed by Diana mir- Path Software to identify the related KEGG pathways.

All miRNAs belonging to the first cluster converged to the adherens junction pathway (4 converging miRNAs, hsa-miR-10a-5p, hsa-miR-10a-3p, hsa-miR-135b-5p, and hsa-miR-200a-3p, for a total of 27 target genes); all three miRNAs of the second cluster focused to ECM-receptor interaction pathway (3 converging miRNAs, hsa-miR-146b-5p, hsa-miR-335-3p, and hsa-miR-335-5p, for a total of 32 target genes); for the third cluster, the analysis identified the TGFβ signaling pathway (3 converging miRNAs, hsa-miR-122-5p, hsa-miR-

576-3p, and hsa-miR-595, for a total of 21 target genes) with the exclusion of hsa-miR-1973, for which no correlated pathways have been found; finally, three of the four miRNAs grouped in the last cluster merged into the cell cycle pathway (3 converging miRNAs, hsa-miR-924, hsa-miR-146a-5p, and hsa-miR-658, for a total of 13 target genes).

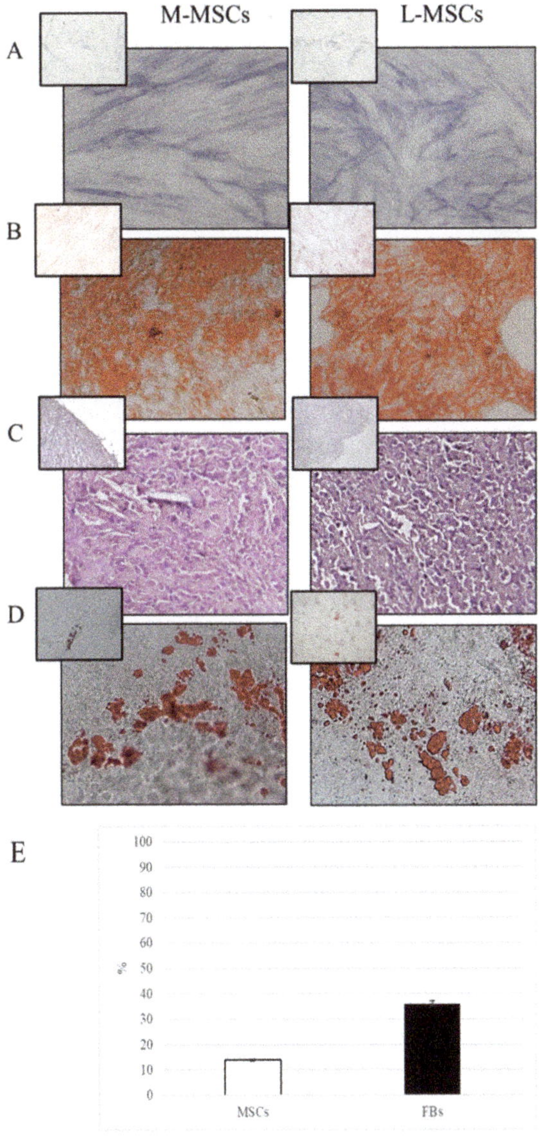

Figure 2. Multilineage differentiation of MSCs from myometrium (M-MSCs) and leiomyoma (L-MSCs) and relative fibroblasts (in the inserts). Representative images of osteogenic differentiation assessment by ALP reaction (**A**) and Alizarin red staining (**B**); chondrogenic differentiation by Safranin-O staining (**C**); adipocyte differentiation by Oil red staining (**D**). Percentage of CD9 positive cells after flow cytometric analysis (**E**). Data are presented as mean ±SD of the 12 samples of L- and M-MSCs and related FBs.

The miRNAs converging to the four pathways (hsa-miR-1973 and hsa-miR-4284 were excluded from the analysis as they did not converge in the pathways identified for their cluster) were differentially expressed in M-MSCs and L-MSCs compared to AF-MSCs and 8/13 of them (hsa-miR-10a-5p; hsa-miR-10a-3p; hsa-miR-135b-5p; hsa-miR-200a-3p; hsa-miR-146b-5p; hsa-miR-335-5p; hsa-miR-122-5p; and hsa-miR-146a-5p) showed a significant difference in both cell types compared to amniotic progenitors (Figure 4B, miRNAs marked with ***). In detail, the four miRNAs (hsa-miR-10a-3p; hsa-miR-10a-5p; hsa-miR-135b-5p; and hsa-miR-200a-3p) that converge at the KEGG pathway "adherens junction" were significantly downregulated in M-MSCs and mainly in L-MSCs compared to AF-MSCs; the three miRNAs (hsa-miR-146b-5p; hsa-miR-335-3p; and hsa-miR-335-5p) targeting genes related to ECM-receptor interaction pathway showed a significant increase in M-MSCs compared to AF-MSCs, whereas their expression was variable in L-MSCs versus AF-MSCs. The expression of the three miRNAs (hsa-miR-122-5p; hsa-miR-576-3p; hsa-miR-595) involved in the TGFβ signaling pathway was significantly higher in L-MSCs than in AF-MSCs, whereas it did not change in 2 of 3 miRNAs (hsa-miR-595; hsa-miR-576-3p) in M-MSCs compared to AF-MSCs; finally, all three miRNAs (hsa-miR-924; hsa-miR-146a-5p; hsa-miR-658) related to cell cycle were significantly downregulated in M-MSCs and only one (hsa-miR-146a-5p) in L-MSCs compared to AF-MSCs.

In the absence of direct functional analysis and in the attempt to strengthen the putative link between differentially expressed miRNAs and altered KEGG pathways, identification of validated protein targets and the evaluation of the protein–protein interaction (PPI) networks were performed using mirTarBase and STRING, respectively.

The validated protein targets are listed in Table 1. All proteins are effectively related to the classified KEGG pathways; the PPI networks clearly show the close connection among the proteins enforcing that, even if mostly putative, these targets are specific and highly interconnected (Figure 5).

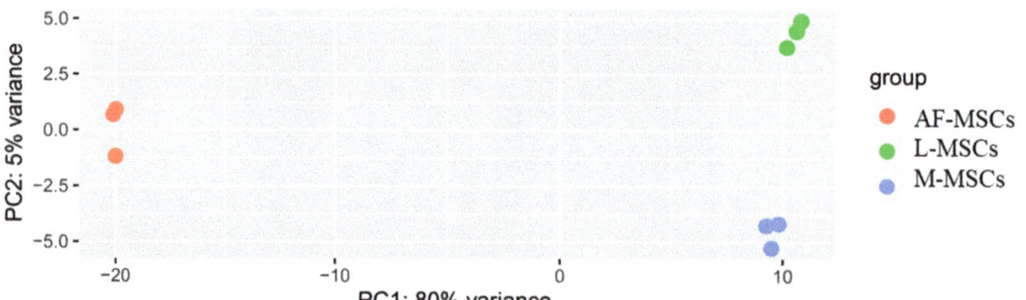

Figure 3. Principal Component Analysis (PCA) shows the variance of samples analysed in the study, based on the full miRNA profile (2646 expressed miRNAs). The analysis was conducted using the DESeq2 package and the values on each axis represent the percentages of variation explained by the principal components. PC1, principal component 1; PC2, principal component 2.

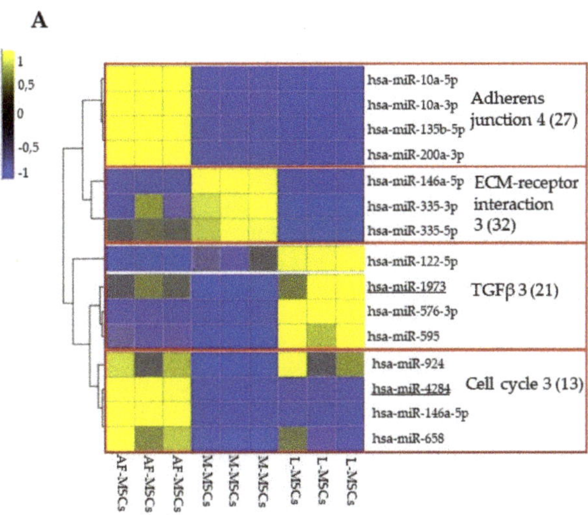

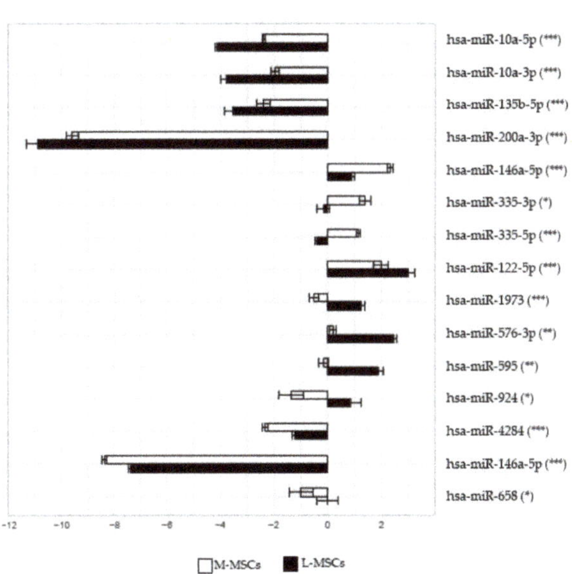

Figure 4. Heat map and expression fold change of 15 miRNAs previously identified as differentially expressed between M-MSCs and L-MSCs based on next generation sequencing analysis. (**A**) Each row represents a different miRNA, and each column represents one sample from the AF-MSCs (*n* = 9), M-MSCs (*n* = 12) or the L-MSCs (*n* = 12) RNA pools. The clusters were obtained by unsupervised hierarchical clustering using Spearman rank correlation and the average linkage method. The key colour illustrates the normalized expression levels (z-scores of normalized counts-cpm) of miRNAs in all samples. The yellow to blue gradient indicates a higher to lower expression. In the right column, the KEGG pathways identified by DIANA-miRPath software as targets of the analysed miRNAs. In brackets, the number of targeted genes per pathways. Underlined miRNAs are not convergent in the indicated KEGG pathway. (**B**) Expression fold changes (log2-transformed) of the 13 selected miRNAs in M-MSCs and L-MSCs as compared to AF-MSCs. Adjusted *p* value < 0.05: *** both comparisons, ** only L-MSCs, * only MSCs.

Table 1. Identification of protein targets by mirTarBase software.

KEGG Pathway	miRNA	TARGET	Validation Methods				
			Strong Evidence			Less Strong Evidence	
			Reporter Assay	Western Blot	qPCR	Microarray	NGS
Adherens junction	hsa-miR-10a-5p	ACTG1			●	●	●
		YES1					●
		CTNND1					●
		MAP3K7	●	●	●	●	
	hsa-miR-200a-3p	CTNNB1	●	●	●		
		TCF7L1	●				
ECM- receptor interaction	hsa-miR-335-3p	COL4A1					●
	hsa-miR-335-5p	CD36				●	
		COL6A1				●	
		COL6A5				●	
		COL6A6				●	
		GP9				●	
		HSPG2				●	
		ITGA1				●	
		ITGA10				●	
		ITGA2				●	
		ITGB4				●	
		ITGB5				●	
		ITGB6				●	
		ITGB8				●	
		LAMA5				●	
		LAMB1				●	
		TNC	●		●	●	
		VTN				●	
		THBS3				●	
		SPP1				●	
TGF-β	hsa-miR-122-5p	NODAL				●	
		RBL1					●
		RPS6KB1					●
		SMURF2				●	
Cell cycle	hsa-miR-146a-5p	CDKN1A	●				
		SMAD4	●	●	●	●	

Dots (●) indicate the methods used for the validation of each target.

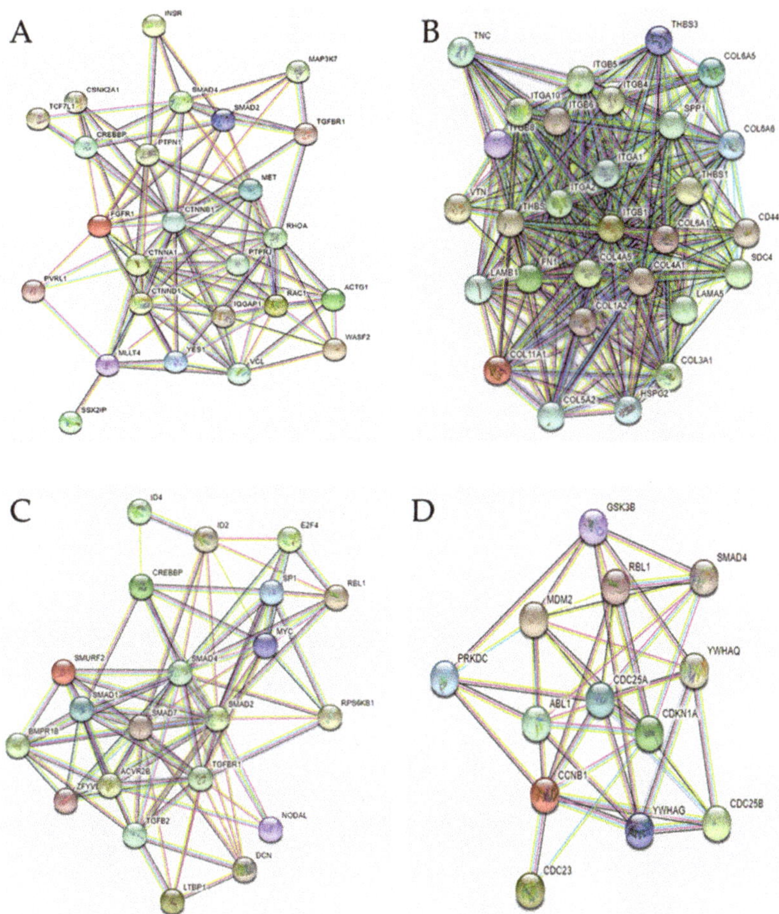

Figure 5. Protein−protein interaction (PPI) network. PPI network has been generated by STRING (Search Tool for the Retrieval of Interacting Genes/Proteins) based on the targeted proteins involved in (**A**) the adherens junction; (**B**) the ECM-receptor interaction pathway; (**C**) the TGFβ; and (**D**) the cell cycle pathway. Nodes are proteins. The thickness of the line is proportional to the strength of the interaction between 2 proteins.

4. Discussion

The expression of 15 miRNAs previously identified as differentially regulated between M-MSCs and L-MSCs [5] was compared with that observed in AF-MSCs in the attempt to elucidate whether dysregulation of the 15 miRNAs occurs: (i) as a result of an early commitment of undifferentiated cells (Figure 1A) or (ii) as an exacerbation of the physiological process of differentiation from M-MSCs to L-MSCs (Figure 1B). AF-MSCs, as a result of their intermediate status between embryonic and adult MSCs, have been considered in this study as a substitute for embryonic cells. However, the regulation of the expression level of the miRNAs across the cell types analysed did not provide conclusive evidence.

The cells isolated from healthy and fibrotic uterine tissues as well as from amniotic fluids were firstly characterized according to the minimal criteria established by the International Society for Cellular Therapy (ISCT) for the identification of human mesenchymal stem cells. In each case, the criteria was met.

After characterization, the choice of AF-MSCs as the most undifferentiated cells, which were used in this study as a substitute for embryonic ones, was validated [9]. Consistent with our hypothesis, AF-MSCs grouped separately from M-MSCs and L-MSCs, whereas the M-MSCs and L-MSCs grouped closer to each other. The expression of the 15 miRNAs in AF-MSCs represents the starting point and will be compared with that detected in M- and L-MSCs.

The heat map and the histogram showed that the expression of the majority of the 15 miRNAs was notably different between AF-MSCs and L-MSCs/M-MSCs (mean of absolute FC 2.6 ± 2.8). The clustering analysis followed by DIANA mir-Path investigation allowed for the identification of four KEGG pathways related to miRNA-clusters defined based on expression: adherens junction, ECM-receptor interaction, TGFβ signaling, and cell cycle. Although the miRNAs targets were not functionally validated, their role was confirmed by mirTarBase and STRING. mirTarBase reported previously validated targets of the selected miRNAs which belong to the identified KEGG pathways. At the same time, STRING showed strong connections among identified putative proteins.

The trend of expression across the cell types of each identified pathway was evaluated in order to determine which hypothesis was more consistent (A, early commitment or B, exacerbation). All four miRNAs (hsa-miR-10a-3p; hsa-miR-10a-5p; hsa-miR-135b-5p; and hsa-miR-200a-3p) converging to the adherens junction pathway showed a stepwise manner along the differentiation process. Within the other three pathways (ECM-receptor interaction, TGFβ, and cell cycle), the miRNAs showed different trends and, therefore, a progressive increase or decrease in expression from AF-MSCs to L-MSCs, i.e., along the hypothetical differentiation process, could not be deduced.

Further, L-MSCs expressed the lowest miRNAs values related to the adherens junction pathway and, most likely as a consequence, the highest level of targeted proteins. As expected, the expression of adhesion molecules is higher in cells residing in connective tissues (where fibroblasts/myofibroblasts develop strong bonds to ECM) than in AF-MSCs [23]. Although the increased expression of proteins related to adherens junction from AF-MSCs to M-MSCs must be considered physiological, its additional shift towards L-MSCs may be connected to the onset of leiomyoma [24,25].

Regarding the ECM-receptor interaction pathway, miRNAs (hsa-miR-146b-5p; hsa-miR-335-3p; and hsa-miR-335-5p) were less expressed in L-MSCs than in M-MSCs. As previously discussed [5], this observation is in line with other studies [23] as well as with the hypothesis that leiomyoma cells, characterized by an aberrant production of ECM, display a strong interaction with the ECM itself. The expression of miRNAs related to the ECM pathway has been previously analyzed in tissues, fibroblasts, and smooth muscle cells derived from myometrium and leiomyoma [26–28], and the results have always highlighted their dysregulation in uterine fibroids. Some previous studies have focused on different miRNAs, such as members of the miR-29 family; however, differentiated cells, instead of mesenchymal cells, were used.

AF-MSCs expressed a discrete number of proteins related to the ECM-receptor interaction; among them, integrins are well known to control many cellular functions through cells/ECM crosstalk, including embryonic development, survival, differentiation, and proliferation [29].

While the expression of the three miRNAs converging in the TGFβ signalling pathway (hsa-miR-122-5p; hsa-miR-576-3p; hsa-miR-595) was weakly altered in M-MSCs compared to AF-MSCs, they were strongly upregulated in L-MSCs.

It has been reported that AF-MSCs express SMAD2/4 (both targeted by hsa-miR-122-5p and hsa-miR-595), as well as NODAL (a target of miRNA hsa-miR-122-5p) and may therefore play a role in regulating self-renewal, similarly as in human embryonic stem cells [30].

The TGFβ pathway is greatly altered in leiomyoma [31–33]; even though it is generally described as upregulated in leiomyoma compared to healthy myometrium (and in this line miRNAs converging to this pathway were expected to be downregulated), the TFGβ pathway consists of many different proteins that act with distinct, and sometimes even opposite, effects. Furthermore, upregulated miRNAs in leiomyoma do not target TGFβIII, which is considered one of the main inducers of aberrant ECM production and the reduced level of ECM degradation in uterine fibroids [32]. Indeed, TGFβIII elevated serum level is among the risk factors for the onset of leiomyoma [34]. Previous studies have reported that TGFβIII is one of the very few growth factors detectable at higher concentrations in uterine fibroids [35]. SMAD7 is one of the target proteins of hsa-miR-595 that, by the interaction with TGF-β/activin type I receptors (targeted by hsa-miR-122-5p), can inhibit TGFβ/activin signaling [36]. Taken together, the lack of TGFβIII and the presence of SMAD7 negatively affected by the high expression of hsa-miR-595, may explain the apparent contradiction between the level of expression of the miRNAs and the upregulation of TGFβ signaling observed in leiomyoma.

The cell cycle pathway is strictly related to TGFβ, which acts as a regulator of cell growth. It inhibits the growth of most cell types while stimulating the growth of fibroblasts [37]. The growth-inhibitory response may be a result of the TGFβRI/TGFβRII/SMAD2/3/4 signaling cascade. SMAD2/3/4 complexes (SMAD 4 is targeted by hsa-miR-146a-5p) activate the transcription of CDK inhibitors (such as CDKN1A targeted by hsa-miR-146a-5p, along with other cell division cycle (CDC) proteins, such as CDC23, CDC25A, and CDC25B targeted by hsa-miR-146a-5p), p15, and p21 [38] causing cell cycle arrest [39]. In AF-MSCs, the high expression of these miRNAs and the consequently weak expression of the target proteins correlate with the maintained strong ability to proliferate and to self-renew; the great variability observed in M-MSCs and L-MSCs samples in the expression of miRNAs belonging to cell cycle pathway reflects the highly heterogeneous proliferative capacity reported by Busnelli [40].

Previous studies on the expression profile of miRNA in fibroid and myometrium tissues have identified a list of involved pathways that encompasses the four considered in this study [41].

Interestingly, even though the four pathways are enclosed in the aforementioned list, the converging miRNAs are different. This contradiction may be explained by the well-established fact that multiple miRNAs target the same genes. However, the finding that the same pathways have been found differentially regulated in progenitor as well as in differentiated cells or tissues confirms their involvement in the onset and development of leiomyoma. This hypothesis is also strengthened by the evidence that the selected miRNAs target more proteins belonging to the same pathways and these proteins show a strong interaction network highlighting the fact that miRNAs act with a convergent and synergistic effects. In this complex scenario, our work is not significant as a further confirmation of the involvement of these pathways but rather because it, for the first time, focuses on progenitor cells that, like the differentiated ones, are subjected to a fine regulation through miRNAs expression. It also opens to the evidence that miRNAs act as triggers for the activation of various pathways during differentiation. The role of miRNAs is also evidently linked to the genetic abnormalities, which, together with hormonal and environmental factors, seem to favor the onset of leiomyoma [42]. About 40% of leiomyomas has non-random and tumor-specific chromosomal abnormalities and, consequently, much attention has been paid to the study of genes located in chromosomal regions affected by recurrent changes, such as the subunit 12 of the mediator complex (MED12), high mobility group AT-Hook 2 (HMGA2), and type I procollagen cooh-terminal proteinase enhancer (PCOLCE).

MED12, implicated as an oncogene in about 70% of uterine leiomyoma [43,44], is a target of hsa-miR-10a-5p, whose expression is gradually reduced from AF-MSCS to L-MSCs, possibly explaining the upregulation of MED12 in leiomyoma. HMGA2 plays a key role in the onset of uterine fibroids and resides in the chromosomal rearrangements affecting the 12q14-15 region that targets the HMGA2 gene. It is usually overexpressed in leiomyoma [45] and this is consistent with the lower expression of the relative hsa-miR-10a-3p found in L-MSCs.

PCOLCE maps to the critical interval on chromosome 7, q22band, and a 60% decrease of its expression was found in fibroid tissues compared to myometrial levels in the same patient [46]. PCOLCE is a target of hsa-miR-122-5p whose expression gradually increases from AF-MSCS up to L-MSCs.

5. Conclusions

In conclusion, the adherent junction, ECM-receptor interaction, TGFβ, and cell cycle pathways drive the onset and the development of leiomyoma, not only in differentiated cells and tissues as previously reported, but also at the mesenchymal level where their alterations are already detectable.

MiRNAs, acting as switches during the differentiation process, can turn these pathways on or off. However, our results do not trace a precise mechanism underlying the involvement of miRNAs and their mode of action (whether alterations in miRNAs and target pathways occur gradually during the differentiation process or result from divergent cellular engagement).

Further functional analyses are needed to better explain the role played of miRNAs during the differentiation of progenitor cells; once again, miRNAs target the major genes linked to the genetic predisposition to leiomyoma. Despite this growing evidence on the involvement of miRNAs, it is not yet possible to establish their role or whether they act as a causal or consequential effect of a phenomenon.

Author Contributions: Conceptualization, M.O. and A.C.; methodology, M.D.V. and R.L.; software, C.D.Q. and M.R.; formal analysis, M.R. and C.D.Q.; data curation, M.D.V.; writing—original draft preparation, M.O. and G.D.C.; writing—review and editing, M.O. and A.C.; supervision, M.O. and A.C. All authors have read and agreed to the published version of the manuscript.

Funding: This research received no external funding.

Institutional Review Board Statement: Institutional ethics committees (2016-0360OR); all patients provided their written informed consent to participate.

Informed Consent Statement: Written informed consent has been obtained from the patient(s) to publish this paper.

Data Availability Statement: The datasets used and/or analysed during the current study are available from the corresponding author on reasonable request.

Conflicts of Interest: The authors declare no conflict of interest.

References

1. Ciavattini, A.; Di Giuseppe, J.; Stortoni, P.; Montik, N.; Giannubilo, S.R.; Litta, P.; Islam, M.; Tranquilli, A.L.; Reis, F.M.; Ciarmela, P. Uterine fibroids: Pathogenesis and interactions with endometrium and endomyometrial junction. *Obstet. Gynecol. Int.* **2013**, *2013*, 173184. [CrossRef] [PubMed]
2. Flake, G.; Andersen, J.; Dixon, D. Etiology and pathogenesis of uterine leiomyomas: A review. *Environ. Health Perspect.* **2003**, *111*, 1037–1054. [CrossRef] [PubMed]
3. Ono, M.; Qiang, W.; Serna, V.A.; Yin, P.; Coon, J.S.; Navarro, A.; Monsivais, D.; Kakinuma, T.; Dyson, M.; Druschitz, S.; et al. Role of stem cells in human uterine leiomyoma growth. *PLoS ONE* **2012**, *7*, e36935. [CrossRef]
4. Orciani, M.; Caffarini, M.; Biagini, A.; Lucarini, G.; Delli Carpini, G.; Berretta, A.; Di Primio, R.; Ciavattini, A. Cronic Inflammation May Enhance Leiomyoma Development by the Involvement of Progenitor Cells. *Stem Cells Int.* **2018**, *2018*, 1716246. [CrossRef]
5. Lazzarini, R.; Caffarini, M.; Delli Carpini, G.; Ciavattini, A.; Di Primio, R.; Orciani, M. From 2646 to 15: Differentially regulated microRNAs between progenitors from normal myometrium and leiomyoma. *Am. J. Obstet. Gynecol.* **2020**, *222*, 596.e1–596.e9. [CrossRef]
6. Kørbling, M.; Estrov, Z. Adult stem cells for tissue repair—A new therapeutic concept? *N. Engl. J. Med.* **2003**, *349*, 570–582. [CrossRef] [PubMed]
7. Dominici, M.; Le Blanc, K.; Mueller, I.; Slaper-Cortenbach, I.; Marini, F.; Krause, D.; Deans, R.J.; Keating, A.; Prockop, D.J.; Horwitz, E.M. Minimal criteria for defining multipotent mesenchymal stromal cells. The International Society for Cellular Therapy position statement. *Cytotherapy* **2006**, *8*, 315–317. [CrossRef] [PubMed]
8. Kozlowska, U.; Krawczenko, A.; Futoma, K.; Jurek, T.; Rorat, M.; Patrzalek, D.; Klimczak, A. Similarities and differences between mesenchymal stem/progenitor cells derived from various human tissues *World J. Stem Cells* **2019**, *11*, 347–374.

9. Borlongan, C.V. Amniotic fluid as a source of engraftable stem cells. *Brain Circ.* **2017**, *3*, 175–179. [CrossRef]

10. Choi, E.; Choi, E.; Hwang, K.C. MicroRNAs as novel regulators of stem cell fate. *World J. Stem Cells* **2013**, *5*, 172–187. [CrossRef]

11. Carbonell, T.; Gomes, A.V. MicroRNAs in the regulation of cellular redox status and its implications in myocardial ischemia-reperfusion injury. *Redox Biol.* **2020**, *36*, 101607. [CrossRef] [PubMed]

12. Orciani, M.; Morabito, C.; Emanuelli, M.; Guarnieri, S.; Sartini, D.; Giannubilo, S.R.; Di Primio, R.; Tranquilli, A.L.; Mariggiò, M.A. Neurogenic potential of mesenchymal-like stem cells from human amniotic fluid: The influence of extracellular growth factors. *J. Biol. Regul. Homeost. Agents* **2011**, *25*, 115–130. [PubMed]

13. Orciani, M.; Emanuelli, M.; Martino, C.; Pugnaloni, A.; Tranquilli, A.L.; Di Primio, R. Potential role of culture mediums for successful isolation and neuronal differentiation of amniotic fluid stem cells. *Int. J. Immunopath. Pharmacol.* **2008**, *21*, 595–602. [CrossRef] [PubMed]

14. Campanati, A.; Orciani, M.; Sorgentoni, G.; Consales, V.; Offidani, A.; Di Primio, R. Pathogenetic characteristics of mesenchymal stem cells in hidradenitis suppurativa. *JAMA Dermatol.* **2018**, *154*, 1184–1190. [CrossRef] [PubMed]

15. Bonifazi, M.; Di Vincenzo, M.; Caffarini, M.; Mei, F.; Salati, M.; Zuccatosta, L.; Refai, M.; Mattioli-Belmonte, M.; Gasparini, S.; Orciani, M. How the Pathological Microenvironment Affects the Behavior of Mesenchymal Stem Cells in the Idiopathic Pulmonary Fibrosis. *Int. J. Mol. Sci.* **2020**, *21*, 8140. [CrossRef]

16. Orciani, M.; Caffarini, M.; Lazzarini, R.; Delli Carpini, G.; Tsiroglou, D.; Di Primio, R.; Ciavattini, A. Mesenchymal Stem Cells from Cervix and Age: New Insights into CIN Regression Rate. *Oxid. Med. Cell. Longev.* **2018**, *2018*, 1545784. [CrossRef]

17. Halfon, S.; Abramov, N.; Grinblat, B.; Ginis, I. Markers distinguishing mesenchymal stem cells from fibroblasts are downregulated with passaging. *Stem Cells Dev.* **2011**, *20*, 53–66. [CrossRef]

18. FastQC Software. Available online: http://www.bioinformatics.babraham.ac.uk/projects/fastqc/ (accessed on 20 April 2020).

19. MiRbase. Available online: https://www.mirbase.org/ (accessed on 27 April 2020).

20. DIANA-mirPath. Available online: http://diana.imis.athena-innovation.gr/DianaTools/index.php?r=site/page&view=software (accessed on 11 May 2020).

21. mirTarBase. Available online: https://mirtarbase.cuhk.edu.cn/ (accessed on 18 May 2020).

22. STRING: Functional Protein Association Networks. Available online: https://string-db.org/ (accessed on 8 June 2020).

23. Malik, M.; Segars, J.; Catherino, W.H. Integrin β1 regulates leiomyoma cytoskeletal integrity and growth. *Matrix Biol.* **2012**, *31*, 389–397. [CrossRef]

24. Islam, M.S.; Ciavattini, A.; Petraglia, F.; Castellucci, M.; Ciarmela, P. Extracellular matrix in uterine leiomyoma pathogenesis: A potential target for future therapeutics. *Hum. Reprod. Update* **2018**, *24*, 59–85. [CrossRef]

25. Leppert, P.C.; Baginski, T.; Prupas, C.; Catherino, W.H.; Pletcher, S.; Segars, J.H. Comparative ultrastructure of collagen fibrils in uterine leiomyoma and normal myometrium. *Fertil. Steril.* **2004**, *82*, 1182–1187. [CrossRef]

26. Kim, Y.J.; Kim, Y.Y.; Shin, J.H.; Kim, H.; Ku, S.Y.; Suh, C.S. Variation in MicroRNA Expression Profile of Uterine Leiomyoma with Endometrial Cavity Distortion and Endometrial Cavity Non-Distortion. *Int. J. Mol. Sci.* **2018**, *19*, 2524. [CrossRef] [PubMed]

27. Chuang, T.D.; Khorram, O. Expression profiling of lncRNAs, miRNAs, and mRNAs and their differential expression in leiomyoma using next-generation RNA sequencing. *Reprod. Sci.* **2018**, *25*, 246–255. [CrossRef] [PubMed]

28. Pan, Q.; Luo, X.; Chegini, N. MicroRNA 21: Response to hormonal therapies and regulatory function in leiomyoma, transformed leiomyoma and leiomyosarcoma cells. *Mol. Hum. Reprod.* **2010**, *16*, 215–227. [CrossRef] [PubMed]

29. Prowse, A.B.J.; Chong, F.; Gray, P.P.; Munro, T.P. Stem cell integrins: Implications for ex-vivo culture and cellular therapies. *Stem Cell Res.* **2011**, *6*, 1–12. [CrossRef] [PubMed]

30. Stefanidis, K.; Pergialiotis, V.; Christakis, D.; Loutradis, D.; Antsaklis, A. Nodal, Nanog, DAZL and SMAD gene expression in human amniotic fluid stem cells. *J. Stem Cells* **2013**, *8*, 17–23.

31. Norian, J.M.; Malik, M.; Parker, C.Y.; Joseph, D.; Leppert, P.C.; Segars, J.H.; Catherino, W.H. Transforming growth factor beta3 regulates the versican variants in the extracellular matrix-rich uterine leiomyomas. *Reprod. Sci.* **2009**, *16*, 1153–1164. [CrossRef]

32. Ciebiera, M.; Włodarczyk, M.; Wrzosek, M.; Męczekalski, B.; Nowicka, G.; Łukaszuk, K.; Ciebiera, M.; Słabuszewska-Jóźwiak, A.; Jakiel, G. Role of Transforming Growth Factor β in Uterine Fibroid Biology. *Int. J. Mol. Sci.* **2017**, *18*, 2435. [CrossRef]

33. Lee, B.S.; Nowak, R.A. Human leiomyoma smooth muscle cells show increased expression of transforming growth factor-β3 (TGF-β3) and altered responses to the antiproliferative effects of TGF-β. *J. Clin. Endocrinol. Metab.* **2001**, *86*, 913–920.

34. Ciebiera, M.; Wlodarczyk, M.; Slabuszewska-Jozwiak, A.; Nowicka, G.; Jakiel, G. Influence of vitamin D and transforming growth factor β3 serum concentrations, obesity, and family history on the risk for uterine fibroids. *Fertil. Steril.* **2016**, *106*, 1787–1792. [CrossRef]

35. Joseph, D.S.; Malik, M.; Nurudeen, S.; Catherino, W.H. Myometrial cells undergo fibrotic transformation under the influence of transforming growth factor β-3. *Fertil. Steril.* **2010**, *93*, 1500–1508. [CrossRef]

36. Nakao, A.; Afrakhte, M.; Moren, A.; Nakayama, T.; Christian, J.L.; Heuchel, R.; Kawabata, M.; Heldin, N.E.; Heldin, C.H.; Dijke, P. Identification of SMAD7, a TGFβ-inducible antagonist of TGF-β signalling. *Nature* **1997**, *389*, 631–635. [CrossRef] [PubMed]

37. Roberts, A.B.; Sporn, M.B. Physiological actions and clinical applications of transforming growth factor-β (TGF-β). *Growth Factors* **1993**, *8*, 1–9. [CrossRef] [PubMed]

38. Derynck, R.; Zhang, Y.E. Smad-dependent and Smad-independent pathways in TGF-β family signalling. *Nature* **2003**, *425*, 577–584. [CrossRef]

39. Huang, S.S.; Huang, J.S. TGF-b Control of Cell Proliferation. *J. Cell Biochem.* **2005**, *96*, 447–462. [CrossRef]

40. Busnelli, M.; Rimoldi, V.; Viganò, P.; Persani, L.; Di Blasio, A.M.; Chini, B. Oxytocin-induced cell growth proliferation in human myometrial cells and leiomyomas. *Fertil. Steril.* **2010**, *94*, 1869–1874. [CrossRef]

41. Zavadil, J.; Ye, H.; Liu, Z.; Wu, J.; Lee, P.; Hernando, E.; Soteropoulos, P.; Toruner, G.A.; Wei, J.J. Profiling and functional analyses of microRNAs and their target gene products in human uterine leiomyomas. *PLoS ONE* **2010**, *5*, e12362. [CrossRef]

42. Medikare, V.; Kandukuri, L.R.; Ananthapur, V.; Deenadayal, M.; Nallari, P. The genetic bases of uterine fibroids; a review. *J. Reprod. Infertil.* **2011**, *12*, 181–191.

43. Mäkinen, N.; Mehine, M.; Tolvanen, J.; Kaasinen, E.; Li, Y.; Lehtonen, H.J.; Gentile, M.; Yan, J.; Enge, M.; Taipale, M.; et al. MED12, the mediator complex subunit 12 gene, is mutated at high frequency in uterine leiomyomas. *Science* **2011**, *334*, 252–255. [CrossRef]

44. Ravegnini, G.; Mariño-Enriquez, A.; Slater, J.; Eilers, G.; Wang, Y.; Zhu, M.; Nucci, M.R.; George, S.; Angelini, S.; Raut, C.P.; et al. MED12 mutations in leiomyosarcoma and extrauterine leiomyoma. *Mod. Pathol.* **2013**, *26*, 743–749. [CrossRef]

45. Klemke, M.; Meyer, A.; Nezhad, M.H.; Bartnitzke, S.; Drieschner, N.; Frantzen, C.; Schmidt, E.H.; Belge, G.; Bullerdiek, J. Overexpression of HMGA2 in uterine leiomyomas points to its general role for the pathogenesis of the disease. *Genes Chromosomes Cancer* **2009**, *48*, 171–178. [CrossRef]

46. Ligon, A.L.; Scott, J.A.; Takahara, K.; Greenspan, D.S.; Morton, C.C. PCOLCE deletion and expression analyses in uterine leiomyomata. *Cancer Genet. Cytogenet.* **2002**, *137*, 133–137. [CrossRef]

 genes

MDPI

Article

Identification of Antitumor *miR-30e-5p* Controlled Genes; Diagnostic and Prognostic Biomarkers for Head and Neck Squamous Cell Carcinoma

Chikashi Minemura [1,†], Shunichi Asai [2,3,†], Ayaka Koma [1], Naoko Kikkawa [2,3], Mayuko Kato [2], Atsushi Kasamatsu [1], Katsuhiro Uzawa [1], Toyoyuki Hanazawa [3] and Naohiko Seki [2,*]

1. Department of Oral Science, Graduate School of Medicine, Chiba University, Chiba 260-8670, Japan; minemura@chiba-u.jp (C.M.); axna4812@chiba-u.jp (A.K.); kasamatsua@faculty.chiba-u.jp (A.K.); uzawak@faculty.chiba-u.jp (K.U.)
2. Department of Functional Genomics, Chiba University Graduate School of Medicine, Chiba 260-8670, Japan; cada5015@chiba-u.jp (S.A.); naoko-k@hospital.chiba-u.jp (N.K.); mayukokato@chiba-u.jp (M.K.)
3. Department of Otorhinolaryngology/Head and Neck Surgery, Chiba University Graduate School of Medicine, Chiba 260-8670, Japan; thanazawa@faculty.chiba-u.jp
* Correspondence: naoseki@faculty.chiba-u.jp; Tel.: +81-43-226-2971; Fax: +81-43-227-3442
† These authors contributed equally to this work.

Citation: Minemura, C.; Asai, S.; Koma, A.; Kikkawa, N.; Kato, M.; Kasamatsu, A.; Uzawa, K.; Hanazawa, T.; Seki, N. Identification of Antitumor *miR-30e-5p* Controlled Genes; Diagnostic and Prognostic Biomarkers for Head and Neck Squamous Cell Carcinoma. *Genes* **2022**, *13*, 1225. https://doi.org/10.3390/genes13071225

Academic Editors: Giuseppe Iacomino and Fabio Lauria

Received: 12 June 2022
Accepted: 7 July 2022
Published: 9 July 2022

Publisher's Note: MDPI stays neutral with regard to jurisdictional claims in published maps and institutional affiliations.

Abstract: Analysis of microRNA (miRNA) expression signatures in head and neck squamous cell carcinoma (HNSCC) has revealed that the *miR-30* family is frequently downregulated in cancer tissues. The Cancer Genome Atlas (TCGA) database confirms that all members of the *miR-30* family (except *miR-30c-5p*) are downregulated in HNSCC tissues. Moreover, low expression of *miR-30e-5p* and *miR-30c-1-3p* significantly predicts shorter survival of HNSCC patients ($p = 0.0081$ and $p = 0.0224$, respectively). In this study, we focused on *miR-30e-5p* to investigate its tumor-suppressive roles and its control of oncogenic genes in HNSCC cells. Transient expression of *miR-30e-5p* significantly attenuated cancer cell migration and invasive abilities in HNSCC cells. Nine genes (*DDIT4*, *FOXD1*, *FXR1*, *FZD2*, *HMGB3*, *MINPP1*, *PAWR*, *PFN2*, and *RTN4R*) were identified as putative targets of *miR-30e-5p* control. Their expression levels significantly predicted shorter survival of HNSCC patients ($p < 0.05$). Among those targets, *FOXD1* expression appeared to be an independent factor predicting patient survival according to multivariate Cox regression analysis ($p = 0.049$). Knockdown assays using siRNAs corresponding to *FOXD1* showed that malignant phenotypes (e.g., cell proliferation, migration, and invasive abilities) of HNSCC cells were significantly suppressed. Overexpression of *FOXD1* was confirmed by immunostaining of HNSCC clinical specimens. Our miRNA-based approach is an effective strategy for the identification of prognostic markers and therapeutic target molecules in HNSCC. Moreover, these findings led to insights into the molecular pathogenesis of HNSCC.

Keywords: microRNA; HNSCC; tumor-suppressor; *miR-30e-5p*; *FOXD1*; TCGA

1. Introduction

A head and neck squamous cell carcinoma (HNSCC) may arise in the oral cavity, hypopharynx, nasopharynx, and larynx. Based on Global Cancer Statistics 2018, HNSCC is the eighth most common cancer in the world [1]. Approximately 890,000 people are diagnosed with HNSCC annually, and 450,000 die from the disease [2]. A notable characteristic of HNSCC is that many patients are already in the advanced stage of disease at the time of diagnosis. Treatment strategies are limited for patients at such an advanced stage [3]. Cisplatin-based treatment is selected for patients for whom surgery is not indicated. However, cancer cells develop drug-resistance in the course of treatment [4]. No effective treatments have been reported for patients with HNSCC who have acquired drug resistance [5]. Thus, there is a pressing need for improved understanding of the molecular pathogenesis of HNSCC using the latest genomic science.

The human genome project showed that a vast number of non-coding RNA molecules (ncRNAs) are transcribed and function in normal and diseased cells [6,7]. MicroRNAs (miR-NAs), which constitute a class of ncRNA, are single strands of RNA, 19 to 23 bases in length. The unique nature of miRNA is that a single miRNA negatively controls a large number of RNA transcripts (both protein coding RNAs and ncRNAs) in a sequence-dependent manner [8,9]. Therefore, dysregulated miRNA expression can disrupt tightly controlled RNA networks in normal cells. The involvement of aberrantly expressed miRNAs has been reported in a variety of human diseases, including cancer [10].

Over the past decade, a large number of studies have shown that aberrantly expressed miRNAs are closely connected to human oncogenesis [11]. Identifying miRNAs with altered expression in cancer cells is a first step in the characterization of their role. The latest RNA-sequence technology is accelerating the creation of miRNA expression signatures in a variety of cancer types [12]. We have established miRNA expression signatures in several types of cancers, including HNSCC [13,14].

The analysis of miRNA signatures has revealed that members of the *miR-30* family are frequently downregulated in several types of cancers, including HNSCC [15,16]. In the human genome, the *miR-30* family is composed of five species, i.e., *miR-30a* to *miR-30e*. *miR-30c* is further subdivided into *miR-30c-1* and *miR-30c-2*. The *miR-30* family is encoded by six genes located on human chromosomes 1, 6, and 8 [17]. The guide strands of the *miR-30* family share the same seed sequence (GUAAACA). The passenger strands of the *miR-30* family are divided into two groups according to their seed sequences. The first group (UUUCAGU) includes *miR-30a-3p*, *miR-30d-3p*, and *miR-30e-3p*. The second group (UGGGAG) encompasses *miR-30b-3p*, *miR-30c-1-3p*, and *miR-30c-2-3p*. We theorized that clarifying the targets/pathways controlled by miRNAs for each cancer type could greatly enhance our understanding of the molecular pathogenesis of cancer cells.

The Cancer Genome Atlas (TCGA, https://www.cancer.gov/tcga), a landmark cancer genomics program, molecularly characterizes over 20,000 primary cancers (and matched normal samples) spanning 33 cancer types. Based on a TCGA analysis, all members of the *miR-30* family (except *miR-30c-5p*) are downregulated in HNSCC tissues. Moreover, the low expression of two miRNAs, *miR-30e-5p* and *miR-30c-1-3p*, significantly predicts shorter survival of patients with HNSCC ($p = 0.0081$ and $p = 0.0224$, respectively).

In this study, we focused on *miR-30e-5p* and investigated its control of genes involved in the molecular pathogenesis of HNSCC. A total of nine genes (*DDIT4*, *FOXD1*, *FXR1*, *FZD2*, *HMGB3*, *MINPP1*, *PAWR*, *PFN2*, and *RTN4R*) were identified as putative targets of *miR-30e-5p*. Their expression level predicted shorter survival of the patients ($p < 0.05$).

Among these genes, we focused on *FOXD1* (Forkhead Box D1), which belongs to the forkhead family of transcription factors. Here, we investigated its functional significance in HNSCC cells.

RNA immunoprecipitation (RIP) assays revealed that *FOXD1* mRNA was incorporated into the RNA-induced silencing complex (RISC). Additionally, dual luciferase reporter assays showed that *miR-30e-5p* was directly bound to the 3′-UTR of *FOXD1*. Functional assays using siRNAs targeting *FOXD1* showed that the knockdown of *FOXD1* expression attenuated HNSCC cell malignant behaviors (migration and invasion). Our approach will reveal the new insights into the involvement of miRNA and the molecular pathogenesis of HNSCC.

2. Materials and Methods

2.1. Analysis of miRNAs and miRNA Target Genes in HNSCC Patients

The sequences of the *miR-30* family were confirmed using miRbase ver. 22.1 (https://www.mirbase.org, accessed on 10 July 2020) [18]. TCGA-HNSC (TCGA, Firehose Legacy) was used to investigate the clinical significance of miRNAs and their target genes. Clinical parameters and gene expression data were obtained from cBioPortal (http://www.cbioportal.org/, data downloaded on 20 August 2020) and OncoLnc (http://www.oncolnc.org/, data downloaded on 20 August 2020) [19–21].

The putative target genes with the *miR-30-5p* binding sites were selected by using TargetScanHuman ver. 7.2 (http://www.targetscan.org/vert_72/, accessed on 22 November 2020) [22].

To identify differentially expressed genes in HNSCC tissues, a newly created microarray data (accession number: GSE180077) was used in this study. Three hypopharyngeal squamous cell carcinoma (HSCC) tissues, three normal hypopharyngeal tissues, and two cervical lymph nodes harvested from one HSCC patient who underwent surgical resection at Chiba University Hospital were subjected to Agilent whole genome microarrays. In this study, we compared gene expressions in cancer tissues with those in normal tissues. The clinical information of this patient was summarized in Table S1.

2.2. HNSCC Cell Lines

Two human HNSCC cell lines were obtained from the RIKEN BioResource Center (Tsukuba, Ibaraki, Japan). Sa3 was harvested from upper gingiva cancer of 63y male. SAS was harvested from oral tongue cancer of 69y female.

2.3. RNA Extraction and Quantitative Real-Time Reverse-Transcription PCR (qRT-PCR)

Total RNA was isolated using TRIzol reagent (Invitrogen, Waltham, MA, USA) according to the manufacturer's protocol. qRT-PCR was performed using the High-Capacity cDNA Reverse Transcription Kit (Applied Biosystems, Waltham, MA, USA) and the StepOnePlus™ Real-Time PCR System (Applied Biosystems). Gene expressions were quantified relatively by the delta-delta Ct method (used *GAPDH* as internal control). Taqman assays used in this study are summarized in Table S2.

2.4. Transfection of miRNAs and siRNAs into HNSCC Cells

The protocols used for transient transfection of miRNAs and siRNAs were described in our previous studies [23–26]. The miRNA precursors and siRNAs used in this report were detailed in Table S2. The miRNAs and siRNAs were transfected into HNSCC cell lines using Opti-MEM (Gibco, Carlsbad, CA, USA) and Lipofectamine™ RNAiMax Transfection Reagent (Invitrogen, Waltham, MA, USA). All miRNA precursors were transfected into HNSCC cell lines at 10 nM, and siRNAs were transfected at 5 nM. Mock transfection consisted of cells without precursors or siRNAs. Control groups were transfected with the negative control precursor.

2.5. RIP Assay

The procedures for the RIP assay have been described previously [23]. Briefly, SAS cells were cultured in 10-cm plates at 80% confluency. Negative control miRNA precursors and *miR-30e-5p* precursors were transfected. After 6 h, immunoprecipitation was performed using the MagCapture™ microRNA Isolation Kit, Human Ago2, obtained from FUJIFILM Wako Pure Chemical Corporation (Wako, Osaka, Japan) according to the manufacturer's protocol. Expression levels of *FOXD1* bound to Ago2 were measured by qRT-PCR. Taqman primers used in the assay are summarized in Table S2.

2.6. Functional Assays of HNSCC Cells (Cell Proliferation, Migration, and Invasion Assays)

The procedures for conducting functional assays (cell proliferation, migration, and invasion assays) with HNSCC cells were described in earlier publications [23–26]. In brief, for proliferation assays, Sa3 or SAS cells were transferred to 96-well plates at 3.0×10^3 cells per well. Cell proliferation was evaluated using the XTT assay kit II (Sigma–Aldrich, St. Louis, MO, USA) 72 h after the transfection procedure. For migration and invasion assays, Sa3 and SAS cells were transfected in 6-well plates at 2.0×10^5 cells per well. After 48 h, transfected Sa3 and SAS cells were added into each chamber at 1.0×10^5 per well. Corning BioCoat™ cell culture chambers (Corning, Corning, NY, USA) were used for migration assays whereas Corning BioCoat Matrigel Invasion Chambers were used for invasion assays. After 48 h, the cells on the lower surface of chamber membranes were stained and counted for analysis. All experiments were performed in triplicate.

2.7. Plasmid Construction and Dual-Luciferase Reporter Assays

The procedures for plasmid construction and the dual-luciferase reporter assays were outlined previously [23–26]. Briefly, a partial wild-type sequence, including the seed sequence of *FOXD1* within the 3′-untranslated region (3′-UTR), was inserted into the psiCHECK-2 vector (C8021; Promega, Madison, WI, USA). Alternatively, a deletion type that was missing the *miR-30e-5p* target site was synthesized. These synthesized vectors (50 ng) were transfected into 1.0×10^5 cells in each well using Lipofectamine 2000 (Invitrogen, Waltham, MA, USA) and Opti-MEM. After 48 h of transfection, dual luciferase reporter assays using the Dual Luciferase Reporter Assay System (Promega, Madison, WI, USA) were conducted. Luminescence data were presented as the *Renilla/Firefly* luciferase activity ratio.

2.8. Immunohistochemistry

The procedures for immunohistochemistry were described in our previous studies [23–26]. The clinical samples were obtained from HNSCC cases who received surgical treatment at Chiba University Hospital. The slides were incubated with primary antibody FOXD1 (1: 50, PA5-27142, Thermo Fisher Scientific, Waltham, MA, USA). The clinical features are shown in Table S3. The reagents used in the analysis are listed in Table S2.

2.9. Gene Set Enrichment Analysis (GSEA)

To analyze the molecular pathways related to *FOXD1* (regulated by *miR-30e-5p*), GSEA was performed. Using TCGA-HNSC data, HNSCC patients were divided into high and low expression groups according to the Z-score of the *FOXD1* expression level. A ranked list of genes was generated by the $\log_2$ ratio comparing the expression levels of each gene between the two groups. The ranked gene list was uploaded into GSEA software [27,28] and we applied the Hallmark gene set in The Molecular Signatures Database [27,29].

2.10. Statistical Analysis

Statistical analyses were determined using JMP Pro 15 (SAS Institute Inc., Cary, NC, USA). Welch's *t*-tests were performed to determine the significance of differences between two groups. Dunnett's tests were applied for comparisons among multiple groups. For correlation analyses, Spearman's test was applied. Survival analyses were analyzed by log-rank test. Each expression levels of target genes, age, disease stage, and pathological grade in TCGA-HNSC were used as variables for Cox's proportional hazards model. A *p*-value less than 0.05 was considered statistically significant.

3. Results

3.1. Expression Levels and the Clinical Significance of the miR-30 Family in HNSCC Clinical Specimens Assessed by TCGA Analysis

The human *miR-30* family consists of 12 members. They include *miR-30a-5p, miR-30a-3p, miR-30b-5p, miR-30b-3p, miR-30c-1-5p, miR-30c-1-3p, miR-30c-2-5p, miR-30c-2-3p, miR-30d-5p, miR-30d-3p, miR-30e-5p,* and *miR-30e-3p.* Seed sequences of *miR-30-5p* are identical. In contrast, two types of seed sequences are present in *miR-30-3p* (Figure 1A). Human *miR-30e* and *miR-30c-1* are located on chromosome 1p34.2, whereas *miR-30c-2* and *miR-30a* are on chromosome 6q13, and *miR-30b* and *miR-30d* are found on chromosome 8q24.22 (Figure 1B).

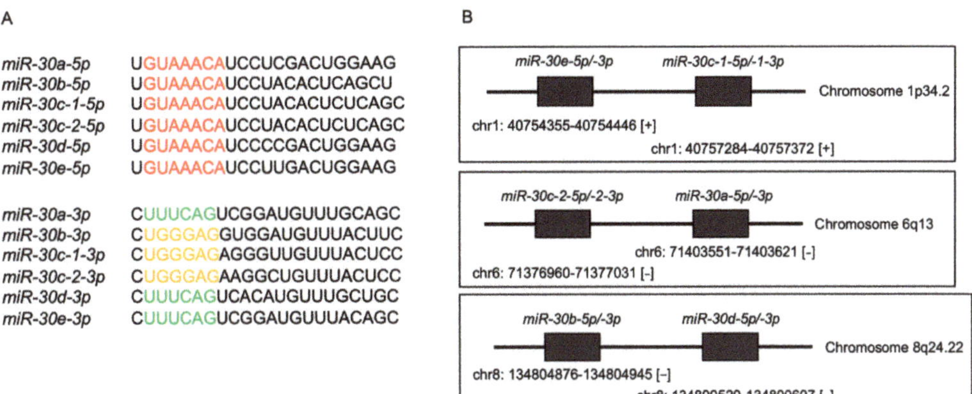

Figure 1. Twelve miRNAs are present in the human genome as members of the *miR-30* family. (**A**) Seed sequences of *miR-30-5p* are identical. In contrast, two types of seed sequences are present in *miR-30-3p*. (**B**) The locations of each microRNA family on chromosome.

Expression levels of all members of the *miR-30* family were validated by TCGA database analyses. All members of the *miR-30* family (except for *miR-30c-5p*) were significantly downregulated in HNSCC tissues ($n = 484$) compared to normal tissues ($n = 44$) (Figure 2).

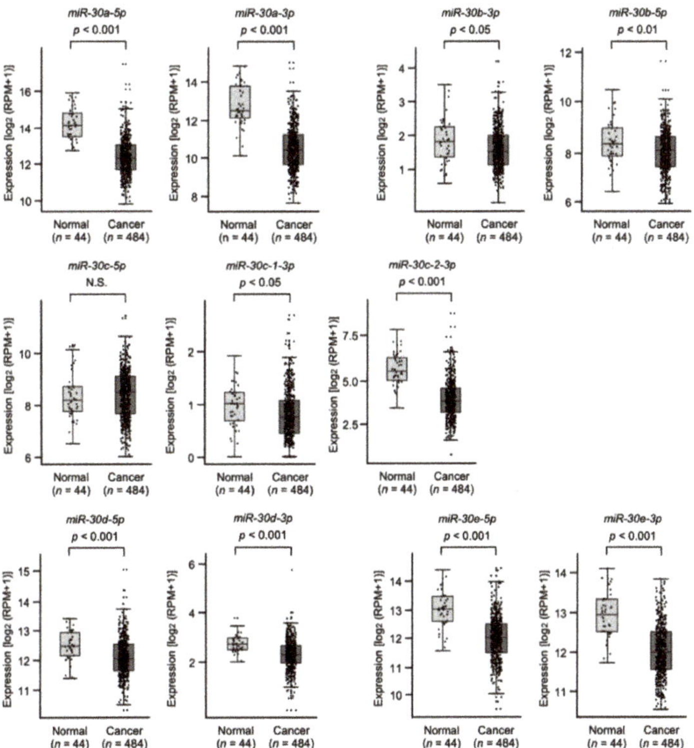

Figure 2. The expression level of the *miR-30* family was analyzed using TCGA-HNSC database. A total of 484 HNSCC tissues and 44 normal epithelial tissues were evaluated (N.S.: not significant).

Moreover, low expression of *miR-30e-5p* and *miR-30c-1-3p* significantly predicted poor prognosis of patients with HNSCC (Figure 3). Expression levels of other members of the *miR-30* family were not related to patient prognosis (Figure 3).

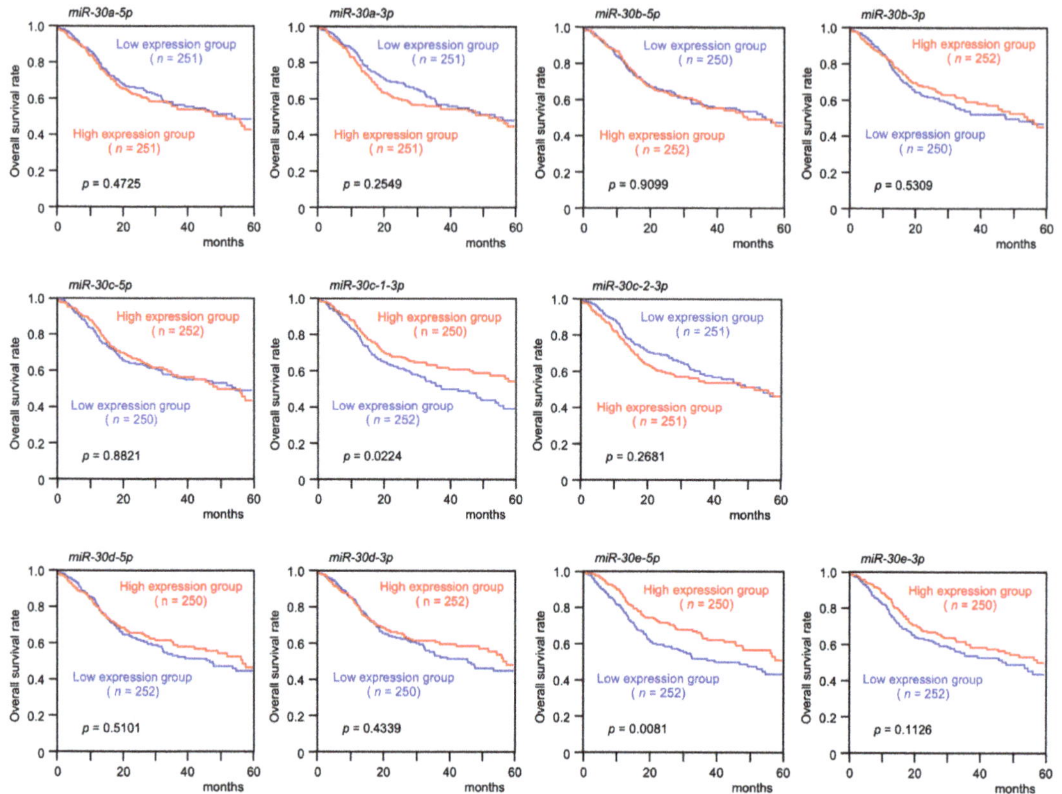

Figure 3. Kaplan–Meier survival analyses of HNSCC patients using data from TCGA-HNSC. Patients were divided into two groups according to the median miRNA expression level: high and low expression groups. The red and blue lines represent the high and low expression groups, respectively.

3.2. Effect of Transient Transfection of miR-30e-5p on HNSCC Cell Proliferation, Migration and Invasion

In this study, we focused on *miR-30e-5p* because its expression was significantly downregulated in HNSCC tissues, and it was closely associated with poor prognosis of the patients, suggesting that *miR-30e-5p* acts as a tumor-suppressive miRNA in HNSCC cells.

The tumor-suppressive roles of *miR-30e-5p* were assessed by transient transfection of *miR-30e-5p* in two cell lines, Sa3 and SAS. Transient transfection of *miR-30e-5p* inhibited HNSCC cell proliferation (Figure 4A). Cancer cell migration and invasive abilities were markedly blocked by *miR-30e-5p* expression in Sa3 and SAS cells (Figure 4B,C). Photographs of typical results from the migration and invasion assays are shown in Figure S1.

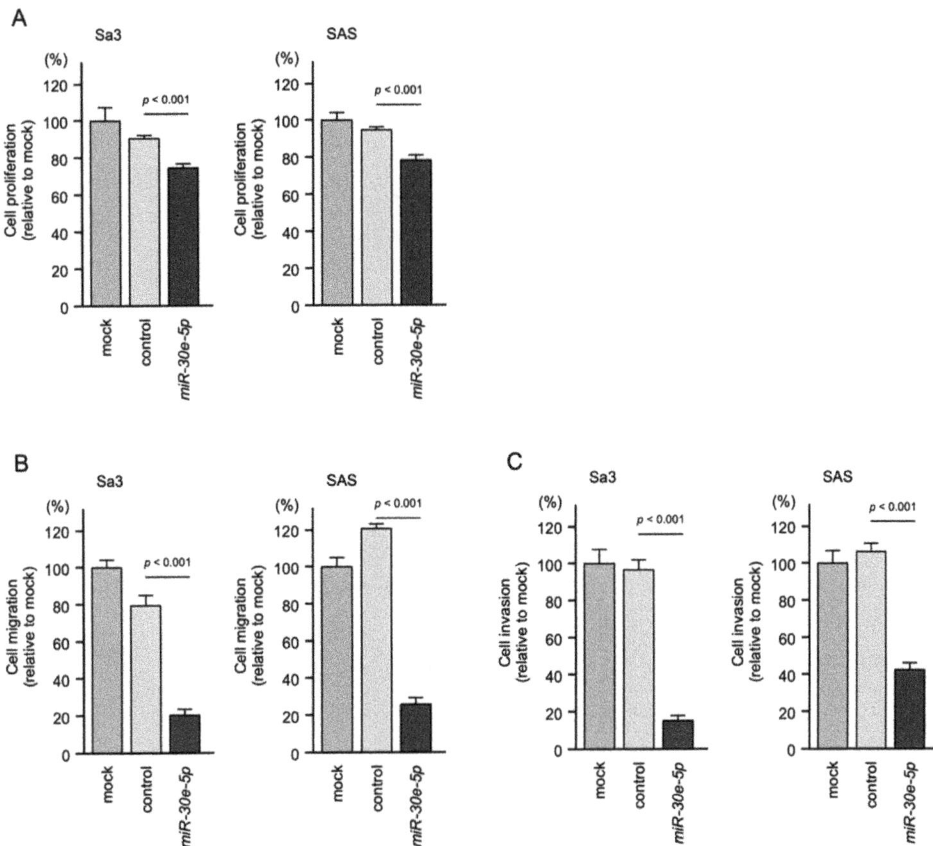

Figure 4. Functional assays of *miR-30e-5p* in HNSCC cell lines (Sa3 and SAS). (**A**) Cell proliferation was assessed using XTT assays 72 h after miRNA transfection. (**B**) Cell migration was assessed using a membrane culture system 48 h after seeding miRNA-transfected cells into the chambers. (**C**) Cell invasion was determined using Matrigel invasion assays 48 h after seeding miRNA-transfected cells into the chambers.

3.3. Screening for Oncogenic Targets of miR-30e-5p in HNSCC

To identify genes modulated by *miR-30e-5p* that were closely involved in HNSCC molecular pathogenesis, we performed in silico database analyses combined with our gene expression data. We created new gene expression data using clinical specimens of hypopharyngeal squamous cell carcinoma (HSCC). Three HSCC tissues, three normal hypopharyngeal tissues, and two cervical lymph nodes harvested from one HSCC patient who underwent surgical resection at Chiba University Hospital were subjected to Agilent whole genome microarrays. In this study, we compared gene expressions in cancer tissues with those in normal tissues. The clinical information of this patient was summarized in Table S1. Expression data were deposited in the GEO database (accession number: GSE180077). Our selection strategy is shown in Figure 5.

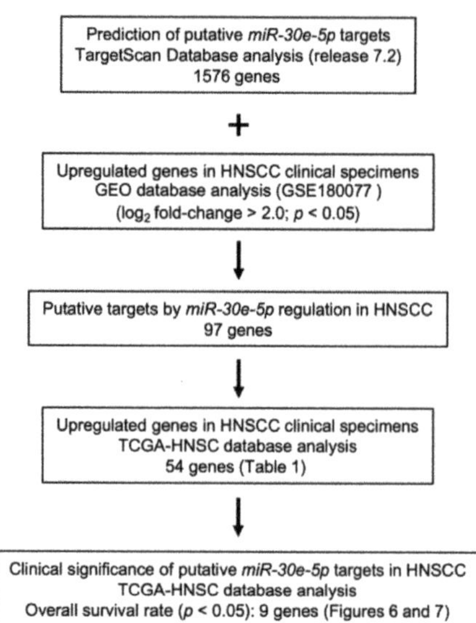

Figure 5. Flow chart of the strategy used to identify putative tumor suppressor genes regulated by *miR-30e-5p*.

TargetScan Human database (release 7.2) provides data on the putative targets of *miR-30e-5p*. A total of 1576 genes are listed. Among these targets, 97 genes were upregulated in HSCC patients. Furthermore, the increased expression of 54 genes was confirmed in TCGA database analyses (Table 1).

Table 1. Candidate target genes regulated by *miR-30-5p*.

Entrez Gene ID	Gene Symbol	Gene Name	Total Binding Sites	GEO [1] p Value	GEO $\log_2$(FC [2])	5y OS [3] p Value
8087	FXR1	Fragile X mental retardation, autosomal homolog 1	1	0.014	3.06	<0.001
5217	PFN2	Profilin 2	1	0.004	2.67	<0.001
54541	DDIT4	DNA-damage-inducible transcript 4	1	0.017	2.70	0.004
2297	FOXD1	Forkhead box D1	1	0.008	3.79	0.008
9562	MINPP1	Multiple inositol-polyphosphate phosphatase 1	1	0.013	2.02	0.019
5074	PAWR	PRKC, apoptosis, WT1, regulator	2	0.003	2.04	0.025
3149	HMGB3	High mobility group box 3	1	0.004	2.17	0.028
2535	FZD2	Frizzled class receptor 2	1	0.006	2.36	0.032
65078	RTN4R	Reticulon 4 receptor	1	0.008	2.20	0.044
115908	CTHRC1	Collagen triple helix repeat containing 1	1	0.010	2.96	0.059
3218	HOXB8	Homeobox B8	1	0.046	3.29	0.077
9143	SYNGR3	Synaptogyrin 3	1	0.042	2.12	0.077
1012	CDH13	Cadherin 13	1	0.013	2.13	0.091
6683	SPAST	Spastin	2	0.012	2.10	0.092
79718	TBL1XR1	Transducin (β)-like 1 X-linked receptor 1	2	0.002	2.63	0.104
114088	TRIM9	Tripartite motif containing 9	1	0.005	4.27	0.113
84733	CBX2	Chromobox homolog 2	1	0.005	3.00	0.133
27	ABL2	ABL proto-oncogene 2, non-receptor tyrosine kinase	1	0.008	2.19	0.155
23657	SLC7A11	solute carrier family 7 (Anionic amino acid transporter light chain, xc-system), member 11	1	0.007	4.00	0.160
154214	RNF217	Ring finger protein 217	1	0.016	2.36	0.186
79712	GTDC1	Glycosyltransferase-like domain containing 1	1	0.004	4.31	0.193

Table 1. *Cont.*

Entrez Gene ID	Gene Symbol	Gene Name	Total Binding Sites	GEO [1] *p* Value	GEO log$_2$(FC [2])	5y OS [3] *p* Value
26059	ERC2	ELKS/RAB6-interacting/CAST family member 2	1	0.027	3.54	0.210
3237	HOXD11	Homeobox D11	1	0.033	4.24	0.214
89796	NAV1	Neuron navigator 1	1	0.007	2.88	0.234
6659	SOX4	SRY (sex determining region Y)-box 4	1	0.005	2.30	0.258
54434	SSH1	Slingshot protein phosphatase 1	1	0.016	2.05	0.280
2048	EPHB2	EPH receptor B2	1	0.013	2.53	0.303
9258	MFHAS1	Malignant fibrous histiocytoma amplified sequence 1	1	0.005	2.27	0.311
54566	EPB41L4B	Erythrocyte membrane protein band 4.1 like 4B	1	0.004	3.06	0.413
8448	DOC2A	Double C2-like domains, α	2	0.021	3.00	0.485
28982	FLVCR1	Feline leukemia virus subgroup C cellular receptor 1	1	0.005	2.43	0.492
55785	FGD6	FYVE, RhoGEF and PH domain containing 6	1	0.014	2.21	0.530
490	ATP2B1	ATPase, Ca++ transporting, plasma membrane 1	1	0.007	2.93	0.556
4644	MYO5A	Myosin VA (heavy chain 12, myoxin)	1	0.003	2.06	0.605
4015	LOX	Lysyl oxidase	1	0.006	4.46	0.613
50805	IRX4	Iroquois homeobox 4	1	0.039	2.67	0.652
23432	GPR161	G protein-coupled receptor 161	1	0.008	2.50	0.662
2729	GCLC	Glutamate-cysteine ligase, catalytic subunit	1	0.004	3.14	0.710
8038	ADAM12	ADAM metallopeptidase domain 12	2	0.003	4.29	0.721
3631	INPP4A	Inositol polyphosphate-4-phosphatase, type I, 107kDa	1	0.003	2.12	0.747
9832	JAKMIP2	Janus kinase and microtubule interacting protein 2	1	0.010	4.19	0.769
121268	RHEBL1	Ras homolog enriched in brain like 1	1	0.004	2.92	0.775
144455	E2F7	E2F transcription factor 7	3	0.021	2.47	0.786
94032	CAMK2N2	Calcium/calmodulin-dependent protein kinase II inhibitor 2	1	0.003	2.95	0.823
84206	MEX3B	Mex-3 RNA binding family member B	2	0.016	2.75	0.880
2887	GRB10	Growth factor receptor-bound protein 10	1	0.008	2.07	0.890
23333	DPY19L1	Dpy-19-like 1 (C. elegans)	1	0.013	3.07	0.893
9435	CHST2	Carbohydrate (N-acetylglucosamine-6-O) sulfotransferase 2	1	0.006	3.30	0.917
54165	DCUN1D1	DCN1, defective in cullin neddylation 1, domain containing 1	2	0.009	2.50	0.938
54714	CNGB3	Cyclic nucleotide gated channel β 3	1	0.008	2.98	0.968
221002	RASGEF1A	RasGEF domain family, member 1A	1	0.005	2.81	0.968
55144	LRRC8D	Leucine rich repeat containing 8 family, member D	1	0.002	2.60	0.978
8534	CHST1	Carbohydrate (keratan sulfate Gal-6) sulfotransferase 1	1	0.005	4.41	0.980
8632	DNAH17	Dynein, axonemal, heavy chain 17	1	0.006	5.26	0.999

[1] Gene Expression Omnibus, [2] Fold Change, [3] 5-year Overall Survival.

3.4. Clinical Significance of miR-30e-5p Targets in Patients with HNSCC Determined by TCGA Analysis

Among the 54 putative target genes, high expression of nine genes (*DDIT4*, *FOXD1*, *FXR1*, *FZD2*, *HMGB3*, *MINPP1*, *PAWR*, *PFN2*, and *RTN4R*) showed statistically significant correlations with the 5-year overall survival frequencies of patients with HNSCC ($p < 0.05$; Figures 6 and 7).

Therefore, univariate analysis for 5-year overall survival was conducted first, and then multivariate analysis was performed for the statistically significant variables ($p < 0.05$). Each expression levels of target genes, age, disease stage, and pathological grade in TCGA-HNSC were used as variables for Cox's proportional hazards model. As a result, the high expression level of *FOXD1* (HR: 1.374, 95% CI: 1.002–1.890, $p = 0.049$), age ($\geq$70) (HR: 1.922, 95% CI: 1.369–2.698, $p < 0.001$) and disease stage (III and IV) (HR: 1.774, 95% CI: 1.159–2.716, $p = 0.008$) were independent prognostic factors (Table 2).

The correlations of the expression levels between these nine genes and *miR-30e-5p* were evaluated by TCGA-HNSC. A Spearman's rank test confirmed weak negative correlations in seven genes (*DDIT4, FOXD1, FZD2, MINPP1, PAWR, PFN2,* and *RTN4R*) ($p < 0.001$, $r < -0.2$, Figure 8), whereas the correlations were not confirmed in *FXR1* and *HMGB3*.

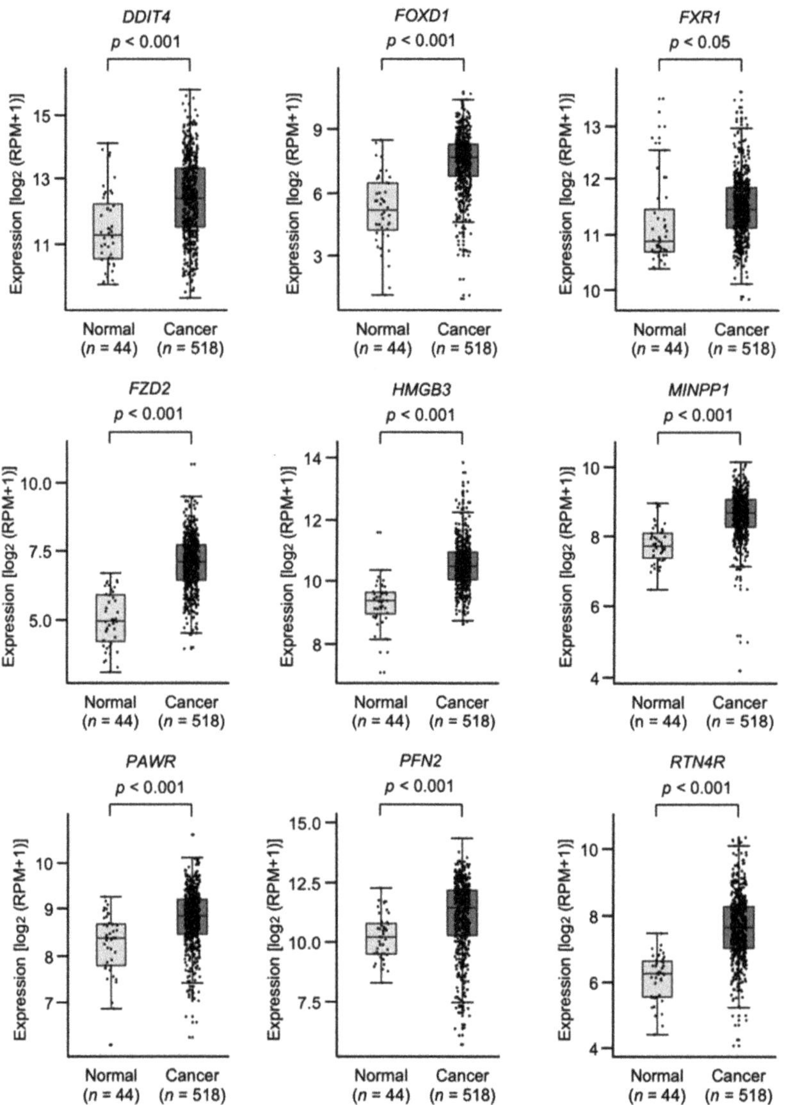

Figure 6. Expression levels of nine target genes (*DDIT4, FOXD1, FXR1, FZD2, HMGB3, MINPP1, PAWR, PFN2,* and *RTN4R*) in HNSCC clinical specimens from TCGA-HNSC. All genes were found to be upregulated in HNSCC tissues ($n = 518$) compared with normal tissues ($n = 44$).

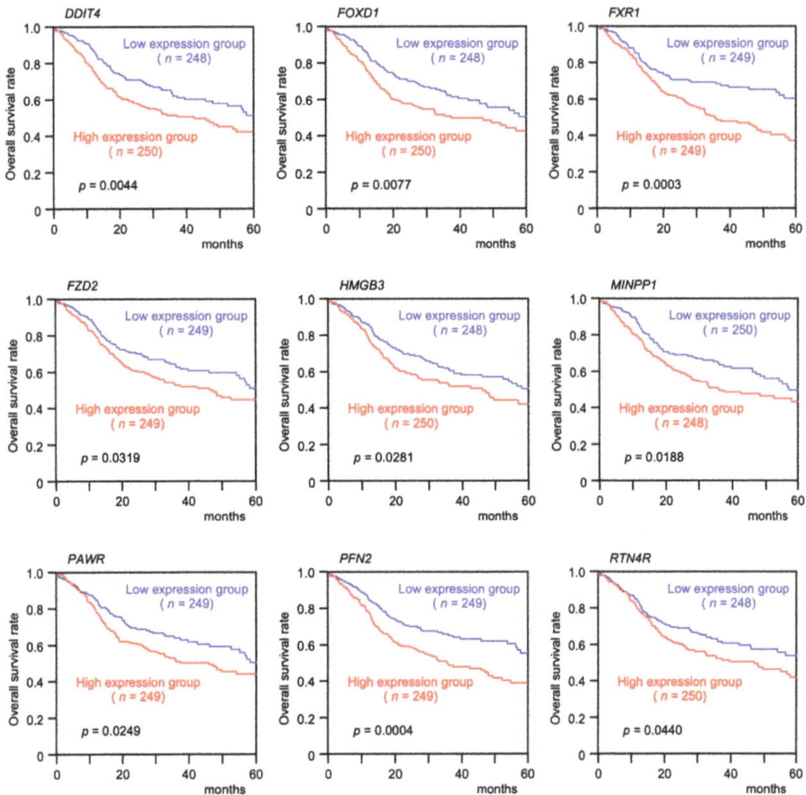

Figure 7. Clinical significance of nine target genes (*DDIT4*, *FOXD1*, *FXR1*, *FZD2*, *HMGB3*, *MINPP1*, *PAWR*, *PFN2*, and *RTN4R*) according to TCGA-HNSC data analysis. Kaplan–Meier curves of the 5-year overall survival rates according to the expression of each gene are presented. High expression levels of all nine genes were significantly predictive of a poorer prognosis in patients with HNSCC. Patients were divided into two groups according to the median gene expression level: high and low expression groups. The red and blue lines represent the high and low expression groups, respectively.

Table 2. Results of Cox regression analysis of overall survivals in five years in TCGA-HNSC.

Variables	Monovariate			Multivariate		
	HR	**95% CI**	*p*-**Value**	**HR**	**95% CI**	*p*-**Value**
DDIT4 (High vs. Low expression)	1.506	1.138–1.994	0.004	1.292	0.939–1.776	0.115
FOXD1 (High vs. Low expression)	1.526	1.153–2.019	0.003	1.374	1.002–1.890	0.049
FXR1 (High vs. Low expression)	1.651	1.242–2.193	0.001	1.303	0.930–1.825	0.124
FZD2 (High vs. Low expression)	1.337	1.011–1.768	0.042	1.069	0.776–1.473	0.684
HMGB3 (High vs. Low expression)	1.413	1.069–1.867	0.015	1.186	0.861–1.633	0.297
MINPP1 (High vs. Low expression)	1.451	1.096–1.921	0.009	1.308	0.940–1.818	0.111
PAWR (High vs. Low expression)	1.438	1.084–1.908	0.012	1.255	0.913–1.727	0.162
PFN2 (High vs. Low expression)	1.642	1.238–2.178	0.001	1.238	0.885–1.731	0.213
RTN4R (High vs. Low expression)	1.326	1.003–1.752	0.048	0.962	0.699–1.323	0.810
Age (≥70 vs. <70)	1.628	1.197–2.213	0.002	1.922	1.369–2.698	<0.001
Disease Stage (III, IV vs. I, II)	1.746	1.158–2.633	0.008	1.774	1.159–2.716	0.008
Pathological Grade (3, 4 vs. 1, 2)	0.901	0.653–1.245	0.529	-	-	-

HR: hazard ratio, CI: confidence interval.

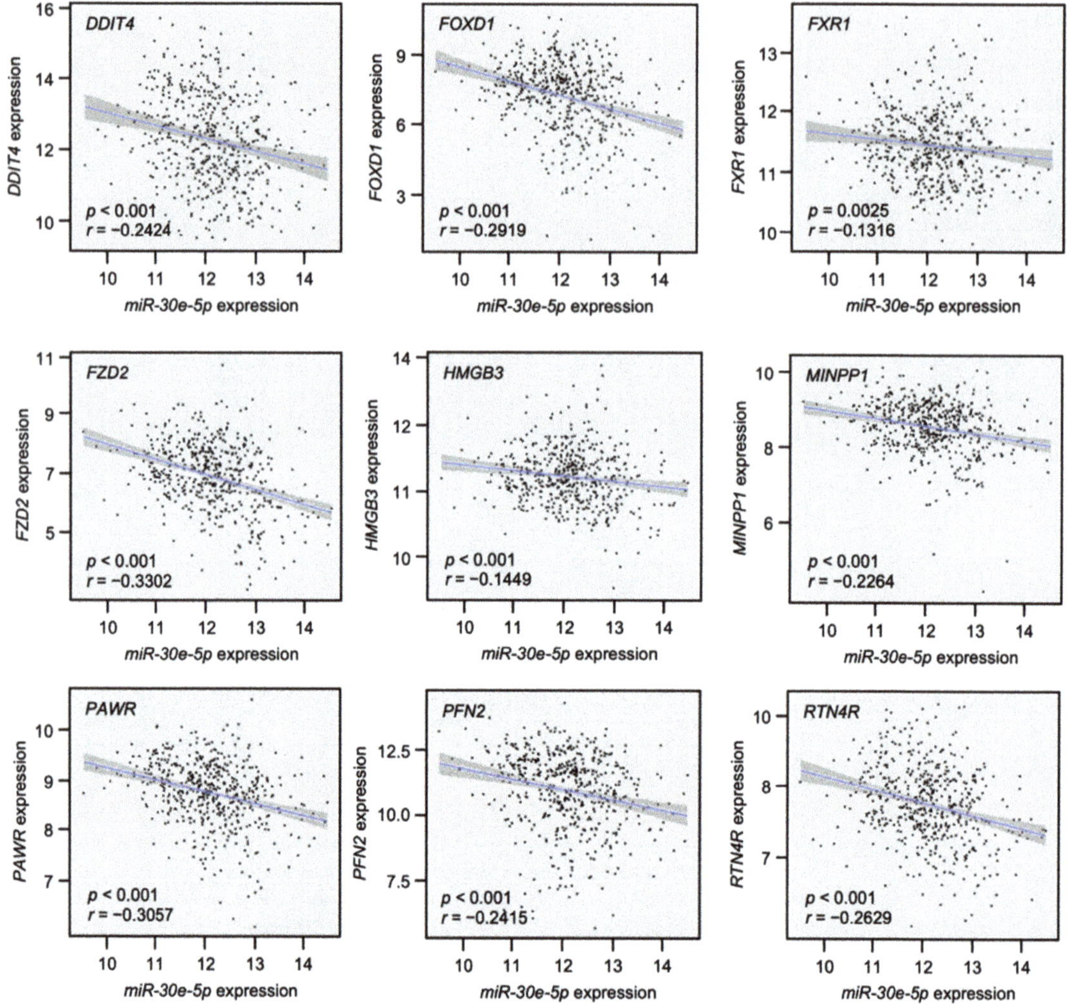

Figure 8. Correlation analysis by TCGA-HNSC for nine genes. Negative correlation of expression levels between *miR-30e-5p* and nine genes in HNSCC clinical specimens.

3.5. Regulated Expression of the Nine Identified Genes by miR-30e-5p in HNSCC Cells

qRT-PCR revealed that the mRNA expression levels of *DDIT4*, *FOXD1*, *MINPP1*, *PAWR*, and *PFN2* were significantly suppressed in *miR-30e-5p*-transfected HNSCC cells. The other genes were not significantly suppressed by *miR-30e-5p* (Figure 9).

Considering the multivariate analysis, the correlation analysis, and the results of qRT-PCR, *FOXD1* was of the greatest interest among the nine targets. Thus, we focused on *FOXD1* as the target gene of *miR-30e-5p* in this study.

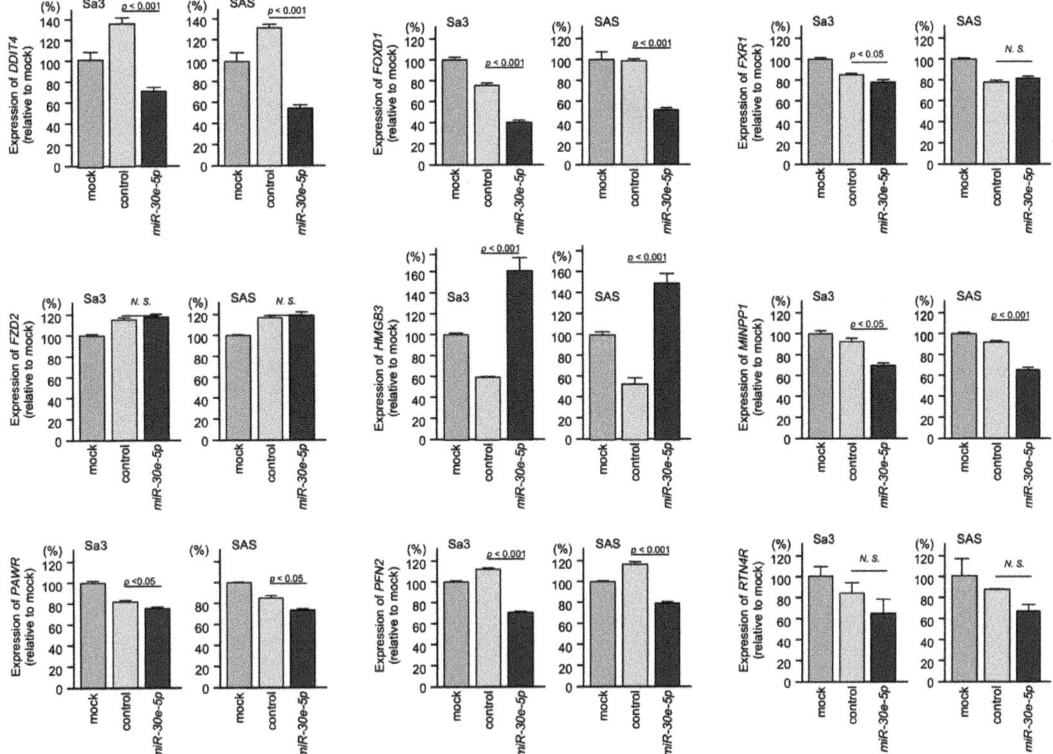

Figure 9. Regulation of the expression of nine genes by *miR-30e-5p* in HNSCC cells. qRT-PCR showing significantly reduced expression of *FOXD1* mRNA (top, middle) 72 h after *miR-30e-5p* transfection in Sa3 and SAS cells (N.S.: not significant compared to control group).

3.6. Incorporation of FOXD1 mRNA into the RNA-Induced Silencing Complex (RISC) and Direct Control of FOXD1 Expression by miR-30e-5p in HNSCC Cells

To confirm incorporation of *FOXD1* mRNA into RISC, RIP assays were performed (Figure 10A,B). The schematic illustration displays the concept of RIP assays. Ago2-bound miRNA and mRNA were isolated by the immunoprecipitation of Ago2, the protein that plays a central role in RISC (Figure 10A). Using samples isolated by immunoprecipitation, qRT-PCR showed that the *FOXD1* mRNA level was significantly higher than those of mock or miR control transfected cells ($p < 0.01$; Figure 10B), suggesting a significant incorporation into RISC.

To confirm that *miR-30e-5p* bound directly to the 3′-UTR of *FOXD1*, a dual-luciferase reporter assay was performed. Luciferase activity was significantly reduced following co-transfection with *miR-30e-5p* and a vector containing the *miR-30e-5p*-binding site in the 3′-UTR of *FOXD1* (Figure 10C). In contrast, co-transfection with a vector containing the *FOXD1* 3′-UTR, in which the *miR-30e-5p*-binding site was deleted, resulted in no change in luciferase activity (Figure 10C).

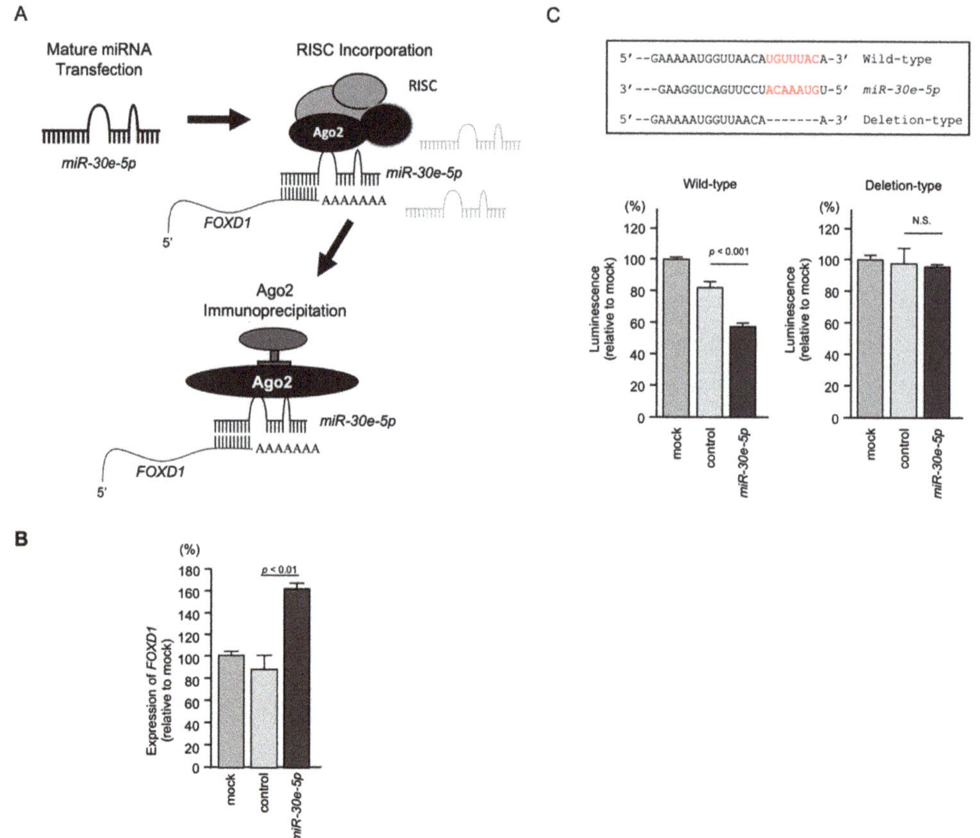

Figure 10. Isolation of RISC-incorporated *FOXD1* mRNA by Ago2 immunoprecipitation. Direct regulation of *FOXD1* expression by *miR-30e-5p* in HNSCC cells. (**A**) Schematic illustration of RIP assay. (**B**) qRT-PCR suggested *FOXD1* mRNA was significantly incorporated into RISC. (**C**) TargetScan database analysis predicting putative *miR-30e-5p*-binding sites in the 3′-UTR of *FOXD1* (upper panel). Dual-luciferase reporter assays showed reduced luminescence activity after co-transfection of the wild-type vector and *miR-30e-5p* in Sa3 cells (lower panel). Normalized data were calculated as the *Renilla/Firefly* luciferase activity ratio (N.S.: not significant compared to control group).

3.7. Expression of FOXD1 in HNSCC Clinical Specimens

HNSCC clinical specimens displayed moderate immunoreactivity in the cytoplasm (Figure 11B,D), whereas normal epithelium showed no expression of *FOXD1* (Figure 11A,C). The clinical features of HNSCC specimens are summarized in Table S3.

3.8. Effects of FOXD1 Knockdown on the Proliferation, Migration, and Invasion of HNSCC Cells

To assess the tumor-promoting effect of *FOXD1* in HNSCC cells, we performed knockdown assays using siRNAs.

First, the inhibitory effects of two different siRNAs targeting *FOXD1* (si*FOXD1*-1 and si*FOXD1*-2) expression were examined. The *FOXD1* mRNA levels were effectively inhibited by each siRNA (Figure S2).

The knockdown of *FOXD1* slightly inhibited Sa3 and SAS cell proliferation (Figure 12A). In contrast, cell migration and invasion were significantly inhibited after si*FOXD1*-1 and si*FOXD1*-2 transfection into Sa3 and SAS cells (Figure 12B,C). Photographs of typical results from the migration and invasion assays are shown in Figure S3.

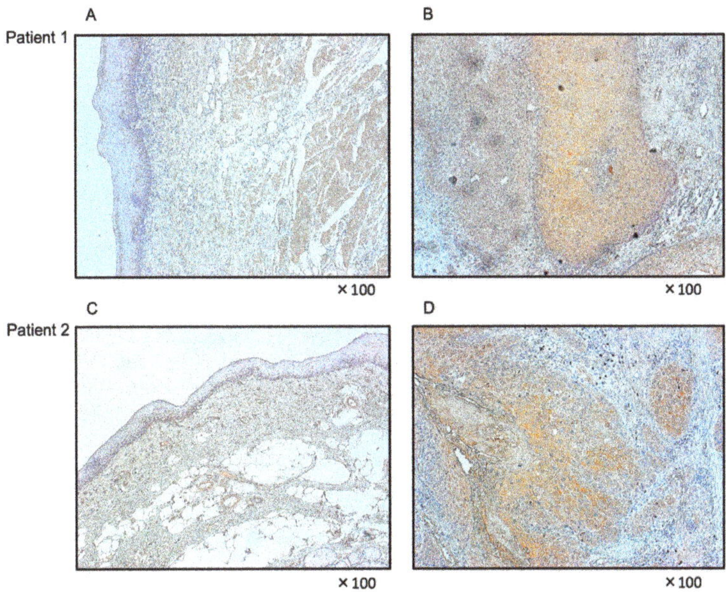

Figure 11. Immunohistochemical staining of *FOXD1* in HNSCC clinical specimens. Weak to moderate immunoreactivity of *FOXD1* was observed in the cancer lesions (**B,D**) whereas negative immunoreactivity was shown in normal mucosa (**A,C**).

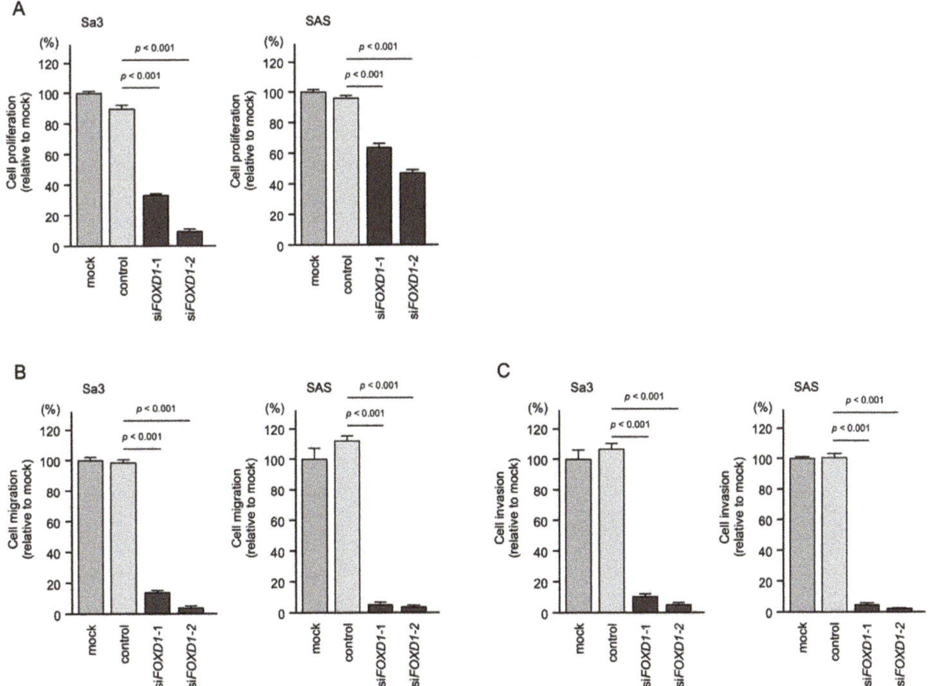

Figure 12. Functional assays of cell proliferation, migration, and invasion following transient transfection of siRNAs targeting *FOXD1* in HNSCC cell lines (Sa3 and SAS cells). (**A**) Cell proliferation

assessed by XTT assay 72 h after siRNA transfection. (**B**) Cell migration assessed using a membrane culture system 48 h after seeding miRNA-transfected cells into the chambers. (**C**) Cell invasion assessed by Matrigel invasion assays 48 h after seeding miRNA-transfected cells into chambers.

3.9. FOXD1-Mediated Molecular Pathways in HNSCC Cells

We performed gene set enrichment analysis (GSEA) to identify genes that were differentially expressed in the high and low expression groups using TCGA-HNSCC data. A GSEA analysis of *FOXD1* showed that the most enriched molecular pathway in the high expression groups of *FOXD1* was "epithelial-mesenchymal transition" (Table 3, Figure 13). Additional pathways (MYC targets, TNFα signaling, Hypoxia, E2F targets, Glycolysis, and DNA repair) were also associated with groups expressing high levels of *FOXD1* (Table 3). The aberrant expression and activation of these pathways were closely associated with the downregulation of *miR-30e-5p* in HNSCC cells and contributed to HNSCC oncogenesis.

Table 3. Results of gene set enrichment analysis.

Enriched Gene Sets in High *FOXD1* Expression Group		
Name	**Normalized Enrichment Score**	**FDR *q*-Value**
Epithelial mesenchymal transition	3.674	*q* < 0.001
MYC targets V1	3.412	*q* < 0.001
TNFα signaling via NFκB	3.262	*q* < 0.001
Hypoxia	3.044	*q* < 0.001
E2F targets	2.988	*q* < 0.001
Glycolysis	2.797	*q* < 0.001
DNA repair	2.602	*q* < 0.001
Unfolded protein response	2.275	0.001
G2M checkpoint	2.248	0.001
TGFβ signaling	2.025	0.001
KRAS signaling up	1.946	0.005
Coagulation	1.848	0.010
Inflammatory response	1.753	0.014
Apical junction	1.739	0.015
Oxidative phosphorylation	1.713	0.015
P53 pathway	1.702	0.016
IL6/JAK/STAT3 signaling	1.617	0.027
Apoptosis	1.601	0.029

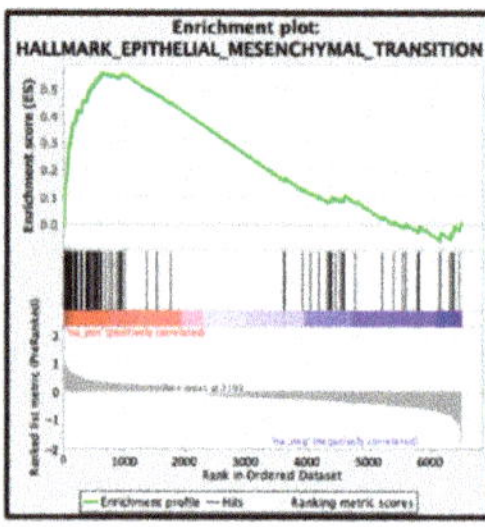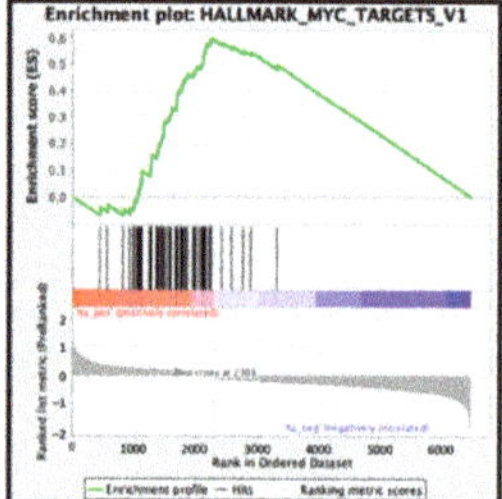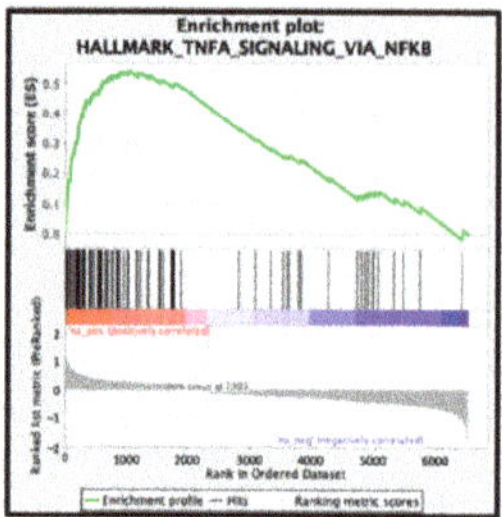

Figure 13. *Cont.*

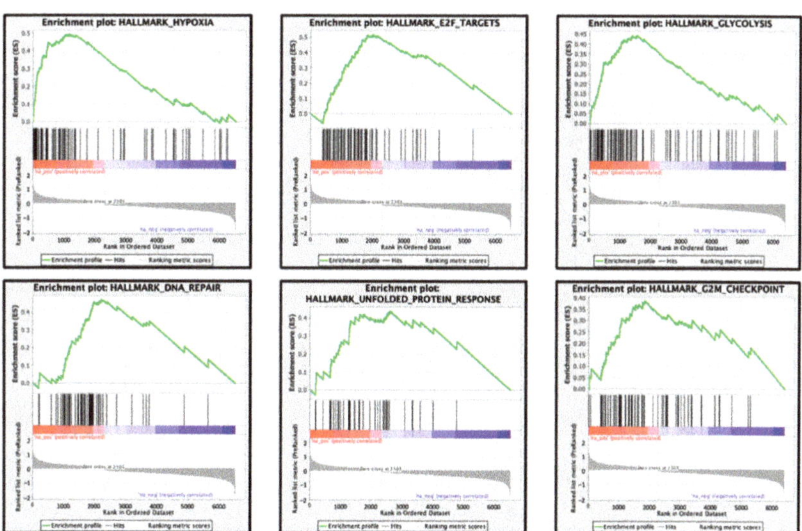

Figure 13. Pathways enriched among the differentially expressed genes in the high *FOXD1* expression group according to gene set enrichment analysis. The nine significantly enriched pathways (top 9) are shown. Most enriched pathway was "Epithelial Mesenchymal Transition".

4. Discussion

In the human genome, numerous miRNAs have very similar sequences and therefore constitute a "family". Because the seed sequences of the miRNA family are the same, each miRNA family controls the expression of the same set of genes. Therefore, aberrant expression of the miRNA family members will cause the disruption of intracellular molecular networks, and these events can initiate the transformation of normal cells to cancer cells. Based on our miRNA signatures, we previously focused on several miRNA families and identified the molecular pathways that were controlled by the *miR-29* family, the *miR-199* family and the *miR-216* family [14,23,30].

Our RNA sequence-based miRNA signatures, including those in HNSCC, showed that some *miR-30* family members were frequently downregulated in cancer tissues. Those observations suggest that these miRNAs control targets with pivotal roles in cancer progression, metastasis, and drug-resistance [31]. Here, we first investigated the expression levels and clinical significance of all members of the *miR-30* family using TCGA database. The family consists of 12 miRNAs: guide and passenger strands of *miR-30a*, *miR-30b*, *miR-30c-1*, *miR-30c-2*, *miR-30d*, and *miR-30e*). Among them, the expression levels of *miR-30e-5p* and *miR-30c-1-3p* were closely associated with HNSCC molecular pathogenesis. Identifying molecular networks controlled by these miRNAs is important for understanding HNSCC oncogenesis.

In this study, we focused on *miR-30e-5p* to investigate the targets that it controls in HNSCC. The seed sequences of the guide strands of the *miR-30a* family are identical, suggesting that these miRNAs may control common targets in a sequence-dependent manner. Previous studies reported that *miR-30e-5p* had tumor-suppressive functions in several types of cancers [32–34]. A previous study showed that *miR-30e-5p* suppressed astrocyte elevated gene-1 (*AEG-1*), an oncogene that contributes to angiogenesis, metastasis, and EMT processes [35]. Another study showed that *miR-30e-5p* was a direct transcriptional target of P53 in colorectal cancer. Expression of *miR-30e-5p* blocked tumor cell migration, invasion, and in vivo metastasis by directly controlling integrin molecules [36].

A single miRNA controls a large number of genes, and targeted genes can differ depending on the type of cancer. Defining miRNA-targeted genes is an important focus in miRNA research. Our target identification strategy successfully identified nine genes that significantly predicted 5-year survival frequencies of HNSCC patients. Unfortunately, none of those genes were independent prognostic factors contributing to 5-year overall survival rates. According to multivariate analysis, a high expression level of *FOXD1* was an independent prognostic factor (HR: 1.374, 95% CI: 1.002–1.890, p = 0.049). Furthermore, the negative correlation between the expression of *FOXD1* and *miR-30e-5p* was confirmed. For those reasons, we focused on *FOXD1* as the target gene in this study. Future studies may examine the other genes in the belief that resultant data will improve our understandings of the molecular pathogenesis of HNSCC.

For example, *FXR1* is an RNA-binding protein that regulates co-transcriptional and post-transcriptional gene expression. It is a member of the Fragile X-mental retardation (*FXR*) family of proteins, which includes *FMR1* and *FXR2* [37]. Recent studies showed that the overexpression of *FRX1* was observed in several types of cancers, including HNSCC, and its expression correlates with poor prognosis of the patients [38]. In oral cancer cells, *FXR1* stabilized *miR-301a-3p* and *miR-301a-3p*, both of which target *p21* [39]. Another study showed that the overexpression of *FXR1* facilitates the bypass of senescence and tumor progression [40]. Aberrant expression of *FXR1*-enhanced HNSCC progression and its expression is closely associated with the molecular pathogenesis of HNSCC.

PFN2 was identified as a *miR-30e-5p* target in this study. *PFN2* is a member of the profilin family, namely, profilin 1, 2, 3, and 4 [41]. Profilins are actin-binding proteins involved in the regulation of cytoskeletal dynamics. Overexpression of *PFN2* was reported in several types of cancers, including HNSCC [42]. Recently, we revealed that *PFN2* was directly regulated by tumor-suppressive *miR-1*/*miR-133* clustered miRNAs in HNSCC cells, and its overexpression promoted cancer cell malignant transformation [25]. Another study showed that overexpression of *PFN2* enhanced the aggressiveness of small cell lung cancer (SCLC), including cell proliferation, migration, and invasion. Knockdown of *PFN2* decreased the cells' aggressive nature. Moreover, a mouse xenograft model demonstrated that the overexpression of *PFN2* dramatically elevated SCLC growth and vasculature formation as well as lung metastasis [43]. Notably, a previous study showed that *miR-30a-5p* suppressed the expression of *PFN2* in lung cancer cell lines [44].

In this study, we confirmed that the overexpression of *FOXD1* facilitated cancer cell malignant transformation, e.g., enhanced migratory and invasive abilities in HNSCC cells. *FOXD1* is a member of the forkhead family of transcription factors [45]. *FOXD1* expression is upregulated in several types of cancers [46]. Moreover, the knockdown of *FOXD1* impairs the colony-forming abilities of oral cancer cells after radiation treatment [47]. Another study showed that upregulated *FOXD1* contributed to melanoma cells' resistance to vemurafenib via the recruited expression of connective tissue growth factor [48]. In gastric cancer (GC) cells, *FOXD1*-AS1 expression induced *FOXD1* translation, and these events enhanced GC cell progression and cisplatin resistance [49]. Importantly, *miR-30a-5p* directly binds to the 3'-UTR of *FOXD1* and suppresses its expression in several types of cancer cells, e.g., lung squamous cell carcinoma, pancreatic ductal adenocarcinoma, osteosarcoma, and ovarian cancer [50–53]. Our present data and previous studies indicate that the overexpression of *FOXD1* is closely involved in aggressive cancer cell transformation in a wide range of cancers, including HNSCC.

These studies show that searching for target genes of tumor-suppressive miRNAs is an attractive strategy for exploring the molecular mechanisms of HNSCC.

5. Conclusions

We showed that the *FOXD1* gene is directly controlled by tumor-suppressive *miR-30e-5p* in HNSCC cells. The aberrant expression of *FOXD1* facilitated cancer cell migration and invasion, and its expression was closely associated with the prognosis of HNSCC patients. Our strategy (analysis of tumor-suppressive miRNAs and their controlled genes) can identify genes that are deeply involved in the molecular pathogenesis of HNSCC.

Supplementary Materials: The following are available online at https://www.mdpi.com/article/10.3390/genes13071225/s1, Figure S1: Photomicrographs of cells subjected to migration and invasion assays. Figure S2: Efficiency of siRNA mediated *FOXD1* knockdown in HNSCC cell lines (Sa3 and SAS cells). Figure S3: Photomicrographs of cells subjected to migration and invasion assays. Table S1: Clinical features of the patient in GSE180077. Table S2: Used reagents in this study. Table S3: Clinical features of HNSCC patients for IHC.

Author Contributions: Conceptualization, N.S., C.M and S.A.; methodology, N.S., C.M. and S.A.; software, S.A.; validation, C.M. and S.A.; formal analysis, C.M. and S.A.; investigation, C.M., S.A. and A.K. (Ayaka Koma); resources, M.K. and N.K.; data curation, C.M. and S.A.; writing—original draft preparation, N.S., C.M. and S.A.; writing—review and editing, A.K. (Atsushi Kasamatsu); visualization, C.M. and S.A.; supervision, T.H. and K.U.; project administration, N.S.; funding acquisition, N.S., A.K. (Atsushi Kasamatsu), M.K., N.K. and K.U. All authors have read and agreed to the published version of the manuscript.

Funding: This study was supported by JSPS KAKENHI (grant numbers: 21K09577, 22K09679, 21K09367, 22H03286, 20H03883, 21KK0159).

Institutional Review Board Statement: The study was conducted according to the Declaration of Helsinki and approved by the Bioethics Committee of Chiba University (approval number: 915, 11 July 2021).

Informed Consent Statement: Informed consent was obtained from all subjects involved in the study.

Data Availability Statement: Our expression data were deposited in the GEO database (accession number: GSE180077).

Acknowledgments: The results shown here are in part based upon data generated by TCGA Research Network: https://www.cancer.gov/tcga (accessed on 20 August 2020).

Conflicts of Interest: The authors declare no conflict of interest.

References

1. Bray, F.; Ferlay, J.; Soerjomataram, I.; Siegel, R.L.; Torre, L.A.; Jemal, A. Global cancer statistics 2018: GLOBOCAN estimates of incidence and mortality worldwide for 36 cancers in 185 countries. *CA Cancer J. Clin.* **2018**, *68*, 394–424. [CrossRef] [PubMed]
2. Johnson, D.E.; Burtness, B.; Leemans, C.R.; Lui, V.W.Y.; Bauman, J.E.; Grandis, J.R. Head and neck squamous cell carcinoma. *Nat. Rev. Dis. Primers* **2020**, *6*, 92. [CrossRef] [PubMed]
3. Cohen, E.E.W.; Bell, R.B.; Bifulco, C.B.; Burtness, B.; Gillison, M.L.; Harrington, K.J.; Le, Q.T.; Lee, N.Y.; Leidner, R.; Lewis, R.L.; et al. The Society for Immunotherapy of Cancer consensus statement on immunotherapy for the treatment of squamous cell carcinoma of the head and neck (HNSCC). *J. Immunother. Cancer* **2019**, *7*, 184. [CrossRef] [PubMed]
4. Chow, L.Q.M. Head and Neck Cancer. *N. Engl. J. Med.* **2020**, *382*, 60–72. [CrossRef]
5. Bonner, J.A.; Harari, P.M.; Giralt, J.; Cohen, R.B.; Jones, C.U.; Sur, R.K.; Raben, D.; Baselga, J.; Spencer, S.A.; Zhu, J.; et al. Radiotherapy plus cetuximab for locoregionally advanced head and neck cancer: 5-year survival data from a phase 3 randomised trial, and relation between cetuximab-induced rash and survival. *Lancet Oncol.* **2010**, *11*, 21–28. [CrossRef]
6. Betel, D.; Wilson, M.; Gabow, A.; Marks, D.S.; Sander, C. The microRNA.org resource: Targets and expression. *Nucleic Acids Res.* **2008**, *36*, D149–D153. [CrossRef]
7. Anfossi, S.; Babayan, A.; Pantel, K.; Calin, G.A. Clinical utility of circulating non-coding RNAs—An update. *Nat. Rev. Clin. Oncol.* **2018**, *15*, 541–563. [CrossRef]
8. Krek, A.; Grun, D.; Poy, M.N.; Wolf, R.; Rosenberg, L.; Epstein, E.J.; MacMenamin, P.; da Piedade, I.; Gunsalus, K.C.; Stoffel, M.; et al. Combinatorial microRNA target predictions. *Nat. Genet.* **2005**, *37*, 495–500. [CrossRef]
9. Gebert, L.F.R.; MacRae, I.J. Regulation of microRNA function in animals. *Nat. Rev. Mol. Cell Biol.* **2019**, *20*, 21–37. [CrossRef]
10. Rupaimoole, R.; Slack, F.J. MicroRNA therapeutics: Towards a new era for the management of cancer and other diseases. *Nat. Rev. Drug Discov.* **2017**, *16*, 203–222. [CrossRef]
11. Goodall, G.J.; Wickramasinghe, V.O. RNA in cancer. *Nat. Rev. Cancer* **2021**, *21*, 22–36. [CrossRef]
12. Mitra, R.; Adams, C.M.; Jiang, W.; Greenawalt, E.; Eischen, C.M. Pan-cancer analysis reveals cooperativity of both strands of microRNA that regulate tumorigenesis and patient survival. *Nat. Commun.* **2020**, *11*, 968. [CrossRef]
13. Koshizuka, K.; Nohata, N.; Hanazawa, T.; Kikkawa, N.; Arai, T.; Okato, A.; Fukumoto, I.; Katada, K.; Okamoto, Y.; Seki, N. Deep sequencing-based microRNA expression signatures in head and neck squamous cell carcinoma: Dual strands of pre-miR-150 as antitumor miRNAs. *Oncotarget* **2017**, *8*, 30288–30304. [CrossRef]
14. Yonemori, K.; Seki, N.; Idichi, T.; Kurahara, H.; Osako, Y.; Koshizuka, K.; Arai, T.; Okato, A.; Kita, Y.; Arigami, T.; et al. The microRNA expression signature of pancreatic ductal adenocarcinoma by RNA sequencing: Anti-tumour functions of the microRNA-216 cluster. *Oncotarget* **2017**, *8*, 70097–70115. [CrossRef]

15. Yang, S.J.; Yang, S.Y.; Wang, D.D.; Chen, X.; Shen, H.Y.; Zhang, X.H.; Zhong, S.L.; Tang, J.H.; Zhao, J.H. The miR-30 family: Versatile players in breast cancer. *Tumor Biol.* **2017**, *39*, 1010428317692204. [CrossRef]
16. Saleh, A.D.; Cheng, H.; Martin, S.E.; Si, H.; Ormanoglu, P.; Carlson, S.; Clavijo, P.E.; Yang, X.; Das, R.; Cornelius, S.; et al. Integrated Genomic and Functional microRNA Analysis Identifies miR-30-5p as a Tumor Suppressor and Potential Therapeutic Nanomedicine in Head and Neck Cancer. *Clin. Cancer Res.* **2019**, *25*, 2860–2873. [CrossRef]
17. Mao, L.; Liu, S.; Hu, L.; Jia, L.; Wang, H.; Guo, M.; Chen, C.; Liu, Y.; Xu, L. miR-30 Family: A Promising Regulator in Development and Disease. *BioMed Res. Int.* **2018**, *2018*, 9623412. [CrossRef]
18. Kozomara, A.; Birgaoanu, M.; Griffiths-Jones, S. miRBase: From microRNA sequences to function. *Nucleic Acids Res.* **2019**, *47*, D155–D162. [CrossRef]
19. Gao, J.; Aksoy, B.A.; Dogrusoz, U.; Dresdner, G.; Gross, B.; Sumer, S.O.; Sun, Y.; Jacobsen, A.; Sinha, R.; Larsson, E.; et al. Integrative analysis of complex cancer genomics and clinical profiles using the cBioPortal. *Sci. Signal.* **2013**, *6*, pl1. [CrossRef]
20. Cerami, E.; Gao, J.; Dogrusoz, U.; Gross, B.E.; Sumer, S.O.; Aksoy, B.A.; Jacobsen, A.; Byrne, C.J.; Heuer, M.L.; Larsson, E.; et al. The cBio cancer genomics portal: An open platform for exploring multidimensional cancer genomics data. *Cancer Discov.* **2012**, *2*, 401–404. [CrossRef]
21. Anaya, J. Linking TCGA survival data to mRNAs, miRNAs, and lncRNAs. *PeerJ Comput. Sci.* **2016**, *2*, e67. [CrossRef]
22. Agarwal, V.; Bell, G.W.; Nam, J.W.; Bartel, D.P. Predicting effective microRNA target sites in mammalian mRNAs. *eLife* **2015**, *4*, e05005. [CrossRef]
23. Koshizuka, K.; Hanazawa, T.; Kikkawa, N.; Arai, T.; Okato, A.; Kurozumi, A.; Kato, M.; Katada, K.; Okamoto, Y.; Seki, N. Regulation of ITGA3 by the anti-tumor miR-199 family inhibits cancer cell migration and invasion in head and neck cancer. *Cancer Sci.* **2017**, *108*, 1681–1692. [CrossRef]
24. Koma, A.; Asai, S.; Minemura, C.; Oshima, S.; Kinoshita, T.; Kikkawa, N.; Koshizuka, K.; Moriya, S.; Kasamatsu, A.; Hanazawa, T.; et al. Impact of Oncogenic Targets by Tumor-Suppressive miR-139-5p and miR-139-3p Regulation in Head and Neck Squamous Cell Carcinoma. *Int. J. Mol. Sci.* **2021**, *22*, 9947. [CrossRef]
25. Asai, S.; Koma, A.; Nohata, N.; Kinoshita, T.; Kikkawa, N.; Kato, M.; Minemura, C.; Uzawa, K.; Hanazawa, T.; Seki, N. Impact of miR-1/miR-133 Clustered miRNAs: PFN2 Facilitates Malignant Phenotypes in Head and Neck Squamous Cell Carcinoma. *Biomedicines* **2022**, *10*, 663. [CrossRef]
26. Minemura, C.; Asai, S.; Koma, A.; Kase-Kato, I.; Tanaka, N.; Kikkawa, N.; Kasamatsu, A.; Yokoe, H.; Hanazawa, T.; Uzawa, K.; et al. Identification of Tumor-Suppressive miR-30e-3p Targets: Involvement of SERPINE1 in the Molecular Pathogenesis of Head and Neck Squamous Cell Carcinoma. *Int. J. Mol. Sci.* **2022**, *23*, 3808. [CrossRef]
27. Mootha, V.K.; Lindgren, C.M.; Eriksson, K.F.; Subramanian, A.; Sihag, S.; Lehar, J.; Puigserver, P.; Carlsson, E.; Ridderstråle, M.; Laurila, E.; et al. PGC-1alpha-responsive genes involved in oxidative phosphorylation are coordinately downregulated in human diabetes. *Nat. Genet.* **2003**, *34*, 267–273. [CrossRef]
28. Subramanian, A.; Tamayo, P.; Mootha, V.K.; Mukherjee, S.; Ebert, B.L.; Gillette, M.A.; Paulovich, A.; Pomeroy, S.L.; Golub, T.R.; Lander, E.S.; et al. Gene set enrichment analysis: A knowledge-based approach for interpreting genome-wide expression profiles. *Proc. Natl. Acad. Sci. USA* **2005**, *102*, 15545–15550. [CrossRef]
29. Liberzon, A.; Subramanian, A.; Pinchback, R.; Thorvaldsdóttir, H.; Tamayo, P.; Mesirov, J.P. Molecular signatures database (MSigDB) 3.0. *Bioinformatics* **2011**, *27*, 1739–1740. [CrossRef]
30. Yamada, Y.; Sugawara, S.; Arai, T.; Kojima, S.; Kato, M.; Okato, A.; Yamazaki, K.; Naya, Y.; Ichikawa, T.; Seki, N. Molecular pathogenesis of renal cell carcinoma: Impact of the anti-tumor miR-29 family on gene regulation. *Int. J. Urol.* **2018**, *25*, 953–965. [CrossRef]
31. Kanthaje, S.; Baikunje, N.; Kandal, I.; Ratnacaram, C.K. Repertoires of MicroRNA-30 family as gate-keepers in lung cancer. *Front. Biosci.* **2021**, *13*, 141–156. [CrossRef]
32. Xu, G.; Cai, J.; Wang, L.; Jiang, L.; Huang, J.; Hu, R.; Ding, F. MicroRNA-30e-5p suppresses non-small cell lung cancer tumorigenesis by regulating USP22-mediated Sirt1/JAK/STAT3 signaling. *Exp. Cell Res.* **2018**, *362*, 268–278. [CrossRef] [PubMed]
33. Ma, Y.X.; Zhang, H.; Li, X.H.; Liu, Y.H. MiR-30e-5p inhibits proliferation and metastasis of nasopharyngeal carcinoma cells by target-ing USP22. *Eur. Rev. Med. Pharmacol. Sci.* **2018**, *22*, 6342–6349. [CrossRef] [PubMed]
34. Zhang, Z.; Qin, H.; Jiang, B.; Chen, W.; Cao, W.; Zhao, X.; Yuan, H.; Qi, W.; Zhuo, D.; Guo, H. miR-30e-5p suppresses cell proliferation and migration in bladder cancer through regulating metadherin. *J. Cell. Biochem.* **2019**, *120*, 15924–15932. [CrossRef]
35. Zhang, S.; Li, G.; Liu, C.; Lu, S.; Jing, Q.; Chen, X.; Zheng, H.; Ma, H.; Zhang, D.; Ren, S.; et al. miR-30e-5p represses angiogenesis and metastasis by directly targeting AEG-1 in squamous cell carcinoma of the head and neck. *Cancer Sci.* **2020**, *111*, 356–368. [CrossRef]
36. Laudato, S.; Patil, N.; Abba, M.L.; Leupold, J.H.; Benner, A.; Gaiser, T.; Marx, A.; Allgayer, H. P53-induced miR-30e-5p inhibits colorectal cancer invasion and metastasis by targeting ITGA6 and ITGB1. *Int. J. Cancer* **2017**, *141*, 1879–1890. [CrossRef]
37. Majumder, M.; Johnson, R.H.; Palanisamy, V. Fragile X-related protein family: A double-edged sword in neurodevelopmental disorders and cancer. *Crit. Rev. Biochem. Mol. Biol.* **2020**, *55*, 409–424. [CrossRef]
38. Qian, J.; Hassanein, M.; Hoeksema, M.D.; Harris, B.K.; Zou, Y.; Chen, H.; Lu, P.; Eisenberg, R.; Wang, J.; Espinosa, A.; et al. The RNA binding protein FXR1 is a new driver in the 3q26-29 amplicon and predicts poor prognosis in human cancers. *Proc. Natl. Acad. Sci. USA* **2015**, *112*, 3469–3474. [CrossRef]

39. Majumder, M.; Palanisamy, V. RNA binding protein FXR1-miR301a-3p axis contributes to p21WAF1 degradation in oral cancer. *PLoS Genet.* **2020**, *16*, e1008580. [CrossRef]
40. Qie, S.; Majumder, M.; Mackiewicz, K.; Howley, B.V.; Peterson, Y.K.; Howe, P.H.; Palanisamy, V.; Diehl, J.A. Fbxo4-mediated degradation of Fxr1 suppresses tumorigenesis in head and neck squamous cell carcinoma. *Nat. Commun.* **2017**, *8*, 1534. [CrossRef]
41. Pinto-Costa, R.; Sousa, M.M. Profilin as a dual regulator of actin and microtubule dynamics. *Cytoskeleton* **2020**, *77*, 76–83. [CrossRef]
42. Zhou, K.; Chen, J.; Wu, J.; Xu, Y.; Wu, Q.; Yue, J.; Song, Y.; Li, S.; Zhou, P.; Tu, W.; et al. Profilin 2 Promotes Proliferation and Metastasis of Head and Neck Cancer Cells by Regulating PI3K/AKT/β-Catenin Signaling Pathway. *Oncol. Res.* **2019**, *27*, 1079–1088. [CrossRef]
43. Cao, Q.; Liu, Y.; Wu, Y.; Hu, C.; Sun, L.; Wang, J.; Li, C.; Guo, M.; Liu, X.; Lv, J.; et al. Profilin 2 promotes growth, metastasis, and angiogenesis of small cell lung cancer through cancer-derived exosomes. *Aging* **2020**, *12*, 25981–25999. [CrossRef]
44. Yan, J.; Ma, C.; Gao, Y. MicroRNA-30a-5p suppresses epithelial-mesenchymal transition by targeting profilin-2 in high invasive non-small cell lung cancer cell lines. *Oncol. Rep.* **2017**, *37*, 3146–3154. [CrossRef]
45. Katoh, M.; Katoh, M. Human FOX gene family (Review). *Int. J. Oncol.* **2004**, *25*, 1495–1500. [CrossRef]
46. Quintero-Ronderos, P.; Laissue, P. The multisystemic functions of FOXD1 in development and disease. *J. Mol. Med.* **2018**, *96*, 725–739. [CrossRef]
47. Lin, C.H.; Lee, H.H.; Chang, W.M.; Lee, F.P.; Chen, L.C.; Lu, L.S.; Lin, Y.F. FOXD1 Repression Potentiates Radiation Effectiveness by Downregulating G3BP2 Expression and Promoting the Activation of TXNIP-Related Pathways in Oral Cancer. *Cancers* **2020**, *12*, 2690. [CrossRef]
48. Sun, Q.; Novak, D.; Hüser, L.; Poelchen, J.; Wu, H.; Granados, K.; Federico, A.; Liu, K.; Steinfass, T.; Vierthaler, M.; et al. FOXD1 promotes dedifferentiation and targeted therapy resistance in melanoma by regulating the expression of connective tissue growth factor. *Int. J. Cancer* **2021**, *149*, 657–674. [CrossRef]
49. Wu, Q.; Ma, J.; Wei, J.; Meng, W.; Wang, Y.; Shi, M. FOXD1-AS1 regulates FOXD1 translation and promotes gastric cancer progression and chemoresistance by activating the PI3K/AKT/mTOR pathway. *Mol. Oncol.* **2021**, *15*, 299–316. [CrossRef]
50. Chen, C.; Tang, J.; Xu, S.; Zhang, W.; Jiang, H. miR-30a-5p Inhibits Proliferation and Migration of Lung Squamous Cell Carcinoma Cells by Targeting FOXD1. *BioMed Res. Int.* **2020**, *2020*, 2547902. [CrossRef]
51. Zhou, L.; Jia, S.; Ding, G.; Zhang, M.; Yu, W.; Wu, Z.; Cao, L. Down-regulation of miR-30a-5p is Associated with Poor Prognosis and Promotes Chemoresistance of Gemcitabine in Pancreatic Ductal Adenocarcinoma. *J. Cancer* **2019**, *10*, 5031–5040. [CrossRef] [PubMed]
52. Tao, J.; Cong, H.; Wang, H.; Zhang, D.; Liu, C.; Chu, H.; Qing, Q.; Wang, K. MiR-30a-5p inhibits osteosarcoma cell proliferation and migration by targeting FOXD1. *Biochem. Biophys. Res. Commun.* **2018**, *503*, 1092–1097. [CrossRef] [PubMed]
53. Wang, Y.; Qiu, C.; Lu, N.; Liu, Z.; Jin, C.; Sun, C.; Bu, H.; Yu, H.; Dongol, S.; Kong, B. FOXD1 is targeted by miR-30a-5p and miR-200a-5p and suppresses the proliferation of human ovarian carcinoma cells by promoting p21 expression in a p53-independent manner. *Int. J. Oncol.* **2018**, *52*, 2130–2142. [CrossRef] [PubMed]

 *genes*

Article

Doxorubicin and Cisplatin Modulate miR-21, miR-106, miR-126, miR-155 and miR-199 Levels in MCF7, MDA-MB-231 and SK-BR-3 Cells That Makes Them Potential Elements of the DNA-Damaging Drug Treatment Response Monitoring in Breast Cancer Cells—A Preliminary Study

Anna Mizielska [1], Iga Dziechciowska [1], Radosław Szczepański [1], Małgorzata Cisek [1], Małgorzata Dąbrowska [1], Jan Ślężak [1], Izabela Kosmalska [1], Marta Rymarczyk [1], Klaudia Wilkowska [1], Barbara Jacczak [1], Ewa Totoń [1], Natalia Lisiak [1], Przemysław Kopczyński [2] and Błażej Rubiś [1,*]

[1] Department of Clinical Chemistry and Molecular Diagnostics, Poznan University of Medical Sciences, Rokietnicka 3, 60-806 Poznan, Poland
[2] Centre for Orthodontic Mini-Implants, Department and Clinic of Maxillofacial Orthopedics and Orthodontics, Poznan University of Medical Sciences, Bukowska 70, 60-812 Poznan, Poland
* Correspondence: blazejr@ump.edu.pl; Tel.: +48-61-641-83-03

Citation: Mizielska, A.; Dziechciowska, I.; Szczepański, R.; Cisek, M.; Dąbrowska, M.; Ślężak, J.; Kosmalska, I.; Rymarczyk, M.; Wilkowska, K.; Jacczak, B.; et al. Doxorubicin and Cisplatin Modulate miR-21, miR-106, miR-126, miR-155 and miR-199 Levels in MCF7, MDA-MB-231 and SK-BR-3 Cells That Makes Them Potential Elements of the DNA-Damaging Drug Treatment Response Monitoring in Breast Cancer Cells—A Preliminary Study. *Genes* **2023**, *14*, 702. https://doi.org/10.3390/genes14030702

Academic Editors: Giuseppe Iacomino and Fabio Lauria

Received: 15 January 2023
Revised: 8 March 2023
Accepted: 9 March 2023
Published: 12 March 2023

Abstract: One of the most innovative medical trends is personalized therapy, based on simple and reproducible methods that detect unique features of cancer cells. One of the good prognostic and diagnostic markers may be the miRNA family. Our work aimed to evaluate changes in selected miRNA levels in various breast cancer cell lines (MCF7, MDA-MB-231, SK-BR-3) treated with doxorubicin or cisplatin. The selection was based on literature data regarding the most commonly altered miRNAs in breast cancer (21-3p, 21-5p, 106a-5p, 126-3p, 126-5p, 155-3p, 155-5p, 199b-3p, 199b-5p, 335-3p, 335-5p). qPCR assessment revealed significant differences in the basal levels of some miRNAs in respective cell lines, with the most striking difference in miR-106a-5p, miR-335-5p and miR-335-3p—all of them were lowest in MCF7, while miR-153p was not detected in SK-BR-3. Additionally, different alterations of selected miRNAs were observed depending on the cell line and the drug. However, regardless of these variables, 21-3p/-5p, 106a, 126-3p, 155-3p and 199b-3p miRNAs were shown to respond either to doxorubicin or to cisplatin treatment. These miRNAs seem to be good candidates for markers of breast cancer cell response to doxorubicin or cisplatin. Especially since some earlier reports suggested their role in affecting pathways and expression of genes associated with the DNA-damage response. However, it must be emphasized that the preliminary study shows effects that may be highly related to the applied drug itself and its concentration. Thus, further examination, including human samples, is required.

Keywords: miRNA; breast cancer; diagnostics; therapy response

1. Introduction

Breast cancer is still among women's most common leading causes of death worldwide [1]. The main challenge is an early diagnosis as well as personalized therapy adjustment. They both require the identification of specific qualitative and quantitative assessments of reliable parameters that could be used as markers and therapy targets, respectively. The main obstacle to therapy efficacy is the resistance of cancer cells to drugs. This is driven by different pathways, including individual genetic characteristics and multi-drug resistance, cell death inhibition (apoptosis suppression), altered drug metabolism, epigenetics, enhanced DNA repair and gene amplification [2]. As reported below (Table 1), all these processes can be controlled by the specific miRNA-controlled expression of certain genes. Another critical characteristic of breast cancer cells (apart from immortality) is the ability to metastasize associated with altered adhesion, migration and invasion. It is mainly related to signaling pathways controlled by different genes or their phosphorylated/dephosphorylated expression products.

Moreover, miRNAs can control genes that regulate adhesion molecules and cell–cell interactions (Table 1). Consequently, it is difficult to identify a specific signature of cancer cells with so many variables. Nevertheless, epigenetic factors can significantly modulate these processes, and microRNA profiling seems to be a promising strategy in diagnostics and therapy response monitoring. Among numerous studies performed so far, some particular miRNAs can be perceived as suitable candidates, especially since most of them are commonly associated with different cancer types, including breast cancer. The idea of miRNA functioning as early biomarkers for the evaluation of drug efficacy and drug safety was proposed not only in cancer but also in other diseases, including multiple sclerosis [3], bipolar disorder [4], diabetes [5] and others [6]. Thus miRNA profiling, supported by bioinformatic analysis, is perceived as a specific and sensitive biomarker for evaluating drug efficacy/resistance and drug safety in patients [7]. Specifically, it was shown that blood-borne miRNA profiles monitored over time have the potential to predict complete pathological responses in breast cancer [8]. Altogether, miRNA assessment shows some potential as a marker in pharmacogenomics (as modulators of pharmacology-related genes). It can be evaluated using low-invasive methods providing high specificity and sensitivity. However, since one miRNA can target different mRNA, a comprehensive profiling study must be performed to obtain an informative and valuable pattern.

Table 1. Contribution of selected miRNAs to breast cancer metabolism.

miRNA	Effect and Pathways Mediated in Breast Cancer	Ref.
miRNA-21 (-3p and -5p)	tumor growth, cancer cells proliferation, metastasis, invasion, sensitivity to chemotherapy, modulation of cancer-related gene expression	[9,10]
miRNA-106a-5p	cancer cell proliferation, colony-forming capacity, migration, invasion, breast cancer cell apoptosis and sensitivity to cisplatin, DNA damage response, suppression of the ATM-associated pathway	[11,12]
miRNA-126 (-3p and -5p)	cancer cell migration, tumor growth, proliferation, invasion and angiogenesis of triple-negative breast cancer cells	[13–15]
miRNA-155 (-3p and -5p)	inflammation, apoptosis, adhesion	[1,16,17]
miRNA-199b (-3p and -5p)	cancer aggressiveness, tumor growth, angiogenesis	[18–20]
miRNA-335 (-3p and -5p)	sensitivity of triple-negative breast cancer cells to paclitaxel, cisplatin and doxorubicin	[21]

2. Materials and Methods

2.1. Cell Culture

Three cell lines representing different molecular subtypes of breast cancer were enrolled in the study, i.e., (i) MCF7 (ER/PR+, HER2low, TP53WT), (ii) MDA MB-231—basal-like subtype, also called triple-negative breast cancer (TNBC; ER/PR-, HER2-, TP53mut), and (iii) SK-BR-3 (ER/PR-, HER2+, TP53mut). MCF7 cell line, in comparison to the MDA-MB-231 cell line, is a poorly aggressive and non-invasive cell line. Overall, it is being considered to have low metastatic potential. SK-BR-3 cells are the least invasive cells out of the three studied, according to previous findings [22]. The MCF7 (HTB-22) and MDA-MB-231 (HTB-26) cells were maintained in RPMI-1640 (Biowest, Nuaillé, France), while SK-BR-3 (HTB-30) cells were cultured in McCoy's 5A (Biowest, Nuaillé, France) media supplemented with 10% fetal bovine serum (FBS) (Biowest, Nuaillé, France) at 37 °C in an atmosphere of 5% CO_2 and saturated humidity. All cell lines were obtained from the American Type Culture Collection (ATCC).

2.2. Studied Drugs

Both studied drugs belong to the DNA-damaging and proapoptotic agents, and both are ABC family substrates (ABCB1 and ABCC3, respectively) [23]. Doxorubicin is commonly used in breast cancer treatment disruption of topoisomerase-II-mediated DNA repair and generation of free radicals and their damage to cellular membranes, DNA and proteins [24]. The mechanism of resistance to doxorubicin results from a reduction in the ability of the drug to accumulate in the nucleus, decreased DNA damage and suppression of the downstream events that transduce the DNA damage signal to apoptosis [25].

In turn, cisplatin effectively blocks breast cancer metastasis and inhibits cancer growth together with paclitaxel in neoadjuvant chemotherapy [26]. Resistance of cancer cells to this drug is associated with decreased drug import, increased drug export, increased drug inactivation by detoxification enzymes, increased DNA damage repair and inactivated cell death signaling [27].

2.3. MTT Cell Survival Assay

Cell survival was determined using MTT assay by assessing the sensitivity of cells subjected to drugs, i.e., doxorubicin or cisplatin, as previously described [28]. Drugs selection was based on common use in breast cancer [29]. Briefly, the cells were seeded at a density of 5×10^3 cells per well in 96-well culture plates, incubated overnight to allow for cell attachment, and the next day either DOX was added at a concentration range of 0, 0.05, 0.1, 0.5, 1, 2 or 5 µM or cis-Pt was administered at a concentration range of 0, 1, 2, 5, 10, 20 or 50 µM (DOX was dissolved in water, and cis-Pt in saline). Cells were treated for 24 h and incubated with 10 µL of MTT reagent (5 mg/mL) (Sigma-Aldrich, St. Louis, MO, USA). The cells were incubated at 37 °C for 4 h, followed by the addition of 100 µL of solubilization buffer (10% SDS in 0.01 M HCl). The absorbance was measured in each well with the Microplate Reader Multiskan FC (Thermo Scientific, Waltham, MA, USA) at two wavelengths of 570 and 690 nm. Each experimental point was determined in biological triplicate (each in 6 technical repeats). IC_{50} (half-maximal (50%) inhibitory concentration) values were calculated using CompuSyn (ComboSyn, Inc., Paramus, NJ, USA) (Table 2), and the standard deviation was calculated using Excel software (Microsoft, Syracuse, NY, USA).

Table 2. IC_{50} values determination for doxorubicin or cisplatin after MCF7, MDA-MB-231 or Sk-BR3 cells treatment for 24 h. IC_{50} values were calculated using CompuSyn program (three biological repeats, six technical repeats each, were performed).

Cell Line	Drug	IC_{50} [µM]	
		DOX	Cis-Pt
MCF7		27.4 ± 0.9	>50
MDA-MB-231		12.9 ± 1.8	>50
SK-BR-3		5.8 ± 0.7	44.3 ± 2.6

2.4. miRNA Isolation

Cells were subjected to treatment with DOX (0.1 µM) or cis-Pt (10 µM) for 24 h. Concentration selection resulted from the survival curve obtained in MTT assay and these specific concentrations were chosen as subcytotoxic but known from previous experiments to provoke apoptosis in a longer incubation time (verified by clonogenic assay, data not shown).

miRNAs were isolated using a miRNA Isolation Kit Cell (BioVendor, Czech Rep.) based on optimized silica membrane column according to manufacturers' instructions. Briefly, dedicated cell lysis buffer (250 µL per sample; 1% β-mercaptoethanol freshly added) was added to cell pellets (1×10^6 cells) followed by vortexing, a brief spin and incubation at 25 °C (room temperature) for 3 min. Next, RCL1 and RCL2 buffers were sequentially added that was accompanied by brief vortexing, spinning and short incubation at 25 °C (room temperature). After centrifugation at $11,000 \times g$ for 3 min, the clear supernatant was transferred to a new 1.5 mL micro-centrifuge tube, 330 µL of isopropanol was added and after short pulse-vortexing for 10 s, all content was transferred to the MR13 Column, followed by incubation at 25 °C for 2 min. After another centrifugation at $11,000 \times g$ for 1 min, 500 µL of buffer CRW1 was added to wash the column, followed by another centrifugation at $11,000 \times g$ for 1 min. After washing the column with 500 µL of Buffer CRW2, the miRNAs were eluted with 30 µL of RNase-Free water (centrifugation at $11,000 \times g$ for 1 min), assessed using spectrophotometer (Eppendorf Biophotometer Plus Spectrophotometer, Hamburg, Germany) and stored at -80 °C for further study.

2.5. qPCR Assessment of miRNAs

Two-Tailed qPCR (BioVendor, Brno, Czech Republic) was performed to assess individual miRNAs according to manufacturers' instructions. The test is based on primers which consist of two hemiprobes, connected by a folded tether. Complementarity of two hemiprobes provides specific binding and specific cDNA synthesis. The cDNA was obtained using the miR-TT-PRI kit containing set of miRNA-specific primers and the Two-Tailed cDNA Synthesis System using a thermocycler (Eppendorf EP Gradient S Thermocycler, Hamburg, Germany) according to the following protocol: 25 °C for 5 min, 50 °C for 15 min, 85 °C for 5 min followed by cooling at 4 °C. For each sample, 600 ng of total RNA was taken for reverse transcription—altogether, it was a combination of three biological repeats (200 ng of each) in one sample. Next, specific PCR primers (PCR Primer F and PCR Primer R) from the miR-TT-PRI kit were added. The qPCR was performed as follows: 95 °C for 30 s, 95 °C for 5 s, 60 °C for 15 s, 72 °C for 10 s and signal detection. After 40 cycles, melting curve analysis was performed and final, relative expression was evaluated based on Cq (quantification cycle) and thermocycler software (Roche LightCycler 480-II PCR, Basel, Switzerland) as previously described [30]. Quantitative qPCR was done with individual reactions for each miRNA target. Validation was performed against U6 (RNU6-1) expression as previously described [31]. Melting temperature analysis was used (SYBRGreen-based) for verification of specific products detection.

2.6. Statistical Analysis

Results were expressed as mean $\pm$ SD. All statistical analyses were carried out using GraphPad Prism 5 (GraphPad Software, Sandiego, CA, USA). Differences were assessed for statistical significance using repeated-measures ANOVA, followed by the post-hoc Dunnett's test method. All experiments were performed in triplicates unless specified otherwise. The threshold for significance was defined as $p < 0.05$ and are indicated by the ($*$, #, •) symbol for $p < 0.05$. qPCR was performed with pooled samples, as commonly acknowledged [32], giving mean target miRNAs levels.

3. Results

3.1. Doxorubicin and Cisplatin Cytotoxicity Evaluation

All three cell lines were subjected to different concentrations of studied drugs to find the range of low toxicity concentrations that could be applied to cells during the evaluation of the association between drug treatment and cell response measured by alterations in selected miRNAs levels. The 24 h time interval selection was based on our previous experience and numerous cytotoxicity assays, as well as more mechanistic assessment of apoptotic pathways, induced by the studied drugs. We observed that longer incubation time (48 and 72 h MTT test were performed, showing more than 40% viability decrease in all three cell lines at the lowest concentrations of both studied drugs; data not included) would provoke cytotoxic effect that might not reflect mechanistic aspect of the specificity of the cancer cells response to the treatment. Additionally, according to our previous studies, many genes are regulated within 24 h after DOX or cis-Pt treatment. Similarly, numerous reports indicate 24 h time interval as a sufficient and optimal time to observe significant alterations in miRNA levels after cancer cells treatment (e.g., [33–36]). For this reason, we were interested in a rapid response that was supposed to be specific. Thus, we decided to verify our hypothesis concerning monitory potential of miRNAs in breast cancer cells in a 24 h time interval that was supposed to provide an early response analysis.

An MTT survival test was involved for DOX (Figure 1A) and cisplatin (Figure 1B) cytotoxicity assessment. Cell treatment was 24 h and, as demonstrated, cytotoxicity effects were concentration dependent. MCF7 and MDA-MB-231 showed a similar survival rate when treated with DOX at 0.05, 0.1, 0.5 or 1 μM (up to ca 20% decrease in survival, relative to control, untreated cells). Simultaneously, treatment of SK-BR-3 cells with the same concentrations of DOX showed significant decrease in survival only at the concentration of 0.5 and 1 μM (up to 20% decrease in survival, relative to control cells). In turn, higher concentration (2 μM) revealed higher resistance of MDA-MB-231 than two other cell lines (survival at 80 vs. 65% in MCF7 and 60% in SK-BR-3), while incubation with 5 μM DOX led to similar survival rate in all three cell lines i.e., 60% (Figure 1A).

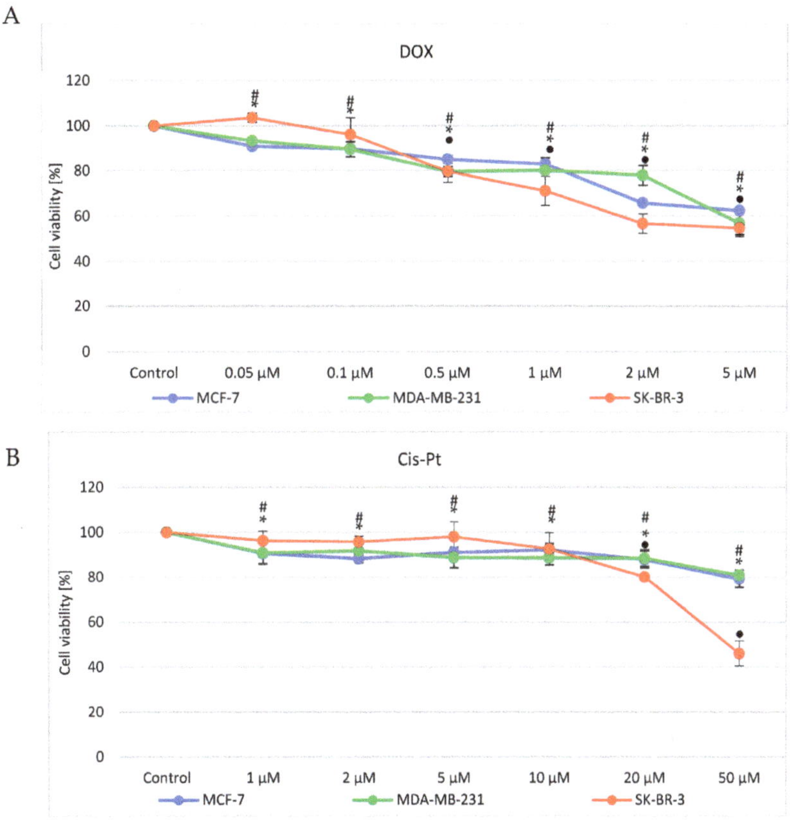

Figure 1. Cytotoxicity assessment and determination of the optimal range of drugs concentration i.e., doxorubicin (**A**) and cisplatin (**B**). MTT assay was performed to find optimal concentrations of doxorubicin and cisplatin for evaluation of miRNAs alterations as a response. Cells were treated for 24 h, followed by colorimetric assessment. All experiments were performed in triplicates (each with six technical repeats). Alterations in survival are relative to control, untreated cells. *, •, # $p < 0.05$.

The same cell lines were treated with cisplatin at the range of concentrations 1, 2, 5, 10, 20 or 50 μM. When cells were subjected to 1, 2, 5 or 10 μM cisplatin, the survival rate of MCF7 and MDA-MB-231 was reduced by circa 10%. At the same time viability of SK-BR-3 was unchanged. Increasing the concentration of cis-Pt (20 or 50 μM) provoked decrease of all three breast cancer cell lines survival, with a more dominant effect observed in SK-BR-3 (50 vs. 20% decrease at 50 µM) (Figure 1B).

Based on the assessment of toxicity of the drugs, IC_{50} values were calculated (Table 1). Primarily, time-course experiments were performed, but since after longer time intervals (48 or 72 h; data not shown), the compounds appeared highly toxic, we decided to subject cells to 24 h treatment only. Low-cytotoxicity concentrations were selected based on the survival curves (these concentrations, however, are known from previous experiments and literature data to induce apoptosis).

3.2. Target miRNAs Selection

First, we used the targetscan algorithm for selection of target miRNAs that were supposed to be evaluated [37]. However, as it gave us enormous, not entirely coherent data, and bearing in mind that miRNAs do not have to be fully complementary to interact with target mRNAs, we performed a selection based on literature data. Additionally, TCGA analysis (GEPIA2 [38], Xena Browser [39] and oncolnc.org (access date: 27 February 2023) [40]) was also performed and discussed below.

Consequently, eleven most commonly breast cancer-associated miRNAs were subjected to identification after doxorubicin (DOX) or cisplatin (cis-Pt) treatment of breast cancer cell lines. Additionally, assessment of a synthetic nonmammalian miR-54-3p was performed as a negative control. The target miRNAs (with short justification) were as follows:

miRNA-21 (-3p and -5p). miR-21-5p was identified as a typical onco-miRNA. When upregulated, it could promote tumor growth, metastasis and invasion and reduce sensitivity to chemotherapy. It modulates the expression of multiple cancer-related target genes and is dysregulated in various tumors [9]. Decreased expression of miR-21 is known to suppress the invasion and proliferation of MCF7 cells [10].

miRNA-106a-5p. In human breast cancer, miR-106a expression was found to be significantly upregulated. It enhanced breast cancer cell proliferation, colony-forming capacity, migration and invasion. Additionally, miR-106a overexpression significantly decreased breast cancer cell apoptosis and sensitivity to cisplatin [11]. Upregulation of miRNA-106a modified DNA damage response and led to the suppression of the ATM gene and formation of its protein product at nuclear foci [12].

miRNA-126 (-3p and -5p). Studies suggest that miR-126-3p acts as either a tumor suppressor or an oncogene in different types of cancer. Upregulation of miR-126-5p can inhibit the migration of the MCF7 breast cancer cell line [13]. Furthermore, overexpression of miR-126-3p significantly reduced tumor size [14]. miRNA-126-3p overexpression inhibited the proliferation, migration, invasion and angiogenesis of triple-negative breast cancer cells (MDA-MB-231 and HCC1937) [15].

miRNA-155 (-3p and -5p). miR-155 was shown to play a crucial role in various physiological and pathological processes, including inflammation and cancer. It was found overexpressed in several solid tumors, including breast, colon, cervical and lung cancers [1] and it is supposed to mediate the pathway controlled by caspase-3 [16]. The mechanism of action of this miRNA is based on downregulation of the cell adhesion molecule 1 (CADM1) that functions as a tumor suppressor [17].

miRNA-199b (-3p and -5p). miR-199b-5p was reported to play a critical role in various types of malignancy. There are studies suggesting that miR-199b-5p downregulation is correlated with aggressive clinical characteristics of breast cancer [18,19]. Overexpression of miR-199b-5p inhibited the formation of capillary-like tubular structures and reduced breast tumor growth and angiogenesis in vivo [20]. Downregulation of miR-199b-5p is correlated with poor prognosis for breast cancer patients.

miRNA-335 (-3p and -5p). The expression of miR-335 depends on the type of cancer. It is downregulated in breast cancers and upregulated in gallbladder carcinoma, endometrial and gastric cancers. Overexpression of miR-335 increases the sensitivity of triple-negative breast cancer cells to paclitaxel, cisplatin and doxorubicin, and improves the effectiveness of chemotherapy. It is also associated with cisplatin sensitivity in ovarian cancer [21].

To summarize the contribution of selected miRNAs to breast cancer metabolism, the justification was collected in Table 2.

3.3. Quantitative Assessment of the Basal Levels of Selected miRNAs

All three cell lines were subjected to quantitative assessment of the basal levels of selected miRNAs using qPCR and relative quantification. All biological experiments were performed in triplicates followed by RT-PCR and qPCR assessment. Consequently, all target miRNAs levels were shown as relative to respective targets in MCF7 cells, used as a calibrator (value "1" assigned to basal level of each miRNA in MCF7). Additionally, the data were divided into three groups i.e., relatively high miRNA levels, low levels and other, relative to results observed in MCF7 cells (selected as reference cell line; Figure 2A,B and Figure 3C, respectively).

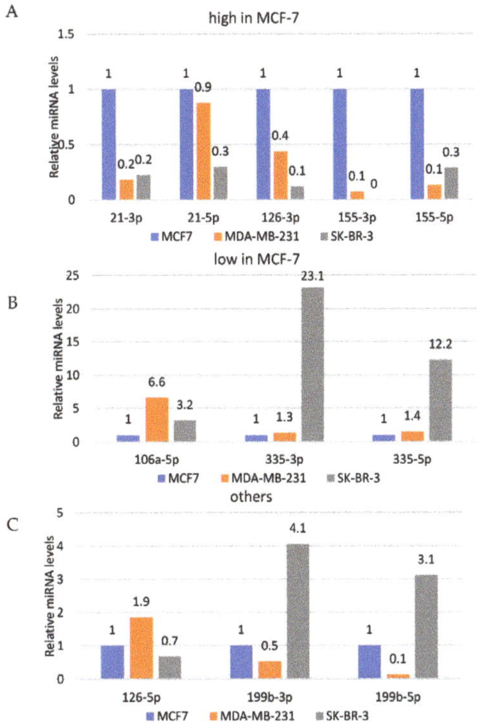

Figure 2. Quantitative assessment of the basal levels of selected miRNAs using qPCR in studied cell lines. miRNAs expression levels were grouped according to their pattern of expression for better representation. The data for all three cell lines (i.e., (**A**), MCF7; (**B**), MDA-MB-231; (**C**), SK-Br-3) were divided into three categories i.e., low, high and others, relative to MCF7 cells, selected as the reference cell line (value "1").

As demonstrated, average basal levels of individual miRNAs were different in different cell lines with no specific pattern. All the studied miRNAs (21-3p, 21-5p, 106a-5p, 126-3p, 126-5p, 155-3p, 155-5p, 199b-3p, 199b-5p, 335-3p, 335-5p) were detected in MCF7 and MDA-MB-231, while SK-BR-3 did not show expression of the miR-155-3p (Figure 2A). Reaction designed for detection of a synthetic nonmammalian cel-miR-54-3p was used as a negative control.

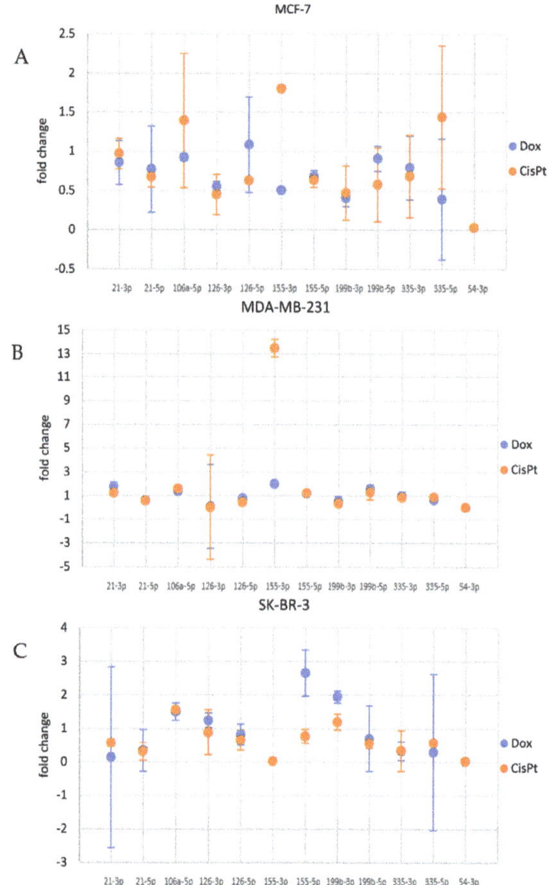

Figure 3. Evaluation of cancer cells response to doxorubicin or cisplatin measured by miRNA levels alterations. All three cell lines ((**A**), MCF7; (**B**), MDA−MB−231; (**C**), SK−BR−3) were subjected to selected miRNAs assessment after treatment with DOX (0.1 µM) or cis-Pt (10 µM) for 24 h. Experiments were performed in triplicates using RT-qPCR assessment. The values represent fold changes, relative to MCF7 cells (used as reference). Values higher than "1" mean induction, while values lower than "1" mean reduction of selected miRNAs levels.

3.4. Cancer Cells Response to Drugs—Evaluation of the Potential Association between Drug Treatment and miRNA Levels

All three cell lines were subjected to selected miRNAs assessment after DOX (0.1 µM) or cis-Pt (10 µM) 24 h treatment. Some alterations of studied miRNAs were observed with no significant pattern but with some consistency between the two applied DNA-damaging drugs within particular cell lines. We decided to focus on the miRNAs that showed at least a 40% change relative to control samples.

3.4.1. miRNA Alterations in MCF7 Cells

The qPCR showed that DOX treatment of MCF7 cells provoked downregulation of 126-3p, 155-3p, 199b-3p and 335-5p miRNAs (Figure 3). The same cells treated with cisplatin (10 µM) revealed induction of miRNAs 155-3p and 335-5p. At the same time, 126-3p and 199b-3p miRNAs were downregulated by cis-Pt (Figure 3A).

3.4.2. miRNA Alterations in MDA-MB-231 Cells

When MDA-MB-231 cells were treated with DOX, an induction of 21-3p, 155-3p and 199b-5p was observed. At the same time, treatment of these cells with DOX provoked downregulation of 126-3p and 199b-3p (Figure 3B). Incubation of MDA-MB-231 cells with cisplatin caused increased accumulation of 106a-5p and 155-3p miRNAs, while 21-5p, 126-3p, 126-5p and 199-3p were downregulated (Figure 3B).

3.4.3. miRNA Alterations in SK-BR-3

Evaluation of miRNAs alterations in SK-BR-3 cells subjected to DOX treatment showed that cells treated with the drug triggered accumulation of 106a-5p, 155-5p and 199b-3p. At the same time 21-3p, 21-5p, 155-3p, 335-3p and 335-5p were downregulated (Figure 3C). When cells were incubated with cisplatin, upregulation of 106a-5p was observed, while 21-3p, 21-5p, 155-3p, 199-5p, 335-3p and 335-5p were downregulated (Figure 3C).

4. Discussion

Epigenetics seems to play a pivotal role in the metabolism of all human cells, including cancer cells. One of the mechanisms involved in regulation of gene expression without changing its sequence is provided by miRNA. Although the whole family of miRNAs can show tissue- and time-specific patterns, we believe that we can not only detect but also modulate these small polymers. First, we need to identify the miRNAs that represent a certain metabolic status, e.g., cancer. Such research has been already conducted, but depending on different study groups (different cancer stage or grade) and different methods involved (ELISA, qPCR, detection in serum, exosomes or cancer cells), it may give varying results. Some studies show alterations in seven miRNAs (miR-10b, miR-21, miR-125b, miR-145, miR-155 miR-191 and miR-382) in serum of breast cancer patients compared to healthy controls [41], while other studies reveal more than 50 different miRNAs altered in breast cancer patients [42]. Gene expression control may become a way for cancer cells to overcome therapeutic strategies, as well as an efficient way to eliminate cancer cells or make them more sensitive to therapeutic agents [43]. Thus, we performed a study that aimed to evaluate the alterations in breast cancer cell lines exposed to anticancer drugs i.e., doxorubicin or platin. First, we performed the assessment of basal levels of eleven miRNAs that are most commonly evoked when breast cancer is studied [1,2,9–12,14,16,18,19,21,44–57]. The data were demonstrated as relative miRNA levels compared to MCF7 cells, indicated as a reference cell line. Since the basal levels were extremely different (data range from 0.1 to 23 arbitrary units) and putting all the data on one graph might be misleading, we divided the results into three groups presented in three independent graphs i.e., relative miRNA levels high, low and other, relative to results observed in MCF7 cells (selected as reference cell line; Figure 2A, 2B and 2C, respectively). Noteworthy, the molecular characteristics of the three studied cell lines significantly differ, which may justify the different basal levels of the assessed miRNAs. However, it is truly difficult to tell if miRNA levels are affected by (or affect) respective features of selected cell lines. Additionally, we should not forget that all three cell lines are derived from three different people and, as commonly known, represent not only different genotypes but also heterogeneous population of cancer cells. However, one of the critical differences between studies cells is the ER/PR/HER2 receptors status. Importantly, these receptors mediate cell proliferation, growth, metabolism and other signaling pathways that, since they are related to the mechanisms affected by the studied miRNAs (Table 2), seem to justify alterations in their basal levels. Similarly, studied cells are characterized by different p53 statuses (i.e., MCF7/wt, MDA-MB-231 and SK-BR-3/mut), which is one of the key players not only in cell proliferation control and apoptosis but also in response to exposition to DNA-damaging compounds. Even with that knowledge, it is difficult to find a pattern regarding basal levels of selected miRNAs in studied cell lines.

After evaluation of the cytotoxic activity of doxorubicin and cisplatin, we performed an analysis of the alterations of target miRNAs in cells subjected to selected low-cytotoxicity concentrations of studied chemotherapeutics in MCF7, MDA-MB-231 and SK-BR-3. Although relative changes were observed in the accumulation of most of the analyzed miRNAs after drug administration, we focused on the changes that showed at least 40% change relative to control samples. Thus, essential alterations were observed in a couple of miRNAs that showed a trend in all cell lines subjected to two DNA-damaging agents.

The miRNA that was considerably altered in two of the three cell lines (MDA-MB-231 and SK-BR-3) was miR-21. It was already suggested that identification of this miRNA in serum of breast cancer patients can be used for breast cancer diagnosis at an early stage of the disease. Although it was not associated with the status of ER, PR and clinical stages [58], it was reported that miR-21 could be related to the development of Multi Drug Resistance (MDR) in breast cancer [59]. Specifically, miR-21 was shown to contribute to breast cancer proliferation and metastasis by targeting LZTFL1 [60]. As reported in colorectal cancer, an increase in miR-21 expression correlated with resistance to fluorouracil therapy due to lowered expression of the repair protein MSH2 [61]. Thus, it is possible that treatment of cells with ABC substrates may provoke alterations in one of the MDR drivers, i.e., miR-21.

It is known that miR-106 is significantly upregulated in human breast cancer, as it can enhance cell proliferation, colony-forming capacity, migration and invasion. Additionally, miR-106a overexpression significantly decreased BC cell apoptosis and sensitivity to cisplatin [62]. It was also shown that upregulation of miRNA-106a modified DNA damage response and led to the suppression of the ATM gene and formation of its protein product at nuclear foci [63]. Thus, it may be concluded that induction of this miRNA in cells treated with DOX or cis-Pt could indicate a response of the cancer cells to the DNA-damaging agent. Consequently, it might enhance the resistance of breast cancer cells to cis-Pt; although we do not know the exact mechanism, we might try to target this miRNA to attenuate the resistance effect. Importantly, mir-106a is known to be involved in DNA damage repair systems and cause sensitization of cancer cells to irradiation by targeting the 3′-UTR of ATM kinase. It was found upregulated in breast cancer cells subjected to DNA damage induction [64]. Importantly, this pathway is associated with checkpoint protein 2 (Chk2), mediating the effects of ATM on DNA damage repair mechanisms and other cellular responses that consequently halt the cell cycle (phosphorylates p53) [12]. We also showed that this miRNA was upregulated in MDA-MB-231 and SK-BR-3 cells as well as in all cell lines after cis-Pt. Lack of significant alteration of miR-106a in MCF7 cells after DOX treatment might be associated with the wild type of p53 in those cells, which has a much shorter half-life than the mutated form [65]. However, such correlation verification would require a certain signaling feedback assessment.

Another altered miRNA was miRNA-155. It was not expressed in SK-BR-3 cells, while it was significantly reduced in MCF7 and induced in MDA-MB-231 after DOX treatment, while induction was observed in both cell lines after cis-Pt treatment. In turn, the mir155-5p was downregulated in MCF7 cells after either DOX or cis-Pt treatment. Administration of DOX provoked induction of this miRNA in the other two cell lines, while cis-Pt treatment showed similar effect in MDA-MB-231 and SK-BR-3 cells. Specifically, miR-155 was found to be overexpressed in breast cancers [66]. It is also known to suppress apoptosis in MDA-MB-453 breast cancer cells by blocking caspase-3 activity [67]. It can also promote loss of genomic integrity in cancer cells by targeting genes involved in microsatellite instability and DNA repair, which strengthens the oncogenic features of this miRNA. It was also shown to decrease chemosensitivity to cisplatin in colon cancer cells and caspase-3 activity induced by cisplatin [45]. Additionally, it was found to be upregulated in the doxorubicin-resistant human lung cancer A549/DOX cell line [46]. As demonstrated previously, miR-155-5p accelerated DNA damage repair, which led to resistance to radiation of esophageal carcinoma cells [68]. However, a contrary observation was made in breast cancer, in which it was revealed that miR-155-5p decreased the efficiency of homologous recombination repair and enhanced sensitivity to radiation by targeting RAD51 directly [69]. This may be due to the different interactions of miRNA-mRNAs in different types of cancer that is also a known fact [70]. Thus, it seems that downregulation of this miRNA, which accompanied drug treatment, might be a good prognostic factor that could show high efficacy for the therapy in breast cancer. miR-155 is one of just a few miRNAs studied in the context of response to DNA-damaging drugs in breast cancer [71].

Another miRNA that was significantly altered in all three cell lines was miR-199. Specifically, miR-199b-5p was downregulated in DOX- as well as cis-Pt-treated MCF7 cells. It was upregulated in DOX-treated MDA-MB-231 cells and cis-Pt-treated SK-BR-3 cells. In turn, 199b-3p was downregulated after cis-Pt treatment in MCF7 and MDA-MB-231 cells, while in SK-BR-3, it was upregulated after DOX treatment. This particular miRNA was shown modulated in ovarian and prostate cancers, osteosarcoma and hepatocellular carcinoma but also in breast cancer. There are studies suggesting that miR-199b-5p is involved in the Notch signaling pathway in osteosarcoma and its downregulation is correlated with aggressive clinical characteristics of breast cancer [18,19]. In fact, the relative level of this miRNA was lower in the most invasive cell line studied in our work, i.e., MDA-MB-231 cells. According to the literature, downregulation of miR-199b-5p is correlated with poor prognosis for breast cancer patients [72]. Thus, modification of this miRNA may show some prognostic value in the context of breast cancer therapy monitoring. However, even if it seems to be a very sensitive marker, it may also be a very unstable marker that requires further profiling in a dynamic environment of cancer cells subjected to drugs. In previous reports, miR-199a-3pwas shown to be induced in response to DNA damage mediated by homologous repair system that suggests involvement of mTOR and c-Met [73]. Similarly, miR-199-5p/3p was shown to target DNA-damage inducible 1 homolog 2, which also implies involvement of this miRNA in response to DNA-damaging agents [74]. This miRNA is known to significantly diminish aggressive progression, including cell oxygen consumption, colony formation and mobility of breast cancer cells [21]. Variable changes of this polymer after DOX or cis-Pt treatment suggest important role of this miRNA in the response of breast cancer cells to therapy.

Another miRNA modulated after DOX or cis-Pt treatment was miR-335-3p (in SK-BR-3 cells). It is known to be associated with p53 in a positive feedback loop to drive cell cycle arrest indicating its important role in proliferation control of cancer cells [75]. The expression level of miR-335 in tissues and cells varies with cancer types, and miR-335 has been proposed as a potential biomarker for the prognosis of cancer. Besides, miR-335 may serve as an oncogene or tumor suppressor via regulating different targets or pathways in tumor initiation, development and metastasis. Furthermore, miR-335 also influences tumor microenvironment and drug sensitivity [21]. Importantly, overexpression of miR-335 was shown to increase the sensitivity of triple-negative breast cancer cells to paclitaxel, cisplatin and doxorubicin, and improve the effect of chemotherapy, as demonstrated in breast cancer patients [76]. We could not see any significant alteration in the expression of either miR-335-3p or miR-335-5p in MDA-MB-231 (triple negative) or MCF7 cells. However, it is difficult to state if alteration in the level of this miRNA after cancer cell exposure to a drug (with a decrease most noticeable in SK-BR-3 cells) contributes to the protection or toxicity mechanism. However, it was suggested that this miRNA could be a tumor suppressor and could serve as a potential therapeutic target for breast cancer treatment [77]. However, again, further studies on the mechanism involved in different cancer types and metabolic conditions or therapy regimen must be evaluated.

4.1. Clinical Relevance

The ultimate goal for scientific studies is providing tools for controlling biological processes to achieve the most wanted outcomes that are length and/or quality of life. This approach requires analysis of the data that may significantly contribute to the metabolism of cell and all human body. These data include genetic code as well as gene expression profiling that is covered by The Cancer Genome Atlas (TCGA) [78]. We wanted to evaluate the clinical relevance of the studied miRNAs in breast cancer data panel. Such assessment was performed using the algorithm available at oncolnc.org [40]. Although numerous studies show potential effect of target miRNAs modulation on cell survival or resistance, a general assessment of the survival relative to low or high selected miRNAs levels did not show any significant contribution of studied miRNAs expression to this parameter (Figure 4). Nine out of eleven of all studied miRNAs were found in the base but, surprisingly, no significant association of any of the miRNAs and patients' survival was found. It may result from still low data numbers in TCGA (high and low—296 cases each group) that may not be sufficient when facing breast cancer, which is a very heterogeneous disease. Especially since, as previously reported, different roles of selected miRNAs in different breast cancer types (ER/PR/HER2 positive vs. negative) were reported [76], presented in Table 2. Although there is no significant difference in the survival time relative to the studied miRNAs' expression, some apparent trends can be recognized, but it may require larger group studies and evaluation of other parameters, such as cancer stage or grade, p53 variant, etc., to obtain conclusive results.

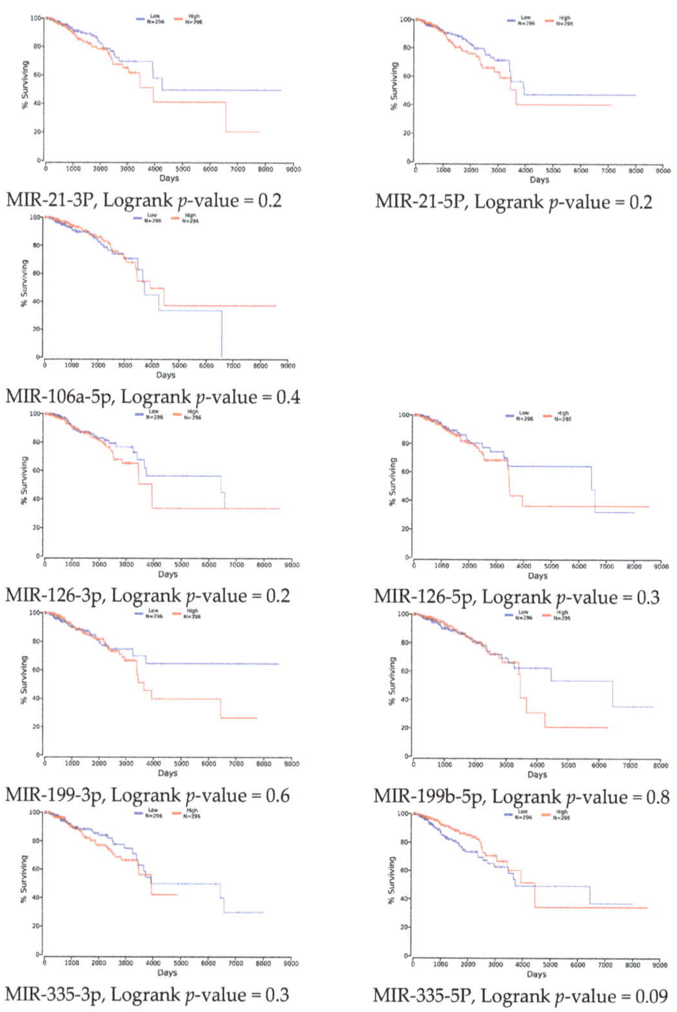

MIR-21-3P, Logrank *p*-value = 0.2   MIR-21-5P, Logrank *p*-value = 0.2

MIR-106a-5p, Logrank *p*-value = 0.4

MIR-126-3p, Logrank *p*-value = 0.2   MIR-126-5p, Logrank *p*-value = 0.3

MIR-199-3p, Logrank *p*-value = 0.6   MIR-199b-5p, Logrank *p*-value = 0.8

MIR-335-3p, Logrank *p*-value = 0.3   MIR-335-5P, Logrank *p*-value = 0.09

Figure 4. Association between studied miRNAs levels and patients' survival estimated with the use of miRNA-seq data submitted in TCGA. Lower and upper percentile were set at 30; high and low expression was estimated in 296 cases each group; data analyzed in [40].

4.2. Study Limitations

There are many studies focused on the assessment of the role of miRNA in cancer development and diagnostics. The potential of miRNA to monitor cancer response to therapy is also raised. However, serious challenges must be faced before more unequivocal conclusions are delivered. First, it must be taken into account that breast cancer (as many other cancer types) reveals high heterogeneity that is driven by many factors, including ethnically diverse backgrounds, age at diagnosis, stage at diagnosis and genetic and non-genetic alterations (including genomic, transcriptomic, proteomic and epigenetic). This diversity of tumor cells' profiles led to distinguishing different classification levels, i.e., based on histology and expression profiles of the molecular markers; estrogen receptor (ER), progesterone receptor (PR) and the overexpression or gene amplification of human epidermal growth factor receptor 2 (HER2). According to the presence or absence of these critical receptors, specific molecular subtypes were selected: Luminal A (ER and PR-positive, HER2-negative, low Ki67), Luminal B (ER and/or PR positive, HER2-positive or high Ki67), HER2-enriched (ER and PR-negative, HER2-positive) and triple-negative (TNBC) (ER, PR, HER2-negative). They all exhibit distinct clinical outcomes and require different treatment strategies [79]. The seminal studies using gene expression profiling have further subdivided breast cancers into molecular and transcriptomic subtypes of prognostic and predictive importance. Thus, we see some limitations of our work that refers to three cell lines only that did not provide representative or coherent data.

Additionally, referring to the subject of the study, i.e., miRNA, we are aware that it is commonly known for its non-specific action and broad target profile. This, in turn, requires more advanced studies involving specific miRNA downregulation or induction to observe their specific role in affecting specific molecular pathways that control selected functional mechanisms in cancer cells. We also encountered some technical issues. As commonly acknowledged, there is no perfect reference gene for miRNA evaluation. Even the U6 reference gene is sometimes questioned as not stable enough. Another issue is drug selection and its concentration and treatment time during the study. All these factors may significantly affect the cancer cell response that is followed by the different responses of cells and, consequently, different alterations in cell metabolism and gene expression. Lastly, some experiments enable observation of early response, while others provide information regarding the prolonged effect. Altogether, it seems that the evaluation of the role of miRNAs in cancer response to therapy is based on the assessment of a very subtle and sensitive to changes parameter. Thus, it may be difficult to obtain conclusive results before reaching a broader context and evaluating samples derived from patients with different cancer types and characteristics. Without a doubt, miRNA downregulation, mRNA and protein profiling and functional studies are required that will show how all these modulations affect cancer cell metabolism and, eventually, the patient's outcomes.

4.3. Potential Mechanism

Evaluation of the biological potential of selected miRNAs in breast cancer patients was performed using targetscan.org (access date: 27 February 2023) [37] and literature data (as shown in Table 2). Based on these analyses, all selected miRNAs could trigger significant effect on the expression of genes contributing to the most critical features of cancer cells i.e., proliferation, adhesion, DNA damage/repair pathways, apoptosis, autophagy, etc. (see Table 2). Identification of potential targets for selected miRNAs showed that miR-21-3p could affect numerous genes but some of them were predominantly associated with cancer homeostasis, such as DNA damage-regulated autophagy modulator 1 (DRAM1), DNA damage-inducible transcript 4-like (DDIT4L) and p53 and DNA damage-regulated 1 (PDRG1). A similar analysis also showed that miR-126-5p is associated with DDIT4L and PDRG1, but also with growth arrest and DNA damage-inducible, gamma (GADD45G). In turn, miR-335-3p levels corresponded with the expression of GADD45A and mediator of DNA-damage checkpoint 1 (MDC1). Literature data were even more abundant but mostly referred to in vitro conditions. Although the role of selected miRNAs in tumor development and response to therapy is well documented, it is still difficult to state if the observed associations are the cause or the result of alterations observed during carcinogenesis or therapeutic agent treatment. For many years, it was thought that the expression of miRNA in cancer cells was primarily reduced. Only a comparison of the miRNA profile of normal and cancer tissues showed significant overexpression of some miRNAs [80]. Depending on the function of miRNAs in the development of tumors, they are classified as suppressor miRNAs (inhibiting the expression of oncogenes or genes that induce apoptosis) and oncogenic miRNAs (activating oncogenesis or inhibiting the expression of suppressor genes) [81]. It should be emphasized that this classification is a significant simplification, because in the case of many miRNAs (e.g., miR-155), the effect of their activity depends on the total activity of regulated genes and tissue type [81]. Although miRNA levels seem very variable, we are still convinced that they can be used as a diagnostic marker and a potential target in modern anticancer therapies. However, due to the enormous number of this short polymers and no need for full complementarity to act, it may be difficult to find conclusive remarks. It seems that different conditions, including time and concentration or a type of therapeutic strategy, significantly affect the observed alteration. Nevertheless, they still seem to be a promising target that reflects not only a disease-associated modulation of the metabolism, but could also reflect the response of cancer cells to therapy that can be monitored. Consequently, targeting specific miRNAs could also be an important element of an efficient therapeutic approach. We suggest that the alterations due to drugs treatment of certain miRNA fractions depend on the breast cancer cell line characteristics. However, these preliminary results require further detailed studies in vitro and in vivo to verify their clinical potential in monitoring and therapy based on miRNAs profiling and targeting.

Author Contributions: A.M.—qPCR; I.D., R.S., M.C., M.D., J.Ś., I.K., M.R., K.W. and B.J.—samples collection, RNA isolation, grant application writing; E.T. and N.L.—cell culture; P.K.—data analysis; B.R.—experiments planning, project coordination, manuscript writing. All authors have read and agreed to the published version of the manuscript.

Funding: The project is financed by the Minister of Science and Higher Education SKN/SP/496721/2021 "Student science clubs create innovations".

Institutional Review Board Statement: Not applicable.

Data Availability Statement: All relevant data are included in the manuscript.

Conflicts of Interest: The authors declare no conflict of interest.

References

1. Faraoni, I.; Antonetti, F.R.; Cardone, J.; Bonmassar, E. miR-155 gene: A typical multifunctional microRNA. *Biochim. Biophys. Acta Mol. Basis Dis.* **2009**, *1792*, 497–505. [CrossRef]
2. Esquela-Kerscher, A.; Slack, F. Oncomirs—microRNAs with a role in cancer. *Nat. Rev. Cancer* **2006**, *6*, 259–269. [CrossRef]
3. Gao, Y.; Han, D.; Feng, J. MicroRNA in multiple sclerosis. *Clin. Chim. Acta* **2021**, *516*, 92–99. [CrossRef] [PubMed]
4. Clausen, A.R.; Durand, S.; Petersen, R.L.; Staunstrup, N.H.; Qvist, P. Circulating miRNAs as Potential Biomarkers for Patient Stratification in Bipolar Disorder: A Combined Review and Data Mining Approach. *Genes* **2022**, *13*, 1038. [CrossRef]
5. Angelescu, M.A.; Andronic, O.; Dima, S.O.; Popescu, I.; Meivar-Levy, I.; Ferber, S.; Lixandru, D. miRNAs as Biomarkers in Diabetes: Moving towards Precision Medicine. *Int. J. Mol. Sci.* **2022**, *23*, 12843. [CrossRef] [PubMed]
6. Koturbash, I.; Tolleson, W.H.; Guo, L.; Yu, D.; Chen, S.; Hong, H.; Mattes, W.; Ning, B. microRNAs as pharmacogenomic biomarkers for drug efficacy and drug safety assessment. *Biomark. Med.* **2015**, *9*, 1153–1176. [CrossRef] [PubMed]
7. Uhr, K.; Prager-van der Smissen, W.J.C.P.; Heine, A.A.J.; Ozturk, B.; Van Jaarsveld, M.T.M.; Boersma, A.W.M.; Jager, A.; Wiemer, E.A.C.; Smid, M.; Foekens, J.A.; et al. MicroRNAs as possible indicators of drug sensitivity in breast cancer cell lines. *PLoS ONE* **2019**, *14*, e0216400. [CrossRef]
8. Kahraman, M.; Röske, A.; Laufer, T.; Fehlmann, T.; Backes, C.; Kern, F.; Kohlhaas, J.; Schrörs, H.; Saiz, A.; Zabler, C.; et al. MicroRNA in diagnosis and therapy monitoring of early-stage triple-negative breast cancer. *Sci. Rep.* **2018**, *8*, 11584. [CrossRef] [PubMed]
9. MiR-21-5p | BioVendor R&D. Available online: https://www.biovendor.com/mir-21-5p (accessed on 12 January 2023).
10. Huang, S.; Fan, W.; Wang, L.; Liu, H.; Wang, X.; Zhao, H.; Jiang, W. Maspin inhibits MCF-7 cell invasion and proliferation by downregulating miR-21 and increasing the expression of its target genes. *Oncol. Lett.* **2020**, *19*, 2621–2628. [CrossRef]

11. You, F.; Luan, H.; Sun, D.; Cui, T.; Ding, P.; Tang, H.; Sun, D. miRNA-106a Promotes Breast Cancer Cell Proliferation, Clonogenicity, Migration, and Invasion through Inhibiting Apoptosis and Chemosensitivity. *DNA Cell Biol.* **2019**, *38*, 198–207. [CrossRef]

12. Szatkowska, M.; Krupa, R. Regulation of DNA Damage Response and Homologous Recombination Repair by microRNA in Human Cells Exposed to Ionizing Radiation. *Cancers* **2020**, *12*, 1838. [CrossRef] [PubMed]

13. Miao, Y.; Lu, J.; Fan, B.; Sun, L. MicroRNA-126-5p Inhibits the Migration of Breast Cancer Cells by Directly Targeting CNOT7. *Technol. Cancer Res. Treat.* **2020**, *19*, 153303382097754. [CrossRef]

14. MiR-126-3p | BioVendor R&D. Available online: https://www.biovendor.com/mir-126-3p (accessed on 12 January 2023).

15. Hong, Z.; Hong, C.; Ma, B.; Wang, Q.; Zhang, X.; Li, L.; Wang, C.; Chen, D. MicroRNA-126-3p inhibits the proliferation, migration, invasion, and angiogenesis of triple-negative breast cancer cells by targeting RGS3. *Oncol. Rep.* **2019**, *42*, 1569–1579. [CrossRef]

16. Ovcharenko, D.; Kelnar, K.; Johnson, C.; Leng, N.; Brown, D. Genome-Scale MicroRNA and Small Interfering RNA Screens Identify Small RNA Modulators of TRAIL-Induced Apoptosis Pathway. *Cancer Res* **2007**, *67*, 10782–10788. [CrossRef]

17. Zhang, G.; Zhong, L.; Luo, H.; Wang, S. MicroRNA-155-3p promotes breast cancer progression through down-regulating CADM1. *OncoTargets Ther.* **2019**, *12*, 7993–8002. [CrossRef] [PubMed]

18. Won, K.Y.; Kim, Y.W.; Kim, H.-S.; Lee, S.K.; Jung, W.-W.; Park, Y.-K. MicroRNA-199b-5p is involved in the Notch signaling pathway in osteosarcoma. *Hum. Pathol.* **2013**, *44*, 1648–1655. [CrossRef]

19. Fang, C.; Wang, F.-B.; Li, Y.; Zeng, X.-T. Down-regulation of miR-199b-5p is correlated with poor prognosis for breast cancer patients. *Biomed. Pharmacother.* **2016**, *84*, 1189–1193. [CrossRef]

20. Lin, X.; Qiu, W.; Xiao, Y.; Ma, J.; Xu, F.; Zhang, K.; Gao, Y.; Chen, Q.; Li, Y.; Li, H.; et al. MiR-199b-5p Suppresses Tumor Angiogenesis Mediated by Vascular Endothelial Cells in Breast Cancer by Targeting ALK1. *Front. Genet.* **2020**, *10*, 1397. [CrossRef] [PubMed]

21. Ye, L.; Wang, F.; Wu, H.; Yang, H.; Yang, Y.; Ma, Y.; Xue, A.; Zhu, J.; Chen, M.; Wang, J.; et al. Functions and Targets of miR-335 in Cancer. *OncoTargets Ther.* **2021**, *14*, 3335–3349. [CrossRef]

22. Sun, N.; Xu, H.N.; Luo, Q.; Li, L.Z. Potential Indexing of the Invasiveness of Breast Cancer Cells by Mitochondrial Redox Ratios. In *Oxygen Transport to Tissue XXXVIII*; Luo, Q., Li, L.Z., Harrison, D.K., Shi, H., Bruley, D.F., Eds.; Advances in Experimental Medicine and Biology; Springer International Publishing: Cham, Switzerland, 2016; Volume 923, pp. 121–127. ISBN 978-3-319-38808-3.

23. Atalay, C.; Gurhan, I.D.; Irkkan, C.; Gündüz, U. Multidrug Resistance in Locally Advanced Breast Cancer. *Tumor Biol.* **2006**, *27*, 309–318. [CrossRef]

24. Thorn, C.F.; Oshiro, C.; Marsh, S.; Hernandez-Boussard, T.; McLeod, H.; Klein, T.E.; Altman, R.B. Doxorubicin Pathways. *Pharm. Genom.* **2011**, *21*, 440–446. [CrossRef]

25. Cox, J.; Weinman, S. Mechanisms of doxorubicin resistance in hepatocellular carcinoma. *Hepatic Oncol.* **2016**, *3*, 57–59. [CrossRef]

26. Wang, H.; Guo, S.; Kim, S.-J.; Shao, F.; Ho, J.W.K.; Wong, K.U.; Miao, Z.; Hao, D.; Zhao, M.; Xu, J.; et al. Cisplatin prevents breast cancer metastasis through blocking early EMT and retards cancer growth together with paclitaxel. *Theranostics* **2021**, *11*, 2442–2459. [CrossRef] [PubMed]

27. Chen, S.-H.; Chang, J.-Y. New Insights into Mechanisms of Cisplatin Resistance: From Tumor Cell to Microenvironment. *Int. J. Mol. Sci.* **2019**, *20*, 4136. [CrossRef] [PubMed]

28. Romaniuk-Drapała, A.; Totoń, E.; Konieczna, N.; Machnik, M.; Barczak, W.; Kowal, D.; Kopczyński, P.; Kaczmarek, M.; Rubiś, B. hTERT Downregulation Attenuates Resistance to DOX, Impairs FAK-Mediated Adhesion, and Leads to Autophagy Induction in Breast Cancer Cells. *Cells* **2021**, *10*, 867. [CrossRef] [PubMed]

29. Di, H.; Wu, H.; Gao, Y.; Li, W.; Zou, D.; Dong, C. Doxorubicin- and cisplatin-loaded nanostructured lipid carriers for breast cancer combination chemotherapy. *Drug Dev. Ind. Pharm.* **2016**, *42*, 2038–2043. [CrossRef]

30. Lisiak, N.M.; Lewicka, I.; Kaczmarek, M.; Kujawski, J.; Bednarczyk-Cwynar, B.; Zaprutko, L.; Rubis, B. Oleanolic Acid's Semisynthetic Derivatives HIMOXOL and Br-HIMOLID Show Proautophagic Potential and Inhibit Migration of HER2-Positive Breast Cancer Cells In Vitro. *Int. J. Mol. Sci.* **2021**, *22*, 11273. [CrossRef]

31. Lou, G.; Ma, N.; Xu, Y.; Jiang, L.; Yang, J.; Wang, C.; Jiao, Y.; Gao, X. Differential distribution of U6 (RNU6-1) expression in human carcinoma tissues demonstrates the requirement for caution in the internal control gene selection for microRNA quantification. *Int. J. Mol. Med.* **2015**, *36*, 1400–1408. [CrossRef]

32. Takele Assefa, A.; Vandesompele, J.; Thas, O. On the utility of RNA sample pooling to optimize cost and statistical power in RNA sequencing experiments. *BMC Genom.* **2020**, *21*, 312. [CrossRef]

33. Xu, M.; Sizova, O.; Wang, L.; Su, D.-M. A Fine-Tune Role of Mir-125a-5p on Foxn1 During Age-Associated Changes in the Thymus. *Aging Dis.* **2017**, *8*, 277–286. [CrossRef]

34. Khan, A.A.; Betel, D.; Miller, M.L.; Sander, C.; Leslie, C.S.; Marks, D.S. Transfection of small RNAs globally perturbs gene regulation by endogenous microRNAs. *Nat. Biotechnol.* **2009**, *27*, 549–555. [CrossRef]

35. Zhao, S.; Li, J.; Feng, J.; Li, Z.; Liu, Q.; Lv, P.; Wang, F.; Gao, H.; Zhang, Y. Identification of Serum miRNA-423-5p Expression Signature in Somatotroph Adenomas. *Int. J. Endocrinol.* **2019**, *2019*, 8516858. [CrossRef]

36. Gutiérrez-Vázquez, C.; Rodríguez-Galán, A.; Fernández-Alfara, M.; Mittelbrunn, M.; Sánchez-Cabo, F.; Martínez-Herrera, D.J.; Ramírez-Huesca, M.; Pascual-Montano, A.; Sánchez-Madrid, F. miRNA profiling during antigen-dependent T cell activation: A role for miR-132-3p. *Sci. Rep.* **2017**, *7*, 3508. [CrossRef]

37. TargetScanHuman 8.0. Available online: https://www.targetscan.org/vert_80/ (accessed on 14 January 2023).

38. GEPIA 2. Available online: http://gepia2.cancer-pku.cn/#index (accessed on 9 February 2023).

39. UCSC Xena. Available online: https://xenabrowser.net/ (accessed on 9 February 2023).

40. OncoLnc. Available online: http://www.oncolnc.org/ (accessed on 1 March 2023).

41. Mar-Aguilar, F.; Mendoza-Ramírez, J.A.; Malagón-Santiago, I.; Espino-Silva, P.K.; Santuario-Facio, S.K.; Ruiz-Flores, P.; Rodríguez-Padilla, C.; Reséndez-Pérez, D. Serum Circulating microRNA Profiling for Identification of Potential Breast Cancer Biomarkers. *Dis. Markers* **2013**, *34*, 163–169. [CrossRef]

42. Loh, H.-Y.; Norman, B.P.; Lai, K.-S.; Rahman, N.M.A.N.A.; Alitheen, N.B.M.; Osman, M.A. The Regulatory Role of MicroRNAs in Breast Cancer. *Int. J. Mol. Sci.* **2019**, *20*, 4940. [CrossRef] [PubMed]

43. Grimaldi, A.M.; Salvatore, M.; Incoronato, M. miRNA-Based Therapeutics in Breast Cancer: A Systematic Review. *Front. Oncol.* **2021**, *11*, 668464. [CrossRef] [PubMed]

44. Gironella, M.; Seux, M.; Xie, M.-J.; Cano, C.; Tomasini, R.; Gommeaux, J.; Garcia, S.; Nowak, J.; Yeung, M.L.; Jeang, K.-T.; et al. Tumor protein 53-induced nuclear protein 1 expression is repressed by miR-155, and its restoration inhibits pancreatic tumor development. *Proc. Natl. Acad. Sci. USA* **2007**, *104*, 16170–16175. [CrossRef]

45. Gao, Y.; Liu, Z.; Ding, Z.; Hou, S.; Li, J.; Jiang, K. MicroRNA-155 increases colon cancer chemoresistance to cisplatin by targeting forkhead box O3. *Oncol. Lett.* **2018**, *15*, 4781–4788. [CrossRef] [PubMed]

46. Lv, L.; An, X.; Li, H.; Ma, L. Effect of miR-155 knockdown on the reversal of doxorubicin resistance in human lung cancer A549/dox cells. *Oncol. Lett.* **2016**, *11*, 1161–1166. [CrossRef] [PubMed]

47. MiR-155-5p | BioVendor R&D. Available online: https://www.biovendor.com/mir-155-5p (accessed on 12 January 2023).

48. Dan, T.; Shastri, A.A.; Palagani, A.; Buraschi, S.; Neill, T.; Savage, J.E.; Kapoor, A.; DeAngelis, T.; Addya, S.; Camphausen, K.; et al. miR-21 Plays a Dual Role in Tumor Formation and Cytotoxic Response in Breast Tumors. *Cancers* **2021**, *13*, 888. [CrossRef]

49. Meng, F.; Henson, R.; Wehbe–Janek, H.; Ghoshal, K.; Jacob, S.T.; Patel, T. MicroRNA-21 Regulates Expression of the PTEN Tumor Suppressor Gene in Human Hepatocellular Cancer. *Gastroenterology* **2007**, *133*, 647–658. [CrossRef] [PubMed]

50. Yu, X.; Chen, Y.; Tian, R.; Li, J.; Li, H.; Lv, T.; Yao, Q. miRNA-21 enhances chemoresistance to cisplatin in epithelial ovarian cancer by negatively regulating PTEN. *Oncol. Lett.* **2017**, *14*, 1807–1810. [CrossRef] [PubMed]

51. Wang, P.; Zou, F.; Zhang, X.; Li, H.; Dulak, A.; Tomko, R.J., Jr.; Lazo, J.S.; Wang, Z.; Zhang, L.; Yu, J. microRNA-21 Negatively Regulates Cdc25A and Cell Cycle Progression in Colon Cancer Cells. *Cancer Res.* **2009**, *69*, 8157–8165. [CrossRef] [PubMed]

52. Hu, H.; Gatti, R.A. MicroRNAs: New players in the DNA damage response. *J. Mol. Cell Biol.* **2011**, *3*, 151–158. [CrossRef]

53. Luo, L.; Xia, L.; Zha, B.; Zuo, C.; Deng, D.; Chen, M.; Hu, L.; He, Y.; Dai, F.; Wu, J.; et al. miR-335-5p targeting ICAM-1 inhibits invasion and metastasis of thyroid cancer cells. *Biomed. Pharmacother.* **2018**, *106*, 983–990. [CrossRef]

54. Tomasetti, M.; Monaco, F.; Manzella, N.; Rohlena, J.; Rohlenova, K.; Staffolani, S.; Gaetani, S.; Ciarapica, V.; Amati, M.; Bracci, M.; et al. MicroRNA-126 induces autophagy by altering cell metabolism in malignant mesothelioma. *Oncotarget* **2016**, *7*, 36338–36352. [CrossRef] [PubMed]

55. Huang, Q.; Ma, Q. MicroRNA-106a inhibits cell proliferation and induces apoptosis in colorectal cancer cells. *Oncol. Lett.* **2018**, *15*, 8941–8944. [CrossRef] [PubMed]

56. Pan, Y.-J.; Wei, L.-L.; Wu, X.-J.; Huo, F.-C.; Mou, J.; Pei, D.-S. MiR-106a-5p inhibits the cell migration and invasion of renal cell carcinoma through targeting PAK5. *Cell Death Dis.* **2017**, *8*, e3155. [CrossRef]

57. Lai, Y.; Quan, J.; Hu, J.; Chen, P.; Xu, J.; Guan, X.; Xu, W.; Lai, Y.; Ni, L. miR-199b-5p serves as a tumor suppressor in renal cell carcinoma. *Exp. Ther. Med.* **2018**, *16*, 436–444. [CrossRef]

58. Gao, J.; Zhang, Q.; Xu, J.; Guo, L.; Li, X. Clinical significance of serum miR-21 in breast cancer compared with CA153 and CEA. *Chin. J. Cancer Res.* **2013**, *25*, 743–748. [CrossRef]

59. Najjary, S.; Mohammadzadeh, R.; Mokhtarzadeh, A.; Mohammadi, A.; Kojabad, A.B.; Baradaran, B. Role of miR-21 as an authentic oncogene in mediating drug resistance in breast cancer. *Gene* **2020**, *738*, 144453. [CrossRef] [PubMed]

60. Wang, H.; Tan, Z.; Hu, H.; Liu, H.; Wu, T.; Zheng, C.; Wang, X.; Luo, Z.; Wang, J.; Liu, S.; et al. microRNA-21 promotes breast cancer proliferation and metastasis by targeting LZTFL1. *BMC Cancer* **2019**, *19*, 738. [CrossRef]

61. Valeri, N.; Gasparini, P.; Braconi, C.; Paone, A.; Lovat, F.; Fabbri, M.; Sumani, K.M.; Alder, H.; Amadori, D.; Patel, T.; et al. MicroRNA-21 induces resistance to 5-fluorouracil by down-regulating human DNA MutS homolog 2 (hMSH2). *Proc. Natl. Acad. Sci. USA* **2010**, *107*, 21098–21103. [CrossRef]

62. You, F.; Li, J.; Zhang, P.; Zhang, H.; Cao, X. miR106a Promotes the Growth of Transplanted Breast Cancer and Decreases the Sensitivity of Transplanted Tumors to Cisplatin. *Cancer Manag. Res.* **2020**, *12*, 233–246. [CrossRef] [PubMed]

63. Mayr, C. Regulation by 3′-Untranslated Regions. *Annu. Rev. Genet.* **2017**, *51*, 171–194. [CrossRef] [PubMed]

64. Kato, M.; Paranjape, T.; Ullrich, R.; Nallur, S.; Gillespie, E.; Keane, K.; Esquela-Kerscher, A.; Weidhaas, J.B.; Slack, F.J. The mir-34 microRNA is required for the DNA damage response *in vivo* in C. elegans and in vitro in human breast cancer cells. *Oncogene* **2009**, *28*, 2419–2424. [CrossRef] [PubMed]

65. Lukashchuk, N.; Vousden, K.H. Ubiquitination and Degradation of Mutant p53. *Mol. Cell. Biol.* **2007**, *27*, 8284–8295. [CrossRef]

66. Habbe, N.; Koorstra, J.-B.M.; Mendell, J.T.; Offerhaus, G.J.; Ryu, J.K.; Feldmann, G.; Mullendore, M.E.; Goggins, M.G.; Hong, S.-M.; Maitra, A. MicroRNA miR-155 is a biomarker of early pancreatic neoplasia. *Cancer Biol. Ther.* **2009**, *8*, 340–346. [CrossRef]

67. Zhang, C.-M.; Zhao, J.; Deng, H.-Y. MiR-155 promotes proliferation of human breast cancer MCF-7 cells through targeting tumor protein 53-induced nuclear protein 1. *J. Biomed. Sci.* **2013**, *20*, 79. [CrossRef]

68. Luo, W.; Zhang, H.; Liang, X.; Xia, R.; Deng, H.; Yi, Q.; Lv, L.; Qian, L. DNA methylation-regulated miR-155-5p depresses sensitivity of esophageal carcinoma cells to radiation and multiple chemotherapeutic drugs via suppression of MAP3K10. *Oncol. Rep.* **2020**, *43*, 1692–1704. [CrossRef]

69. Gasparini, P.; Lovat, F.; Fassan, M.; Casadei, L.; Cascione, L.; Jacob, N.K.; Carasi, S.; Palmieri, D.; Costinean, S.; Shapiro, C.L.; et al. Protective role of miR-155 in breast cancer through *RAD51* targeting impairs homologous recombination after irradiation. *Proc. Natl. Acad. Sci. USA* **2014**, *111*, 4536–4541. [CrossRef]

70. Chiu, Y.-C.; Tsai, M.-H.; Chou, W.-C.; Liu, Y.-C.; Kuo, Y.-Y.; Hou, H.-A.; Lu, T.-P.; Lai, L.-C.; Chen, Y.; Tien, H.-F.; et al. Prognostic significance of NPM1 mutation-modulated microRNA−mRNA regulation in acute myeloid leukemia. *Leukemia* **2016**, *30*, 274–284. [CrossRef]

71. Cao, W.; Gao, W.; Liu, Z.; Hao, W.; Li, X.; Sun, Y.; Tong, L.; Tang, B. Visualizing miR-155 To Monitor Breast Tumorigenesis and Response to Chemotherapeutic Drugs by a Self-Assembled Photoacoustic Nanoprobe. *Anal. Chem.* **2018**, *90*, 9125–9131. [CrossRef]

72. Wu, A.; Chen, Y.; Liu, Y.; Lai, Y.; Liu, D. miR-199b-5p inhibits triple negative breast cancer cell proliferation, migration and invasion by targeting DDR1. *Oncol. Lett.* **2018**, *16*, 4889–4896. [CrossRef]

73. Fornari, F.; Milazzo, M.; Chieco, P.; Negrini, M.; Calin, G.A.; Grazi, G.L.; Pollutri, D.; Croce, C.M.; Bolondi, L.; Gramantieri, L. MiR-199a-3p Regulates mTOR and c-Met to Influence the Doxorubicin Sensitivity of Human Hepatocarcinoma Cells. *Cancer Res.* **2010**, *70*, 5184–5193. [CrossRef] [PubMed]

74. Tanaka, N.; Minemura, C.; Asai, S.; Kikkawa, N.; Kinoshita, T.; Oshima, S.; Koma, A.; Kasamatsu, A.; Hanazawa, T.; Uzawa, K.; et al. Identification of *miR-199-5p* and *miR-199-3p* Target Genes: Paxillin Facilities Cancer Cell Aggressiveness in Head and Neck Squamous Cell Carcinoma. *Genes* **2021**, *12*, 1910. [CrossRef]

75. Scarola, M.; Schoeftner, S.; Schneider, C.; Benetti, R. miR-335 Directly Targets Rb1 (pRb/p105) in a Proximal Connection to p53-Dependent Stress Response. *Cancer Res* **2010**, *70*, 6925–6933. [CrossRef] [PubMed]

76. Hao, J.; Lai, M.; Liu, C. Expression of miR-335 in triple-negative breast cancer and its effect on chemosensitivity. *J. BUON* **2019**, *24*, 1526–1531. [PubMed]

77. Hajibabaei, S.; Sotoodehnejadnematalahi, F.; Nafissi, N.; Zeinali, S.; Azizi, M. Aberrant promoter hypermethylation of miR-335 and miR-145 is involved in breast cancer PD-L1 overexpression. *Sci. Rep.* **2023**, *13*, 1003. [CrossRef]

78. The Cancer Genome Atlas Program-NCI. Available online: https://www.cancer.gov/about-nci/organization/ccg/research/structural-genomics/tcga (accessed on 1 March 2023).

79. Mavrommati, I.; Johnson, F.; Echeverria, G.V.; Natrajan, R. Subclonal heterogeneity and evolution in breast cancer. *NPJ Breast Cancer* **2021**, *7*, 155. [CrossRef] [PubMed]

80. Chen, C.-Z. MicroRNAs as Oncogenes and Tumor Suppressors. *N. Engl. J. Med.* **2005**, *353*, 1768–1771. [CrossRef] [PubMed]

81. Chen, Y.; Fu, L.L.; Wen, X.; Liu, B.; Huang, J.; Wang, J.H.; Wei, Y.Q. Oncogenic and tumor suppressive roles of microRNAs in apoptosis and autophagy. *Apoptosis* **2014**, *19*, 1177–1189. [CrossRef] [PubMed]

Article

microRNA Expression Profile of Purified Alveolar Epithelial Type II Cells

Stefan Dehmel [1,2,†], Katharina J. Weiss [1,3,†], Natalia El-Merhie [4,†], Jens Callegari [5,6], Birte Konrad [1], Kathrin Mutze [5], Oliver Eickelberg [7], Melanie Königshoff [5,7] and Susanne Krauss-Etschmann [4,8,*]

1   Institute for Lung Biology and Disease, Ludwig-Maximilians University Hospital Munich,
    Asklepios Clinic Gauting and Helmholtz Zentrum München, Comprehensive Pneumology Center Munich,
    Max-Lebsche-Platz 31, 81377 Munich, Germany
2   Helmholtz Zentrum München, Department Strategy, Programs, Resources,
    Helmholtz Zentrum München German Research Center for Environmental Health,
    Ingolstädter Landstraße 1, 85764 Neuherberg, Germany
3   Dr. von Hauner Children's Hospital, Ludwig-Maximilians-University, 80337 Munich, Germany
4   Early Life Origins of Chronic Lung Disease, Research Center Borstel, Leibniz Lung Center,
    Member of the German Center for Lung Research (DZL) and the Airway Research Center North (ARCN),
    23845 Borstel, Germany
5   Helmholtz Zentrum Munich, Lung Repair and Regeneration, Comprehensive Pneumology Center,
    Member of the German Center for Lung Research, 81377 Munich, Germany
6   Evangelisches Krankenhaus Bergisch Gladbach, Ferrenbergstraße, 51465 Bergisch Gladbach, Germany
7   Division of Pulmonary, Allergy and Critical Care Medicine, Department of
    Medicine, University of Pittsburgh Medical Center, Pittsburgh, PA 15261, USA
8   Institute for Experimental Medicine, Christian-Albrechts-Universität zu Kiel, 24118 Kiel, Germany
*   Correspondence: skrauss-etschmann@fz-borstel.de
†   These authors contributed equally to this work.

Citation: Dehmel, S.; Weiss, K.J.; El-Merhie, N.; Callegari, J.; Konrad, B.; Mutze, K.; Eickelberg, O.; Königshoff, M.; Krauss-Etschmann, S. microRNA Expression Profile of Purified Alveolar Epithelial Type II Cells. *Genes* **2022**, *13*, 1420. https://doi.org/10.3390/genes13081420

Academic Editors: Giuseppe Iacomino and Fabio Lauria

Received: 12 June 2022
Accepted: 6 August 2022
Published: 10 August 2022

Publisher's Note: MDPI stays neutral with regard to jurisdictional claims in published maps and institutional affiliations.

Abstract: Alveolar type II (ATII) cells are essential for the maintenance of the alveolar homeostasis. However, knowledge of the expression of the miRNAs and miRNA-regulated networks which control homeostasis and coordinate diverse functions of murine ATII cells is limited. Therefore, we asked how miRNAs expressed in ATII cells might contribute to the regulation of signaling pathways. We purified "untouched by antibodies" ATII cells using a flow cytometric sorting method with a highly autofluorescent population of lung cells. TaqMan® miRNA low-density arrays were performed on sorted cells and intersected with miRNA profiles of ATII cells isolated according to a previously published protocol. Of 293 miRNAs expressed in both ATII preparations, 111 showed equal abundances. The target mRNAs of bona fide ATII miRNAs were used for pathway enrichment analysis. This analysis identified nine signaling pathways with known functions in fibrosis and/or epithelial-to-mesenchymal transition (EMT). In particular, a subset of 19 miRNAs was found to target 21 components of the TGF-β signaling pathway. Three of these miRNAs (miR-16-5p, -17-5p and -30c-5p) were down-modulated by TGF-β1 stimulation in human A549 cells, and concomitant up-regulation of associated mRNA targets (BMPR2, JUN, RUNX2) was observed. These results suggest an important role for miRNAs in maintaining the homeostasis of the TGF-β signaling pathway in ATII cells under physiological conditions.

Keywords: alveolar epithelial type II cells; type II pneumocytes; ATII; AECII; flow cytometry; autofluorescence; miRNAs; pathway analysis; TGF-beta; homeostasis; EMT

1. Introduction

MicroRNAs (miRNAs) are small, endogenous non-coding RNA molecules that regulate the expression of ~60% of the human genome by post-transcriptional inhibition of target mRNAs [1–3]. Since their discovery in 1993, interest in miRNAs has increased, uncovering their importance in physiological processes, such as metabolism, growth, cell signaling, inflammation and cell differentiation, as well as their implication in the pathogenesis of several diseases [4–7].

It has been shown that miRNAs play an important role in lung development and in the maintenance of pulmonary homeostasis, which is vital for the preservation of normal lung function and health [8,9]. Moreover, miRNAs could regulate the interplay of epithelial cells with other cell types through targeting of multiple pulmonary pathways [10–14]. Taking into consideration that miRNAs affect the expression of a large part of the genome, the detrimental dysregulation of miRNAs disrupts lung homeostasis and initiates disease pathogenesis. Indeed, there is emerging evidence that altered miRNA expression in respiratory diseases modulates disease phenotypes and ultimately disease progression. For example, members of the miR-200 and miR-29 families are down-regulated in models of idiopathic pulmonary fibrosis (IPF) [15,16], and miR-154 has been suggested to promote fibrosis in interstitial lung disease [17]. Furthermore, the dysregulation of miRNAs was associated with an increased asthma exacerbation risk [18], as well as with the pathogenesis of COPD [19,20]. Therefore, expression of miRNAs maintains tissue homeostasis, whereas when miRNA expression is dysregulated, pathological changes occur. It has been suggested that some alveolar epithelial type II (ATII) cell-derived miRNAs could play a role in the maintenance of alveolar homeostasis in response to injury [21].

Two types of epithelial cells—alveolar epithelial type I (ATI) and type II (ATII) cells— form the alveolus. ATI cells cover ~95% of the alveolar surface to mediate gas exchange and maintain barrier integrity [22,23]. In lung injury, dying ATI cells slough off, leading to increased permeability [23–25]. ATII cells, in turn, orchestrate re-epithelialization and function as progenitors for dead ATI cells, thus restoring barrier function and gas exchange [26–31]. In mature lungs, proliferation and turnover of cells is relatively low, with an estimate of 28–35 days for ATII cells. However, this kinetics is enhanced in response to lung injury [32–34]. On the other hand, failure to repair injured alveolar epithelium is associated with progression and initiation of many pulmonary diseases [35,36]. Studies reported that epithelial destruction and ATII cell apoptosis are critical hallmarks in many pulmonary diseases [37–40]. ATII cells, which cover only ~5% of the alveolar surface [41], maintain the homeostasis of the alveolus [42–45]. This later role is carried out by surfactant proteins SP-A, SP-B, SP-C, and SP-D, which are secreted by lamellar bodies within ATII cells, the only lung epithelial cells which produce and secrete all four surfactant proteins [46,47]. Moreover, surfactant proteins A and D play an additional role in host defense and regulation of immune responses [48–50]. Therefore, any deficiencies or mutations in surfactant protein synthesis result in the disruption of lung homeostasis [51–56]. Taken together, it is clear that ATII cells exert several important biological functions and thus are critical for the maintenance of alveolar homeostasis and promoting pulmonary health [32,57,58]. However, this contrasts with the limited knowledge of the expression of the miRNAs and miRNA-regulated networks which control homeostasis and coordinate diverse functions of murine ATII cells.

The goal of this study was, therefore, to identify a set of miRNAs that are critical for maintenance of ATII cell homeostasis. These miRNAs could be further used to design miRNA-based therapeutics that target their function. The goal of this study was to identify miRNAs expressed by murine ATII cells under normal, non-pathologic conditions and to elucidate potential miRNA-controlled pathways of ATII cell homeostasis. We assumed that every method for isolating ATII cells will bias at least some microRNAs to some extent. To circumvent this problem, we decided to use ATII cells obtained by two different isolation procedures (panning and sorting) and to use the cut set of expressed microRNAs expressed by both sATII and pATII. We aimed to identify miRNAs that are expressed in all kinds of putative ATII cell subsets and not in a subset that might be enriched by a single method. To this end, we used two different methods (panning and sorting) for the purification of ATII cells to avoid bias in the miRNA composition introduced by a single method. For this purpose, a three-step approach was followed: first, a protocol for the isolation of highly pure murine ATII cells was developed using fluorescence-activated cell sorting (FACS) for further miRNA profiling. Second, we intersected miRNA profiles from our FACS-based procedure with those obtained by a previously published protocol relying on negative selection of ATII cells in antibody-coated plastic dishes [57]. Third, we used this dataset for in silico pathway enrichment analysis of ATII miRNA targets. In silico target prediction tools for miRNAs are highly prone to false-positive results. We therefore restricted the pathway analyses to miRNA–target pairs that have been confirmed previously and are available through the Ingenuity® software. Finally, we corroborated our findings in human epithelial alveolar cell line A549. The study limitations include the fact that impurities in the panned cell fraction were mostly due to contamination with CD31- and CD45-postive cells, while the sorted cells had a very low number of contaminating cells of unknown composition, which might have been due to ATII progenitor cells.

2. Materials and Methods

2.1. Animals

Female C57BL/6NCrl wild type 6–12-week-old mice (5 mice per group) were maintained under specific pathogen-free conditions in individually ventilated cages. Mice were fed fortified rodent chow and water *ad libitum*. All animal experiments were approved by the Animal Ethics Committee of the government of Upper Bavaria, and all animal studies were conducted in compliance with the guidelines of the Institutional Animal Care and Use Committee of the Helmholtz Center Munich, Bavaria, Germany.

2.2. Preparation of Single-Cell Suspensions and Cell Sorting

Single-cell suspensions were prepared as described previously [58] from the whole lungs of female C57BL/6 mice (6–12 w). The mice were anesthetized by intraperitoneal injection of MMF (Medetomidine 0.5 μg/g, Midazolam 5.0 μg/g, Fentanyl 0.05 μg/g) and 60 μL of heparin (for blood coagulation inhibition (5 IU/μL, Ratiopharm, Ulm, Germany)). Lungs were perfused via the right ventricle with 10 mL PBS and 1.5 mL Dispase (BD, CA) instilled over a tracheal catheter. This was followed by a 0.3 mL instillation of pre-warmed to 42 °C low-melt agarose (1%) (Invitrogen, Darmstadt, Germany). Lungs were removed and incubated for 45 min in 2.5 mL Dispase at room temperature. Then, lungs were transferred to a culture dish containing 5 mL medium (DMEM/F12 (1:1) (Gibco, Darmstadt, Germany) supplemented with 0.04 mg/mL DNase I (AppliChem, Darmstadt, Germany), 3.6 mg/mL D-(+)-Glucose (AppliChem, Germany) and 1% Penicillin/Streptomycin (PAA, Cölbe, Austria)), and divided into separate lobes. The lobes were then sequentially transferred into a new culture dish containing 8 ml of medium, where the tissue was gently teased apart with forceps. The resulting cell suspension was homogenized and transferred into a 50 mL conical tube. The cell suspension was serially filtered through 100, 20 and 10 μm nylon meshes and then centrifuged at $200\times g$ for 10 min at 15 °C. The supernatant was discarded, and the cell pellet was resuspended in medium (DMEM/F12 (1:1), (Gibco) containing 3.6 mg/mL D-(+)-Glucose (AppliChem), 1% Penicillin/Streptomycin (PAA) and 2% FBS Gold (PAA)).

Thereafter, single-cell lung suspensions from 3–4 mice were incubated on ice with rat anti-mouse CD45-APC (IgG2b, κ; BD Pharmingen) and rat anti-mouse CD31-APC (IgG2a, κ; BD Pharmingen). The antibodies are listed in Table 1. Cells were then washed and resuspended to a final concentration of 10×10^6/mL in DMEM/F12 (1:1) (Gibco, Germany) containing 2% FBS Gold (PAA, Cölbe, Austria). After serial filtration through 100, 40 and 35 μm cell strainers (BD Biosciences, Heidelberg, Germany), cells were sorted on a FACS Aria II (BD Biosciences). Cell doublets were excluded according to FSC-H to FSC-A and FSC-W to FSC-A characteristics. ATII cells were identified as the CD45/CD31-negative and autofluorescence (FITC channel)-high population. Cells were sorted using an 85 μm nozzle tip at 45 psi sheath fluid pressure. Cells isolated by this procedure were designated as sATII. For RNA isolation, sorted cells were immediately pelleted and stored at −80 °C. Cell sorting was performed immediately after cell extraction.

Table 1. Antibodies used in this study.

Antibodies for Flow Cytometry and Cell Sorting (ITC: Isotype Control):					
Antigen	Host	Isotype	Fluorochrome	Clone	Company
CD31	Rat	IgG2a, k	APC	MEC 13.3	BD Pharmingen
ITC for CD31	Rat	IgG2a, k	APC	R35-95	BD Pharmingen
CD31	Rat	IgG2a, k	PE	390	BioLegend
ITC for CD31	Rat	IgG2a, k	PE	RTK2758	BioLegend
CD45	Rat	IgG2b, k	APC	30-F11	BD Pharmingen
ITC for CD45	Rat	IgG2b, k	APC	A95-1	BD Pharmingen
CD74	Rat	IgG2b, k	FITC	In-1	BD Pharmingen
ITC for CD74	Rat	IgG2b, k	FITC	A95-1	BD Pharmingen
Primary Antibodies for Immunofluorescence Staining:					
Antigen	Host	Isotype	Clone	Company	
Pan-cytokeratin	Goat	IgG1	C-11	Abcam	
E-Cadherin	Mouse	IgG2a, k	36/E-Cadherin	BD Pharmingen	
Alpha-SMA	Mouse	IgG2a	1A4	Sigma	
CD31	Rabbit	IgG	Polyclonal	Abcam	
Pro-SPC	Rabbit	IgG	Polyclonal	Chemicon/Millipore	
CCSP	Rabbit	IgG	Polyclonal	Upstate/Millipore	
CD45	Rat	IgG2b, k	30-F11	BD Pharmingen	
Secondary Antibodies for Immunofluorescence Staining:					
Antigen	Host	Isotype	Fluorochrome	Company	
Rabbit-IgG (H+L)	Goat	IgG	Alexa Fluor 555	Invitrogen	
Mouse-IgG (H+L)	Goat	IgG	Alexa Fluor 555	Inivtrogen	
Rat-IgG (H+L)	Goat	IgG	Alexa Fluor 555	Inivtrogen	
Goat-IgG (H+L)	Donkey	IgG	Alexa Fluor	Inivtrogen	

* The CD31-PE antibody used recognizes a different epitope of CD31 than the CD31-APC antibody used for sorting.

2.3. Flow Cytometry and Immunofluorescence Staining for Purity Assessment

Cells were stained with antibodies for 20 minutes on ice and expression markers were analyzed with a BD LSR II flow cytometer (BD Biosciences). For immunofluorescence staining, 1×10^5 cells in 200 µL/chamber were used, and for flow cytometry, 1×10^5 in 50 µL antibody were used. For intracellular staining, cells were fixed and permeabilized with IntraPrep (Beckman Coulter, Krefeld, Germany) according to the manufacturer's protocol. For immunofluorescence staining, cells were centrifuged for 5 min at $200 \times g$ (4 °C) on culture slides using a Rotina 420R centrifuge (Hettich, Tuttlingen, Germany) and dried overnight. Cytospins were fixed with acetone:methanol (1:1) (AppliChem), blocked with 5% BSA (Sigma-Aldrich, Schnelldorf, Germany) in PBS and stained with primary and secondary antibodies (Table 1) diluted in 0.1% BSA in PBS. Cells were then fixed with 4% PFA (Microcos, Germany) and mounted with ProLong® Gold antifade reagent with DAPI (Invitrogen). The antibodies used in this study are listed in Table 1. For analysis of dead cells, Propidium iodide (Sigma-Aldrich) was added for 10 min at 4 °C prior to the analysis by flow cytometry.

2.4. Isolation of ATII Cells (Panning)

Lung single-cell suspensions were prepared and primary ATII cell isolation by panning (designated pATII) was performed as described by Königshoff et al. [57]. Briefly, culture dishes coated with CD45 and CD16/32 antibodies (Table 1) (15 µL of each antibody/10 mL DMEM per culture dish) were incubated overnight at 4 °C and thereafter washed with 5 mL DMEM twice. Then, 5 mL of single-cell suspension was added to the coated dishes and incubated for 35 min at 37 °C in order to remove lymphocytes and macrophages. To allow the adherence of fibroblasts, the unattached cells were collected, transferred to new uncoated dishes and incubated for 35 min at 37 °C. The supernatant was then pooled and centrifuged at 15 °C for 10 min at 200 g. The pellet was stored at −80 °C for further RNA isolation and primary ATII cells were resuspended and processed for flow cytometry.

2.5. Papanicolaou Staining

PAP staining was performed as described by Dobbs LG [59]. In brief, cells were centrifuged on coverslips and dried overnight. The following day, cells were stained with hematoxylin and thereafter dipped in lithium carbonate solution. After incubation with increasing concentrations of ethanol, cells were immersed in xylene:ethanol 1:1 and thereafter rinsed with xylene. Afterwards, cells were embedded in Entellan.

2.6. TGF-β1 Stimulation of A549 Cells

Human alveolar epithelial A549 cells (ATCC, CCL-185™) were cultured in DMEM/F12 (Gibco) medium supplemented with 10% FBS Gold (PAA). The cells were seeded into 6-well plates at a density of 2×10^5 cells/well and incubated overnight at 37 °C in a humidified atmosphere at 5% CO_2. Prior to the treatment, the cells were starved for 24 h in DMEM/F12 media containing 0.1% FBS and then treated for 72 h with either vehicle control (0.1% BSA in 4 mM HCl) or recombinant human TGF-β1 (2 ng/mL) (R&D Systems, USA). Thereafter, cells were lysed with Qiazol (Qiagen, Hilden, Germany) and the lysates were stored at −20 °C until RNA isolation. All stimulations were carried out in triplicate and repeated independently three times.

2.7. RNA Isolation

Total RNA, including miRNAs, was isolated from primary ATII and A549 cells using an miRNeasy miRNA purification kit (Qiagen) according to the manufacturer's instructions. Total RNA concentration was quantified by absorbance at 26 0nm with a NanoDrop 1000 spectrophotometer (Thermo Scientific), and RNA integrity was assessed by agarose gel electrophoresis.

2.8. Reverse Transcription and Quantitative PCR of mRNAs

Reverse transcription was performed using random hexamers and MuLV reverse transcriptase according to the manufacturer's instructions (Life Technologies, Darmstadt, Germany), with 350ng total RNA as input. Relative quantification of mRNA expression was performed using LightCycler® 480 SYBR Green I Master Mix (Roche, Mannheim, Germany) with the LightCycler® 480 II system (Roche). All primers had an amplification efficiency of $\geq$92.5% and Cq values were corrected for inter-run variations. The primer sequences are listed in Table 2. Cq values above 35 were regarded as not expressed. Transcript abundance was calculated using the ΔΔCq method [60]. For sATII cells, *Hprt* was used as a reference gene and *Sftpc* mRNA expression served as a calibrator. For A549 cells, the arithmetic mean of the Cq values for *HPRT1* and *RNA18S5* served as a normalizer. Outliers were excluded using a modified Z-score [61]. T-bars, representing the range of expression levels due to sample variation, were calculated as $2\text{\textasciicircum}-(\Delta\Delta\text{Cq} \pm \text{S})$. S (standard deviation of the ΔCq value) was calculated by the formula $S = (s_1\text{\textasciicircum}2 + s_2\text{\textasciicircum}2)\text{\textasciicircum}1/2$, where s_1 and s_2 are the SEMs of the Cq(*target*) and Cq(*reference*) values from four (sATII) or three (A549) independent experiments, with three technical replicates for each. Statistical significance was calculated using ΔCq values and unpaired *t*-tests (GraphPad Prism).

Table 2. Primers used for RT-qPCR.

Gene Symbol	Species	NCBI GenBank Accession	Primers (5′->3′)	Product Size (bp)
Acta2	Mmu	NM_007392	Fwd: GCTGGTGATGATGCTCCCA Rev: GCCCATTCCAACCATTACTCC	81
Aqp5	Mmu	NM_009701	Fwd: CCTTATCCATTGGCTTGTCG Rev: CTGAACCGATTCATGACCAC	115
Cd74	Mmu	NM_001042605	Fwd: GATGGCTACTCCCTTGCTGA Rev: TGGGTCATGTTGCCGTACT	93
Cdh1	Mmu	NM_009864	Fwd: CCATCCTCGGAATCCTTGG Rev: TTTGACCACCGTTCTCCTCC	89
Hprt	Mmu	NM_013556	Fwd: CCTAAGATGAGCGCAAGTTGAA Rev: CCACAGGACTAGAACACCTGCTAA	86
Pecam1	Mmu	NM_008816	Fwd: ATCGGCAAAGTGGTCAAGAG Rev: GGCATGTCCTTTTATGATCTCAG	111
Ptprc	Mmu	NM_001111316	Fwd: GTCCCTACTTGCCTATGTCAATG Rev: CCGGGAGGTTTTCATTCC	115
Sftpa1	Mmu	NM_023134	Fwd: GGAGAGCCTGGAGAAAGGGGGC Rev: ATCCTTGCAAGCTGAGGACTCCC	124
Sftpc	Mmu	NM_011359	Fwd: AGCAAAGAGGTCCTGATGGA Rev: GAGCAGAGCCCCTACAATCA	153
Tjp1	Mmu	NM_009386	Fwd: ACGAGATGCTGGGACTGACC Rev: AACCGCATTTGGCGTTACAT	112
ACTA2	HSA	NM_001141945	Fwd: GGCTCTGGGCTCTGTAAGG Rev: TTTGCTCTGTGCTTCGTCAC	147
BCL2	HSA	NM_000633	Fwd: CTGAGTACCTGAACCGGCA Rev: GAGAAATCAAACAGAGGCCG	106
BMPR2	HSA	NM_001204	Fwd: TGCCCTCCTGATTCTTGG Rev: CATAGCCGTTCTTGATTCTGC	130
CDH1	HSA	NM_004360	Fwd: ATACACTCTCTTCTCTCACGCTGTGT Rev: CATTCTGATCGGTTACCGTGATC	89
FN1	HSA	NM_212482	Fwd: CCGACCAGAAGTTTGGGTTCT Rev: CAATGCGGTACATGACCCCT	81
HPRT1	HSA	NM_000194	Fwd: TTGTTGTAGGATATGCCCTTGAC Rev: TCTCATCTTAGGCTTTGTATTTTGC	105
JUN	HSA	NM_002228	Fwd: CAGAGAGACAGACTTGAGAACTTGAC Rev: GACGCAACCCAGTCCAAC	100
MAP2K4	HSA	NM_003010	Fwd: GGCCAAAGTATAAAGAGCTTCTGA Rev: CAGCGATATCAATCGACATACAT	145
RNA18S5	HSA	NR_003286	Fwd: GCAATTATTCCCCATGAACG Rev: AGGGCCTCACTAAACCATCC	125
RUNX2	HSA	NM_001024630	Fwd: TAGATGGACCTCGGGAACC Rev: GAGGCGGTCAGAGAACAAAC	77
SMAD3	HSA	NM_005902	Fwd: GTCAAGAGCCTGGTCAAGAAAC Rev: GATGGGACACCTGCAACC	136
SNAI1	HSA	NM_005985	Fwd: CTTCTCTAGGCCCTGGCTG Rev: AGGTTGGAGCGGTCAGC	105
TGFBR2	HSA	NM_001024847	Fwd: TCTGTGGATGACCTGGCTAAC Rev: TCATTTCCCAGAGCACCAG	148
TJP1	HSA	NM_003257	Fwd: GAGGAAACAGCTATATGGGAACAAC Rev: TGACGTTTCCCCACTCTGAAA	120
VIM	HSA	NM_003380	Fwd: AGATGGCCCTTGACATTGAG Rev: TGAGTGGGTATCAACCAGAGG	146

2.9. Reverse Transcription and Quantitative PCR of miRNAs

Quantification of miRNAs was performed using TaqMan® miRNA assays (Life Technologies) and the TaqMan® miRNA reverse transcription kit (Life Technologies), according to manufacturer's instructions. MiR quantitation was performed on a LightCycler® 480 II (Roche) instrument using TaqMan® Universal Master Mix II, no UNG (Life Technologies). TaqMan® miRNA assays used were: hsa-miR-16-5p (Assay ID 000391), hsa-miR-17-5p (Assay ID 002308), hsa-miR-24-3p (Assay ID 000402), hsa-miR-30c-5p, (Assay ID 000419) and RNU6B (Assay ID 001093). Relative transcript abundance levels and statistical significance were calculated as described for mRNAs, with the difference that RNU6B served as a reference gene and ΔCq values of vehicle control-treated A549 cells were used as a calibrator.

2.10. Analysis of TaqMan® Real-Time PCR miRNA Array Data

Total RNA concentration was quantified using a NanoDrop 1000 instrument (Thermo Scientific), and RNA integrity was assessed with a Bioanalyzer 2100 instrument (Agilent, Stuttgart, Germany). Samples with OD260/280 ratio of $\geq$1.85 and with an RNA integrity number (RIN) of $\geq$6.5 were used for miR array studies. For miR cDNA synthesis, 135 ng total RNA was reverse transcribed using stem-loop MegaPlex RT primers (rodent pool sets A+B v3.0) and an miRNA reverse transcription kit on a PeqStar 96 thermal cycler (Peqlab, Erlangen, Germany). Pre-amplification of the RT product was performed using TaqMan® PreAmp Master Mix and PreAmp Primer Mix (rodent pool sets A + B v3.0) on the 7900HT Fast RT-qPCR system. For miR expression profiling, TaqMan® Array Rodent MiRNA A+B Card Sets v3.0 containing 641 TaqMan® assays detecting mature murine miRNAs present in miRBase v15 were used65. Quantitative real-time PCR was performed on a 7900HT Fast RT-qPCR system using TaqMan® Universal PCR Master Mix. Raw cycle threshold (Cq) values were determined using Sequence Detection Software (SDS) v2.4 and SDS RQ Manager 1.2.1 (Life Technologies, Germany) with automatic settings for baseline and threshold. MiRNA assays with replicate differences larger than one Cq were filtered out and miRNAs with Cq > 32 were regarded as not detectable and excluded from the analysis. Global mean normalization was used to determine normalized relative quantities (NRQs) 66. MiRNAs with I NRQ fold differences I of $\leq$1.5 (sorted vs. panned ATII cells) were regarded as similarly expressed in both cell preparations. Pathway enrichment analysis of ATII miR targets was carried out using the Ingenuity® software. Target mRNAs were filtered using the Ingenuity® miRNA target filter to contain only those mRNAs with previously confirmed miR seed–target interactions. The identified target mRNAs were then associated with the canonical pathway library contained in the Ingenuity® Knowledge Base. The significance of the association between the dataset and a given canonical pathway was measured in two ways: (1) as a ratio of the number of molecules from the dataset that map to the pathway divided by the total number of molecules that map to the canonical pathway; and (2) Fisher's exact test, with Benjamini–Hochberg (BH) correction for multiple testing, was used to calculate the p-value determining the probability that the association between the genes in the dataset could be explained by chance alone.

3. Results

3.1. Isolation by Sorting and Assessment of Purity of Primary Murine ATII Cells

We developed a method for the isolation of highly purified "untouched by antibodies" primary ATII cells from murine lungs based on their autofluorescence [62,63] (Figure 1A). Critical steps included cell sorting of the ATII cell population based on its autofluorescence parameters measured in the FITC channel. ATII cells were isolated as negative for lineage markers of hematopoietic (CD45) and endothelial (CD31) cells (Figure 1A, left panels) defined as CD45/CD31-APC[negative] and autofluorescence-FITC[high]. FSC served to remove doublet cells, while the APC channel was used as a dump channel for CD31[positive] endothelial cells and CD45[positive] leukocytes (mainly alveolar macrophages) using APC-conjugated antibodies for both of these antigens. Flow cytometric re-analysis showed that sorted cells remained highly autofluorescent (Figure 1A, right panels). Since ATII cells express MHC class II antigens and the associated invariant chain polypeptide CD74 [63–65], we investigated CD74 expression of the sorted cells as an indicator of ATII cell purity after the removal of CD45[positive] cells, which are also known to express CD74 in the murine lung (Figure 1B).

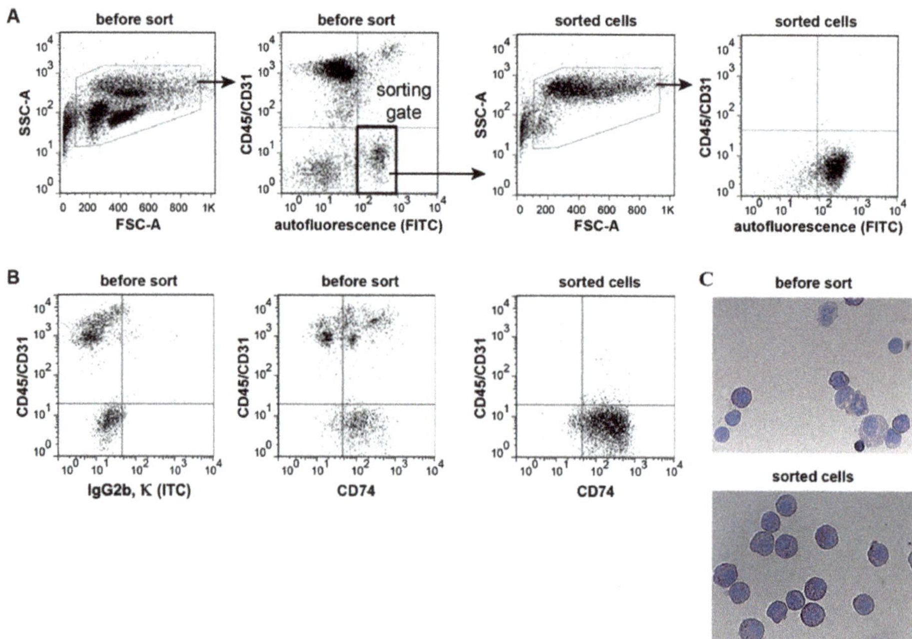

Figure 1. FACS strategy and purity of sorted ATII cells. (**A**) ATII cells were sorted based on high autofluorescence (FITC-channel) and absence of CD45 and CD31 surface expression (left panels). CD45 and CD31 expression levels were measured in the same channel (APC) and the gate used for sorting is highlighted as a thick rectangle. Sorted cells were re-analysed using the same gating strategy (right panels). Removal of doublets based on FSC characteristics not shown. Dot plots are representative of four independent experiments. (**B**) Representative dot plots of cells stained for CD45, CD31 and intracellular CD74 before and after sorting (ITC: isotype control). (**C**) Light microscopic images of Papanicolaou-stained cytospin preparations of cells before and after sorting (×400). ATII cells show characteristic dark blue inclusions in the cytoplasm (n = 4, mean ± SEM).

Papanicolaou staining (Figure 1C) showed that nearly all of the sorted cells have dark blue inclusions in the cytoplasm (lamellar bodies)—a characteristic feature of ATII cells [59].

3.2. Confirmation of Epithelial and ATII Identity of the Sorted Cell Population

To further corroborate the identity of sorted ATII cells, cytocentrifuge preparations were stained with characteristic ATII and non-ATII phenotypic markers. Sorted cells were highly positive for pro-surfactant protein C (proSP-C) and epithelial cell marker proteins E-cadherin and cytokeratin (Figure 2). On the other hand, the expression of leukocyte marker CD45, endothelial marker CD31 and smooth muscle cell marker α-SMA was not detected in sorted cells. Some very few cells were found to express the Club cell secretory protein (CCSP) after sorting.

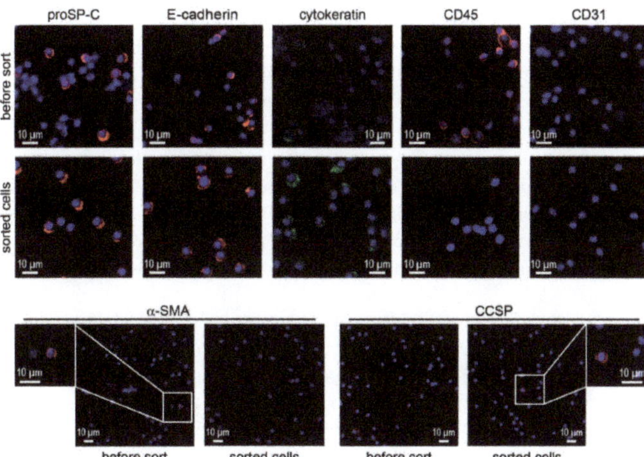

Figure 2. Immunofluorescence for phenotypic markers on cytocentrifuge preparations of lung cell suspensions (before sorting) and sorted cells. Cytocentrifuge preparations of whole lung cell suspensions and sorted cells were stained for phenotypic markers associated with ATII cells (proSP-C, E-cadherin, cytokeratin), leukocytes (CD45), endothelial cells (CD31), smooth muscle cells (α-SMA) and Club cells (CCSP). Scale bars represent 10 µm.

3.3. Viability, Purity and Phenotypes of ATII Cells Isolated by Sorting and Panning

Sorting and panning methods were compared based on cell viability (PI exclusion using flow cytometry) and purity (expression of phenotypic markers using flow cytometry and qRT-PCRs).

Viable cells were analyzed by flow cytometry as PI-negative. Propidium iodide (PI) exclusion (Figure 3A,B) confirmed the high viability of cell populations before isolation (PI: above 98%) and after isolation (PI: above 96%) for sATII and pATII, with a slightly higher viability (96.7%) for pATII.

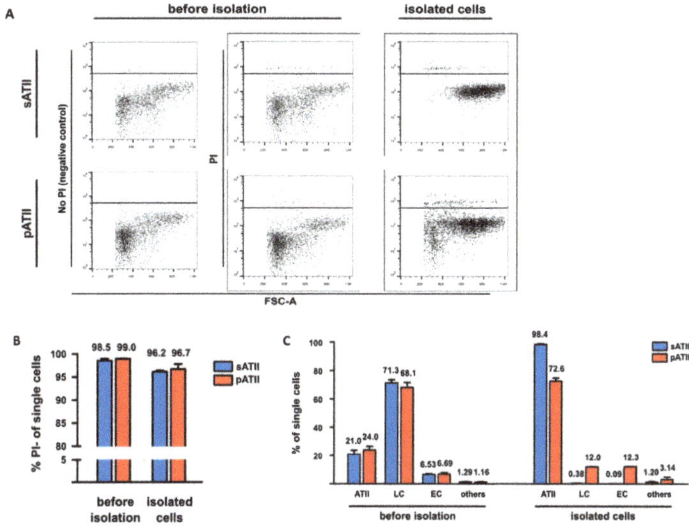

Figure 3. Viability, phenotypic markers and phenotypic expression of sATII and pATII cells before and after isolation. (**A**) Whole lung suspensions without PI staining were used as negative controls

(left panels). Viable cells were identified for sATII (upper row) and pATII (lower row) in the whole lung suspension (before isolation) and isolated cells were identified by PI exclusion (middle and right panels). (**B**) Flow cytometric analysis of the viability of sATII and pATII cell populations before and after isolation as determined by propidium iodide (PI) negativity. (**C**) Purity of sATII and pATII cell preparation before and after isolation. ATII cells were defined as CD45/31^{neg} CD74^{pos} cells, leukocytes as CD45/31-APC^{pos} cells without CD31-PE^{pos} cells and endothelial cells as CD31-PE$^{pos.}$ cells. Note that the CD31-PE antibody used recognizes a different epitope of CD31 than the CD31-APC antibody used for sorting. Each value is the mean of four independent experiments for sATII and two independent experiments for pATII. T-bars show the standard errors of the means (SEMs).

To assess the purity (phenotypic markers) and to determine the fractions of non ATII cells, such as endothelial cells and leukocytes, sATII and pATII isolated cells were stained with a PE-conjugated antibody recognizing a different CD31 epitope [66] to the CD31-APC antibody used for sorting (Figure 3C). ATII cells were defined as CD45^{neg}/CD31^{neg}/CD74^{pos} cells, and their phenotype was confirmed by qRT-PCR (*Sftpc*). We observed that the percentage of ATII cells in sATII and pATII cell populations increased from 21.0% and 24.0% before the isolation to 98.4% and 72.6% in the isolated cells, respectively. However, the percentage of leukocytes (defined as CD45$^{pos.}$ (APC) minus CD31$^{pos.}$ (PE) cells) and endothelial cells (CD31$^{pos.}$ (PE) cells) in the sATII population decreased from 71.3% and 6.53% before sorting to 0.38% and 0.09% after the cell sorting, respectively. By comparison, the percentage of leukocytes in the pATII cells decreased from 68.1% before isolation to 12.0% after isolation, while endothelial cells in pATII showed a relative increase from 6.69% before sorting to 12.3% in sorted cells (Figure 3B) in four independent experiments. Cells not expressing CD45, CD31 or CD74 were labeled as other.

3.4. MiRNA Expression Profiling of ATII Cells and Pathway Enrichment Analysis of Downstream mRNA Targets

ATII cells obtained by sorting (sATII, n = 2 biological replicates) and panning (pATII, n = 2) were used for miRNA profiling, which showed an expression of 293 miRNAs at detectable levels. Of these, 111 miRNAs were expressed at similar levels ($|FC| \leq 1.5x$) in both sATII and pATII preparations, and hence were termed ATII miRNAs (Figure 4) and further used for pathway enrichment analysis. To identify target mRNAs from these 111 miRNAs, we used Ingenuity$^{®}$'s [67] miR–target filter restricted to experimentally observed miR–target interactions. By this means, we identified 40 ATII miRNAs with 662 previously validated mRNA interactions in the cut set of 111 ATII miRNAs. Of note, 38 of these miRNAs were associated with 343 mRNAs present in the canonical pathway library of Ingenuity$^{®}$ (see Figure 4 for an overview of the workflow).

Significant enrichment (adj. *p*-value < 0.001) of 343 target mRNAs was observed in 143 signaling pathways and 2 metabolic pathways (nicotinate and nicotinamide metabolism, inositol phosphate metabolism). These pathways were assigned to 20 categories (Table 3). From the top 20 significant signaling pathways, 9 (marked with an asterisk*) have already been associated with fibrosis and/or EMT (e.g., PI3K/Akt, PTEN, IGF-1 and TGF-β) [68–75] (Figure 5). Another 9 of the top 20 pathways have been associated with cancer. Taken together, these results suggest an important role for ATII miRNAs in controlling cellular growth, proliferation and development.

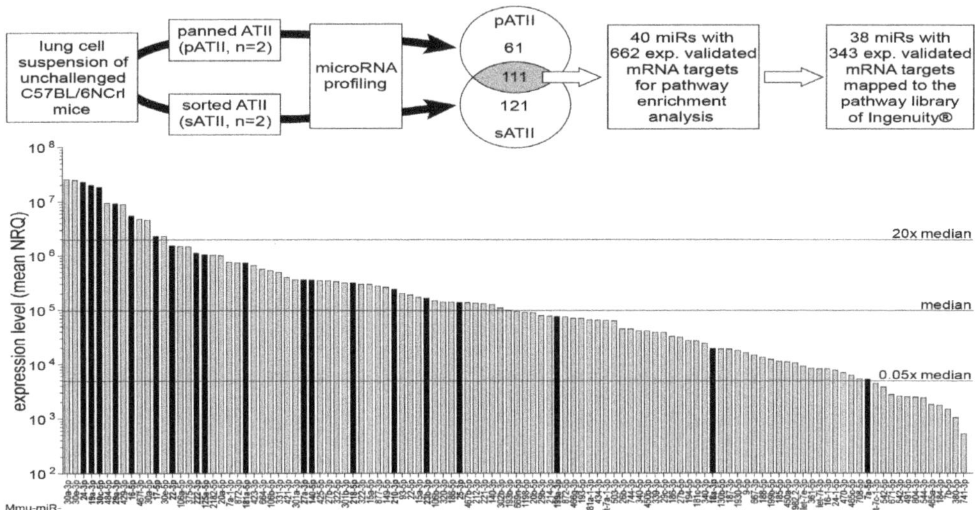

Figure 4. Workflow and miRNA expression profile of ATII cells. Lung single-cell suspensions were generated from unchallenged female C57BL/6NCrl mice. ATII cells were isolated either by negative selection (panning, pATII, n = 2) or by cell sorting (sATII, n = 2). Both cell preparations were subjected to miRNA profiling using TaqMan® array microfluidic cards (Life Technologies). In pATII and sATII, respectively, 61 and 121 miRNAs had fold differences larger than 1.5 (FC > 1.5). A cut set of 111 miRNAs with similar expression levels (| fold difference | ≤1.5) in pATII and sATII was identified and is represented as a bar diagram. For 40 of these bona fide ATII miRNAs, experimentally observed interactions with 662 mRNA targets were available in the Ingenuity® database and this information was used for pathway enrichment analysis, resulting in 38 miRNAs targeting 343 mRNAs in 145 pathways. Black bars indicate 19 miRNA targeting components of the TGF-β signaling pathway. The majority (16 out of 19) of these miRNAs were expressed above median level. A complete list of the miRNAs expressed in ATII cells is available as a spreadsheet file (see Table S2 supplementary material). n = 2 represents a biological replicate.

Table 3. Categories of pathways with significant ATII miR–target enrichment.

Pathway Category	Pathways per Category	Examples of Pathways within Category
Cancer	30	Small and non-small cell lung cancer, p53
Cellular growth, proliferation and development	28	PI3K/Akt, ILK, TGF-β, Integrin, FAK, mTOR
Cytokine signaling	27	Chemokine, IL-6, IL-8, IL-9, IL-10, IL-15, IL-17, IL-22, TNFR1
Cellular immune response	22	CXCR4, HMGB1, NF-κB, dendritic cell maturation
Growth factor signaling	21	IGF-1, EGF, GM-CSF, VEGF, FGF, PDGF
Apoptosis signaling	16	PTEN, death receptor, 14-3-3, JAK/Stat, tight junction signaling
Cell cycle regulation	13	G1/S checkpoint regulation, G2/M DNA damage checkpoint regulation
Intracellular and second messenger	13	Glucocorticoid receptor, ERK/MAPK, Rac, Rho, Gα12/13, PAK
Neurotransmitters and other nervous system signaling	13	Neuregulin, ErbB, Ephrin receptor, axonal guidance
Organismal growth and development	13	Stem cell pluripotency, HGF, BMP, Wnt/β-catenin
Disease-specific pathways	9	Hepatic fibrosis, rheumatoid arthritis, Huntington's disease
Cardiovascular signaling	7	Cardiac hypertrophy, atherosclerosis, thrombin signaling
Cellular stress and injury	6	HMGB1, HIF1α, p70S6K
Humoral immune response	5	CD40, IL-4, B cell receptor signaling
Nuclear receptor signaling	5	PPARα/RXRα activation, PPAR, RAR activation, VDR/RXR activation
Pathogen-influenced	3	LPS-stimulated MAPK signaling
Transcriptional regulation	2	Role of NANOG and Oct4 in mammalian embryonic stem cell pluripotency
Xenobiotic metabolism	1	Aryl hydrocarbon receptor signaling
Metabolism of cofactors and vitamins	1	Nicotinate and nicotinamide metabolism
Metabolism of complex lipids	1	Inositol phosphate metabolism

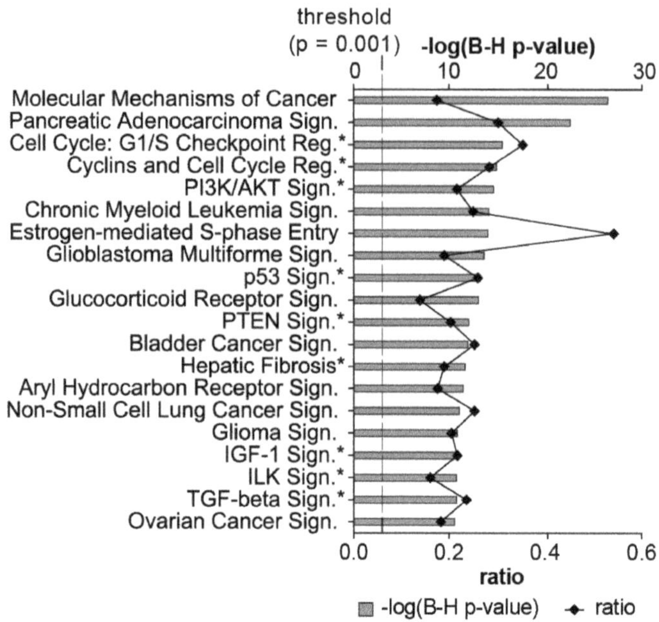

Figure 5. Top 20 enriched signaling pathways targeted by miRNAs expressed in ATII cells. The 111 miRNAs expressed at similar levels in pATII and sATII were used as inputs for the miRNA target filter module in Ingenuity®. The top 20 signaling pathways associated with the dataset are shown. The significance of this association is expressed by the probability (grey bars) that the association between the targets and the pathway is not due to chance (BH-adjusted *p*-value, Fisher's exact test). The degree of miRNA interaction within a certain pathway was calculated as the ratio of the number of targets that map to a given pathway to the total number of molecules within the pathway (black line). The dashed line indicates the significance threshold at *p* = 0.001. An asterisk highlights signaling pathways that have been associated previously with fibrosis and/or EMT.

3.5. Key Upstream Regulators of Target mRNAs

In the next step, we looked in silico for potential upstream regulators of all 662 target mRNAs, thereby identifying three miRNAs (16-5p, 30c-5p and 302d-3p) and two growth factors (TGFβ1 and EGF) as the top upstream regulators of our target mRNAs (see Table S1). Intriguingly, miR-16-5p and miR-30c-5p showed, also, very high expression levels (above 20× median) in the ATII expression profile (Figure 4, Table S2—Supplementary Material). TGF-β as well as EGF signaling pathways showed significant enrichment of ATII miRNA targets. Since deregulation of TGF-β signaling plays a crucial role in chronic lung diseases [76–78], we further focused on the investigation of ATII miRNAs in the TGF-β signaling pathway.

The canonical TGF-β signaling pathway of the Ingenuity® pathway library consists of 89 molecules. Nineteen ATII miRNAs (16 of which were expressed above median level; see Figure 4 and Table S2—Supplementary Material) were found to target 21 TGF-β signaling components located at several levels in the pathway, from ligands to transcription factors and target genes (Figure 6 and Table 4). Eleven molecules within the TGF-β pathway were targeted by up to four miRNAs, and ten miRNAs targeted more than one TGF-β signaling molecule (Table 4). Overall, these findings indicate a tight regulation of this pathway in ATII cells by miRNAs and the fact that they are expressed under physiological conditions indicates a role for these miRNAs in the homeostasis of the TGF-β pathway.

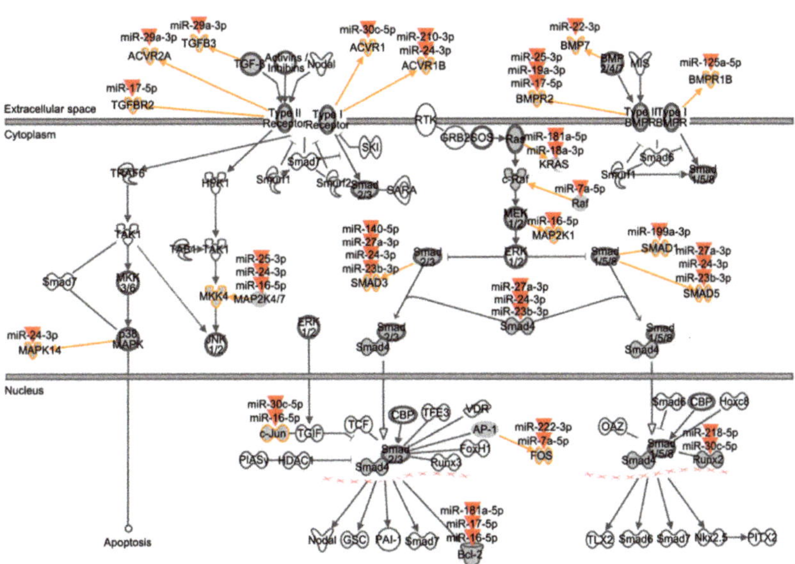

Figure 6. Mapping of ATII miRNAs to TGF-β signaling pathway components. This figure represents the canonical TGF-β signaling pathway from the Ingenuity® pathway library. Orange arrows and outlines indicate ATII miRNA targeted members within molecule families (dark grey icons). Red arrowheads symbolize ATII miRNAs. For example, TGFBR2 is a member of the type II TGF-β receptors, and its mRNA is targeted by miR-17-5p. Compare Table 4 for an overview of interactions.

Table 4. Mapping of ATII-expressed miRNAs to TGF-β pathway signaling components.

miRNA	miRBase MIMAT ID	Number of Targets	Pubmed ID for Exp. Obs. Interaction	mRNA Target	Transduction Level, Molecular Type
Mmu-miR-22-3p	0000531	1	19011694	*Bmp7*	Extracellular ligand, growth factor
Mmu-miR-29a-3p	0000535	2	19342382	*Tgfb3*	
Mmu-miR-30c-5p	0000514	3	18258830	*Acvr1*	
Mmu-miR-24-3p	0000219	6	17906079	*Acvr1b*	
Mmu-miR-210-3p	0000658	1	19520079	*Acvr1b*	
Mmu-miR-29a-3p	0000535	2	19342382	*Acvr2a*	
Mmu-miR-125a-5p	0000135	1	19738052	*Bmpr1b*	Plasma membrane receptor, kinase
Mmu-miR-19a-3p	0000651	1	19390056	*Bmpr2*	
Mmu-miR-25-3p	0000652	2	19390056	*Bmpr2*	
Mmu-miR-17-5p	0000649	3	19390056	*Bmpr2*	
Mmu-miR-17-5p	0000649	3	20709030	*Tgfbr2*	
Mmu-miR-18a-3p	0004626	1	19372139	*Kras*	Cytoplasmatic signaling, enzyme
Mmu-miR-181a-5p	0000210	2	20080834	*Kras*	
Mmu-miR-16-5p	0000527	4	20065103	*Map2k1*	
Mmu-miR-16-5p	0000527	4	19861690	*Map2k4*	Cytoplasmatic signaling, kinase
Mmu-miR-24-3p	0000219	6	19861690	*Map2k4*	
Mmu-miR-25-3p	0000652	2	19861690	*Map2k4*	
Mmu-miR-24-3p	0000219	6	19502786	*Mapk14*	
Mmu-miR-7a-5p	0000677	2	19072608	*Raf1*	
Mmu-miR-199a-3p	0000230	1	19251704	*Smad1*	
Mmu-miR-23b-3p	0000125	3	19582816	*Smad3*	
Mmu-miR-24-3p	0000219	6	19582816	*Smad3*	
Mmu-miR-27a-3p	0000537	3	19582816	*Smad3*	
Mmu-miR-140-5p	0000151	1	20071455	*Smad3*	
Mmu-miR-23b-3p	0000125	3	19582816	*Smad4*	
Mmu-miR-24-3p	0000219	6	19582816	*Smad4*	
Mmu-miR-27a-3p	0000537	3	19582816	*Smad4*	
Mmu-miR-23b-3p	0000125	3	19582816	*Smad5*	Transcription factor
Mmu-miR-24-3p	0000219	6	19582816	*Smad5*	
Mmu-miR-27a-3p	0000537	3	19582816	*Smad5*	

Table 4. *Cont.*

miRNA	miRBase MIMAT ID	Number of Targets	Pubmed ID for Exp. Obs. Interaction	mRNA Target	Transduction Level, Molecular Type
Mmu-miR-7a-5p	0000677	2	17028171	*Fos*	
Mmu-miR-222-3p	0000670	1	20299489	*Fos*	
Mmu-miR-16-5p	0000527	4	18362358	*Jun*	
Mmu-miR-30c-5p	0000514	3	18668040	*Jun*	
Mmu-miR-30c-5p	0000514	3	21628588	*Runx2*	
Mmu-miR-218-5p	0000663	1	21628588	*Runx2*	
Mmu-miR-16-5p	0000527	4	18449891	*Bcl2*	Transcription factor
Mmu-miR-17-5p	0000649	3	19666108	*Bcl2*	target,
Mmu-miR-181a-5p	0000210	2	20204284	*Bcl2*	transporter

3.6. mRNA Targets Located within the TGF-β Signaling Pathway

3.6.1. Effect of TGF-β1 Treatment on miRNA Expression in A549 Cells

Our in silico analyses indicated that highly expressed ATII miRNAs inhibit TGF-β signaling. Hence, we hypothesized that TGF-β stimulation might induce a down-regulation of these inhibitory miRNAs and lead to de-repression of this pathway. Since primary ATII cells gradually lose their phenotypes in in vitro culture, we studied the effects of TGF-β stimulation on these inhibitory miRNAs in a human alveolar epithelial A549 cell line.

Initially, we confirmed TGF-β pathway activation by showing down-regulation of E-Cadherin (*CDH1*) mRNA and up-regulation of the EMT markers vimentin (*VIM*), fibronectin (*FN1*) and snail family zinc finger 1 (*SNAI1*) in A549 cells upon TGF-β1 stimulation (Figure 7A).

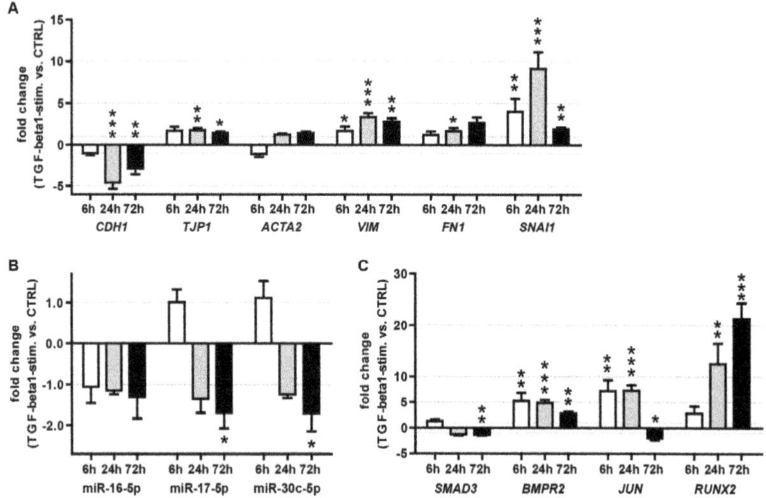

Figure 7. Effect of TGF-β1 treatment on the expression of EMT markers (**A**), miRNAs (**B**) and TGF-β pathway miRNA targets (**C**) in A549 cells. Shown are the mean fold changes (TGF-β1 vs. vehicle control) at 6, 24 and 72 h after stimulation with human recombinant TGF-β1. The arithmetic mean of *RNA18S5* and *HPRT1* mRNA expression served as a normalizer for mRNA quantitation. The small nuclear RNA *RNU6B* served as a reference gene for miRNA quantitation. The results were derived from three independent experiments, with each time point measured in triplicate. T-bars indicate maximum fold changes based on SEMs for target and reference gene expression. Unpaired *t*-test, vs. control treatment: *: $p < 0.05$, **: $p < 0.01$, ***: $p < 0.001$.

To focus on miRNAs with the highest potential relevance for TGF-β regulation, we selected miRNAs with high expression levels (above 20-fold of the median of all expressed miRNAs) (Figure 4) and with $\geq$3 targets in the TGF-β pathway (Figure 6, Table 4). This resulted in a selection of four miRNAs: miR-16-5p, -17-5p, -24-3p and -30c-5p.

MiRs 17-5p and 30c-5p showed a significant reduction at 72 h (~1.8-fold) after TGF-β1 stimulation compared to the vehicle control, while miR-16-5p and miR-24-3p remained unchanged (Figure 7B).

3.6.2. Effect of TGF-β1 Treatment on Target mRNA Expression in A549 Cells

Due to these results, we speculated that the putative miRNA-based repression of the TGF-β pathway under healthy conditions is released upon stimulation with TGF-β. Moreover, this mechanism, which sustains the homeostasis of TGF-β signaling, might be conserved in humans. Therefore, we also investigated the expression patterns of ATII miRNA targets within the canonical TGF-β pathway upon stimulation with TGF-β1 in A549 cells. While the expression of *BCL2, MAP2K4, TGFBR2* (data not shown) and *SMAD3* mRNAs was not, or was only mildly, affected by TGF-β1 stimulation. The expression of *BMPR2* (~5.5-fold, ~5-fold and ~2.5-fold for 6 h, 24 h and 72 h treatment, respectively), *JUN* (~15-fold 6 h and 24 h treatment) and *RUNX2* (~12-fold and ~22-fold for 24 h and 72 h treatment, respectively) mRNAs increased drastically within the investigated time frame (Figure 7C). This could indicate that de-repression by down-modulation of miRNAs might, at least in part, play a role in the activation of the TGF-β signaling pathway.

4. Discussion

Aberrant miRNA expression has been implicated in the pathogenesis of various ATII-associated diseases [15–18]. Nonetheless, until now, miRNA expression in healthy controls and respiratory diseases has been mainly studied in cell lines and whole lung samples. Few studies have analysed the expression of various miRNAs in primary ATII cells [19,79–81]; however, a complete miRNA expression profile of primary ATII cells has gone unaccounted for. We therefore aimed to provide intact ATII cells for miRNA profiling and to analyse the expression of miRNAs in primary "untouched by antibodies" ATII cells from healthy C57BL/6 mice (commonly used as a model animal for ATII-relevant diseases). We termed the cells "intact ATII cells" because these cells are "untouched by antibodies"; they were isolated by taking advantage of the autofluorescence of this cell type and by staining of the surface markers CD45 of leukocytes and CD31 of endothelial cells. The aim was to establish a preparation method that will provide intact cells for prospective miRNA profiling. Therefore, the isolated cells had to have three main properties: (1) "untouched by antibodies", (2) high viability and (3) high purity. Thereafter, we aimed to study the regulated pathways of the target mRNAs which could provide an insight into the role of miRNAs in healthy ATII cells.

Studies utilising freshly isolated primary ATII cells are necessary to understand molecular pathways regulating diverse functions of this cell type. However, this research remains highly elusive since the isolation of highly pure, viable and proliferative ATII cells for functional studies is fraught with challenges [82–85]. First, ATII cells in vitro undergo phenotypic change to resemble ATI cells [61,86,87]. Second, no cell line exists to complement these studies and represent the broad extent of known ATII properties. Therefore, an efficient method for ATII cell isolation is required. Many different isolation methods for ATII cells from mice have been described, including magnetic bead separation [58,88], panning [57,82] and cell-sorting [67,89,90]. However, it is always challenging to isolate highly pure ATII cells. First, extracellular ATII-specific markers for mice are rare and, second, positive selection of ATII cells using ATII-specific markers (CD74 and EpCAM^{high}/T1α^{neg}) [64,91] could affect certain cellular pathways and therefore change the activation status of purified cells [86,87,92]. It is assumed that EpCAM is involved in diverse intracellular processes, such as cell signaling, migration, differentiation and proliferation [92]. Monoclonal antibodies to EpCAM were described to induce antibody-dependent cellular cytotoxicity in colorectal cancer therapy [86]. The antibody to CD74, which was recently documented as an ATII-specific marker [64], stimulated the cleavage of the CD74 cystolic fragment, inducing NF-ƙB activation [87]. Therefore, a positive selection of ATII cells carries the risk of activating cellular pathways.

We therefore developed a new method for the isolation of ATII cells based on the autofluorescence characteristics of cell populations, thus allowing the isolation of "untouched by antibodies" ATII cells with high viability and purity. It was reported that ATII cells characterized as $CD45^{neg}/CD31^{neg}/Sca-1^{neg}/CCSP^{neg}$ showed high autofluorescence [62]. The presence of a $CD45^{pos.}$ population in our preparations with a slightly higher autofluorescence than that of ATII cells might have indicated macrophages [63]. Autofluorescence arises from endogenous fluorophores which are present in cells and extracellular matrix [89,90]. However, it is not very clear which endogenous fluorophore causes ATII cell autofluorescence. High metabolic activity due to surfactant production and consequently the presence of large amount of metabolic enzymes, such as NAD(P)H and flavins, might contribute to ATII cell autofluorescence [93]. Moreover, porphyrins, present in hemoglobin that has been found in primary ATII cells, exhibit natural autofluorescence [94,95]. However, no study to date has used autofluorescence for the isolation of ATII cells. We used autofluorescence and staining of surface markers of the other cell types to isolate "untouched by antibodies" highly pure sATII cells by FACS. The population of isolated and FACS-sorted ATII cells showed high purity in terms of CD45- and CD31-negativity and CD74-positivity. The few contaminating cells were within the $CD45/CD31^{neg}$ $CD74^{neg.}$ populations. The absence of CD31- and CD45-expressing cells and the expression of epithelial marker proteins (i.e., cytokeratin and E-cadherin) on nearly all sorted cells, as demonstrated by immunofluorescence, confirmed the ATII phenotype of the sorted cells. Expression of CCSP in a few sorted cells might have been due to the presence of Club cells, which have high autofluorescence and co-express proSP-C and CCSP after enzymatic dissociation and sorting [58,96]. Moreover, bronchioalveolar stem cells (BASCs), which develop into bronchiolar and alveolar epithelial cells, also co-express CCSP and proSP-C [19,96]. Nonetheless, they are unlikely to have contaminated the sorted population, as they exhibit low autofluorescence [96]. Another possibility is the presence of ATII progenitor cells that express CCSP and are highly autofluorescent due to their high metabolic activity [97,98]. The existence of endothelial cells post-pATII cell isolation protocol was most likely due to the fact that the "panning" protocol does not use antibodies to deplete endothelial cells.

The purity of the isolated sATII and pATII cells was also confirmed by the high expression and abundance of ATII epithelial and phenotypic markers. Isolated sATII and pATII cells expressed moderate levels of Aqp5, which was in accordance with other studies, showing that murine ATII cells express AQP5, unlike human and rat lung cells, where AQP5 is exclusively ATI-specific [99–103].

In order to dissect the functional role of miRNAs in ATII cells under normal, physiological conditions, miRNA profiling was performed. We used a cut set of miRNAs expressed at similar levels in ATII cells isolated by two different methods. This approach was used in order to identify miRNAs that are common to all ATII cells and not only restricted to distinct ATII cell subsets that are enriched by one of the isolation methods. Furthermore, this method reduces the activation of pathways which could be triggered during the isolation process and thus minimizes the changes in miRNA expression. Enrichment of target mRNAs with binding sites for cut set of ATII miRNAs in distinct pathways argues for biological relevance of a given pathway in these cells. This way, miRNAs can serve as a tool to detect or prioritize important pathways that might be overlooked in primarily mRNA-based identification strategies.

A cut set of 111 ATII miRNAs was used for pathway enrichment analysis in order to identify miRNA-regulated pathways involved in ATII cell homeostasis. Of 145 classified pathways with statistically significant target enrichment, only two pathways regulate metabolic processes. This could indicate that under normal physiological conditions, miRNAs in ATII cells may not play an important role in metabolic pathway regulation. However, since our approach was limited to miRNA–target interactions that were experimentally observed to date, it is possible that more miRNA targets and relevant pathways may be identified in ATII cells in the future.

The role of miRNAs was further confirmed by pathway analysis, which revealed that top network functions of the ATII miRNA target gene set were associated with pathways related to "cancer" and to "fibrosis and/or EMT". These findings are in agreement with those of a study by Fujino and colleagues, who showed that ATII cells isolated from human biopsy samples expressed genes enriched for positive regulation of cell differentiation and lung development [91]. Further, these results are in accordance with a recent study by Zacharias and colleagues [104]. Specific molecular pathways that have been already associated with these functions in ATII cells include TGF-β, Wnt/β-Catenin and growth factor signaling (e.g., EGF, HGF, KGF) [59,93,105,106]. In line with these studies, we have found that TGF-β and EGF are among the top five upstream regulators. Among miRNAs expressed above the median level, 16 miRNAs have their targets in the canonical TGF-β pathway, such that two of these miRNAs are within the top five upstream regulators. In this context, TGF-β is a potent EMT inducer that functions in cellular proliferation and differentiation, as well as in apoptosis, and therefore plays a crucial role in the regulation of epithelial homeostasis [107–109]. Similarly, the EGF protein family promotes EMT by stimulation of alveolar epithelial cell proliferation and migration [107,110]. Therefore, it has been suggested that there is cross-talk between these growth factors within the TGF-β pathway; however, the exact mechanism is unknown [105,106,111,112].

Several reports have shown that miRNAs are able to regulate these pathways. We found that miR-30a-3p/5p, miR-30c-5p and miR-30e-3p/5p were amongst those with the highest expression in ATII cells. Along this line, miR-30c-5p was among the top upstream regulators with three targets within the TGF-β pathway, while miRs 30a/e-3p were the most abundant. Down-regulation of miR-30 was observed in lung samples from IPF and NSCLC patients [13,79]. Transfection of hepatocyte cell line AML12 with these miRNAs resulted in decreased TGF-β1-induced EMT, while TGF-β1 treatment resulted in down-regulation of these miRNAs [113]. Moreover, Zhou and colleagues further showed that miR-30a down-regulates TGF-β1-induced EMT and peritoneal dialysis-related peritoneal fibrosis through down-regulation of snail1 [114]. Therefore, the high expression of three miR-30 family members in the present study could suggest that this family plays a crucial role in suppressing EMT in ATII cells under normal physiological conditions. Four members of the miR17~92 cluster (miR-19a, -17, -20a and 18a) revealed high to moderate expression in ATII cells in our study, with miR-17-5p having three targets within the canonical TGF-beta signaling pathway. The activation of this cluster in neuroblastoma cells was reported to regulate TGF-β signaling components [115]. In addition, it was reported that this cluster regulates cell proliferation and collagen synthesis by targeting the TGF-β pathway [116]. Further, the current study shows that the expression of miR-16-5p is among the top upstream regulators, with four targets within the TGF-β pathway, thus suggesting a crucial role for this miRNA in ATII homeostasis. It was shown that overexpression of miR-16 inhibited EMT-mediated factors Snail and Twist in vitro in a prostate cancer cell line [117]. It was reported that p53, a tumor suppressor, induces miR-16, whereas the down-regulation of p53 leads to EMT-related stem cell phenotypes [118,119], suggesting that miRNAs are regulators of the p53-controlled epithelial phenotype in ATII cells under normal physiological conditions. Hence, miRNAs are important in protection from fibrosis and cancer progression and in the maintenance of the ATII cell phenotype.

The miR-200 family member, miR-429, was also strongly expressed in our ATII cell preparation. ATII cells play an important role in the pathogenesis of IPF due to loss of their regenerative capacity [120,121]. Moreover, ATII cells isolated from IPF patients demonstrate impairment in their transdifferentiation into ATI cells [122], which triggers dysfunction in epithelial–mesenchymal transition (EMT) in the alveolar epithelium and leads to fibrosis [83,123]. miR-200 family members were reported to control these pathways, such that they were shown to be down-regulated in the lungs of IPF patients as well as in mice with experimental pulmonary fibrosis [15]. Interestingly, it was described that miR-200 family can restore normal regenerative function in exhausted senescent IPF pneumocytes by induction of transdifferentiation of primary human IPF ATII cells into ATI cells [81]. Moimas et al. [81] have demonstrated that upon transfection of IPF ATII cells with synthetic mimics of the entire miR-200 family, i.e., with miR-200b-3p and miR-200c-3p, they were able to restore the capability of exhausted senescent IPF ATII cells to transdifferentiate into ATI cells. It was shown that miR-429 reversed EMT in metastatic ovarian cancer cells [113] and was down-regulated in TGF-β1 treated MDCK cells [124]. The miR-200 family is a well-known inhibitor of TGF-β-induced EMT, and it is highly expressed in almost all epithelial cell types, except cells of mesenchymal origin [125–127]. Studies have shown that low expression of miR-200 family members is associated with poor prognosis in cancers, such as ovarian, gastric and thyroid cancers and many more [128–131]. Moreover, Tellez and colleagues further reported that miR-200 family members were repressed in immortalized human bronchial epithelial cells during EMT induced by tobacco carcinogens [132]. However, the overexpression of miR-200 family members in a lung adenocarcinoma mouse model restricted the cancer cells to an epithelial phenotype and stopped metastases [133]. In line with this, various functional studies showed that the down-regulation of miR-200 induced EMT, whereas its overexpression provoked mesenchymal-to-epithelial transition (MET) and inhibited cancer cell motility by repression of ZEB1 and ZEB2 [127,134–136]. Thus, both transcriptional factors could be involved in the TGF-β- pathway via a negative loop with miR-200. Therefore, the expression of miR-30- and miR-200 family members shows that ATII miRNAs play an important role in maintaining epithelial homeostasis. The finding that miRNAs targeting TGF-β signaling components are down-regulated by TGF-β is in accordance with our findings for down-modulation of miR-17-5p and -30c-5p in TGF-β1-treated A549 cells and supports the idea that under physiological conditions this pathway is at least partially controlled by miRNAs.

Recent studies have indicated that downstream molecules of the TGF-β signaling pathway interfere with miRNA expression either by regulating their transcriptional or post-transcriptional processing via interaction with components of the miRNA biogenesis machinery or by modulating epigenetic marks on miRNA promoters [137,138]. Additionally, miRNAs target components of the TGF-β signaling pathway, resulting in a complex network of signaling loops that contribute to the modulation of this pathway (reviewed in [139,140]). Since primary ATII cells gradually lose their phenotypes during in vitro culture, we decided to study the effects of TGF-β stimulation in the human alveolar epithelial cell line A549 that features hallmark characteristics of ATII cells [141].

MiR-16-5p and miR-30c-5p are top upstream regulators of the investigated set of target genes and have four (MAP2K1, MAP2K4, JUN and BCL2) and three (ACVR1, JUN and RUNX2) experimentally observed targets in the TGF-β signaling pathway, respectively. MiR-17-5p has three experimentally observed targets in the TGF-β signaling pathway (TGFBR2, BMPR2 and BCL2) and is a member of the miR-17~92 cluster, which has been associated with inhibition of TGF-β signaling. Most importantly, all of these ATII miRNAs have been associated with inhibition of TGF-β signaling [113,115,142–145]. Additionally, Corcoran et al. [146] reported that a set of miRNAs expressed in ATII cells is down-regulated in A549 cells upon TGF-β stimulation. These findings are in agreement with our findings regarding the down-regulation of miR-30c and miR-17-5p in A549 cells. Furthermore, it was shown that TGF-β-induced target gene expression is tightly controlled through down-regulation of miRNAs via TGFβ-induced transcription factors, such as AP-1, SMAD3/4 and NF-κB [146]. Of note, we found that TGF-β1 is a predicted upstream regulator for the set of validated mRNA targets interacting with the ATII-expressed miRNAs, of which 19 had experimentally observed targets in the canonical TGF-β pathway of the Ingenuity® database. This finding supports the idea of a complex interaction between TGF-β signaling and miR regulation in ATII cells already under normal conditions, hence underlining the importance of miRNAs for sustaining TGF-β pathway homeostasis. Additionally, Pandit et al. showed the down-regulation of 18 miRNAs—7 of which were also expressed in our ATII cells—in lung tissues of patients with idiopathic pulmonary fibrosis, indicating that disturbance of this mechanism might contribute to disease progression [79]. Finally, homeostatic down-modulation of TGF-β signaling by miRNAs expressed under normal conditions might contribute to the inhibition of ATII to ATI trans-differentiation [147].

In summary, these findings suggest that autofluorescence characteristics of murine lung cells can be exploited to isolate highly pure, untouched ATII cells and that miRNAs expressed in ATII cells contribute to cellular homeostasis by the modulation of proliferation and cell-activation pathways. Based on our data for miRNA expression in ATII cells, under normal conditions, and enrichment of miRNA targets in the TGF-β pathway, we hypothesize that miRNAs might represent valuable tools for the early detection of pathological conditions, such as fibrotic lung diseases and lung cancer.

Supplementary Materials: The following are available online at https://www.mdpi.com/article/10.3390/genes13081420/s1, Table S1: Upstream regulators. Table S2: ATII miR expression profile.

Author Contributions: S.D. participated in the study design, carried out the microRNA expression profiling, pathway analysis, performed the qPCR analysis and drafted the manuscript. K.J.W. established and carried out the flow cytometric isolation, immunophenotypic characterization of ATII cells and helped to draft the manuscript. N.E.-M. participated in critical data interpretation and drafting and writing of the manuscript. J.C. prepared the panned ATII cells. B.K. participated in the isolation and Papanicolaou-staining of sorted ATII cells. K.M. carried out the TGF-β stimulation of A549 cells. O.E. helped to draft the manuscript. M.K. participated in the design of the study. S.K.-E. conceived the manuscript and designed and supervised the study. All authors contributed to the writing of the manuscript. All authors have read and agreed to the published version of the manuscript.

Funding: This work is part of a PhD thesis by Katharina Singer and was supported in part by a grant of the FöFoLe-program of the Medical Faculty of the Ludwig Maximilian University of Munich.

Informed Consent Statement: Not applicable.

Acknowledgments: NEM and SKE are members of the Leibniz Science Campus "Evolutionary Medicine of the Lung".

Conflicts of Interest: The authors declare no conflict of interest.

References

1. Ambros, V. The functions of animal microRNAs. *Nature* **2004**, *431*, 350–355. [CrossRef]
2. Lewis, B.P.; Shih, I.H.; Jones-Rhoades, M.W.; Bartel, D.P.; Burge, C.B. Prediction of mammalian microRNA targets. *Cell* **2003**, *115*, 787–798. [CrossRef]
3. Berezikov, E. Evolution of microRNA diversity and regulation in animals. *Nat. Rev. Genet.* **2011**, *12*, 846–860. [CrossRef] [PubMed]
4. Sayed, D.; Abdellatif, M. MicroRNAs in development and disease. *Physiol. Rev.* **2011**, *91*, 827–887. [CrossRef] [PubMed]
5. Bartel, D.P. MicroRNAs: Genomics, biogenesis, mechanism, and function. *Cell* **2004**, *116*, 281–297. [CrossRef]
6. Kozomara, A.; Griffiths-Jones, S. miRBase: Integrating microRNA annotation and deep-sequencing data. *Nucleic Acids Res.* **2011**, *39*, D152–D157. [CrossRef]
7. Brown, D.; Rahman, M.; Nana-Sinkam, S.P. MicroRNAs in respiratory disease. A clinician's overview. *Ann. Am. Thorac. Soc.* **2014**, *11*, 1277–1285. [CrossRef]
8. Williams, A.E.; Moschos, S.A.; Perry, M.M.; Barnes, P.J.; Lindsay, M.A. Maternally imprinted microRNAs are differentially expressed during mouse and human lung development. *Dev. Dyn. Off. Publ. Am. Assoc. Anat.* **2007**, *236*, 572–580. [CrossRef]
9. Williams, A.E.; Perry, M.M.; Moschos, S.A.; Lindsay, M.A. microRNA expression in the aging mouse lung. *BMC Genom.* **2007**, *8*, 172. [CrossRef]
10. Wang, Y.; Frank, D.B.; Morley, M.P.; Zhou, S.; Wang, X.; Lu, M.M.; Lazar, M.A.; Morrisey, E.E. HDAC3-Dependent Epigenetic Pathway Controls Lung Alveolar Epithelial Cell Remodeling and Spreading via miR-17-92 and TGF-beta Signaling Regulation. *Dev. Cell* **2016**, *36*, 303–315. [CrossRef]
11. Mathis, C.; Poussin, C.; Weisensee, D.; Gebel, S.; Hengstermann, A.; Sewer, A.; Belcastro, V.; Xiang, Y.; Ansari, S.; Wagner, S.; et al. Human bronchial epithelial cells exposed in vitro to cigarette smoke at the air-liquid interface resemble bronchial epithelium from human smokers. *Am. J. Physiol. Lung Cell. Mol. Physiol.* **2013**, *304*, L489–L503. [CrossRef]
12. Huang, Y.; Crawford, M.; Higuita-Castro, N.; Nana-Sinkam, P.; Ghadiali, S.N. miR-146a regulates mechanotransduction and pressure-induced inflammation in small airway epithelium. *FASEB J. Off. Publ. Fed. Am. Soc. Exp. Biol.* **2012**, *26*, 3351–3364. [CrossRef]

13. Zhang, Y.; Huang, X.R.; Wei, L.H.; Chung, A.C.; Yu, C.M.; Lan, H.Y. miR-29b as a therapeutic agent for angiotensin II-induced cardiac fibrosis by targeting TGF-beta/Smad3 signaling. *Mol. Ther. J. Am. Soc. Gene Ther.* **2014**, *22*, 974–985. [CrossRef]

14. Wang, Y.; Liu, J.; Chen, J.; Feng, T.; Guo, Q. MiR-29 mediates TGFbeta 1-induced extracellular matrix synthesis through activation of Wnt/beta -catenin pathway in human pulmonary fibroblasts. *Technol. Health Care Off. J. Eur. Soc. Eng. Med.* **2015**, *23* (Suppl. 1), S119–S125.

15. Yang, S.; Banerjee, S.; de Freitas, A.; Sanders, Y.Y.; Ding, Q.; Matalon, S.; Thannickal, V.J.; Abraham, E.; Liu, G. Participation of miR-200 in pulmonary fibrosis. *Am. J. Pathol.* **2012**, *180*, 484–493. [CrossRef]

16. Xie, T.; Liang, J.; Geng, Y.; Liu, N.; Kurkciyan, A.; Kulur, V.; Leng, D.; Deng, N.; Liu, Z.; Song, J.; et al. MicroRNA-29c Prevents Pulmonary Fibrosis by Regulating Epithelial Cell Renewal and Apoptosis. *Am. J. Respir. Cell Mol. Biol.* **2017**, *57*, 721–732. [CrossRef]

17. Milosevic, J.; Pandit, K.; Magister, M.; Rabinovich, E.; Ellwanger, D.C.; Yu, G.; Vuga, L.J.; Weksler, B.; Benos, P.V.; Gibson, K.F.; et al. Profibrotic role of miR-154 in pulmonary fibrosis. *Am. J. Respir. Cell Mol. Biol.* **2012**, *47*, 879–887. [CrossRef]

18. Kho, A.T.; McGeachie, M.J.; Moore, K.G.; Sylvia, J.M.; Weiss, S.T.; Tantisira, K.G. Circulating microRNAs and prediction of asthma exacerbation in childhood asthma. *Respir. Res.* **2018**, *19*, 128. [CrossRef]

19. Ezzie, M.E.; Crawford, M.; Cho, J.H.; Orellana, R.; Zhang, S.; Gelinas, R.; Batte, K.; Yu, L.; Nuovo, G.; Galas, D.; et al. Gene expression networks in COPD: microRNA and mRNA regulation. *Thorax* **2012**, *67*, 122–131. [CrossRef]

20. Conickx, G.; Mestdagh, P.; Avila Cobos, F.; Verhamme, F.M.; Maes, T.; Vanaudenaerde, B.M.; Seys, L.J.; Lahousse, L.; Kim, R.Y.; Hsu, A.C.; et al. MicroRNA Profiling Reveals a Role for MicroRNA-218-5p in the Pathogenesis of Chronic Obstructive Pulmonary Disease. *Am. J. Respir. Crit. Care Med.* **2017**, *195*, 43–56. [CrossRef]

21. Quan, Y.; Wang, Z.; Gong, L.; Peng, X.; Richard, M.A.; Zhang, J.; Fornage, M.; Alcorn, J.L.; Wang, D. Exosome miR-371b-5p promotes proliferation of lung alveolar progenitor type II cells by using PTEN to orchestrate the PI3K/Akt signaling. *Stem Cell Res. Ther.* **2017**, *8*, 138. [CrossRef]

22. Mason, R.J. Biology of alveolar type II cells. *Respirology* **2006**, *11*, S12–S15. [CrossRef]

23. Weibel, E.R. On the tricks alveolar epithelial cells play to make a good lung. *Am. J. Respir. Crit. Care Med.* **2015**, *191*, 504–513. [CrossRef]

24. Kapanci, Y.; Weibel, E.R.; Kaplan, H.P.; Robinson, F.R. Pathogenesis and reversibility of the pulmonary lesions of oxygen toxicity in monkeys. II. Ultrastructural and morphometric studies. *Lab. Investig. A J. Tech. Methods Pathol.* **1969**, *20*, 101–118.

25. Hirai, K.I.; Witschi, H.; Cote, M.G. Electron microscopy of butylated hydroxytoluene-induced lung damage in mice. *Exp. Mol. Pathol.* **1977**, *27*, 295–308. [CrossRef]

26. Zemans, R.L.; Briones, N.; Campbell, M.; McClendon, J.; Young, S.K.; Suzuki, T.; Yang, I.V.; De Langhe, S.; Reynolds, S.D.; Mason, R.J.; et al. Neutrophil transmigration triggers repair of the lung epithelium via beta-catenin signaling. *Proc. Natl. Acad. Sci. USA* **2011**, *108*, 15990–15995. [CrossRef] [PubMed]

27. Tanjore, H.; Degryse, A.L.; Crossno, P.F.; Xu, X.C.; McConaha, M.E.; Jones, B.R.; Polosukhin, V.V.; Bryant, A.J.; Cheng, D.S.; Newcomb, D.C.; et al. beta-catenin in the alveolar epithelium protects from lung fibrosis after intratracheal bleomycin. *Am. J. Respir. Crit. Care Med.* **2013**, *187*, 630–639. [CrossRef]

28. Aso, Y.; Yoneda, K.; Kikkawa, Y. Morphologic and biochemical study of pulmonary changes induced by bleomycin in mice. *Lab. Investig. A J. Tech. Methods Pathol.* **1976**, *35*, 558–568.

29. Zhang, Y.; Goss, A.M.; Cohen, E.D.; Kadzik, R.; Lepore, J.J.; Muthukumaraswamy, K.; Yang, J.; DeMayo, F.J.; Whitsett, J.A.; Parmacek, M.S.; et al. A Gata6-Wnt pathway required for epithelial stem cell development and airway regeneration. *Nat. Genet.* **2008**, *40*, 862–870. [CrossRef]

30. Aoshiba, K.; Yokohori, N.; Nagai, A. Alveolar wall apoptosis causes lung destruction and emphysematous changes. *Am. J. Respir. Cell Mol. Biol.* **2003**, *28*, 555–562. [CrossRef]

31. Evans, M.J.; Cabral, L.J.; Stephens, R.J.; Freeman, G. Transformation of alveolar type 2 cells to type 1 cells following exposure to NO_2. *Exp. Mol. Pathol.* **1975**, *22*, 142–150. [CrossRef]

32. Bowden, D.H.; Davies, E.; Wyatt, J.P. Cytodynamics of pulmonary alveolar cells in the mouse. *Arch. Pathol.* **1968**, *86*, 667–670. [PubMed]

33. Kauffman, S.L. Cell proliferation in the mammalian lung. *Int. Rev. Exp. Pathol.* **1980**, *22*, 131–191.

34. Adamson, I.Y.; Bowden, D.H. The type 2 cell as progenitor of alveolar epithelial regeneration. A cytodynamic study in mice after exposure to oxygen. *Lab. Investig. A J. Tech. Methods Pathol.* **1974**, *30*, 35–42.

35. Tesfaigzi, J.; Wood, M.B.; Johnson, N.F.; Nikula, K.J. Apoptosis is a pathway responsible for the resolution of endotoxin-induced alveolar type II cell hyperplasia in the rat. *Int. J. Exp. Pathol.* **1998**, *79*, 303–311. [CrossRef]

36. Bardales, R.H.; Xie, S.S.; Schaefer, R.F.; Hsu, S.M. Apoptosis is a major pathway responsible for the resolution of type II pneumocytes in acute lung injury. *Am. J. Pathol.* **1996**, *149*, 845–852.

37. Motz, G.T.; Eppert, B.L.; Wesselkamper, S.C.; Flury, J.L.; Borchers, M.T. Chronic cigarette smoke exposure generates pathogenic T cells capable of driving COPD-like disease in $Rag2^{-/-}$ mice. *Am. J. Respir. Crit. Care Med.* **2010**, *181*, 1223–1233. [CrossRef]

38. Yokohori, N.; Aoshiba, K.; Nagai, A.; Respiratory Failure Research Group in, J. Increased levels of cell death and proliferation in alveolar wall cells in patients with pulmonary emphysema. *Chest* **2004**, *125*, 626–632. [CrossRef]

39. Tsuji, T.; Aoshiba, K.; Nagai, A. Alveolar cell senescence in patients with pulmonary emphysema. *Am. J. Respir. Crit. Care Med.* **2006**, *174*, 886–893. [CrossRef]

40. Bhandary, Y.P.; Shetty, S.K.; Marudamuthu, A.S.; Gyetko, M.R.; Idell, S.; Gharaee-Kermani, M.; Shetty, R.S.; Starcher, B.C.; Shetty, S. Regulation of alveolar epithelial cell apoptosis and pulmonary fibrosis by coordinate expression of components of the fibrinolytic system. *Am. J. Physiol. Lung Cell. Mol. Physiol.* **2012**, *302*, L463–L473. [CrossRef]

41. Crapo, J.D.; Barry, B.E.; Gehr, P.; Bachofen, M.; Weibel, E.R. Cell number and cell characteristics of the normal human lung. *Am. Rev. Respir. Dis.* **1982**, *126*, 332–337.

42. Ridsdale, R.; Post, M. Surfactant lipid synthesis and lamellar body formation in glycogen-laden type II cells. *Am. J. Physiol. Lung Cell. Mol. Physiol.* **2004**, *287*, L743–L751. [CrossRef]

43. Evans, M.J.; Hackney, J.D. Cell proliferation in lungs of mice exposed to elevated concentrations of oxygen. *Aerosp. Med.* **1972**, *43*, 620–622.

44. Liu, Y.; Sadikot, R.T.; Adami, G.R.; Kalinichenko, V.V.; Pendyala, S.; Natarajan, V.; Zhao, Y.Y.; Malik, A.B. FoxM1 mediates the progenitor function of type II epithelial cells in repairing alveolar injury induced by Pseudomonas aeruginosa. *J. Exp. Med.* **2011**, *208*, 1473–1484. [CrossRef]

45. Jansing, N.L.; McClendon, J.; Henson, P.M.; Tuder, R.M.; Hyde, D.M.; Zemans, R.L. Unbiased Quantitation of Alveolar Type II to Alveolar Type I Cell Transdifferentiation during Repair after Lung Injury in Mice. *Am. J. Respir. Cell Mol. Biol.* **2017**, *57*, 519–526. [CrossRef]

46. Andreeva, A.V.; Kutuzov, M.A.; Voyno-Yasenetskaya, T.A. Regulation of surfactant secretion in alveolar type II cells. *Am. J. Physiol. Lung Cell. Mol. Physiol.* **2007**, *293*, L259–L271. [CrossRef] [PubMed]

47. Phelps, D.S.; Floros, J. Localization of pulmonary surfactant proteins using immunohistochemistry and tissue in situ hybridization. *Exp. Lung Res.* **1991**, *17*, 985–995. [CrossRef] [PubMed]

48. Ariki, S.; Kojima, T.; Gasa, S.; Saito, A.; Nishitani, C.; Takahashi, M.; Shimizu, T.; Kurimura, Y.; Sawada, N.; Fujii, N.; et al. Pulmonary collectins play distinct roles in host defense against Mycobacterium avium. *J. Immunol.* **2011**, *187*, 2586–2594. [CrossRef]

49. Carreto-Binaghi, L.E.; Aliouat, M.E.; Taylor, M.L. Surfactant proteins, SP-A and SP-D, in respiratory fungal infections: Their role in the inflammatory response. *Respir. Res.* **2016**, *17*, 66. [CrossRef]

50. Ikegami, M.; Scoville, E.A.; Grant, S.; Korfhagen, T.; Brondyk, W.; Scheule, R.K.; Whitsett, J.A. Surfactant protein-D and surfactant inhibit endotoxin-induced pulmonary inflammation. *Chest* **2007**, *132*, 1447–1454. [CrossRef]

51. Nogee, L.M.; Garnier, G.; Dietz, H.C.; Singer, L.; Murphy, A.M.; deMello, D.E.; Colten, H.R. A mutation in the surfactant protein B gene responsible for fatal neonatal respiratory disease in multiple kindreds. *J. Clin. Investig.* **1994**, *93*, 1860–1863. [CrossRef]

52. Nogee, L.M.; de Mello, D.E.; Dehner, L.P.; Colten, H.R. Brief report: Deficiency of pulmonary surfactant protein B in congenital alveolar proteinosis. *N. Engl. J. Med.* **1993**, *328*, 406–410. [CrossRef]

53. Glasser, S.W.; Detmer, E.A.; Ikegami, M.; Na, C.L.; Stahlman, M.T.; Whitsett, J.A. Pneumonitis and emphysema in sp-C gene targeted mice. *J. Biol. Chem.* **2003**, *278*, 14291–14298. [CrossRef]

54. Glasser, S.W.; Witt, T.L.; Senft, A.P.; Baatz, J.E.; Folger, D.; Maxfield, M.D.; Akinbi, H.T.; Newton, D.A.; Prows, D.R.; Korfhagen, T.R. Surfactant protein C-deficient mice are susceptible to respiratory syncytial virus infection. *Am. J. Physiol. Lung Cell. Mol. Physiol.* **2009**, *297*, L64–L72. [CrossRef] [PubMed]

55. Glasser, S.W.; Senft, A.P.; Maxfield, M.D.; Ruetschilling, T.L.; Baatz, J.E.; Page, K.; Korfhagen, T.R. Genetic replacement of surfactant protein-C reduces respiratory syncytial virus induced lung injury. *Respir. Res.* **2013**, *14*, 19. [CrossRef] [PubMed]

56. Maitra, M.; Cano, C.A.; Garcia, C.K. Mutant surfactant A2 proteins associated with familial pulmonary fibrosis and lung cancer induce TGF-beta1 secretion. *Proc. Natl. Acad. Sci. USA* **2012**, *109*, 21064–21069. [CrossRef]

57. Konigshoff, M.; Kramer, M.; Balsara, N.; Wilhelm, J.; Amarie, O.V.; Jahn, A.; Rose, F.; Fink, L.; Seeger, W.; Schaefer, L.; et al. WNT1-inducible signaling protein-1 mediates pulmonary fibrosis in mice and is upregulated in humans with idiopathic pulmonary fibrosis. *J. Clin. Investig.* **2009**, *119*, 772–787. [CrossRef]

58. Corti, M.; Brody, A.R.; Harrison, J.H. Isolation and primary culture of murine alveolar type II cells. *Am. J. Respir. Cell Mol. Biol.* **1996**, *14*, 309–315. [CrossRef]

59. Dobbs, L.G. Isolation and culture of alveolar type II cells. *Am. J. Physiol.* **1990**, *258 Pt 1*, L134–L147. [CrossRef]

60. Pfaffl, M.W. A new mathematical model for relative quantification in real-time RT-PCR. *Nucleic Acids Res* **2001**, *29*, e45. [CrossRef]

61. Iglewicz, B.; Hoaglin, D. How to detect and handle outliers. In *The ASQC Basic References in Quality Control: Statistical Techniques*; Mykytka, E.F., Ed.; Asq Press: Milwaukee, WI, USA, 1993; Volume 16.

62. Kim, C.F.; Jackson, E.L.; Woolfenden, A.E.; Lawrence, S.; Babar, I.; Vogel, S.; Crowley, D.; Bronson, R.T.; Jacks, T. Identification of bronchioalveolar stem cells in normal lung and lung cancer. *Cell* **2005**, *121*, 823–835. [CrossRef]

63. Cunningham, A.C.; Milne, D.S.; Wilkes, J.; Dark, J.H.; Tetley, T.D.; Kirby, J.A. Constitutive expression of MHC and adhesion molecules by alveolar epithelial cells (type II pneumocytes) isolated from human lung and comparison with immunocytochemical findings. *J. Cell Sci.* **1994**, *107 Pt 2*, 443–449. [CrossRef]

64. Marsh, L.M.; Cakarova, L.; Kwapiszewska, G.; von Wulffen, W.; Herold, S.; Seeger, W.; Lohmeyer, J. Surface expression of CD74 by type II alveolar epithelial cells: A potential mechanism for macrophage migration inhibitory factor-induced epithelial repair. *Am. J. Physiol. Lung Cell. Mol. Physiol.* **2009**, *296*, L442–L452. [CrossRef]

65. Gereke, M.; Jung, S.; Buer, J.; Bruder, D. Alveolar type II epithelial cells present antigen to CD4(+) T cells and induce Foxp3(+) regulatory T cells. *Am. J. Respir. Crit. Care Med.* **2009**, *179*, 344–355. [CrossRef]

66. Chacko, A.M.; Nayak, M.; Greineder, C.F.; Delisser, H.M.; Muzykantov, V.R. Collaborative enhancement of antibody binding to distinct PECAM-1 epitopes modulates endothelial targeting. *PLoS ONE* **2012**, *7*, e34958. [CrossRef]
67. Ingenuity® Systems, Summer Release 2012. Available online: www.ingenuity.com (accessed on 11 June 2022).
68. Choi, J.E.; Lee, S.S.; Sunde, D.A.; Huizar, I.; Haugk, K.L.; Thannickal, V.J.; Vittal, R.; Plymate, S.R.; Schnapp, L.M. Insulin-like growth factor-I receptor blockade improves outcome in mouse model of lung injury. *Am. J. Respir. Crit. Care Med.* **2009**, *179*, 212–219. [CrossRef]
69. Kavvadas, P.; Kypreou, K.P.; Protopapadakis, E.; Prodromidi, E.; Sideras, P.; Charonis, A.S. Integrin-linked kinase (ILK) in pulmonary fibrosis. *Virchows Arch. Int. J. Pathol.* **2010**, *457*, 563–575. [CrossRef] [PubMed]
70. Le Cras, T.D.; Korfhagen, T.R.; Davidson, C.; Schmidt, S.; Fenchel, M.; Ikegami, M.; Whitsett, J.A.; Hardie, W.D. Inhibition of PI3K by PX-866 prevents transforming growth factor-alpha-induced pulmonary fibrosis. *Am. J. Pathol.* **2010**, *176*, 679–686. [CrossRef]
71. Nakashima, N.; Kuwano, K.; Maeyama, T.; Hagimoto, N.; Yoshimi, M.; Hamada, N.; Yamada, M.; Nakanishi, Y. The p53-Mdm2 association in epithelial cells in idiopathic pulmonary fibrosis and non-specific interstitial pneumonia. *J. Clin. Pathol.* **2005**, *58*, 583–589. [CrossRef]
72. Son, G.; Hines, I.N.; Lindquist, J.; Schrum, L.W.; Rippe, R.A. Inhibition of phosphatidylinositol 3-kinase signaling in hepatic stellate cells blocks the progression of hepatic fibrosis. *Hepatology* **2009**, *50*, 1512–1523. [CrossRef]
73. Xia, H.; Khalil, W.; Kahm, J.; Jessurun, J.; Kleidon, J.; Henke, C.A. Pathologic caveolin-1 regulation of PTEN in idiopathic pulmonary fibrosis. *Am. J. Pathol.* **2010**, *176*, 2626–2637. [CrossRef] [PubMed]
74. Watts, K.L.; Cottrell, E.; Hoban, P.R.; Spiteri, M.A. RhoA signaling modulates cyclin D1 expression in human lung fibroblasts; implications for idiopathic pulmonary fibrosis. *Respir. Res.* **2006**, *7*, 88. [CrossRef] [PubMed]
75. Willis, B.C.; Liebler, J.M.; Luby-Phelps, K.; Nicholson, A.G.; Crandall, E.D.; du Bois, R.M.; Borok, Z. Induction of epithelial-mesenchymal transition in alveolar epithelial cells by transforming growth factor-beta1: Potential role in idiopathic pulmonary fibrosis. *Am. J. Pathol.* **2005**, *166*, 1321–1332. [CrossRef]
76. Willis, B.C.; Borok, Z. TGF-beta-induced EMT: Mechanisms and implications for fibrotic lung disease. *Am. J. Physiol. Lung Cell. Mol. Physiol.* **2007**, *293*, L525–L534. [CrossRef]
77. Jones, C.P.; Gregory, L.G.; Causton, B.; Campbell, G.A.; Lloyd, C.M. Activin A and TGF-beta promote T(H)9 cell-mediated pulmonary allergic pathology. *J. Allergy Clin. Immunol.* **2012**, *129*, 1000–1010.e1003. [CrossRef]
78. Konigshoff, M.; Kneidinger, N.; Eickelberg, O. TGF-beta signaling in COPD: Deciphering genetic and cellular susceptibilities for future therapeutic regimen. *Swiss Med. Wkly.* **2009**, *139*, 554–563.
79. Pandit, K.V.; Corcoran, D.; Yousef, H.; Yarlagadda, M.; Tzouvelekis, A.; Gibson, K.F.; Konishi, K.; Yousem, S.A.; Singh, M.; Handley, D.; et al. Inhibition and role of let-7d in idiopathic pulmonary fibrosis. *Am. J. Respir. Crit. Care Med.* **2010**, *182*, 220–229. [CrossRef]
80. Yamada, M.; Kubo, H.; Ota, C.; Takahashi, T.; Tando, Y.; Suzuki, T.; Fujino, N.; Makiguchi, T.; Takagi, K.; Suzuki, T.; et al. The increase of microRNA-21 during lung fibrosis and its contribution to epithelial-mesenchymal transition in pulmonary epithelial cells. *Respir. Res.* **2013**, *14*, 95. [CrossRef]
81. Moimas, S.; Salton, F.; Kosmider, B.; Ring, N.; Volpe, M.C.; Bahmed, K.; Braga, L.; Rehman, M.; Vodret, S.; Graziani, M.L.; et al. miR-200 family members reduce senescence and restore idiopathic pulmonary fibrosis type II alveolar epithelial cell transdifferentiation. *ERJ Open Res.* **2019**, *5*, 00138-2019. [CrossRef]
82. Rice, W.R.; Conkright, J.J.; Na, C.L.; Ikegami, M.; Shannon, J.M.; Weaver, T.E. Maintenance of the mouse type II cell phenotype in vitro. *Am. J. Physiol. Lung Cell. Mol. Physiol.* **2002**, *283*, L256–L264. [CrossRef]
83. Kim, K.K.; Wei, Y.; Szekeres, C.; Kugler, M.C.; Wolters, P.J.; Hill, M.L.; Frank, J.A.; Brumwell, A.N.; Wheeler, S.E.; Kreidberg, J.A.; et al. Epithelial cell alpha3beta1 integrin links beta-catenin and Smad signaling to promote myofibroblast formation and pulmonary fibrosis. *J. Clin. Investig.* **2009**, *119*, 213–224. [PubMed]
84. Rock, J.R.; Barkauskas, C.E.; Cronce, M.J.; Xue, Y.; Harris, J.R.; Liang, J.; Noble, P.W.; Hogan, B.L. Multiple stromal populations contribute to pulmonary fibrosis without evidence for epithelial to mesenchymal transition. *Proc. Natl. Acad. Sci. USA* **2011**, *108*, E1475–E1483. [CrossRef] [PubMed]
85. Barkauskas, C.E.; Cronce, M.J.; Rackley, C.R.; Bowie, E.J.; Keene, D.R.; Stripp, B.R.; Randell, S.H.; Noble, P.W.; Hogan, B.L. Type 2 alveolar cells are stem cells in adult lung. *J. Clin. Investig.* **2013**, *123*, 3025–3036. [CrossRef] [PubMed]
86. Flieger, D.; Hoff, A.S.; Sauerbruch, T.; Schmidt-Wolf, I.G. Influence of cytokines, monoclonal antibodies and chemotherapeutic drugs on epithelial cell adhesion molecule (EpCAM) and LewisY antigen expression. *Clin. Exp. Immunol.* **2001**, *123*, 9–14. [CrossRef]
87. Starlets, D.; Gore, Y.; Binsky, I.; Haran, M.; Harpaz, N.; Shvidel, L.; Becker-Herman, S.; Berrebi, A.; Shachar, I. Cell-surface CD74 initiates a signaling cascade leading to cell proliferation and survival. *Blood* **2006**, *107*, 4807–4816. [CrossRef]
88. Messier, E.M.; Mason, R.J.; Kosmider, B. Efficient and rapid isolation and purification of mouse alveolar type II epithelial cells. *Exp. Lung Res.* **2012**, *38*, 363–373. [CrossRef]
89. Monici, M. Cell and tissue autofluorescence research and diagnostic applications. *Biotechnol. Annu. Rev.* **2005**, *11*, 227–256.
90. Heikal, A.A. Intracellular coenzymes as natural biomarkers for metabolic activities and mitochondrial anomalies. *Biomark. Med.* **2010**, *4*, 241–263. [CrossRef]

91. Fujino, N.; Kubo, H.; Ota, C.; Suzuki, T.; Suzuki, S.; Yamada, M.; Takahashi, T.; He, M.; Suzuki, T.; Kondo, T.; et al. A novel method for isolating individual cellular components from the adult human distal lung. *Am. J. Respir. Cell Mol. Biol.* **2012**, *46*, 422–430. [CrossRef]

92. Trzpis, M.; McLaughlin, P.M.; de Leij, L.M.; Harmsen, M.C. Epithelial cell adhesion molecule: More than a carcinoma marker and adhesion molecule. *Am. J. Pathol.* **2007**, *171*, 386–395. [CrossRef]

93. Fehrenbach, H. Alveolar epithelial type II cell: Defender of the alveolus revisited. *Respir. Res.* **2001**, *2*, 33–46. [CrossRef]

94. Bhaskaran, M.; Chen, H.; Chen, Z.; Liu, L. Hemoglobin is expressed in alveolar epithelial type II cells. *Biochem. Biophys. Res. Commun.* **2005**, *333*, 1348–1352. [CrossRef]

95. Grek, C.L.; Newton, D.A.; Spyropoulos, D.D.; Baatz, J.E. Hypoxia up-regulates expression of hemoglobin in alveolar epithelial cells. *Am. J. Respir. Cell Mol. Biol.* **2011**, *44*, 439–447. [CrossRef]

96. Teisanu, R.M.; Lagasse, E.; Whitesides, J.F.; Stripp, B.R. Prospective isolation of bronchiolar stem cells based upon immunophenotypic and autofluorescence characteristics. *Stem Cells* **2009**, *27*, 612–622. [CrossRef]

97. Ghaedi, M.; Le, A.V.; Hatachi, G.; Beloiartsev, A.; Rocco, K.; Sivarapatna, A.; Mendez, J.J.; Baevova, P.; Dyal, R.N.; Leiby, K.L.; et al. Bioengineered lungs generated from human iPSCs-derived epithelial cells on native extracellular matrix. *J. Tissue Eng. Regen. Med.* **2018**, *12*, e1623–e1635. [CrossRef]

98. Ghaedi, M.; Calle, E.A.; Mendez, J.J.; Gard, A.L.; Balestrini, J.; Booth, A.; Bove, P.F.; Gui, L.; White, E.S.; Niklason, L.E. Human iPS cell-derived alveolar epithelium repopulates lung extracellular matrix. *J. Clin. Investig.* **2013**, *123*, 4950–4962. [CrossRef]

99. Matsuzaki, T.; Hata, H.; Ozawa, H.; Takata, K. Immunohistochemical localization of the aquaporins AQP1, AQP3, AQP4, and AQP5 in the mouse respiratory system. *Acta Histochem. Cytochem.* **2009**, *42*, 159–169. [CrossRef]

100. Krane, C.M.; Fortner, C.N.; Hand, A.R.; McGraw, D.W.; Lorenz, J.N.; Wert, S.E.; Towne, J.E.; Paul, R.J.; Whitsett, J.A.; Menon, A.G. Aquaporin 5-deficient mouse lungs are hyperresponsive to cholinergic stimulation. *Proc. Natl. Acad. Sci. USA* **2001**, *98*, 14114–14119. [CrossRef]

101. Kreda, S.M.; Gynn, M.C.; Fenstermacher, D.A.; Boucher, R.C.; Gabriel, S.E. Expression and localization of epithelial aquaporins in the adult human lung. *Am. J. Respir. Cell Mol. Biol.* **2001**, *24*, 224–234. [CrossRef]

102. Flodby, P.; Borok, Z.; Banfalvi, A.; Zhou, B.; Gao, D.; Minoo, P.; Ann, D.K.; Morrisey, E.E.; Crandall, E.D. Directed expression of Cre in alveolar epithelial type 1 cells. *Am. J. Respir. Cell Mol. Biol.* **2010**, *43*, 173–178. [CrossRef]

103. King, L.S.; Nielsen, S.; Agre, P. Aquaporins in complex tissues. I. Developmental patterns in respiratory and glandular tissues of rat. *Am. J. Physiol.* **1997**, *273*, C1541–C1548. [CrossRef]

104. Zacharias, W.J.; Frank, D.B.; Zepp, J.A.; Morley, M.P.; Alkhaleel, F.A.; Kong, J.; Zhou, S.; Cantu, E.; Morrisey, E.E. Regeneration of the lung alveolus by an evolutionarily conserved epithelial progenitor. *Nature* **2018**, *555*, 251–255. [CrossRef]

105. Deharvengt, S.; Marmarelis, M.; Korc, M. Concomitant targeting of EGF receptor, TGF-beta and SRC points to a novel therapeutic approach in pancreatic cancer. *PLoS ONE* **2012**, *7*, e39684. [CrossRef]

106. Tian, Y.C.; Chen, Y.C.; Chang, C.T.; Hung, C.C.; Wu, M.S.; Phillips, A.; Yang, C.W. Epidermal growth factor and transforming growth factor-beta1 enhance HK-2 cell migration through a synergistic increase of matrix metalloproteinase and sustained activation of ERK signaling pathway. *Exp. Cell Res.* **2007**, *313*, 2367–2377. [CrossRef]

107. Crosby, L.M.; Waters, C.M. Epithelial repair mechanisms in the lung. *Am. J. Physiol. Lung Cell. Mol. Physiol.* **2010**, *298*, L715–L731. [CrossRef]

108. Yu, H.; Konigshoff, M.; Jayachandran, A.; Handley, D.; Seeger, W.; Kaminski, N.; Eickelberg, O. Transgelin is a direct target of TGF-beta/Smad3-dependent epithelial cell migration in lung fibrosis. *FASEB J. Off. Publ. Fed. Am. Soc. Exp. Biol.* **2008**, *22*, 1778–1789.

109. Annes, J.P.; Munger, J.S.; Rifkin, D.B. Making sense of latent TGFbeta activation. *J. Cell Sci.* **2003**, *116 Pt 2*, 217–224. [CrossRef]

110. Alipio, Z.A.; Jones, N.; Liao, W.; Yang, J.; Kulkarni, S.; Sree Kumar, K.; Hauer-Jensen, M.; Ward, D.C.; Ma, Y.; Fink, L.M. Epithelial to mesenchymal transition (EMT) induced by bleomycin or TFG(b1)/EGF in murine induced pluripotent stem cell-derived alveolar Type II-like cells. *Differ. Res. Biol. Divers.* **2011**, *82*, 89–98.

111. Ouyang, H.; Gore, J.; Deitz, S.; Korc, M. microRNA-10b enhances pancreatic cancer cell invasion by suppressing TIP30 expression and promoting EGF and TGF-beta actions. *Oncogene* **2017**, *36*, 4952. [CrossRef]

112. Han, M.; Wang, F.; Gu, Y.; Pei, X.; Guo, G.; Yu, C.; Li, L.; Zhu, M.; Xiong, Y.; Wang, Y. MicroRNA-21 induces breast cancer cell invasion and migration by suppressing smad7 via EGF and TGF-beta pathways. *Oncol. Rep.* **2016**, *35*, 73–80. [CrossRef]

113. Zhang, J.; Zhang, H.; Liu, J.; Tu, X.; Zang, Y.; Zhu, J.; Chen, J.; Dong, L.; Zhang, J. miR-30 inhibits TGF-beta1-induced epithelial-to-mesenchymal transition in hepatocyte by targeting Snail1. *Biochem. Biophys. Res. Commun.* **2012**, *417*, 1100–1105. [CrossRef] [PubMed]

114. Zhou, Q.; Yang, M.; Lan, H.; Yu, X. miR-30a negatively regulates TGF-beta1-induced epithelial-mesenchymal transition and peritoneal fibrosis by targeting Snai1. *Am. J. Pathol.* **2013**, *183*, 808–819. [CrossRef] [PubMed]

115. Mestdagh, P.; Bostrom, A.K.; Impens, F.; Fredlund, E.; Van Peer, G.; De Antonellis, P.; von Stedingk, K.; Ghesquiere, B.; Schulte, S.; Dews, M.; et al. The miR-17-92 microRNA cluster regulates multiple components of the TGF-beta pathway in neuroblastoma. *Mol. Cell* **2010**, *40*, 762–773. [CrossRef] [PubMed]

116. Li, L.; Shi, J.Y.; Zhu, G.Q.; Shi, B. MiR-17-92 cluster regulates cell proliferation and collagen synthesis by targeting TGFB pathway in mouse palatal mesenchymal cells. *J. Cell. Biochem.* **2012**, *113*, 1235–1244. [CrossRef] [PubMed]

117. Jin, W.; Chen, F.; Wang, K.; Song, Y.; Fei, X.; Wu, B. miR-15a/miR-16 cluster inhibits invasion of prostate cancer cells by suppressing TGF-beta signaling pathway. *Biomed. Pharmacother.* **2018**, *104*, 637–644. [CrossRef] [PubMed]

118. Shi, L.; Jackstadt, R.; Siemens, H.; Li, H.; Kirchner, T.; Hermeking, H. p53-induced miR-15a/16-1 and AP4 form a double-negative feedback loop to regulate epithelial-mesenchymal transition and metastasis in colorectal cancer. *Cancer Res.* **2014**, *74*, 532–542. [CrossRef] [PubMed]

119. De Craene, B.; Berx, G. Regulatory networks defining EMT during cancer initiation and progression. *Nat. Rev. Cancer* **2013**, *13*, 97–110. [CrossRef]

120. Naikawadi, R.P.; Disayabutr, S.; Mallavia, B.; Donne, M.L.; Green, G.; La, J.L.; Rock, J.R.; Looney, M.R.; Wolters, P.J. Telomere dysfunction in alveolar epithelial cells causes lung remodeling and fibrosis. *JCI Insight* **2016**, *1*, e86704. [CrossRef]

121. Selman, M.; Lopez-Otin, C.; Pardo, A. Age-driven developmental drift in the pathogenesis of idiopathic pulmonary fibrosis. *Eur. Respir. J.* **2016**, *48*, 538–552. [CrossRef]

122. Liang, J.; Zhang, Y.; Xie, T.; Liu, N.; Chen, H.; Geng, Y.; Kurkciyan, A.; Mena, J.M.; Stripp, B.R.; Jiang, D.; et al. Hyaluronan and TLR4 promote surfactant-protein-C-positive alveolar progenitor cell renewal and prevent severe pulmonary fibrosis in mice. *Nat. Med.* **2016**, *22*, 1285–1293. [CrossRef]

123. Marmai, C.; Sutherland, R.E.; Kim, K.K.; Dolganov, G.M.; Fang, X.; Kim, S.S.; Jiang, S.; Golden, J.A.; Hoopes, C.W.; Matthay, M.A.; et al. Alveolar epithelial cells express mesenchymal proteins in patients with idiopathic pulmonary fibrosis. *Am. J. Physiol. Lung Cell. Mol. Physiol.* **2011**, *301*, L71–L78. [CrossRef]

124. Chen, J.; Wang, L.; Matyunina, L.V.; Hill, C.G.; McDonald, J.F. Overexpression of miR-429 induces mesenchymal-to-epithelial transition (MET) in metastatic ovarian cancer cells. *Gynecol. Oncol.* **2011**, *121*, 200–205. [CrossRef]

125. Lu, J.; Getz, G.; Miska, E.A.; Alvarez-Saavedra, E.; Lamb, J.; Peck, D.; Sweet-Cordero, A.; Ebert, B.L.; Mak, R.H.; Ferrando, A.A.; et al. MicroRNA expression profiles classify human cancers. *Nature* **2005**, *435*, 834–838. [CrossRef]

126. Hill, L.; Browne, G.; Tulchinsky, E. ZEB/miR-200 feedback loop: At the crossroads of signal transduction in cancer. *Int. J. Cancer* **2013**, *132*, 745–754. [CrossRef]

127. Gregory, P.A.; Bert, A.G.; Paterson, E.L.; Barry, S.C.; Tsykin, A.; Farshid, G.; Vadas, M.A.; Khew-Goodall, Y.; Goodall, G.J. The miR-200 family and miR-205 regulate epithelial to mesenchymal transition by targeting ZEB1 and SIP1. *Nat. Cell Biol.* **2008**, *10*, 593–601. [CrossRef]

128. Hu, X.; Macdonald, D.M.; Huettner, P.C.; Feng, Z.; El Naqa, I.M.; Schwarz, J.K.; Mutch, D.G.; Grigsby, P.W.; Powell, S.N.; Wang, X. A miR-200 microRNA cluster as prognostic marker in advanced ovarian cancer. *Gynecol. Oncol.* **2009**, *114*, 457–464. [CrossRef]

129. Kurashige, J.; Kamohara, H.; Watanabe, M.; Hiyoshi, Y.; Iwatsuki, M.; Tanaka, Y.; Kinoshita, K.; Saito, S.; Baba, Y.; Baba, H. MicroRNA-200b regulates cell proliferation, invasion, and migration by directly targeting ZEB2 in gastric carcinoma. *Ann. Surg. Oncol.* **2012**, *19* (Suppl. 3), S656–S664. [CrossRef]

130. Zidar, N.; Bostjancic, E.; Gale, N.; Kojc, N.; Poljak, M.; Glavac, D.; Cardesa, A. Down-regulation of microRNAs of the miR-200 family and miR-205, and an altered expression of classic and desmosomal cadherins in spindle cell carcinoma of the head and neck–hallmark of epithelial-mesenchymal transition. *Hum. Pathol.* **2011**, *42*, 482–488. [CrossRef]

131. Braun, J.; Hoang-Vu, C.; Dralle, H.; Huttelmaier, S. Downregulation of microRNAs directs the EMT and invasive potential of anaplastic thyroid carcinomas. *Oncogene* **2010**, *29*, 4237–4244. [CrossRef]

132. Tellez, C.S.; Juri, D.E.; Do, K.; Bernauer, A.M.; Thomas, C.L.; Damiani, L.A.; Tessema, M.; Leng, S.; Belinsky, S.A. EMT and stem cell-like properties associated with miR-205 and miR-200 epigenetic silencing are early manifestations during carcinogen-induced transformation of human lung epithelial cells. *Cancer Res.* **2011**, *71*, 3087–3097. [CrossRef]

133. Gibbons, D.L.; Lin, W.; Creighton, C.J.; Rizvi, Z.H.; Gregory, P.A.; Goodall, G.J.; Thilaganathan, N.; Du, L.; Zhang, Y.; Pertsemlidis, A.; et al. Contextual extracellular cues promote tumor cell EMT and metastasis by regulating miR-200 family expression. *Genes Dev.* **2009**, *23*, 2140–2151. [CrossRef] [PubMed]

134. Park, S.M.; Gaur, A.B.; Lengyel, E.; Peter, M.E. The miR-200 family determines the epithelial phenotype of cancer cells by targeting the E-cadherin repressors ZEB1 and ZEB2. *Genes Dev.* **2008**, *22*, 894–907. [CrossRef] [PubMed]

135. Korpal, M.; Lee, E.S.; Hu, G.; Kang, Y. The miR-200 family inhibits epithelial-mesenchymal transition and cancer cell migration by direct targeting of E-cadherin transcriptional repressors ZEB1 and ZEB2. *J. Biol. Chem.* **2008**, *283*, 14910–14914. [CrossRef] [PubMed]

136. Yang, X.; Hu, Q.; Hu, L.X.; Lin, X.R.; Liu, J.Q.; Lin, X.; Dinglin, X.X.; Zeng, J.Y.; Hu, H.; Luo, M.L.; et al. miR-200b regulates epithelial-mesenchymal transition of chemo-resistant breast cancer cells by targeting FN1. *Discov. Med.* **2017**, *24*, 75–85.

137. Ebert, M.S.; Sharp, P.A. Roles for microRNAs in conferring robustness to biological processes. *Cell* **2012**, *149*, 515–524. [CrossRef]

138. Janakiraman, H.; House, R.P.; Gangaraju, V.K.; Diehl, J.A.; Howe, P.H.; Palanisamy, V. The Long (lncRNA) and Short (miRNA) of It: TGFbeta-Mediated Control of RNA-Binding Proteins and Noncoding RNAs. *Mol. Cancer Res.* **2018**, *16*, 567–579. [CrossRef]

139. Blahna, M.T.; Hata, A. Smad-mediated regulation of microRNA biosynthesis. *FEBS Lett.* **2012**, *586*, 1906–1912. [CrossRef]

140. Butz, H.; Racz, K.; Hunyady, L.; Patocs, A. Crosstalk between TGF-beta signaling and the microRNA machinery. *Trends Pharmacol. Sci.* **2012**, *33*, 382–393. [CrossRef]

141. Wang, D.; Haviland, D.L.; Burns, A.R.; Zsigmond, E.; Wetsel, R.A. A pure population of lung alveolar epithelial type II cells derived from human embryonic stem cells. *Proc. Natl. Acad. Sci. USA* **2007**, *104*, 4449–4454. [CrossRef]

142. Wang, J.; Huang, W.; Xu, R.; Nie, Y.; Cao, X.; Meng, J.; Xu, X.; Hu, S.; Zheng, Z. MicroRNA-24 regulates cardiac fibrosis after myocardial infarction. *J. Cell. Mol. Med.* **2012**, *16*, 2150–2160. [CrossRef]

143. Xie, T.; Liang, J.; Guo, R.; Liu, N.; Noble, P.W.; Jiang, D. Comprehensive microRNA analysis in bleomycin-induced pulmonary fibrosis identifies multiple sites of molecular regulation. *Physiol. Genom.* **2011**, *43*, 479–487. [CrossRef]
144. Pandit, K.V.; Milosevic, J.; Kaminski, N. MicroRNAs in idiopathic pulmonary fibrosis. *Transl. Res. J. Lab. Clin. Med.* **2011**, *157*, 191–199. [CrossRef]
145. Dews, M.; Fox, J.L.; Hultine, S.; Sundaram, P.; Wang, W.; Liu, Y.Y.; Furth, E.; Enders, G.H.; El-Deiry, W.; Schelter, J.M.; et al. The myc-miR-17~92 axis blunts TGF{beta} signaling and production of multiple TGF{beta}-dependent antiangiogenic factors. *Cancer Res.* **2010**, *70*, 8233–8246. [CrossRef]
146. Corcoran, D.L. Transcriptional Regulation of microRNA Genes and the Regulatory Networks in Which They Participate. Ph.D. Thesis, University of Pittsburgh, Pittsburgh, PA, USA, 2008. Available online: http://d-scholarship.pitt.edu/8802/8801/DLCorcoran_etd_081208.pdf (accessed on 11 June 2022).
147. Zhao, L.; Yee, M.; O'Reilly, M.A. Transdifferentiation of alveolar epithelial type II to type I cells is controlled by opposing TGF-beta and BMP signaling. *Am. J. Physiol. Lung Cell. Mol. Physiol.* **2013**, *305*, L409–L418. [CrossRef]

 *genes*

Article

Circulating Transcriptional Profile Modulation in Response to Metabolic Unbalance Due to Long-Term Exercise in Equine Athletes: A Pilot Study

Katia Cappelli [1,2], Samanta Mecocci [1], Stefano Capomaccio [1,2,*], Francesca Beccati [1,2], Andrea Rosario Palumbo [1], Alessia Tognoloni [1], Marco Pepe [1,2] and Elisabetta Chiaradia [1,2]

[1] Department of Veterinary Medicine, University of Perugia, 06126 Perugia, Italy; katia.cappelli@unipg.it (K.C.); samanta.mecocci@studenti.unipg.it (S.M.); francesca.beccati@unipg.it (F.B.); fotodarp@gmail.com (A.R.P.); alessia.tognoloni@studenti.unipg.it (A.T.); marco.pepe@unipg.it (M.P.); elisabetta.chiaradia@unipg.it (E.C.)

[2] Sports Horse Research Center, University of Perugia, 06126 Perugia, Italy

* Correspondence: stefano.capomaccio@unipg.it; Tel.: +39-0755857765

Citation: Cappelli, K.; Mecocci, S.; Capomaccio, S.; Beccati, F.; Palumbo, A.R.; Tognoloni, A.; Pepe, M.; Chiaradia, E. Circulating Transcriptional Profile Modulation in Response to Metabolic Unbalance Due to Long-Term Exercise in Equine Athletes: A Pilot Study. *Genes* **2021**, *12*, 1965. https://doi.org/10.3390/genes12121965

Academic Editors: Giuseppe Iacomino, Fabio Lauria and Samantha A. Brooks

Received: 15 October 2021
Accepted: 6 December 2021
Published: 9 December 2021

Publisher's Note: MDPI stays neutral with regard to jurisdictional claims in published maps and institutional affiliations.

Abstract: Physical exercise has been associated with the modulation of micro RNAs (miRNAs), actively released in body fluids and recognized as accurate biomarkers. The aim of this study was to measure serum miRNA profiles in 18 horses taking part in endurance competitions, which represents a good model to test metabolic responses to moderate intensity prolonged efforts. Serum levels of miRNAs of eight horses that were eliminated due to metabolic unbalance (Non Performer-NP) were compared to those of 10 horses that finished an endurance competition in excellent metabolic condition (Performer-P). Circulating miRNA (ci-miRNA) profiles in serum were analyzed through sequencing, and differential gene expression analysis was assessed comparing NP versus P groups. Target and pathway analysis revealed the up regulation of a set of miRNAs (of mir-211 mir-451, mir-106b, mir-15b, mir-101-1, mir-18a, mir-20a) involved in the modulation of myogenesis, cardiac and skeletal muscle remodeling, angiogenesis, ventricular contractility, and in the regulation of gene expression. Our preliminary data open new scenarios in the definition of metabolic adaptations to the establishment of efficient training programs and the validation of athletes' elimination from competitions.

Keywords: mi-RNA; physical exercise; gene expression; stress

1. Introduction

While regular physical activity is known to be beneficial for health, prolonged or high-intensity exercise can cause stress due to duration and inability of humans and horses to adapt physiologically to these conditions [1–5].

Competitive sports are known to be demanding and stressful for both human and animal athletes. In particular, endurance is one of the most challenging among equestrian disciplines; endurance horses are susceptible to metabolic imbalance due to dehydration, acid balance and electrolyte abnormalities, substrate depletion and heat accumulation, which can result in life-threatening conditions [6]. Indeed, equine endurance competitions, which have recently gained popularity worldwide, are governed by the rules established by National and International Equestrian Federations (FEI), which implement strict regulations to safeguard and ensure the welfare of animals [7].

Endurance horses are subjected to specific training that induces the physiological adaptations required for carrying out prolonged moderate intensity exercise on different types of ground surfaces and under different weather conditions, by modulating their energy metabolism towards aerobic conditions [8]. However, when horses are poorly trained, or high speeds are required, anaerobic pathways may be recruited. Endurance athletes also develop hypervolemia, which, when coupled with splanchnic vasoconstriction physiologically triggered by stress hormones, enables maintaining of good central blood pressure and satisfactory perfusion of the main organs. Prolonged physical exercise can also induce changes in coagulation systems, immune modulation, and vascular integrity, and is known to have negative effects on health and welfare [9]. Indeed, the homeostatic disruptions that occur during long competitions have been suggested as the main causes of certain pathological conditions such as myopathies, colic, laminitis, diaphragmatic flutter, cardiac arrhythmias and massive rhabdomyolysis [7]. Therefore, it is essential to find early biomarkers that can promptly identify subjects at risk and prevent dysmetabolism.

Evidence has been found that alterations in circulating miRNA (ci-miRNA) expression are induced by physical exercise and endurance [10–12].

Among the RNAs, micro-RNAs (miRNAs) are highly conserved regulatory molecules that play active roles in cell differentiation, proliferation and metabolism. MiRNAs drive post-transcriptional down-regulation and bind to mRNAs, with a one-to-many (and vice versa) relationship with their targets, in that a gene can be regulated by different miRNAs, and the same miRNA can regulate different genes [13].

MiRNAs can be released into the bloodstream within apoptotic bodies, extracellular vesicles (i.e., exosomes), high/low-density lipoproteins (HDL and LDL) or as active protein complexes (RNA-binding proteins) [14,15]. Stable ci-miRNAs can be found in plasma and/or in serum, many of which are tissue-specific and signatures of certain physiological and/or pathological conditions [12]. Physical exercise has recently been associated with the modulation of small noncoding RNAs in the bloodstream depending on type and duration of the physical activity [16]. Therefore, ci-miRNAs have been proposed as biomarkers for evaluating human athletic performance and were recently used in our previous study on endurance riding [17].

The aim of this research is to acquire a deeper molecular knowledge of responses to prolonged moderate-intensity exercise in endurance horses and to identify ci-miRNAs related to nonphysiological responses. To this aim, the ci-miRNA transcriptional profiles of horses that successfully finished an endurance race were compared using high-throughput sequencing and those with severe metabolic disorders after a competition were eliminated. Our hypothesis is that differentially modulated ci-miRNA in horses after a race elimination could provide clues to a reduced adaptation to exercise stress leading to low performance syndromes and diseases. We therefore tried to obtain a transcriptional picture of the multiorgan response and gain a better understanding of the molecular basis of this phenomenon.

2. Materials and Methods

2.1. Ethics Statement

The study protocol was reviewed and approved by institutional Ethics Committee of University of Perugia (license No. 2019-32). All procedures were performed in accordance with the approved guidelines. Owner informed consent and the approval of the Ground Jury President and the Veterinary Commission President of the event were obtained before initiation of study procedures with the animals.

In order to correctly report research on live animals, the manuscript was prepared following the ARRIVE guidelines (https://arriveguidelines.org/, accessed on 8 December 2021).

2.2. Sample Collection

For this study, eighteen Arabian horses, thirteen females, four geldings and one male, aged from 6 to 13 years old (median 9 years old) engaged in 80–160 km national and international competitions held in the same season of 2018 in Italy were recruited. During competition at least every 40 km there is a compulsory veterinary inspection (vet gate) to determine if the horse is fit to continue as indicated by FEI Endurance Rules (available at: https://inside.fei.org/fei/disc/endurance/rules, accessed on 8 December 2021) including irregular gait (i.e., lameness) or metabolic reasons, such as failure to recovery maximum heart rate (i.e., 64 beats per minutes) cardiac arrhythmia, clinical signs of metabolic instability, excessive fatigue, heat stroke, colic, myopathy, severe dehydration or excessively high temperature by evaluation of heart, gut sound, mucous membrane colour, capillary refill time and muscle tone.

The horses enrolled were divided into two groups based on veterinary gates inspection results: Performer (P) was composed of 10 subjects that successfully completed the competition at free speed and Non Performer (NP) constituted the remaining eight subjects that were eliminated at different stages of the competition (Table S1).

Peripheral blood was collected from the jugular vein using a vacutainer with and without anticoagulant at the end of race or at the intermediate veterinary examination that resulted in elimination of the horse from the competition within 30′. The serum samples were obtained by centrifugation at $400 \times g$ in a bench-top centrifuge for 15 min, immediately after collection and then stored at $-80\ °C$ until biochemical tests and miRNA isolation.

2.3. Blood Count and Biochemical Analysis

A complete blood count with leukocyte differential assessment was performed using a laser haematology analyser (Sysmex XT-2000iV; Sysmex, Kobe, Japan). The analysis included white blood cells (WBCs), red blood cells (RBCs), mean corpuscular volume (MCV), haemoglobin (Hb), platelets (PLTs) and mean platelet volume (MPV), haematocrit (HCT), mean corpuscular haemoglobin (MCH), mean corpuscular haemoglobin concentration (MCHC) and red blood cell distribution width (RDW). Selected biochemical parameters were also analyzed with an Hitachi 904 automated biochemistry analyzer (Boehringer Mannheim, Baden-Wurttemberg, Germany) including urea, creatinine, total bilirubin (Tbil), aspartate aminotransferase (AST), γ-glutamyltransferase (GGT), creatine kinase (CK), lactate dehydrogenase (LDH), total proteins (TPs) and albumin (Alb). For blood count and biochemical analysis, Student's *t*-test was applied to identify differences between continuous data between group NP and group P.

2.4. RNA Extraction and Library Preparation

Total RNA extraction was carried out using the commercial miRNeasy Serum/Plasma Advanced kit (Qiagen, Venlo, The Netherlands) following manufacturer's instructions. For better evaluation, the miRNA differences between the two groups, spike-in sequences were added to the lysis buffer at the beginning of the RNA extraction procedure (2.5 μL of Spike-in solution per 600 μL of serum for each subject) using the QIAseq miRNA Library QC Spike-in kit (Qiagen, Venlo, The Netherlands), which provides 52 spike-in phosphorylated at 5′, an essential feature for the library preparation [18]. Adding these plant origin sequences allowed us to obtain quality control for the sequencing process and to normalize the results in terms of the number of sequenced fragments for each miRNA. The RNA quantity and quality was evaluated with NanoDrop 2000 (Thermo Fisher Scientific, Waltham, MA, USA) spectrophotometer and microfluidic electrophoresis respectively (Bioanalyzer 2100 Agilent Technologies, Santa Clara, CA, USA). The TruSeq Small RNA Library Illumina (Illumina Inc., 5200 Illumina Way, San Diego, CA, USA) kit was used for library construction following the manufacturer's instructions, and fragments were sequenced on a NextSeq500 Illumina (Illumina Inc., 5200 Illumina Way, San Diego, CA, USA) instrument with 75bp single-end chemistry.

2.5. Bioinformatic Analysis

Raw reads in fastq format were quality checked with FastQC (https://www.bioinf ormatics.babraham.ac.uk/projects/fastqc/, accessed on 8 December 2021) and trimmed with Trim Galore 0.5.0 software for the removal of low-quality sequences and adapters (https://www.bioinformatics.babraham.ac.uk/projects/trim_galore/, accessed on 8 December 2021). The Bowtie2 [19] aligner of the Tuxedo suite was used to align cleaned reads adopting a three-step alignment strategy: (i) on the spike-in set used in the experiment; (ii) on the mirbase 22 hairpin database (Release 22.1) (http://www.mirbase.org, accessed on 8 December 2021) for the unmapped reads on spike-in and (iii) on the horse reference genome (equcab3) [20], for the unmapped on miRNA database (Figure 1). After the alignment procedure, the dataset was normalized for spike-ins data through the RUVSeq [21] R package. Briefly, RUVSeq uses spike-in reads to extrapolate correction coefficients for the samples that are integrated, among other confounding effects (i.e., sex), into the differential gene expression (DGE) evaluation analysis model implemented in edgeR [22]. A PCA plot from expression data is available in Figure S1. Expression levels were measured as count per million (CPM) and parameters set for DGE comparing the NP to the P group with an absolute log2-fold change (logFC) > 1.5 and an adjusted *p*-value for multiple testing correction applying the Benjamini-Hochberg method (FDR) < 0.05.

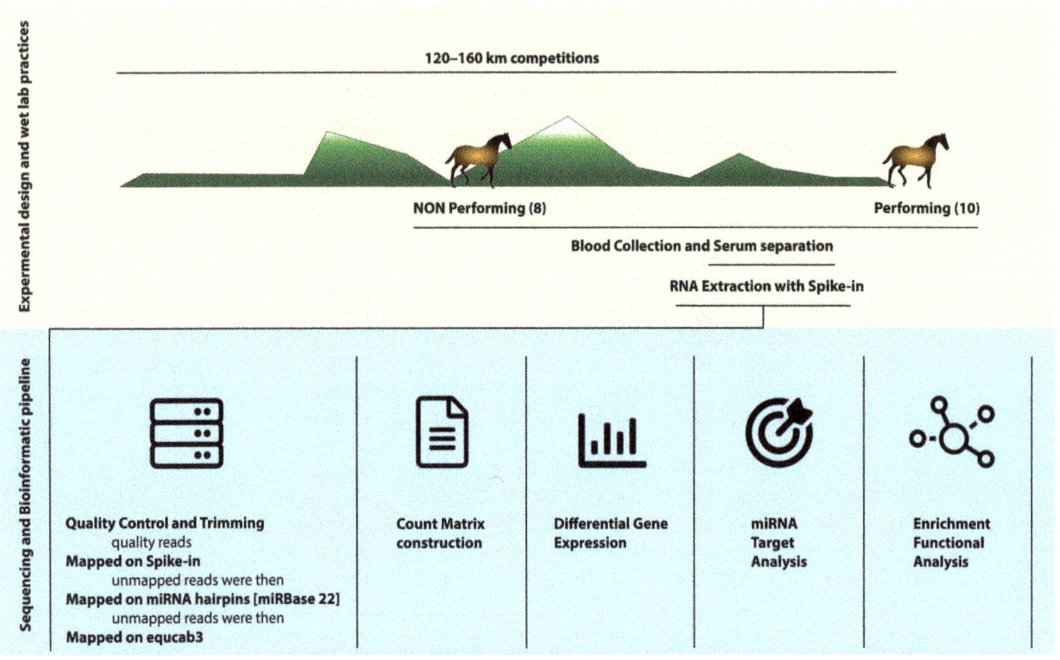

Figure 1. Experimental design (cream background) and bioinformatic pipeline (cadet blue background).

Target Genes and Enrichment Analysis

Human or murine orthologue of all the differentially expressed miRNAs, according to the DGE analysis from edgeR, were retrieved using miRbase (http://www.mirbase.org, accessed on 8 December 2021) software and used to identify putative genes (predicted and/or validated) targeted by these miRNAs. The most frequently sequenced human or murine miRNA form IDs (3p or 5p) were selected as input in the miRWalk 3.0 webtool (http://mirwalk.umm.uni-heidelberg.de, accessed on 8 December 2021). A unique list of predicted and validated targets was identified for each miRNA, specifying also the miRNA site of action with respect to the targeted mRNA: 3'UTR, 5'UTR or CDS (coding region). Only the genes targeted by five or more miRNAs were selected for the downstream analyses.

The Cytoscape 3.7.1 suite [23] was used to build a Protein-Protein Interaction Network (PPI) using the IMEx database, which contains nonredundant information derived from the major public protein databases. The clusterMaker 2.0 a Cytoscape application [24] with the "gLay" option was used to highlight different clusters within the network based on the number and type of connections between the nodes. Clusters with a number of interactions greater than 30 were inspected for Gene Ontology (GO) enrichment analysis carried out for the related biological process through the BiNGO application [25]. Results were filtered with a corrected FDR < 0.05 (Benjamini Hockberg correction). A summary of the experimental design and the bioinformatic pipeline is depicted in Figure 1.

3. Results

3.1. Hematology and Clinical Chemistry Analyses

The hematology and clinical chemistry assays evidenced values of hematocrit ($p < 0.01$), hemoglobin ($p < 0.01$), total protein, albumin ($p < 0.01$), urea ($p < 0.05$) and creatinine ($p < 0.01$) significantly higher in the Non Performer (NP) group compared to that of Performer (P) group whereas classical exercise biochemical markers such as creatine phosphokinase (CPK), aspartate-transaminase (AST) and lactic-dehydrogenase (LDH) values were not significantly different between the two groups, even after normalizing for the kilometers runs (Table S2).

3.2. Sequencing Statistics

The sequencing depth average was ~18,380,000 reads; of these, the 77.72% passed the trimming step. Two samples (P2 and P3) showed an unusually higher percentage (81% and 61% respectively) of discarded reads due to poor quality and/or length. Only the reads mapped exactly one time on miRBase 22 or EquCab3 annotated as sequences referable to miRNA were used for downstream analyses. Sequencing results and alignment rates are summarized in (Table S3).

3.3. Differential Expression Analysis of miRNA

After the statistical analysis with edgeR, starting from a cleaned dataset of 288 miRNAs, a total of seven differentially expressed (FDR < 0.05) were found. All of these were upregulated (logFC ≥ +1.5) in the NP group compared to P group (Table 1).

Table 1. Differentially expressed miRNAs (NP vs. P).

miRNA	logFC	logCPM	FDR
eca-mir-211	4.9223101	0.1280119	0.0015486
eca-mir-15b	4.6239495	3.7887133	0.0027720
eca-mir-451	2.1155572	9.8706133	0.0092029
eca-mir-18a	5.4076681	1.2325327	0.0092029
eca-mir-20a	3.2246991	2.7407456	0.0175767
eca-mir-106b	2.4277831	4.8328315	0.0213777
eca-mir-101-1	2.0112158	7.7859690	0.0213777

3.4. miRNA-Target Evaluation and Pathway Analysis

Targets of each miRNA, for all the binding sites (3′-UTR, 5′-UTR and CDS of the target) were individualised from mirWalk analysis (Table 2). All the target lists were merged in order to produce a unique series of 16 genes that were selected according to the number of miRNAs they targeted, setting the threshold of at least five (Table 3).

These genes were used to create a PPI network to identify clusters of proteins retrieved from the IMEx database. The total network contained 481 nodes and 657 edges; the major protein clusters, i.e., those with more than 30 interactions, are reported in Figure 2, where proteins are indicated of central nodes that correspond to target genes which have the highest number of interactions with other proteins correlated to similar biological functions. Gene Ontology (GO) enrichment analysis with BiNGO was performed on these major clusters (Table S4). The most enriched GO terms for the TTN cluster related to cardiac function and the apoptotic process such as "muscle structure development", "muscle organ development", "cytoskeleton organization", "muscle contraction", "cardiac myofibril assembly", "regulation of apoptotic process" and "regulation of programmed cell death". Regarding AGK cluster central node "mitochondrial transport", "lactate biosynthetic process" and "negative regulation of glucocorticoid secretion" were among the main enriched terms. "Response to stimulus", "response to stress", "regulation of immune system process" together with "cell cycle", "cell migration" and "cytoskeleton organization" were the main enriched GO terms for the cluster with MACF1 central node, while USP49 seemed to be related to protein folding. Terms related to translation regulation, such as "miRNA mediated inhibition of translation", "regulation of translation, ncRNA-mediated" and "negative regulation of translation, ncRNA-mediated", were among the main enriched biological processes found for the SYNE1 central nodes cluster. Complete data from GO analysis for all the PPI clusters are reported in Table S4.

Table 2. Target gene consistencies of each miRNA detailed with respect to the binding site.

miRNA ID	Target 3′-UTR	Target 5′-UTR	Target CDS	Total Target
mir-101-3p	315	43	322	680
mir-106b-5p	1442	263	1435	3140
mir-15b-5p	1102	285	1623	3010
mir-18a-5p	1550	356	2242	4148
mir-20a-5p	1167	183	1038	2388
mir-211-5p	2161	844	2626	5631
mir-451-5p	171	44	409	624

Table 3. Target genes selected on the miRNA consensus analysis (recognized by at least five differentially expressed miRNAs).

Target Genes	miRNAs	# of miRNA
TTN	mir-101; mir-106b; mir-15b; mir-18a; mir-20a; mir-211	6
MACF1	mir-101; mir-106b; mir-15b; mir-18a; mir-20a	5
SYNE1	mir-101; mir-15b; mir-18a; mir-20a; mir-211	5
TDRD12	mir-106b; mir-15b; mir-18a; mir-20a; mir-211	5
USP49	mir-101; mir-106b; mir-15b; mir-18a; mir-211	5
ACTR8	mir-101; mir-106b; mir-15b; mir-18a; mir-211	5
AGK	mir-101; mir-106b; mir-15b; mir-20a; mir-211	5
CACNA1B	mir-106b; mir-15b; mir-18a; mir-20a; mir-211	5
COL6A5	mir-101; mir-106b; mir-15b; mir-18a; mir-211	5
FAM227A	mir-106b; mir-15b; mir-18a; mir-20a; mir-211	5
HECW2	mir-106b; mir-15b; mir-18a; mir-20a; mir-211	5
PGM2L1	mir-106b; mir-15b; mir-18a; mir-20a; mir-211	5
PKD1L2	mir-101; mir-106b; mir-18a; mir-20a; mir-211	5
RIMBP2	mir-106b; mir-15b; mir-18a; mir-20a; mir-211	5
VPS13A	mir-101; mir-15b; mir-18a; mir-20a; mir-211	5
ZBTB37	mir-101; mir-106b; mir-18a; mir-20a; mir-211	5

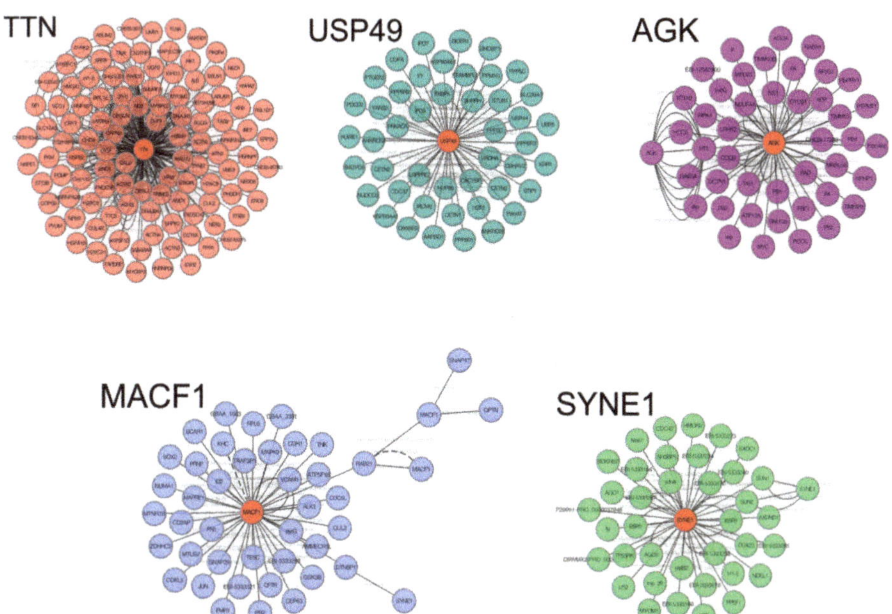

Figure 2. Cluster of proteins derived from clusterMaker 2.0 cytoscape application. Central nodes (red) indicate most relevant proteins. Clusters in boxes refer to networks between mouse proteins.

4. Discussion

The molecular mechanisms underlying successful adaptation to stress associated with prolonged or high-intensity exercise remain poorly understood. In this study, the miRNA expression profiles of equine athletes competing in national and international endurance races were analysed. The differences observed in the number of circulating miRNAs between the nonperforming (NP group, eliminated by the competition) and performing equine athletes (P group, those who successfully finished the competition) were investigated.

Comparing the hematological and biochemical profiles of the two experimental groups, NP horses showed signs of severe dehydration, as they had higher values of hematocrit, hemoglobin, creatinine, total plasma proteins, albumins, and increased protein catabolism indicated by higher serum urea concentration. High levels of muscle enzyme markers (including for example creatine phosphokinase (CPK), aspartate-transaminase (AST) and lactic-dehydrogenase (LDH)) were found, as expected, in both groups' serum (Table S2). However, no significant differences were observed among P and NP horses in enzymes. This confirmed the low correlation between these serum enzymes levels and the poor performance of equine athletes as suggested by other authors [26]. This evidence indicates that deeper knowledge is needed about molecular events modulating the response to physical stress that can also lead to metabolic alterations and low performance. miRNAs are ideal molecules for this purpose due to their properties and their fast response to physiological stress [12,16]. The results show that besides skeletal muscle tissues (SMTs), other tissues may be involved in an equine athlete's adaptation to exercise-induced stress, since physiological stress induced by exercise triggers multiorgan responses in skeletal muscle, vasculature, heart and lung [27]. Garciarena et al. suggested that endurance training can improve cardiac function by "transforming" pathological cardiac hypertrophy into a physiological state [28] during which miRNAs have been described as key players [29].

However, it is important to note that NP horses were eliminated from the competition due to metabolic disorders including long recovery times.

As a whole, the results seem to indicate that the NP horses failed to adapt to exercise-induced stress. We observed, indeed, an increase in mir-101 plasma levels in these horses, which is in contrast to what was observed in trained horses [30]. Mirna-101 plays an important role in cardiac hypertrophy [31], and can be decreased by endurance training [32].

Interestingly, our results also showed high levels of some mir-17-92 cluster members, namely mir-18a, mir-20a and mir-106, which have been linked to myocardial ischemia/reperfusion injuries [31].

Increases in mir-20 and mir-106 levels have been also related to stress [33]. They are both involved in glucose metabolism regulation, while mir-106 has been linked to oxidative stress [33]. This has been proposed as a potential biomarker for recovery after exercise, as it was down-regulated following exhaustive endurance exercise and restored to normal levels within 2 h after the competition [34,35]. Moreover, over-expressed mir-106b has been associated with skeletal muscle mitochondrial dysfunction and insulin resistance, which are affected by exercise [34]. The high levels of mir-106b and mir-20a in our NP athletes also suggests endothelial function modulation failure [36] and the promotion of angiogenesis, that usually occurs during endurance training. Indeed, mir-106 exhibits anti-angiogenic properties, whereas mir-20a, also involved in angiogenesis, was found to quantitatively correlate with peak exercise capacity, cardiorespiratory fitness [37] and endothelial repair capacity [38]. However, increased expression of miR-20a was only observed at rest after sustained training and in response to hypertension-induced heart failure [39].

Mir-15b which was found to be increased in NP horses, is known to be an anti-angiogenic miRNA [40,41]. These miRNAs suppress vascular endothelial growth factor (VEGF) [40,41], basic fibroblast growth factor (bFGF) VEGF receptor 2 and FGF receptor 1 expression) [42]. The mir-15 family modulates cardiac hypertrophy, and its members are up-regulated during myocardial disorders [43]. Mir-451 is one of the most abundant miRNAs found in red blood cells [44,45] and was increased in NP horses. It was found to be increased in low responders to resistance exercise training [46] and has been correlated with coronary artery disease [47] in the occurrence of ischemic stroke.

Most of the modulated miRNAs in NP horses identified protein-coding genes as the targets involved in cardiac and skeletal muscle cytoskeleton, which is the main coordinator of muscle contraction [48]. More specifically, some proteins such as TTN (titin), SYNE1 (nesprin), and MACF1 (Microtubule Actin Cross-linking Factor 1), which are involved in nuclear and cytosolic cytoskeleton organization, were highlighted as being crucial cluster nodes by the PPI network analysis (Figure 2). Titin is involved in cardiac and skeletal muscle remodeling and modulates the elastic properties of the sarcomere as well as contractile muscle properties [49]. Furthermore, titin regulates myocardial passive stiffness and myocardial and ventricular function, mediates nuclear signaling and modulates muscle response to mechanical stress [50]. Moreover, the expression of various titin isoforms has been associated with sarcomere length, which correlates with muscle fascicle length and may enhance running performance.

Other cytoskeleton alterations are suggested by MACF, which is known to be a stress-induced regulator of cardiomyocyte microtubule distribution [48] essential for ventricular adaptation to hemodynamic overload [51], and to cardiac response to exercise. Muscle contraction could also be affected by changes in Ca^{2+} influx in our NP horses since *CACNA1B* (Voltage-dependent N-type calcium channel subunit alpha-1B), the gene target of some modulated miRNAs, was repressed and previous studies reported that many of the genes encoding ion channels are differentially methylated in horses and humans following exercise [52]. Moreover AGK (acylglycerol kinase) protein, a central cluster node evidenced by the PPI network analysis (Figure 2), is involved in the metabolism of mitochondrial phospholipids and in the stability of SLC25A4 (ADP/ATP translocase 1). Interesting, *SLC25A4* knockout mice exhibit phenotypes of hypertrophic cardiomyopathy, exercise intolerance, and lacticacidemia [53].

USP49, another PPI network central node, is involved in splicing alterations as it is essential for the cotranscriptional splicing [54]. Moreover, one of the target genes, *TDRD12*, is a gene expression enhancer via the production of secondary piRNAs. Since these small RNAs are key molecules in the transposable elements activation/expression [55], it is intriguing to think that one of the possible outcomes of this down regulation could be a genome plasticity response induced by exercise [3,56].

Overall, analysis of differentially expressed miRNAs and their target genes in NP horses identified genes were associated with the modulation of the various steps of gene expression essential for adaptation processes, namely "response to stimulus", "regulation of translation", "regulation of apoptotic process" and "muscle structure development", underlying cardiac and skeletal muscle plasticity that occurs during prolonged physical activity and training [57].

5. Conclusions

For the NP horses, the results suggest a specific ci-miRNA profile pointing towards molecular mechanisms and metabolic pathways underlying the inability of tissues/organs to adapt to stress induced by prolonged physical exercise and training. Modulation of myogenesis, cardiac and skeletal muscle remodeling, angiogenesis, ventricular contractility and the regulation of gene expression appear to be the most involved processes.

It has to be said, though, that intrinsic limitations such as sample size and heterogeneity of recruited subjects due to the unpredictability of on-field collection, make this study a preliminary investigation. Validation with alternative techniques and a larger cohort of subjects, and possibly time course sampling, will be valuable in this intriguing research field.

Supplementary Materials: The following materials are available online at https://www.mdpi.com/article/10.3390/genes12121965/s1. Figure S1: Representation of the samples after the Principal Component analysis on miRNA features. Table S1: Details of the distance travelled, average speed and details of the veterinary inspection reported in the vet card at the time of completion (group P) or elimination (group NP). Table S2: Hematology and clinical chemistry values differences between groups P and NP. Table S3: Statistics of sequencing and mapping. Table S4: Multiple tables describing enrichment analysisfnot.

Author Contributions: Conceptualization, K.C. and E.C.; data curation, S.M. and S.C.; formal analysis, S.M., S.C., A.R.P. and A.T.; funding acquisition, K.C.; methodology, S.M., S.C., F.B., A.R.P., A.T. and M.P.; resources, F.B. and M.P.; visualization, S.C.; writing—review & editing, K.C., S.M., S.C., F.B. and E.C. All authors have read and agreed to the published version of the manuscript.

Funding: This research was funded by Dipartimento di Medicina Veterinaria, University of Perugia, grant number CAPPRDB2018.

Institutional Review Board Statement: This study protocol was reviewed and approved by institutional Ethics Committee of University of Perugia (license No. 2019-32). All procedures were performed in accordance with the approved guidelines. Owner informed consent and the approval of the Ground Jury President and the Veterinary Commission President of the event were obtained before initiation of study procedures with the animals.

Informed Consent Statement: Not applicable.

Data Availability Statement: Raw sequence data are available in SRA. BioProject ID collecting all samples is the following: PRJNA726388.

Acknowledgments: The authors would like to thank Gianluca Alunni for his valuable and reliable technical help and DVM Nicola Pilati for his help in sampling procedures.

Conflicts of Interest: The authors declare no conflict of interest. The funders had no role in the design of the study, in the collection, analyses, or interpretation of data, in the writing of the manuscript, or in the decision to publish the results.

References

1. Dhabhar, F.S. Effects of stress on immune function: The good, the bad, and the beautiful. *Immunol. Res.* **2014**, *58*, 193–210. [CrossRef] [PubMed]
2. Morton, J.P.; Kayani, A.C.; McArdle, A.; Drust, B. The Exercise-Induced stress response of skeletal muscle, with specific emphasis on humans. *Sports Med.* **2009**, *39*, 643–662. [CrossRef]
3. Cappelli, K.; Mecocci, S.; Gioiosa, S.; Giontella, A.; Silvestrelli, M.; Cherchi, R.; Valentini, A.; Chillemi, G.; Capomaccio, S. Gallop racing shifts mature mRNA towards introns: Does exercise-induced stress enhance genome plasticity? *Genes* **2020**, *11*, 410. [CrossRef] [PubMed]
4. Cappelli, K.; Amadori, M.; Mecocci, S.; Miglio, A.; Antognoni, M.T.; Razzuoli, E. Immune response in young thoroughbred racehorses under training. *Animals* **2020**, *10*, 1809. [CrossRef] [PubMed]
5. Cappelli, K.; Felicetti, M.; Capomaccio, S.; Nocelli, C.; Silvestrelli, M.; Verini-Supplizi, A. Effect of training status on immune defence related gene expression in Thoroughbred: Are genes ready for the sprint? *Vet. J.* **2013**, *195*, 373–376. [CrossRef]
6. Noakes, T.D. Physiological models to understand exercise fatigue and the adaptations that predict or enhance athletic performance. *Scand. J. Med. Sci. Sports* **2000**, *10*, 123–145. [CrossRef] [PubMed]
7. Nagy, A.; Dyson, S.J.; Murray, J.K. A veterinary review of endurance riding as an international competitive sport. *Vet. J.* **2012**, *194*, 288–293. [CrossRef]
8. Amory, H.; Votion, D.M.; Fraipont, A.; Goachet, A.G.; Robert, C.; Farnir, F.; Van Erck, E. Altered systolic left ventricular function in horses completing a long distance endurance race. *Equine Vet. J.* **2010**, *42*, 216–219. [CrossRef] [PubMed]
9. Scoppetta, F.; Tartaglia, M.; Renzone, G.; Avellini, L.; Gaiti, A.; Scaloni, A.; Chiaradia, E. Plasma protein changes in horse after prolonged physical exercise: A proteomic study. *J. Proteom.* **2012**, *75*, 4494–4504. [CrossRef]
10. Mooren, F.C.; Viereck, J.; Krüger, K.; Thum, T. Circulating micrornas as potential biomarkers of aerobic exercise capacity. *Am. J. Physiol. Heart Circ. Physiol.* **2014**, *306*, H557–H563. [CrossRef] [PubMed]
11. Polakovičová, M.; Musil, P.; Laczo, E.; Hamar, D.; Kyselovič, J. Circulating MicroRNAs as potential biomarkers of exercise response. *Int. J. Mol. Sci.* **2016**, *17*, 1553. [CrossRef] [PubMed]
12. Xu, T.; Liu, Q.; Yao, J.; Dai, Y.; Wang, H.; Xiao, J. Circulating microRNAs in response to exercise. *Scand. J. Med. Sci. Sports* **2015**, *25*, e149–e154. [CrossRef] [PubMed]
13. Lombardi, G.; Perego, S.; Sansoni, V.; Banfi, G. Circulating miRNA as fine regulators of the physiological responses to physical activity: Pre-analytical warnings for a novel class of biomarkers. *Clin. Biochem.* **2016**, *49*, 1331–1339. [CrossRef] [PubMed]
14. Turchinovich, A.; Weiz, L.; Langheinz, A.; Burwinkel, B. Characterization of extracellular circulating microRNA. *Nucleic Acids Res.* **2011**, *39*, 7223–7233. [CrossRef] [PubMed]
15. Valadi, H.; Ekström, K.; Bossios, A.; Sjöstrand, M.; Lee, J.J.; Lötvall, J.O. Exosome-mediated transfer of mRNAs and microRNAs is a novel mechanism of genetic exchange between cells. *Nat. Cell Biol.* **2007**, *9*, 654–659. [CrossRef]
16. Makarova, J.A.; Maltseva, D.V.; Galatenko, V.V.; Abbasi, A.; Maximenko, D.G.; Grigoriev, A.I.; Tonevitsky, A.G.; Northoff, H. Exercise immunology meets MiRNAs. *Exerc. Immunol. Meets MiRNAs* **2014**, *20*, 135–164.
17. Cappelli, K.; Capomaccio, S.; Viglino, A.; Silvestrelli, M.; Beccati, F.; Moscati, L.; Chiaradia, E. Circulating miRNAs as putative biomarkers of exercise adaptation in endurance horses. *Front. Physiol.* **2018**, *9*, 429. [CrossRef] [PubMed]
18. Head, S.R.; Kiyomi Komori, H.; LaMere, S.A.; Whisenant, T.; Van Nieuwerburgh, F.; Salomon, D.R.; Ordoukhanian, P. Library construction for next-generation sequencing: Overviews and challenges. *Biotechniques* **2014**, *56*, 61–77. [CrossRef]
19. Langmead, B.; Salzberg, S.L. Fast gapped-read alignment with Bowtie 2. *Nat. Methods* **2012**, *9*, 357–359. [CrossRef]
20. Kalbfleisch, T.S.; Rice, E.S.; DePriest, M.S.; Walenz, B.P.; Hestand, M.S.; Vermeesch, J.R.; O'Connell, B.L.; Fiddes, I.T.; Vershinina, A.O.; Saremi, N.F.; et al. Improved reference genome for the domestic horse increases assembly contiguity and composition. *Commun. Biol.* **2018**, *1*, 1–8. [CrossRef]
21. Risso, D.; Ngai, J.; Speed, T.P.; Dudoit, S. Normalization of RNA-seq data using factor analysis of control genes or samples. *Nat. Biotechnol.* **2014**, *32*, 896–902. [CrossRef] [PubMed]
22. Robinson, M.D.; McCarthy, D.J.; Smyth, G.K. edgeR: A Bioconductor package for differential expression analysis of digital gene expression data. *Bioinformatics* **2009**, *26*, 139–140. [CrossRef] [PubMed]
23. Shannon, P.; Markiel, A.; Ozier, O.; Baliga, N.S.; Wang, J.T.; Ramage, D.; Amin, N.; Schwikowski, B.; Ideker, T. Cytoscape: A software Environment for integrated models of biomolecular interaction networks. *Genome Res.* **2003**, *13*, 2498–2504. [CrossRef] [PubMed]
24. Morris, J.H.; Apeltsin, L.; Newman, A.M.; Baumbach, J.; Wittkop, T.; Su, G.; Bader, G.D.; Ferrin, T.E. ClusterMaker: A multi-algorithm clustering plugin for Cytoscape. *BMC Bioinform.* **2011**, *12*, 436. [CrossRef] [PubMed]
25. Maere, S.; Heymans, K.; Kuiper, M. BiNGO: A Cytoscape plugin to assess overrepresentation of Gene Ontology categories in Biological Networks. *Bioinformatics* **2005**, *21*, 3448–3449. [CrossRef]
26. Fielding, C.L.; Magdesian, K.G.; Rhodes, D.M.; Meier, C.A.; Higgins, J.C. Clinical and biochemical abnormalities in endurance horses eliminated from competition for medical complications and requiring emergency medical treatment: 30 cases (2005–2006): Retrospective study. *J. Vet. Emerg. Crit. Care* **2009**, *19*, 473–478. [CrossRef] [PubMed]
27. Vega, R.B.; Konhilas, J.P.; Kelly, D.P.; Leinwand, L.A. Molecular Mechanisms Underlying Cardiac Adaptation to Exercise. *Cell Metab.* **2017**, *25*, 1012–1026. [CrossRef]

28. Garciarena, C.D.; Pinilla, O.A.; Nolly, M.B.; Laguens, R.P.; Escudero, E.M.; Cingolani, H.E.; Ennis, I.L. Endurance training in the spontaneously hypertensive rat conversion of pathological into physiological cardiac hypertrophy. *Hypertension* **2009**, *53*, 708–714. [CrossRef]

29. Feng, H.J.; Ouyang, W.; Liu, J.H.; Sun, Y.G.; Hu, R.; Huang, L.H.; Xian, J.L.; Jing, C.F.; Zhou, M.J. Global microRNA profiles and signaling pathways in the development of cardiac hypertrophy. *Braz. J. Med. Biol. Res.* **2014**, *47*, 361–368. [CrossRef]

30. Faraldi, M.; Gomarasca, M.; Sansoni, V.; Perego, S.; Banfi, G.; Lombardi, G. Normalization strategies differently affect circulating miRNA profile associated with the training status. *Sci. Rep.* **2019**, *9*, 1584. [CrossRef]

31. Das, A.; Samidurai, A.; Salloum, F.N. Deciphering Non-coding RNAs in Cardiovascular Health and Disease. *Front. Cardiovasc. Med.* **2018**, *5*, 73. [CrossRef] [PubMed]

32. Keller, P.; Vollaard, N.B.J.; Gustafsson, T.; Gallagher, I.J.; Sundberg, C.J.; Rankinen, T.; Britton, S.L.; Bouchard, C.; Koch, L.G.; Timmons, J.A. A transcriptional map of the impact of endurance exercise training on skeletal muscle phenotype. *J. Appl. Physiol.* **2011**, *110*, 46–59. [CrossRef] [PubMed]

33. Solich, J.; Kuśmider, M.; Faron-Górecka, A.; Pabian, P.; Kolasa, M.; Zemła, B.; Dziedzicka-Wasylewska, M. Serum Level of miR-1 and miR-155 as Potential Biomarkers of Stress-Resilience of NET-KO and SWR/J Mice. *Cells* **2020**, *9*, 917. [CrossRef] [PubMed]

34. Håkansson, K.E.J.; Sollie, O.; Simons, K.H.; Quax, P.H.A.; Jensen, J.; Nossent, A.Y. Circulating Small Non-coding RNAs as Biomarkers for Recovery After Exhaustive or Repetitive Exercise. *Front. Physiol.* **2018**, *9*, 1136. [CrossRef] [PubMed]

35. Nielsen, S.; Åkerström, T.; Rinnov, A.; Yfanti, C.; Scheele, C.; Pedersen, B.K.; Laye, M.J. The miRNA plasma signature in response to acute aerobic exercise and endurance training. *PLoS ONE* **2014**, *9*, e87308. [CrossRef]

36. Brown, M.D.; Hudlicka, O. Modulation of physiological angiogenesis in skeletal muscle by mechanical forces: Involvement of VEGF and metalloproteinases. *Angiogenesis* **2003**, *6*, 1–14. [CrossRef]

37. Baggish, A.L.; Hale, A.; Weiner, R.B.; Lewis, G.D.; Systrom, D.; Wang, F.; Wang, T.J.; Chan, S.Y. Dynamic regulation of circulating microRNA during acute exhaustive exercise and sustained aerobic exercise training. *J. Physiol.* **2011**, *589*, 3983–3994. [CrossRef]

38. Wang, D.; Wang, Y.; Ma, J.; Wang, W.; Sun, B.; Zheng, T.; Wei, M.; Sun, Y. MicroRNA-20a participates in the aerobic exercise-based prevention of coronary artery disease by targeting PTEN. *Biomed. Pharmacother.* **2017**, *95*, 756–763. [CrossRef] [PubMed]

39. Dickinson, B.A.; Semus, H.M.; Montgomery, R.L.; Stack, C.; Latimer, P.A.; Lewton, S.M.; Lynch, J.M.; Hullinger, T.G.; Seto, A.G.; Van Rooij, E. Plasma microRNAs serve as biomarkers of therapeutic efficacy and disease progression in hypertension-induced heart failure. *Eur. J. Heart Fail.* **2013**, *15*, 650–659. [CrossRef] [PubMed]

40. Hua, Z.; Lv, Q.; Ye, W.; Wong, C.K.A.; Cai, G.; Gu, D.; Ji, Y.; Zhao, C.; Wang, J.; Yang, B.B.; et al. Mirna-directed regulation of VEGF and other angiogenic under hypoxia. *PLoS ONE* **2006**, *1*, e116. [CrossRef] [PubMed]

41. Triozzi, P.L.; Achberger, S.; Aldrich, W.; Singh, A.D.; Grane, R.; Borden, E.C. The association of blood angioregulatory microRNA levels with circulating endothelial cells and angiogenic proteins in patients receiving dacarbazine and interferon. *J. Transl. Med.* **2012**, *10*, 241. [CrossRef] [PubMed]

42. Caporali, A.; Emanueli, C. MicroRNA-503 and the Extended MicroRNA-16 Family in Angiogenesis. *Trends Cardiovasc. Med.* **2011**, *21*, 162–166. [CrossRef]

43. Tijsen, A.J.; Van der Made, I.; Van den Hoogenhof, M.M.; Wijnen, W.J.; Van Deel, E.D.; de Groot, N.E.; Alekseev, S.; Fluiter, K.; Schroen, B.; Goumans, M.-J.; et al. The microRNA-15 family inhibits the TGFβ-pathway in the heart. *Cardiovasc. Res.* **2014**, *104*, 61–71. [CrossRef] [PubMed]

44. Kirschner, M.B.; Edelman, J.J.B.; Kao, S.C.H.; Vallely, M.P.; Van Zandwijk, N.; Reid, G. The impact of hemolysis on cell-free microRNA biomarkers. *Front. Genet.* **2013**, *4*, 94. [CrossRef] [PubMed]

45. Doss, J.F.; Corcoran, D.L.; Jima, D.D.; Telen, M.J.; Dave, S.S.; Chi, J.T. A comprehensive joint analysis of the long and short RNA transcriptomes of human erythrocytes. *BMC Genom.* **2015**, *16*, 952. [CrossRef]

46. Davidsen, P.K.; Gallagher, I.J.; Hartman, J.W.; Tarnopolsky, M.A.; Dela, F.; Helge, J.W.; Timmons, J.A.; Phillips, S.M. High responders to resistance exercise training demonstrate differential regulation of skeletal muscle microRNA expression. *J. Appl. Physiol.* **2011**, *110*, 309–317. [CrossRef] [PubMed]

47. Ren, J.; Zhang, J.; Xu, N.; Han, G.; Geng, Q.; Song, J.; Li, S.; Zhao, J.; Chen, H. Signature of circulating MicroRNAs As potential biomarkers in vulnerable coronary artery disease. *PLoS ONE* **2013**, *8*, e80738. [CrossRef] [PubMed]

48. Henderson, C.A.; Gomez, C.G.; Novak, S.M.; Mi-Mi, L.; Gregorio, C.C. Overview of the muscle cytoskeleton. *Compr. Physiol.* **2017**, *7*, 891–944. [PubMed]

49. Anderson, B.R.; Granzier, H.L. Titin-based tension in the cardiac sarcomere: Molecular origin and physiological adaptations. *Prog. Biophys. Mol. Biol.* **2012**, *110*, 204–217. [CrossRef]

50. Lewinter, M.M.; Granzier, H.L. Cardiac titin and heart disease. *J. Cardiovasc. Pharmacol.* **2014**, *63*, 207–212. [CrossRef] [PubMed]

51. Fassett, J.T.; Xu, X.; Kwak, D.; Wang, H.; Liu, X.; Hu, X.; Bache, R.J.; Chen, Y. Microtubule Actin Cross-Linking Factor 1 Regulates Cardiomyocyte Microtubule Distribution and Adaptation to Hemodynamic Overload. *PLoS ONE* **2013**, *8*, e73887. [CrossRef] [PubMed]

52. Allen, D.G.; Lamb, G.D.; Westerblad, H. Impaired calcium release during fatigue. *J. Appl. Physiol.* **2008**, *104*, 296–305. [CrossRef] [PubMed]

53. Chu, B.; Hong, Z.; Zheng, X. Acylglycerol Kinase-Targeted Therapies in Oncology. *Front. Cell Dev. Biol.* **2021**, *9*, 1948. [CrossRef] [PubMed]

54. Zhang, Z.; Jones, A.; Joo, H.Y.; Zhou, D.; Cao, Y.; Chen, S.; Erdjument-Bromage, H.; Renfrow, M.; He, H.; Tempst, P.; et al. USP49 deubiquitinates histone H2B and regulates cotranscriptional pre-mRNA splicing. *Genes Dev.* **2013**, *27*, 1581–1595. [CrossRef] [PubMed]
55. Todd, C.D.; Deniz, Ö.; Taylor, D.; Branco, M.R. Functional evaluation of transposable elements as enhancers in mouse embryonic and trophoblast stem cells. *eLife* **2019**, *8*, e44344. [CrossRef]
56. Capomaccio, S.; Verini-Supplizi, A.; Galla, G.; Vitulo, N.; Barcaccia, G.; Felicetti, M.; Silvestrelli, M.; Cappelli, K. Transcription of LINE-derived sequences in exercise-induced stress in horses. *Anim. Genet.* **2010**, *41*, 23–27. [CrossRef] [PubMed]
57. Choi, S.; Liu, X.; Li, P.; Akimoto, T.; Lee, S.Y.; Zhang, M.; Yan, Z. Transcriptional profiling in mouse skeletal muscle following a single bout of voluntary running: Evidence of increased cell proliferation. *J. Appl. Physiol.* **2005**, *99*, 2406–2415. [CrossRef]

 *genes*

Review

Considerations and Suggestions for the Reliable Analysis of miRNA in Plasma Using qRT-PCR

Eunmi Ban and Eun Joo Song *

Graduate School of Pharmaceutical Sciences, College of Pharmacy, Ewha Womans University, Seoul 03760, Korea; emban@korea.com
* Correspondence: esong@ewha.ac.kr

Abstract: MicroRNAs (miRNAs) are promising molecules that can regulate gene expression, and their expression level and type have been associated with early diagnosis, targeted therapy, and prognosis of various diseases. Therefore, analysis of miRNA in the plasma or serum is useful for the discovery of biomarkers and the diagnosis of implicated diseases to achieve potentially unprecedented progress in early treatment. Numerous methods to improve sensitivity have recently been proposed and confirmed to be valuable in miRNA detection. Specifically, quantitative reverse-transcription polymerase chain reaction (qRT-PCR) is an effective and common method for sensitive and specific analysis of miRNA from biological fluids, such as plasma or serum. Despite this, the application of qRT-PCR is limited, as it can be affected by various contaminants. Therefore, extraction studies have been frequently conducted to maximize the extracted miRNA amount while simultaneously minimizing contaminants. Moreover, studies have evaluated extraction efficiency and normalization of the extracted sample. However, variability in results among laboratories still exists. In this review, we aimed to summarize the factors influencing the qualification and quantification of miRNAs in the plasma using qRT-PCR. Factors influencing reliable analysis of miRNA using qRT-PCR are described in detail. Additionally, we aimed to describe the importance of evaluating extraction and normalization for reliable miRNA analysis and to explore how miRNA detection accuracy, especially from plasma, can be improved.

Keywords: qRT-PCR; plasma; miRNA; amplification efficiency

Citation: Ban, E.; Song, E.J. Considerations and Suggestions for the Reliable Analysis of miRNA in Plasma Using qRT-PCR. *Genes* **2022**, *13*, 328. https://doi.org/10.3390/genes13020328

Academic Editors: Giuseppe Iacomino and Fabio Lauria

Received: 5 January 2022
Accepted: 8 February 2022
Published: 10 February 2022

Publisher's Note: MDPI stays neutral with regard to jurisdictional claims in published maps and institutional affiliations.

1. Introduction

Circulating microRNAs (miRNAs) are highly stable extracellular molecules that circulate in the bloodstream [1,2]. These circulating miRNAs are approximately 22 nucleotides in length and play an important role in gene regulation by binding to and repressing the activity of specific target messenger RNAs (mRNAs). Profiles of miRNAs in plasma and serum have been found to be altered in cancer and other disease states [3,4]. Numerous studies have reported that specific miRNA expression profiles are associated with pathological conditions such as cardiovascular disease [5], cancer [6] and other diseases [7], which may provide diagnostic and therapeutic value as biomarkers. In previous studies, elevated plasma expression levels of miRNA-499 [8], miRNA-122 [9] and miRNA-155 [10] are known to be associated with AMI, liver injury, and inflammation, respectively. Meanwhile, the plasma expression levels of miRNA-34 [11] and miRNA-23a [12] are known to decrease in solid tumors and lung cancer, respectively.

Therefore, analyses of circulating miRNAs are important for the discovery and study of disease biomarkers that may aid in disease risk assessment, diagnosis, prognosis, and monitoring of treatment responses (Figure 1).

Currently, miRNA levels in biological fluids, tissues, and cells are measured after extraction by commercial RNA extraction kits, such as chloroform–phenol-based extraction [13,14], magnetic bead extraction [15], and column-based extraction [16], followed by microarray [17,18], Northern blotting [19,20], and quantitative reverse-transcription polymerase chain reaction (qRT-PCR) analysis [21,22]. Among these methods, qRT-PCR is widely preferred over other detection methods because of its high sensitivity and specificity for detecting low levels of circulating miRNAs in plasma and serum (Figure 2).

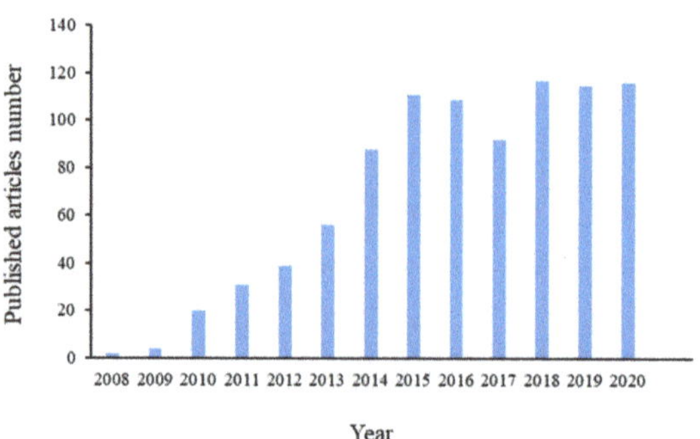

Figure 1. The number of PubMed search results regarding articles reporting on analyses of circulating miRNAs.

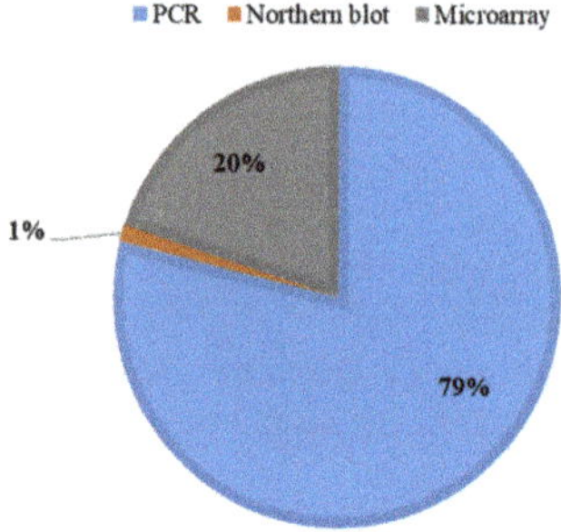

Figure 2. Methods for endogenous miRNA analysis in plasma or serum compared in terms of the proportions of articles reporting their use among PubMed search results from the past 10 years.

However, in previous studies, plasma miRNA levels have varied according to the laboratories performing the measurements [23,24]. This can mainly be due to differences in sample processing, measurements and data analysis [25–28]. In general, accurate miRNA measurement using qRT-PCR requires a high-quality sample, especially because of the low concentration of miRNA in plasma. Therefore, to increase the extraction efficiency and consistency of miRNAs analysis obtained from plasma and serum, numerous efforts over several years have focused on various highly sensitive and specific methods of miRNA extraction from plasma [29–31]. Additionally, similar efforts targeting data normalization [32,33] and the optimization and assessment of PCR conditions have been pursued to improve the accuracy of PCR measurements [34,35]. However, few reviews have focused on the importance and necessity of evaluating PCR conditions for optimizing extraction conditions.

In this review, which focuses on quantity and quality, we describe factors influencing miRNA measurement in plasma and serum using qRT-PCR. Additionally, we review the advantages of assessing PCR efficiency and normalization to obtain reliable and accurate PCR-based results of miRNA analysis in plasma. This evaluation calls attention to the importance of the assessment of PCR efficiency for optimizing PCR and miRNA extraction conditions.

2. Inconsistent Measurement of miRNA Extracted in Plasma Using qRT-PCR

Since miRNA analysis using qRT-PCR greatly depends on the quality of the miRNA extract, the results of miRNA analysis are different depending on the sampling procedure applied to the same sample. Many researchers have reported this inconsistency in miRNA analysis results, and various efforts are being made to analyze the factors that cause analysis inconsistency during the sample extraction step [31,36–38]. Brunet-Vega et al. compared the Cq levels of miRNAs extracted from the same plasma samples using five commercially available miRNA extraction kits [37]. They reported that the levels of the tested miRNAs were similar, but the levels of the spiked-in exogenous miRNAs were different. For this reason, the analyzed endogenous miRNAs were measured differently from the spike-in exogenous miRNAs. Another research group presented differences in the recovery and levels of miRNAs tested using two different miRNA extraction processes [31]. They also showed that the levels of tested miRNAs were affected by various sample treatments during the sample extraction process. Poel et al. also presented the effect of different carriers and pretreatment times on miRNA extraction recovery and showed inconsistent results between studies [38]. These studies recommended the need for standardization of protocols, including sample handling and extraction processes, to reduce the mismatch results of miRNAs in plasma between laboratories and between assays to perform reliable biomarker screening and discovery of miRNAs in plasma samples.

3. Factors Inhibiting Accurate miRNA Measurement in Plasma Using qRT-PCR

miRNA levels in plasma are low—10- and 100-times less than the concentrations in cells and tissues, respectively [39]. Therefore, the reliable and accurate analysis of miRNAs in plasma is a major issue, despite significant developments in the field. Among various analytical methods, qRT-PCR is usually used to analyze circulating miRNA levels in biological fluids, including plasma and serum, owing to its high specificity. However, qRT-PCR analysis can be compromised by various materials, such as matrix and extraction residual reagents in samples, and consequently, miRNA analysis results can vary depending on the purity of the extracted sample. Specifically, the effect of the interference on miRNA analysis in plasma is larger in cells and tissues because of the low abundance of miRNAs in plasma. Therefore, many studies have investigated the interference of miRNA analysis in plasma using qRT-PCR, which is primarily caused by sample components and the residual reagents extracted [36,40].

3.1. Sample Matrix

It is established that interference of qRT-PCR analysis is caused by various components present within the sample matrix. Therefore, for reliable qRT-PCR analysis, a high-purity sample from which components, such as proteins, lipids, and carbohydrates have been removed, must be prepared. The abundance of these confounding components varies significantly according to dietary status, anticoagulant type, as well as sampling and storage conditions [41,42]. Besides matrix components, hemolysis can be a major cause of variation in miRNA levels. Several miRNAs are found in large amounts in red blood cells (RBCs), and they are released from RBCs as a result of hemolysis, thereby increasing the level of certain miRNAs in the blood. However, hemolysis is more difficult to control than other conditions because it occurs frequently during blood sampling. Many authors have reported that miRNA qRT-PCR analysis results differ according to the degree of hemolysis of the sample. Myklebust et al. showed that qRT-PCR miRNA measurement is influenced by hemolysis [43]; in their study, the plasma miRNA concentration increased as the hemolyzed proportion of the sample increased, but the degree of increase depended on the miRNA type. Specifically, miR-16 is one of the most abundant miRNAs in RBCs [44], and many studies have shown that hemolysis may increase miR-16 levels in plasma. This is especially important because, due to its high abundance relative to other miRNAs, miRNA-16 is used as an endogenous reference gene to normalize the data after qRT-PCR analysis. Thus, hemolysis must be taken into consideration for accurate screening of blood miRNA levels.

Feng et al. also demonstrated the effect of the sample matrix on miRNA analysis using qRT-PCR [36]. They showed variability in miRNA levels among matrices with varying compositions, including in terms of types or levels of anticoagulant molecules, plasma protein, and lipids, and hemolysis was analyzed using qRT-PCR. The authors showed that the factors influencing the sample matrix (leading to variability in miRNA analysis in plasma using qRT-PCR) are primarily associated with dietary status, anticoagulant selection, and plasma sample storage conditions. Mompeón et al. also demonstrated the effect of hemolysis on miRNA analysis using qRT-PCR, observing that miRNA levels differed between serum and plasma [45]. In addition to differences in sample components and conditions, there may be differences in the degree of interference with qRT-PCR analysis due to extraction efficiency. Variations in analysis results may also be associated with the purity of the prepared sample, which depends on the sample extraction method. For this reason, various sample extraction kits and methods have been developed, and some of these have been reported to reduce intralaboratory and interlaboratory variability. Several research groups have presented and compared commercial RNA extraction kits, and studies related to the standardization of miRNA extraction methods from biological fluids have been conducted [46–48]. Column-based extraction kits often obtain high-quality miRNA extracts, and they are associated with lower extraction variation chloroform–phenol-based kits. However, chloroform–phenol-based kits, such as Trizol, are also known for favorable extraction recovery and costs.

3.2. Residual Reagents

For reliable qRT-PCR analysis, it is necessary to minimize any interfering factors using an extraction approach. For this reason, different extraction methods have been developed and applied to measure miRNA in cells, plasma, serum, and tissues. Currently, commercial miRNA extraction kits are largely divided into chloroform–phenol-based reagent kits and column-based extraction kits. Both methods include phenol extraction, which is a long-established approach to extracting nucleotides from biological samples. However, with phenol extraction, residual solvents, including phenol, remain in the final purified RNA sample after extraction and interfere with qRT-PCR-based miRNA analysis as a contaminant [49]. Among this interference caused by extraction reagents, residual phenol not only interferes with PCR analysis, but it can also cause errors in the quantification of RNA extracted from plasma. Specifically, the interference by residual phenol on miRNA analysis in plasma or serum is severe (relative to interference in cells and tissues) because of the low abundance of miRNAs in plasma and serum. For this reason, many researchers have investigated the issue of residual phenol. Spectrometric overestimations caused by residual phenol from extracted RNA yields have frequently led to inaccurate and variable plasma miRNA measurements. This problem is exacerbated by the fact that the wavelengths of RNA and phenol are similar, and the level of RNA in plasma is low compared with residual phenol. Several companies have, therefore, developed nanodrop systems to measure residual phenol concentrations in extracted samples to prevent mismeasurement of the amount or concentration of extracted RNAs by spectrometry [50,51]. Additionally, researchers have developed new systems that are not based on absorbance but that use specific fluorescent dyes for small RNAs to reliably measure extracted miRNAs [52,53]. For circulating biomarker detection analysis, accuracy could be optimized via the use of equal volume inputs rather than the same amount of RNA [54]. In plasma miRNA analysis by qRT-PCR, instead of direct RNA measurements, extraction recovery and analyzed samples are evaluated and normalized using spiked exogenous miRNA.

4. Important Considerations for Reliable miRNA Analysis Using qRT-PCR

Plasma miRNA analysis is an important area of biological and clinical research that is gaining increasing recognition. However, low plasma miRNA levels are associated with miRNA measurement errors and consequent inaccurate analyses [32]. These errors are mainly caused by interference in the sample matrix. Therefore, researchers have attempted to evaluate interference and develop normalization methods to minimize errors caused by such interference.

4.1. Amount of miRNA

Given that highly purified samples are needed for successful qRT-PCR analysis, many researchers have concentrated on developing methods that emphasize extracting high-quality RNA rather than high yields. Column-based extraction kits are often used for RNA extraction from various sample types. However, high quantities of RNA are also needed for reliable miRNA analysis, but it is challenging to efficiently extract RNAs from serum and plasma [55]. For this reason, chloroform–phenol-based extraction is still used to extract miRNAs from plasma, although several researchers have discussed the problems associated with chloroform–phenol-based extraction methods in PCR analysis. Various extraction kits have been compared and investigated [53]. In addition to extraction kits, various trials have been conducted, including comparisons between different modifications of extraction processes (such as the addition of carriers to trigger precipitation or the modification of incubation conditions) to increase extraction recovery without loss of quality. Some research teams have optimized carrier types, concentrations, and incubation conditions to maximize plasma miRNA extraction [56,57].

4.2. Normalization

Similar to other analytical methods, qRT-PCR analysis is subject to variations or errors, including in association with elements such as sample handling and volume measurements. Specifically, qRT-PCR plasma miRNA analysis can be greatly influenced by nutritional status, anticoagulant type, and plasma storage conditions. Therefore, normalization is important to mitigate variations in qRT-PCR analysis. One of methods for normalization of miRNA analysis by qRT-PCR is global normalization, which uses the calculated mean of all miRNAs in a given sample as the normalizer. This method is highly recommended when dealing with large scale miRNA expression profiling studies where several hundreds of miRNAs are analyzed [58]. However, global normalization cannot be applied for small-scale studies. Another popular method is normalization through reference genes. U6 is a small nuclear RNA commonly used as an endogenous internal control to normalize miRNA expression levels in different biological samples, including plasma. However, there is evidence that U6 plasma levels vary under certain conditions [59]. Therefore, various studies have been conducted to identify more reliable reference genes for normalizing endogenous plasma miRNA levels [60,61]. However, evaluations of different reported reference genes have yielded inconsistent findings [33,62–64]. Consequently, efforts have led to the identification of an appropriate endogenous miRNA for normalization, accounting for differences in plasma according to various factors, including disease status, sex, and age. External references, such as cel-miR-39-1 for normalization, are also frequently used to correct for extraction recovery and measurement. Zhang et al. showed the accuracy of normalization by reference gene candidates using exogenous miRNA (spiked-in cel-miR-39) as a target (Figure 3).

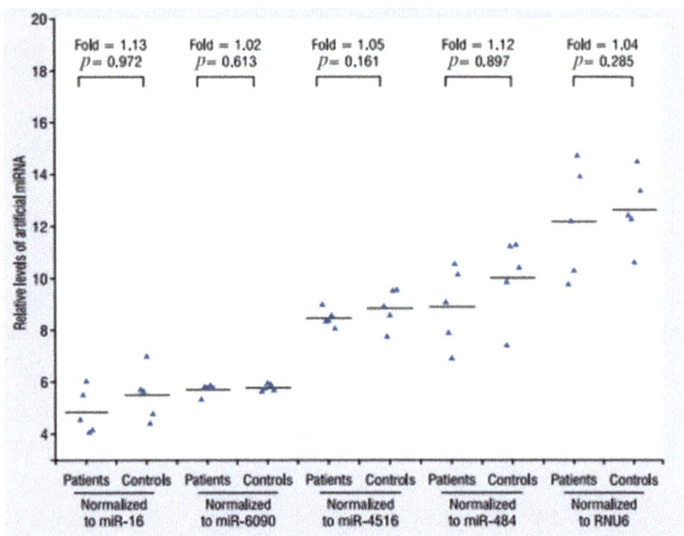

Figure 3. Effects of normalization by different reference gene methods on the expression levels of miRNAs from plasma samples of stable coronary artery disease patients and healthy controls ($n = 5$ in each group). For each group, 2 µL of exogenous cel-miR-39 was spiked into 300 µL plasma. The levels of cel-miR-39 were assessed by qRT-PCR and were normalized to miR-16, miR-6090, miR-4516, miR-484, and RNU6 [62].

Such an approach can eliminate multiple deviations of the experimental process, yielding more robust results. However, the approach also makes the experimental procedure cumbersome, and clinical applications more inconvenient. For example, cel-miR-39 was spiked into serum immediately before RNA extraction, allowing for the control of technical variation. However, the cel-miR-39 recovery was variable, ranging from 1% to 56%, thereby demonstrating the inherent need to take technical variability into account when performing absolute quantification [65]. Additionally, extraction kit-dependent differences in isolation yields across exogenous cel-miRs were reported. Nevertheless, the use of an exogenous cel-miR for normalization and correction presents less variability than strategies based on the concentration of endogenous components, such as the frequently used miR-16-5p [66]. Therefore, the normalization or correction strategy and, to a lesser extent, postanalytical concerns strongly limit the clinical implementation of miRNAs. To date, researchers are yet to establish a robust method of miRNA quantification for qRT-PCR that is clinically easy to implement and universally accepted.

4.3. Amplification Efficiency

In each cycle, qRT-PCR automatically detects the PCR amplification of a specific gene target. In PCR analysis, the number of target sequence molecules should double during each replication cycle, corresponding to 100% amplification efficiency. However, in practice, inappropriate reaction conditions and polymerase inhibition affect primer template annealing, resulting in decreased amplification efficiency and potentially leading to inaccurate conclusions. The assessment of factors affecting amplification efficiency provides information regarding inappropriate or suboptimal reaction conditions, as well as the presence of contaminants interfering with accurate qRT-PCR analysis. Therefore, qRT-PCR assays result in significant uncertainty due to variations in amplification and extraction efficiency [67,68]. For these reasons, several studies have investigated amplification efficiency to improve the accuracy and reliability of qRT-PCR analysis [23,37,69–71]. Brunet-Vega et al. demonstrated the necessity of exogenous genes through circulating miRNA profiling analysis using a commercial RNA extraction kit and exogenous genes to control for technical factors affecting final miRNA levels. Additionally, the observation that PCR efficiency reduces the variability of miRNAs circulating between samples should be validated because miRNA analysis in plasma using PCR can be affected by samples and PCR components [37]. Lebuhn et al. also demonstrated interlaboratory variation in qRT-PCR miRNA analysis in terms of amplification efficiency according to qRT-PCR-related factors, including interlaboratory differences in extraction steps [23]. Zununi Vahed and our team have successfully optimized extraction conditions through evaluations of amplification efficiency for reliable endogenous miRNA analysis using qRT-PCR [69,70]. Table 1 shows extraction method-dependent differences in amplification efficiency and quantification cycle (Ct) values of extracted miRNA.

Table 1. Mean quantification cycle (Ct), PCR efficiency, and correlation–coefficient (R^2) values of miR-21 isolation from cell lines, urine, and plasma by different methods.

Method	Body Fluids				Cell Lines				Urine Sediments			
	Ct	E (%)	R^2	Slope	Ct	E (%)	R^2	Slope	Ct	E (%)	R^2	Slope
KCH$_3$COOH	31.1 ± 0.4	103.54	0.995	−3.24	17.5 ± 0.07	99.5	0.992	−3.33	23.0 ± 0.3	100	0.998	−3.32
PEG4000	33.2 ± 1.0	111.5	0.993	−3.074	20.0 ± 0.13	95.49	0.996	−3.44	25.7 ± 0.45	98	1	−3.37
PEG6000	36.8 ± 0.2	91.99	0.977	−3.53	18.3 ± 0.32	116	0.983	−2.99	28.1 ± 0.74	86	0.976	−3.683
LiCl8M	34.8 ± 0.5	94.17	0.982	−3.47	21.8 ± 0.49	100.46	0.97	−3.31	31.7 ± 0.02	108	0.961	−3.145
Ethanol+LiCl	33.3 ± 0.07	99.46	0.994	−3.34	20.9 ± 0.9	105	0.998	−3.189	25.3 ± 0.62	114	0.993	−3.024
Ethanol	35.0 ± 0.09	120.02	0.979	−2.92	22.4 ± 0.03	98.03	0.991	−3.37	37.8 ± 0.63	105	0.982	−3.189

Data from three biological replicates of cell lines (HT-29 and HUVEC), body fluids (plasma), and urine samples. (Reproduced from [69]).

Svec et al. reported on factors associated with effective amplification efficiency [71]. Given that polymerase inhibition is caused by contaminants transferred from the RNA isolation process or sample matrix, for factors related to PCR reaction conditions, extraction conditions should be evaluated and optimized through assessments of amplification efficiency to reduce contaminants interfering with accurate qRT-PCR analysis. These studies have revealed that contaminants have a greater effect on qRT-PCR-based miRNA analysis from plasma than samples such as cells and tissues due to low levels of plasma miRNAs. Therefore, potential sources of interference in extracted plasma samples must be identified and corrected based on amplification efficiency before conducting qRT-PCR analysis. Reaction conditions such as annealing and primer conditions must first be evaluated and optimized in terms of amplification efficiency to ensure accurate analysis. Importantly, based on the evaluation of amplification efficiency, the specificity and sensitivity of qRT-PCR results can differ by primer type and concentration [72].

5. Discussion

Clinical and pharmaceutical research about plasma or serum miRNAs is becoming increasingly important. Consequently, endogenous plasma miRNA analysis has also become critical. Analysis of endogenous plasma miRNA is conducted using qRT-PCR, but such analyses have shown high variability between different laboratories and individuals [24]. One explanation for this is that miRNA levels in plasma are low, and qRT-PCR analysis is consequently affected by extraction and sample components. Therefore, several studies have investigated reproducible techniques and adjustments applied to miRNA analysis, such as sample extraction and normalization techniques. The high sensitivity of qRT-PCR as an analytical tool is matched by its sensitivity to interference by various factors. Therefore, the optimization of an effective extraction method is a major consideration for reliable PCR analysis, and many published articles report on extraction methods to minimize sample interference. Similarly, normalization and optimization of PCR conditions in terms of amplification efficiency have also been investigated, with consideration of issues, such as hemolysis, as major causes of interference. Specifically, normalization is heavily emphasized as an important factor facilitating reliable evaluation of plasma miRNA. Normally, reference genes are used to normalize endogenous miRNA, while exogenous miRNAs, such as cel-39-1, can also be added to samples to compensate for differences in extraction efficiency between samples [37]. However, these do not reflect extraction recovery because extraction efficiency differs between endogenous and exogenous miRNA, as does the effect of amplification efficiency on the environment of the extracted sample. Therefore, in addition to exogenous miRNA, appropriate reference genes are needed to normalize extracted endogenous miRNA levels, and reference gene selection must be prioritized because some reference genes can differ depending on the sample condition and type. With normalization, the evaluation of amplification efficiency is also crucial for the reliability of qRT-PCR studies [37,73,74]. Sreedharan et al. demonstrated improvements in the reliability of expression data through primer-dependent improvements in amplification efficiency [73]. Variations in primer concentration and annealing temperature, as well as primer design, can affect amplification efficiency and consequently affect the reliability of expression data. Optimization of PCR and extraction conditions through assessments of amplification efficiency might be important determinants of accurate and reliable qRT-PCR analysis of endogenous plasma miRNA.

This review describes the considerable variation and poor reproducibility of qRT-PCR-based plasma miRNA analysis associated with incomplete optimization of extraction and RT-PCR conditions through amplification efficiency and normalization. In the context of evaluating amplification, the use of exogenous and endogenous reference genes for normalization is necessary for the reliable and reproducible quantification of circulating miRNAs in plasma. These factors should be considered when translating the analysis of circulating miRNAs from plasma and serum into validated biomarker-based tests for routine clinical use. However, despite advances, such as the standardization of extraction processes and normalization for reliable qRT-PCR analysis of plasma miRNA, issues remain regarding the accuracy of qRT-PCR analysis due to individual differences in matrix composition. Therefore, as shown in Figure 4, we propose that the optimization of extraction conditions and the evaluation and identification of dependable reference genes (based on assessments of amplification efficiency) are necessary to ensure reliable and robust qRT-PCR-based miRNA analysis necessity for future applications of circulating miRNAs.

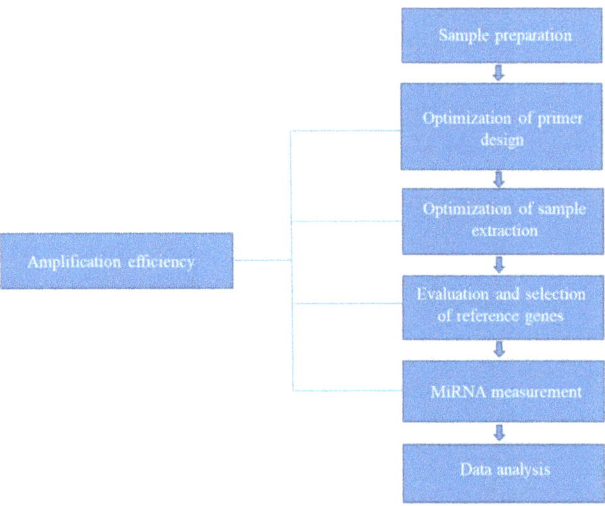

Figure 4. Suggested flowchart for qRT-PCR analysis of plasma miRNA.

6. Conclusions

This review presented factors influencing measurements of miRNAs in plasma/serum including assessment of PCR efficiency and normalization to obtain reliable and accurate

PCR-based results of miRNA analysis in plasma. It could suggest the necessity of the assessment of PCR efficiency for the optimization of PCR and miRNA extraction conditions. In this review, the effect of factors related to extraction efficiency such as sample matrix, residual solvent after extraction process and RNA amount was described among various factors influencing measurements of miRNAs using qRT-PCR. These factors may cause inhibitors of qRT-PCR analysis and consequently can lead to inaccurate qRT-PCR analysis. The necessity of amplification efficiency and normalization as another considerable part was reported for reliable and reproducible quantification of circulating miRNAs in plasma using qRT-PCR. From this review, we conclude that the optimization of extraction conditions and selection of reliable reference genes based on assessment of the amplification efficiency should be prioritized for achieving a reliable qRT-PCR-based miRNA analysis in plasma/serum.

Author Contributions: Conceptualization, E.B. and E.J.S.; investigation, E.B.; writing—original draft preparation, E.B.; writing—review and editing, E.B. and E.J.S.; supervision, E.J.S.; project administration, E.J.S.; funding acquisition, E.J.S. All authors have read and agreed to the published version of the manuscript.

Funding: This work was supported by a National Research Foundation of Korea (NRF) grant funded by the Ministry of Science and ICT (2020R1A4A4079494 and 2019R1A2C2004052), as well as a Korea Basic Science Institute (National Research Facilities and Equipment Center) grant funded by the Ministry of Education (2021R1A6C101A442).

Institutional Review Board Statement: Not applicable.

Informed Consent Statement: Not applicable.

Data Availability Statement: Not applicable.

Conflicts of Interest: The authors declare no conflict of interest.

References

1. Lindner, K.; Haier, J.; Wang, D.I.; Watson, D.J.; Hussey, R. Hummel, Circulating microRNAs: Emerging biomarkers for diagnosis and prog K. nosis in patients with gastrointestinal cancers. *Clin. Sci.* **2015**, *28*, 1–15. [CrossRef] [PubMed]
2. Mitchell, P.S.; Parkin, R.K.; Kroh, E.M.; Fritz, B.R.; Wyman, S.K.; Pogosova-Agadjanyan, E.L.; Peterson, A.; Noteboom, J.; O'Briant, K.C.; Allen, A.; et al. Circulating microRNAs as stable blood-based markers for cancer detection. *Proc. Natl. Acad. Sci. USA* **2008**, *105*, 10513–10518. [CrossRef] [PubMed]
3. Kawaguchi, T.; Komatsu, S.; Ichikawa, D.; Tsujiura, M.; Takeshita, H.; Hirajima, S.; Miyamae, M.; Okajima, W.; Ohashi, T.; Imamura, T.; et al. Circulating MicroRNAs: A Next-Generation Clinical Biomarker for Digestive System Cancers. *Int. J. Mol. Sci.* **2016**, *17*, 1459. [CrossRef] [PubMed]
4. Liao, T.-L.; Chen, Y.-M.; Hsieh, C.-W.; Chen, H.-H.; Lee, H.-C.; Hung, W.-T.; Tang, K.-T.; Chen, D.-Y. Upregulation of circulating microRNA-134 in adult-onset Still's disease and its use as potential biomarker. *Sci. Rep.* **2017**, *7*, 4214. [CrossRef] [PubMed]
5. Zhou, S.-S.; Jin, J.-P.; Wang, J.-Q.; Zhang, Z.-G.; Freedman, J.H.; Zheng, Y.; Cai, L. miRNAS in cardiovascular diseases: Potential biomarkers, therapeutic targets and challenges. *Acta Pharmacol. Sin.* **2018**, *39*, 1073–1084. [CrossRef]
6. Wu, Y.; Li, Q.; Zhang, R.; Dai, X.; Chen, W.; Xing, D. Circulating microRNAs: Biomarkers of disease. *Clin. Chim. Acta* **2021**, *516*, 46–54. [CrossRef] [PubMed]
7. Backes, C.; Meese, E.; Keller, A. Specific miRNA Disease Biomarkers in Blood, Serum and Plasma: Challenges and Prospects. *Mol. Diagn. Ther.* **2016**, *20*, 509–518. [CrossRef]
8. Corsten, M.F.; Dennert, R.; Jochems, S.; Kuznetsova, T.; Devaux, Y.; Hofstra, L.; Wagner, D.R.; Staessen, J.A.; Heymans, S.; Schroen, B. MicroRNA-208b and MicroRNA-499 Reflect Myocardial Damage in Cardiovascular Disease. *Circ. Cardiovasc. Genet.* **2010**, *3*, 499–506. [CrossRef]
9. Bandiera, S.; Pfeffer, S.; Baumert, T.F.; Zeisel, M.B. miR-122—A key factor and therapeutic target in liver disease. *J. Hepatol.* **2015**, *62*, 448–457. [CrossRef]
10. Mahesh, G.; Biswas, R. MicroRNA-155: A Master Regulator of Inflammation. *J. Interferon. Cytokine Res.* **2019**, *39*, 321–330. [CrossRef] [PubMed]
11. Zhang, L.; Liao, Y.; Tang, L. MicroRNA-34 family: A potential tumor suppressor and therapeutic candidate in cancer. *J. Exp. Clin. Cancer Res.* **2019**, *38*, 53. [CrossRef] [PubMed]
12. Hetta, H.F.; Zahran, A.M.; Shafik, E.A.; El-Mahdy, R.I.; Mohamed, N.A.; Nabil, E.E.; Esmaeel, H.M.; Alkady, O.A.; Elkady, A.; Mohareb, D.A.; et al. Circulating miRNA-21 and miRNA-23a Expression Signature as Potential Biomarkers for Early Detection of Non-Small-Cell Lung Cancer. *Microrna* **2019**, *8*, 206–215. [CrossRef] [PubMed]

13. Ban, E.; Chae, D.K.; Song, E.J. Simultaneous detection of multiple microRNAs for expression profiles of microRNAs in lung cancer cell lines by capillary electrophoresis with dual laser-induced fluorescence. *J. Chromatogr. A* **2013**, *1315*, 195–199. [CrossRef]

14. Kroh, E.; Parkin, R.; Mitchell, P.; Tewari, M. Analysis of circulating microRNA biomarkers in plasma and serum using quantitative reverse transcription-PCR (qRT-PCR). *Methods* **2010**, *50*, 298–301. [CrossRef]

15. Chen, S.; Shiesh, S.C.; Lee, G.B.; Chen, C. Two-step magnetic bead-based (2MBB) techniques for immunocapture of extracellular vesicles and quantification of microRNAs for cardiovascular diseases: A pilot study. *PLoS ONE* **2020**, *15*, e0229610. [CrossRef]

16. Guo, Y.; Vickers, K.; Xiong, Y.; Zhao, S.; Sheng, Q.; Zhang, P.; Zhou, W.; Flynn, C.R. Comprehensive evaluation of extracellular small RNA isolation methods from serum in high throughput sequencing. *BMC Genom.* **2017**, *18*, 50. [CrossRef] [PubMed]

17. Zhu, W.; Su, X.; Gao, X.; Dai, Z.; Zou, X. A label-free and PCR-free electrochemical assay for multiplexed microRNA profiles by ligase chain reaction coupling with quantum dots barcodes. *Biosens. Bioelectron.* **2014**, *53*, 414–419. [CrossRef] [PubMed]

18. Oxnard, G.R.; Paweletz, C.P.; Kuang, Y.; Mach, S.L.; O'Connell, A.; Messineo, M.M.; Luke, J.J.; Butaney, M.; Kirschmeier, P.; Jackman, D.M.; et al. Noninvasive Detection of Response and Resistance in EGFR-Mutant Lung Cancer Using Quantitative Next-Generation Genotyping of Cell-Free Plasma DNA. *Clin. Cancer Res.* **2014**, *20*, 1698–1705. [CrossRef] [PubMed]

19. Pall, G.S.; Hamilton, A.J. Improved northern blot method for enhanced detection of small RNA. *Nat. Protoc.* **2008**, *3*, 1077–1084. [CrossRef] [PubMed]

20. Huang, Q.; Mao, Z.; Li, S.; Hu, J.; Zhu, Y. A non-radioactive method for small RNA detection by northern blotting. *Rice* **2014**, *7*, 26. [CrossRef]

21. Xia, L.; Xia, W.; Li, S.; Li, W.; Chu, B. Identification and expression of small non-coding RNA, L10-Leader, in different growth phases of Streptococcus mutans. *Nucleic Acid Ther.* **2012**, *22*, 177–186. [CrossRef] [PubMed]

22. Gharbi, S.; Khateri, S.; Soroush, M.R.; Shamsara, M.; Naeli, P.; Mowla, S.J. MicroRNA expression in serum samples of sulfur mustard veterans as a diagnostic gateway to improve care. *PLoS ONE* **2018**, *13*, e0194530. [CrossRef] [PubMed]

23. Lebuhn, M.; Derenkó, J.; Rademacher, A.; Helbig, S.; Munk, B.; Pechtl, A.; Stolze, Y.; Prowe, S.; Schwarz, W.H.; Schlüter, A.; et al. DNA and RNA Extraction and Quantitative Real-Time PCR-Based Assays for Biogas Biocenoses in an Interlaboratory Comparison. *Bioengineering* **2016**, *3*, 7. [CrossRef] [PubMed]

24. Kloten, V.; Neumann, M.H.; Di Pasquale, F.; Sprenger-Haussels, M.; Shaffer, J.M.; Schlumpberger, M.; Herdean, A.; Betsou, F.; Ammerlaan, W.; Hällström, T.A.; et al. Multicenter Evaluation of Circulating Plasma MicroRNA Extraction Technologies for the Development of Clinically Feasible Reverse Transcription Quantitative PCR and Next-Generation Sequencing Analytical Work Flows. *Clin. Chem.* **2019**, *65*, 1132–1140. [CrossRef] [PubMed]

25. Kuang, J.; Yan, X.; Genders, A.J.; Granata, C.; Bishop, D.J. An overview of technical considerations when using quantitative real-time PCR analysis of gene expression in human exercise research. *PLoS ONE* **2016**, *13*, e0196438. [CrossRef]

26. Iguchi, T.; Niino, N.; Tamai, S.; Sakurai, K.; Mori, K. Absolute Quantification of Plasma MicroRNA Levels in Cynomolgus Monkeys, Using Quantitative Real-time Reverse Transcription PCR. *J. Vis. Exp.* **2018**, *132*, e56850. [CrossRef]

27. Binderup, H.G.; Madsen, J.S.; Heegaard, N.H.H.; Houlind, K.; Andersen, R.F.; Brasen, C.L. Quantification of microRNA levels in plasma—Impact of preanalytical and analytical conditions. *PLoS ONE* **2018**, *13*, e0201069. [CrossRef]

28. Parker, V.L.; Cushen, B.F.; Gavriil, E.; Marshall, B.; Waite, S.; Pacey, A.; Heath, P.R. Comparison and optimisation of microRNA extraction from the plasma of healthy pregnant women. *Mol. Med. Rep.* **2021**, *23*, 258. [CrossRef]

29. Androvic, P.; Romanyuk, N.; Urdzikova-Machova, L.; Rohlova, E.; Kubista, M.; Valihrach, L. Two-tailed RT-qPCR panel for quality control of circulating microRNA studies. *Sci. Rep.* **2019**, *9*, 4255. [CrossRef]

30. Liu, K.; Tong, H.; Li, T.; Wang, X.; Chen, Y. Research progress in molecular biology related quantitative methods of MicroRNA. *Am. J. Transl. Res.* **2020**, *12*, 3198–3321.

31. Fauth, M.; Hegewald, A.B.; Schmitz, L.; Krone, D.J.; Saul, M.J. Validation of extracellular miRNA quantification in blood samples using RT-qPCR. *FASEB BioAdvances* **2019**, *1*, 481–492. [CrossRef] [PubMed]

32. Gevaert, A.B.; Witvrouwen, L.; Vrints, C.J.; Heidbuchel, H.; Van Craenenbroeck, E.M.; Van Laere, S.J. MicroRNA profiling in plasma samples using qPCR arrays: Recommendations for correct analysis and interpretation. *PLoS ONE* **2018**, *13*, e0193173. [CrossRef]

33. Faraldi, M.; Gomarasca, M.; Sansoni, V.; Perego, S.; Banf, G.; Lombardi, G. Normalization strategies differently affect circulating miRNA profile associated with the training status. *Sci. Rep.* **2019**, *9*, 1584. [CrossRef]

34. Mei, Q.; Li, X.; Meng, Y.; Wu, Z.; Guo, M.; Zhao, Y.; Fu, X.; Han, W. A Facile and Specific Assay for Quantifying MicroRNA by an Optimized RT-qPCR Approach. *PLoS ONE* **2012**, *7*, e46890. [CrossRef] [PubMed]

35. Stein, E.V.; Duewer, D.L.; Farkas, N.; Romsos, E.L.; Wang, L.; Cole, K.D. Steps to achieve quantitative measurements of microRNA using two step droplet digital PCR. *PLoS ONE* **2017**, *12*, e0188085. [CrossRef] [PubMed]

36. Feng, X.; Liu, Y.; Wan, N. Plasma microRNA detection standardization test. *J. Clin. Lab. Anal.* **2020**, *34*, e23058. [CrossRef] [PubMed]

37. Brunet-Vega, A.; Pericay, C.; Quílez, M.E.; Ramírez-Lázaro, M.J.; Calvet, X.; Lario, S. Variability in microRNA recovery from plasma: Comparison of five commercial kits. *Anal. Biochem.* **2015**, *488*, 28–35. [CrossRef]

38. Poel, D.; Buffart, T.E.; Oosterling-Jansen, J.; Verheul, H.M.; Voortman, J. Evaluation of several methodological challenges in circulating miRNA qPCR studies in patients with head and neck cancer. *Exp. Mol. Med.* **2018**, *50*, e454. [CrossRef] [PubMed]

39. Pritchard, C.; Cheng, H.; Tewari, M. MicroRNA profiling: Approaches and considerations. *Nat. Rev. Genet.* **2012**, *13*, 358–369. [CrossRef] [PubMed]

40. Dellett, M.; Simpson, M.D.A. Considerations for optimization of microRNA PCR assays for molecular diagnosis. *Expert Rev. Mol. Diagn.* **2016**, *16*, 407–414. [CrossRef]

41. Mooney, C.; Raoof, R.; El-Naggar, H.; Sanz-Rodriguez, A.; Jimenez-Mateos, E.M.; Henshal, D.C. High Throughput qPCR Expression Profiling of Circulating MicroRNAs Reveals Minimal Sex- and Sample Timing-Related Variation in Plasma of Healthy Volunteers. *PLoS ONE* **2015**, *10*, e0145316. [CrossRef] [PubMed]

42. Mussbacher, M.; Krammer, T.L.; Heber, S.; Schrottmaier, W.C.; Zeibig, S.; Holthoff, H.-P.; Pereyra, D.; Starlinger, P.; Hackl, M.; Assinger, A. Impact of Anticoagulation and Sample Processing on the Quantification of Human Blood-Derived microRNA Signatures. *Cells* **2020**, *9*, 1915. [CrossRef] [PubMed]

43. Myklebust, M.P.; Rosenlund, B.; Gjengstø, P.; Bercea, B.S.; Karlsdottir, Á.; Brydøy, M.; Dahl, O. Quantitative PCR Measurement of miR-371a-3p and miR-372-p is influenced by hemolysis. *Front. Genet.* **2019**, *10*, 463. [CrossRef] [PubMed]

44. Merkerova, M.; Belickova, M.; Bruchova, H. Differential expression of microRNAs in hematopoietic cell lineages. *Eur. J. Haematol.* **2008**, *81*, 304–310. [CrossRef]

45. Mompeón, A.; Ortega-Paz, L.; Vidal-Gómez, X.; Costa, T.J.; Pérez-Cremades, D.; Garcia-Blas, S.; Brugaletta, S.; Sanchis, J.; Sabate, M.; Novella, S.; et al. Disparate miRNA expression in serum and plasma of patients with acute myocardial infarction: A systematic and paired comparative analysis. *Sci. Rep.* **2020**, *10*, 5373. [CrossRef] [PubMed]

46. Bryzgunova, O.; Konoshenko, M.; Zaporozhchenko, I.; Yakovlev, A.; Laktionov, P. Isolation of Cell-Free miRNA from Biological Fluids: Influencing Factors and Methods. *Diagnostics* **2021**, *11*, 865. [CrossRef]

47. Farina, N.H.; Wood, M.E.; Perrapato, S.D.; Francklyn, C.S.; Stein, G.S.; Stein, J.L.; Lian, J.B. Standardizing analysis of circulating microRNA: Clinical and biological relevance. *J. Cell Biochem.* **2014**, *115*, 805–811. [CrossRef] [PubMed]

48. Khan, J.; Lieberman, J.A.; Lockwood, C.M. Variability in, variability out: Best practice recommendations to standardize pre-analytical variables in the detection of circulating and tissue microRNAs. *Clin. Chem. Lab. Med.* **2017**, *55*, 608–621. [CrossRef] [PubMed]

49. Schrader, C.; Schielke, A.; Ellerbroek, L.; Johne, R. PCR inhibitors—Occurrence, properties and removal. *J. Appl. Microbiol.* **2012**, *113*, 1014–1026. [CrossRef] [PubMed]

50. Unger, C.; Lokmer, N.; Lehmann, D.; Axmann, L.M. Detection of phenol contamination in RNA samples and its impact on qRT-PCR results. *Anal. Biochem.* **2019**, *571*, 49–52. [CrossRef] [PubMed]

51. Lee, J.T.; Cheung, K.M.; Leung, V.Y. Correction for concentration overestimation of nucleic acids with phenol. *Anal. Biochem.* **2014**, *465*, 179–186. [CrossRef] [PubMed]

52. Garcia-Elias, A.; Alloza, L.; Puigdecanet, E.; Nonell, L.; Tajes, M.; Curado, J.; Enjuanes, C.; Diaz, O.; Bruguera, J.; Marti-Almor, J.; et al. Defining quantification methods and optimizing protocols for microarray hybridization of circulating microRNAs. *Sci. Rep.* **2017**, *7*, 7725. [CrossRef]

53. Wright, K.; de Silva, K.; Purdie, A.C.; Plain, K.M. Comparison of methods for miRNA isolation and quantification from ovine plasma. *Sci. Rep.* **2020**, *10*, 825. [CrossRef]

54. El-Khoury, V.; Pierson, S.; Kaoma, T.; Bernardin, F.; Berchem, G. Assessing cellular and circulating miRNA recovery: The impact of the RNA isolation method and the quantity of input material. *Sci. Rep.* **2016**, *6*, 19529. [CrossRef] [PubMed]

55. Niu, Y.; Zhang, L.; Qiu, H.; Wu, Y.; Wang, Z.; Zai, Y.; Liu, L.; Qu, J.; Kang, K.; Gou, D. An improved method for detecting circulating microRNAs with S-Poly(T) Plus real-time PCR. *Sci. Rep.* **2015**, *5*, 15100. [CrossRef] [PubMed]

56. Ban, E.; Chae, D.K.; Yoo, Y.S.; Song, E.J. An improvement of miRNA extraction efficiency in human plasma. *Anal. Bioanal. Chem.* **2017**, *409*, 6397–6404. [CrossRef] [PubMed]

57. Duy, J.; Koehler, J.W.; Honko, A.N.; Minogue, T.D. Optimized microRNA purification fromTRIzol-treated plasma. *BMC Genomics* **2015**, *16*, 95. [CrossRef] [PubMed]

58. Mestdagh, P.; Van Vlierberghe, P.; De Weer, A.; Muth, D.; Westermann, F.; Speleman, F.; Vandesompele, J. A novel and universal method for microRNA RT-qPCR data normalization. *Genome Biol.* **2009**, *10*, R64. [CrossRef] [PubMed]

59. Xiang, M.; Zeng, Y.; Yang, R.; Xu, H.; Chen, Z.; Zhong, J.; Xie, H.; Xu, Y.; Zeng, X. U6 is not a suitable endogenous control for the quantification of circulating microRNAs. *Biochem. Biophys. Res. Commun.* **2014**, *454*, 210–214. [CrossRef] [PubMed]

60. Donati, S.; Ciuffi, S.; Brandi, M.L. Human Circulating miRNAs Real-time qRT-PCR-based Analysis: An Overview of Endogenous Reference Genes Used for Data Normalization. *Int. J. Mol. Sci.* **2019**, *20*, 4353. [CrossRef]

61. Madadi, S.; Schwarzenbach, H.; Lorenzen, J.; Soleimani, M. MicroRNA expression studies: Challenge of selecting reliable reference controls for data normalization. *Cell Mol. Life Sci.* **2019**, *76*, 3497–3514. [CrossRef]

62. Zhang, Y.; Tang, W.; Peng, L.; Tang, J.; Yuan, Z. Identification and validation of microRNAs as endogenous controls for quantitative polymerase chain reaction in plasma for stable coronary artery disease. *Cardiol. J.* **2016**, *6*, 694–703. [CrossRef]

63. Wang, X.; Zhang, X.; Yuan, J.; Wu, J.; Deng, X.; Peng, J.; Wang, S.; Yang, C.; Ge, J.; Zou, Y. Evaluation of the performance of serum miRNAs as normalizers in microRNA studies focused on cardiovascular disease. *J. Thorac. Dis.* **2018**, *10*, 2599–2607. [CrossRef]

64. Schwarzenbach, H.; da Silva, A.M.; Calin, G.; Pantel, K. Which is the accurate data normalization strategy for microRNA quantification? *Clin. Chem.* **2015**, *61*, 1333–1342. [CrossRef]

65. Sanders, I.; Holdenrieder, S.; Walgenbach-Brünagel, G.; von Ruecker, A.; Kristiansen, G.; Müller, S.C.; Ellinger, J. Evaluation of reference genes for the analysis of serum miRNA in patients with prostate cancer, bladder cancer and renal cell carcinoma. *Int. J. Urol.* **2012**, *19*, 1017–1025. [CrossRef]

66. Vigneron, N.; Meryet-Figuière, M.; Guttin, A.; Issartel, J.P.; Lambert, B.; Briand, M.; Louis, M.H.; Vernon, M.; Lebailly, P.; Lecluse, Y.; et al. Towards a new standardized method for circulating miRNAs profiling in clinical studies: Interest of the exogenous normalization to improve miRNA signature accuracy. *Mol. Oncol.* **2016**, *10*, 981–992. [CrossRef]

67. Ramshani, Z.; Zhang, C.; Richards, K.; Chen, L.; Xu, G.; Stiles, B.L.; Hill, R.; Senapati, S.; Go, D.B.; Chang, H.-C. Extracellular vesicle microRNA quantification from plasma using an integrated microfluidic device. *Commun. Biol.* **2019**, *2*, 189. [CrossRef]

68. Butz, H.; Patócs, A. *Circulating microRNAs in Disease Diagnostics and Their Potential Biological Relevance*; Igaz, P., Ed.; Springer: Basel, Switzerland, 2015; pp. 55–71.

69. Zununi Vahed, S.; Barzegari, A.; Rahbar Saadat, Y.; Mohammadi, S.; Samadi, N. A microRNA isolation method from clinical samples. *BioImpacts* **2016**, *6*, 25–31. [CrossRef]

70. Ban, E.; Kwon, H.; Seo, H.S.; Yoo, Y.S.; Song, E.J. Screening of miRNAs in plasma as a diagnostic biomarker for cardiac disease based on optimization of extraction and qRT-PCR condition assay through amplification efficiency. *BMC Biotechnol.* **2021**, *21*, 50. [CrossRef]

71. Svec, D.; Tichopad, A.; Novosadova, V.; Pfaffl, M.W.; Kubista, M. How good is a PCR efficiency estimate: Recommendations for precise and robust qPCR efficiency assessments. *Biomol. Detect. Quantif.* **2015**, *3*, 9–16. [CrossRef]

72. Balcells, I.; Citrera, S.; Busk, P.K. Specific and sensitive quantificative RT-PCR of miRNAs with DNA primers. *BMC Biotechnol.* **2011**, *11*, 70. [CrossRef]

73. Sreedharan, S.P.; Kumar, A.; Giridhar, P. Primer design and amplification efficiencies are crucial for reliability of quantitative PCR studies of caffeine biosynthetic N-methyltransferases in coffee. *3 Biotech.* **2018**, *8*, 467. [CrossRef]

74. Cirera, S.; Andersen-Ranberg, E.U.; Langkilde, S.; Aaquist, M.; Greda, H. Challenges and standardization of microRNA profiling in serum and cerebrospinal fluid in dogs suffering from non-infectious infammatory CNS disease. *Acta Vet. Scand.* **2019**, *61*, 57. [CrossRef]

Protocol

Molecular Investigation of miRNA Biomarkers as Chemoresistance Regulators in Melanoma: A Protocol for Systematic Review and Meta-Analysis

Peter Shaw [1,2], Greg Raymond [3], Katherine S. Tzou [4], Siddhartha Baxi [5], Ravishankar Ram Mani [6], Suresh Kumar Govind [7], Harish C. Chandramoorthy [8], Palanisamy Sivanandy [9], Mogana Rajagopal [10], Suja Samiappan [11], Sunil Krishnan [4] and Rama Jayaraj [12,*]

[1] Department of Artificial Intelligence, Nanjing University of Information Science and Technology (NUIST), Nanjing 211544, China; 100001@nuist.edu.cn
[2] Menzies School of Health Research, Darwin 0810, Australia
[3] Northern Territory Medical Program, CDU Campus, Flinders University, Ellengowan Drive, Darwin 0909, Australia; greg.raymond@flinders.edu.au
[4] Department of Radiation Oncology, Mayo Clinic Florida, 4500 San Pablo Road S, Jacksonville, FL 32224, USA; tzou.katherine@mayo.edu (K.S.T.); Krishnan.Sunil@mayo.edu (S.K.)
[5] Genesis Care Gold Coast Radiation Oncologist, John Flynn Hospital, 42 Inland Drive, Tugun 4224, Australia; Siddhartha.baxi@genesiscare.com
[6] Department of Pharmaceutical Biology, Faculty of Pharmaceutical Sciences, UCSI University Kuala Lumpur (South Wing), No.1, Jalan Menara Gading, UCSI Heights, Cheras, Kuala Lumpur 56000, Malaysia; Ravishankar@ucsiuniversity.edu.my
[7] Department of Parasitology, Faculty of Medicine, University of Malaya, Kuala Lumpur 50603, Malaysia; suresh@um.edu.my
[8] Stem Cells and Regenerative Medicine Unit, Department of Microbiology and Clinical Parasitology, College of Medicine, King Khalid University, Abha 61421, Saudi Arabia; hshkonda@kku.edu.sa
[9] Department of Pharmacy Practice, School of Pharmacy, International Medical University, Bukit Jalil, Kuala Lumpur 57000, Malaysia; PalanisamySivanandy@imu.edu.my
[10] Faculty of Pharmaceutical Sciences, UCSI University Kuala Lumpur (South Wing), No.1, Jalan Menara Gading, UCSI Heights, Cheras, Kuala Lumpur 56000, Malaysia; mogana@ucsiuniversity.edu.my
[11] Department of Biochemistry, Bharathiar University, Coimbatore 641046, India; sujaramalingam08@gmail.com
[12] Northern Territory Institute of Research and Training, Darwin 0909, Australia
* Correspondence: jramamoorthi@gmail.com

Citation: Shaw, P.; Raymond, G.; Tzou, K.S.; Baxi, S.; Mani, R.R.; Kumar Govind, S.; Chandramoorthy, H.C.; Sivanandy, P.; Rajagopal, M.; Samiappan, S.; et al. Molecular Investigation of miRNA Biomarkers as Chemoresistance Regulators in Melanoma: A Protocol for Systematic Review and Meta-Analysis. *Genes* **2022**, *13*, 115. https://doi.org/10.3390/genes13010115

Academic Editors: Rajiv Kumar, Giuseppe Iacomino and Fabio Lauria

Received: 6 November 2021
Accepted: 5 January 2022
Published: 8 January 2022

Publisher's Note: MDPI stays neutral with regard to jurisdictional claims in published maps and institutional affiliations.

Abstract: Introduction: Melanoma is a global disease that is predominant in Western countries. However, reliable data resources and comprehensive studies on the theragnostic efficiency of miRNAs in melanoma are scarce. Hence, a decisive study or comprehensive review is required to collate the evidence for profiling miRNAs as a theragnostic marker. This protocol details a comprehensive systematic review and meta-analysis on the impact of miRNAs on chemoresistance and their association with theragnosis in melanoma. Methods and analysis: The articles will be retrieved from online bibliographic databases, including Cochrane Review, EMBASE, MEDLINE, PubMed, Scopus, Science Direct, and Web of Science, with different permutations of 'keywords'. To obtain full-text papers of relevant research, a stated search method will be used, along with selection criteria. The Preferred Reporting Items for Systematic Reviews and Meta-Analysis for Protocols 2015 (PRISMA-P) standards were used to create this study protocol. The hazard ratio (HR) with a 95% confidence interval will be analyzed using Comprehensive Meta-Analysis (CMA) software 3.0. (CI). The pooled effect size will be calculated using a random or fixed-effects meta-analysis model. Cochran's Q test and the I2 statistic will be used to determine heterogeneity. Egger's bias indicator test, Orwin's and the classic fail-safe N tests, the Begg and Mazumdar rank collection test, and Duval and Tweedie's trim and fill calculation will all be used to determine publication bias. The overall standard deviation will be evaluated using Z-statistics. Subgroup analyses will be performed according to the melanoma participants' clinicopathological and biological characteristics and methodological factors if sufficient studies and retrieved data are identified and available. The source of heterogeneity will be assessed using a meta-regression analysis. A pairwise matrix could be developed using either a pairwise correlation or expression associations of miRNA with patients' survival for the same studies.

Keywords: chemoresistance; chemosensitivity; melanoma; meta-analysis; miRNAs; protocol; systematic review

---

1. Introduction

1.1. Epidemiology

Skin cancers are uncommon malignancies globally and do not rank among the top ten common cancers [1]. Despite melanoma not being the leading cause of cancer mediated deaths, deaths from melanoma are rising, and it has a vastly inferior prognosis compared to other common types of cancer. The three primary types of skin cancer are basal cell carcinoma, squamous cell carcinoma, and malignant melanoma. Squamous and basal cell carcinoma together are referred to as non-melanoma skin cancers [2]. Among the known skin cancers, the most commonly occurring type is basal cell carcinoma [3]. Despite the fact the global incidence of melanoma (1.6 percent) is lower than that of non-melanoma skin cancers (6.2 percent), melanoma is still considered a progressive disease and is the deadliest form of skin cancer [4]. About 75% of the skin-cancer associated deaths are due to melanoma [5]. It is a rare type of skin cancer that progresses to other parts of the body. Its risk factors include exposure of skin to ultraviolet (UV) light and other factors, such as genetics [6,7]. Although surgery is the standard treatment after diagnosis, prevention strategies are prioritized, as the best way to avoid developing melanoma is by limiting direct exposure to sunlight and, therefore, overexposure to UV light [8].

1.2. Rationale

1.2.1. The Importance of This Study

Chemoresistance continues to be a significant impediment to cancer treatment in medical oncology. Resistance may occur due to prior exposure or even as a result of cancer therapy itself [9,10]. Thus, research on treatment strategies, such as the multimodality approach involving surgery, chemotherapy, radiotherapy, and immune/biotherapy, is currently being conducted in order to circumvent the issue of the development of chemoresistance [11]. miRNAs' involvement in melanoma chemoresistance has not been effectively explored [12–14]. Recent and emerging studies on this topic have generated sufficient clinical data to make a more feasible approach to perform a meta-analysis and systematic review on melanoma patients' chemoresistance [15–17].

1.2.2. What Will the Study's Approach Be to This Problem?

The suggested study has the potential to provide a comprehensive picture of chemoresistance in melanoma and its relationship to miRNA expression. Non-coding RNAs called microRNAs influence gene expression. miRNAs are small RNAs with a length of 19–25 nucleotides that inhibit or degrade genes at the post-transcriptional phase [18]. Several studies have focused on miRNAs' impact on chemotherapeutic resistance in cancers, including breast cancer [19], cervical cancer [20], colorectal cancer [21], gastric cancer [22], lung cancer [23], oral cancer [24], ovarian cancer [25], pancreatic cancer [26] prostate cancer [27], and skin cancer [28]. With a 5-year survival rate of 92 percent, surgery remains the best choice for curing localized, invasive melanoma [29]. The molecular basis of melanoma resistance to chemotherapy is thought to be multifactorial, with a defective drug transport system, an altered apoptotic pathway, apoptosis deregulation, and changes in the enzymatic systems that mediate cellular metabolic machinery all contributing to chemotherapy complications [30].

There have also been several meta-analyses and systematic reviews considering the link between miRNAs and chemoresistance [31,32]. However, topics, such as the clinical outcome predictions of miRNAs in cancer [33], the miRNA prognostic signatures cross-validated in metastatic melanoma [34], and the correlation between DNA repair gene polymorphism and cutaneous melanoma, still require further investigation [35]. Understanding the impact of changes in chemoresistance-related biological processes could aid in the development of new therapeutic approaches for malignant melanoma treatment [36].

1.2.3. How Will It Help?

The role played by microRNAs in chemoresistance is found to be complex, and linking distinct miRNAs to different genetic pathways is still in its infancy. Clinical samples can benefit from miRNA profiling and can allow for the distinguishing of cancerous cells from normal cells and could be a useful tool for classifying poorly differentiated tumors. Providing a detailed systematic review may aid oncologists, gastroenterologists, and clinical researchers to expand their understanding of the theragnostic and predictive role of miRNAs and the potential implementation of these biomarkers for future clinical practice. Therefore, studies analyzing the effects of miRNA expression on chemoresistance and sensitivity in melanoma patients, as well as studies exploring the effects of miRNAs on chemotherapy via in vitro experimentation, will be included in our review. Our study will provide a network of chemoresistance mechanisms and drug regulatory pathways in conjunction with the different chemotherapy drugs commonly utilized in melanoma. Thus, it is hoped that identifying the specific miRNAs and the associated pathways of chemoresistance in melanoma may help in the development of future therapeutics by indicating how miRNAs' profiles could predict the efficacy of chemotherapy and chemoresistance. The results obtained from the meta-analysis will ideally help improve clinical treatment and prognosis [37,38]. This protocol and the study following it should act as a reference for future studies regarding the prognosis and diagnosis of melanoma using microRNAs and, thereby, help in the proliferation of literature in this field.

2. Methods

2.1. Study Design

An all-encompassing search approach will be carried out using the bibliographic databases Cochrane Review, MEDLINE, EMBASE, PubMed, Science Direct, Scopus, and Web of Science for the last ten years. Previous studies evaluating the role of miRNAs on chemoresistance and sensitivity in melanoma will be identified. Additional studies will be extracted from the included studies' reference lists through a manual search and also from the review of literature articles that discuss melanoma chemoresistance. The publication language will be limited to English, including officially translated materials, with no restrictions on the publication date or status.

2.1.1. Eligibility Criteria

Inclusion Criteria

As a key inclusion criterion, investigations must examine at the impact of miRNA expression in melanoma patients and cell lines.

Other norms will include:

- Studies that deal with resistance in melanoma.
- Studies published until December 2021.
- Reporting of miRNA profiling platforms.
- Studies with appropriate patient data with therapeutic measures.
- Studies reporting the genes and/or pathways involved in chemoresistance or chemosensitivity.
- miRNA expression analysis using in vitro assays.
- Studies reporting the patient's survival with 95% CI (confidence interval) values in hazard ratio (HR) or Kaplan–Meier (KM) curves for quantitative synthesis or meta-analysis.

Exclusion Criteria

The following will be excluded from the study:

- Letters to the editor, fact sheets, conference proceedings, unpublished materials, review articles, case studies, and studies conducted solely in patients or in vitro.
- Studies examining patient data from bioinformatic datasets.
- Duplicate publications from the same study will be treated as one study.
- Studies using non-human data.

Search Strategy and Study Selection

Databases will be used to identify the literature that is related to miRNAs, drug resistance, and melanoma. The search terms used should be in all combinations of "miRNA" or "microRNA" AND "Drug resistance" or "Chemosensitivity" or "Chemoresistance" AND "Melanoma" (Table 1). Additional relevant articles will be identified by manually examining the retrieved articles. Potentially relevant articles will be carefully collated for further processing. The studies will initially be chosen based on the individual judgement of two authors upon the reading of the titles and abstracts of the articles. Full-text articles will be scrutinized if the titles and abstracts are uncertain. All authors will be contacted for pertinent information. Any disagreement will be solved by discussion amongst the two authors. Any major differences will involve a team decision or third reviewer to make a decision.

Table 1. Search terms.

Search Number	Parameter
1	Melanoma "[Topic]" OR miRNA "[Topic]"
2	Melanoma "[Topic]" OR miRNA "[Topic]" OR patient "[Topic]" OR clinical study "[Topic]"
3	Melanoma "[Topic]" OR miRNA "[Topic]" OR microRNA "[Topic]" AND resistance "[Topic]" OR patient "[Topic]" OR clinical study "[Topic]"
4	Melanoma "[Topic]" OR miRNA "[Topic]" OR microRNA "[Topic]" AND chemoresistance (Chemoresist*) "[Topic]" OR patient "[Topic]" OR clinical study "[Topic]"
5	Melanoma "[Topic]" OR miRNA "[Topic]" OR microRNA "[Topic]" AND chemosensitivity (Chemosens*) "[Topic]" OR patient "[Topic]" OR clinical study "[Topic]"
6	1 AND 2 AND 3 AND 4 AND 5

* The search terms "Chemosensitivity" or "Chemoresistance" will be substituted by wildcards, such as "Chemosens*" or "Chemoresist*".

Data Extraction and Management

The studies in the selection criteria will be evaluated individually and the respective authors will be contacted by the authors. to gather any missing information. The data extraction form will collect bibliographic and demographic information, as well as clinico-pathological and biological aspects of melanoma patients if relevant data and information are available. Data from the included studies will be reviewed by three authors and cross-checked by the corresponding author. The corresponding authors of the selected articles will be contacted for further clarifications.

2.1.2. Data Collection Process

1. From the studies, five major categories of data will be extracted: The study characteristics, including the author, geographic region, year of publication, study period, sample size, study design, sampling procedures, validity of confirmative diagnosis, method of data collection, and number of melanoma cancer cases/patients, as well as the International Classification of Disease (ICD) Code for the anatomical site of cancer under study.
2. Clinical, pathological, and biological attributes, including comorbidity, risk factors, tumor histology (squamous, adenocarcinoma, clear cell, and undifferentiated), pathological grades (1, 2, and 3), tumor size, negative and positive lymph node metastasis, positive and negative vascular involvement, the lymphocyte infiltration (if any), histology grade (well, moderate, poor, and undetermined), P16 (positive and negative), deep stromal invasion (%), and specific body sites, such as the face (the temporal, frontal, periorbital, infraorbital, buccal, zygomatic, mental, or perioral region), nose, lip, ear, scalp, trunk, neck, and extremities [39].
3. miRNA expression in melanoma patients and their responses towards their treatment.
4. Hazard ratio (HR) and 95% confidence interval (CI) estimates of overall survival (OS), disease-free survival (DFS), and other endpoint measures.
5. In vitro and in vivo studies.

Outcomes and Prioritization

The primary outcome is to evaluate the role of the miRNAs associated with chemoresistance in melanoma patients.

Secondary outcomes will be used to correlate variations in primary outcomes with clinicopathological and biological parameters.

Quality Assessment of Included Studies

The Dutch Cochrane Centre's Meta-Analysis Of Observational Studies in Epidemiology (MOOSE) guidelines [40] will be used to assess the quality of the included studies, and the following information will be extracted:

i.   Information about the patient's tissue collection.
ii.   Location of the study.
iii. Gender.
iv.   Age.
v.   Exposure to sunlight.
vi.   Ulceration status.
vii. miRNA analysis in melanoma patients.
viii. List of melanoma cell lines used.
ix.   Tumor stage.
x.   Lymph node status.
xi.   miRNA profiling platform.
xii. The form of therapy used.
xiii. Genes and/or pathways involved in resistance.

All the mentioned criteria will be required for the study to be qualified for the systematic review. The Newcastle-Ottawa scale (NOS) will also be used to assess the methodological quality of cohort studies [41].

Assessment of Risk of Bias in Individual Studies

The authors will assess the risk of bias based on parameters, such as the number of patients studied, the year of publication, the mode of disease diagnosis, geographical demarcation, and the length of the study. A predetermined checklist incorporating questions from eight categories from the Dutch Cochrane Centre's Meta-Analysis Of Observational Studies in Epidemiology (MOOSE) guidelines will be used to assess the quality of the studies [40]. Six elements will be included in the tool's reporting: background, search strategy, techniques, results, discussion, and conclusions. The reporting elements of the checklist are based on epidemiological concepts and will be provided even if individual studies lack strong empirical evidence [42–45].

Publication Bias

A significant concern in meta-analysis is the risk of publication bias [46–50]. To understand publication bias, Egger's and Begg's bias indicator tests, as well as the inverted funnel plot, will be used [51]. Trim and fill calculations by Duval and Tweedie will also be evaluated [52]. To investigate the effect size of statistically non-significant and unpublished studies, the classic and Orwin's fail-safe N tests will be used [53–58]. All the authors will assess publication bias individually. Team decisions will be involved in case of disagreements.

2.1.3. Statistical Analysis
Meta-Analysis

Meta-regression analysis will be used to study the heterogeneity between the involved studies. Potential influences, such as the number of patients, year of publication, study period, research location, kind of study, and diagnostic process will be investigated for heterogeneity using the Higgins I-squared statistic [59] and Cochran's Q test [60].

The hazard ratio (HR) will be analyzed using the comprehensive meta-analysis software (CMA) 3.0 with a 95% CI (confidence interval). Fixed model effects will be used in significant studies and, if studies are not significant, random model effects will be used. Z-statistics will be used to calculate the overall standard deviation.

Subgroup Analyses

Subgroup analyses will be performed based on the melanoma participants' clinical, pathological, and biological characteristics, as well as methodological aspects, if adequate studies and recovered data are discovered and accessible. Our research team intends to look into particular subgroup analyses based on clinical and pathological factors and biological information, such as comorbidity, risk factors, tumor histology (squamous, adenocarcinoma, clear cell, and undifferentiated), pathological grades (1, 2, and 3), tumor size, negative and positive lymph node metastasis, negative and positive vascular involvement, histology grade (well, moderate, poor, and undetermined), P16 (positive and negative), deep stromal invasion (percentage), and specific body sites, such as the face (the temporal, frontal, periorbital, infraorbital, buccal, zygomatic, mental, or perioral region), nose, lip, ear, neck, scalp, trunk, and other parameters.

Meta-Regression

A meta-regression analysis will be used to determine the source of heterogeneity. A P-value of less than 0.05 will be considered significant for heterogeneity. Gender distribution, data collection methods, research quality, sample size, and sampling procedure will all be evaluated. In order to weigh every study by calculating R^2 with the proposed quantity variance, a random-effects model will be employed. A meta-regression analysis will be used to explain the heterogeneity of cancer research in relation to one or more study variables, with a large ratio of studies required for a genuine regression. For each deviation, a ratio of at least 10 is recommended [61–64].

Network-Centric Model Analysis

A pairwise matrix could be developed using either pairwise correlation or expression associations of miRNA with patients' survival for the same papers listed in this protocol [65,66]. A clique-centric pattern search using cluster editing could then be used to identify pathways/systems [67,68]. The details of this analysis toolkit are summarized in Figure 1. The strength of this approach stems from the fact that that it provides more insight into the upregulated and downregulated expression of the miRNAs and the melanoma cancers that are not possible with other statistical methods such as principle component analysis.

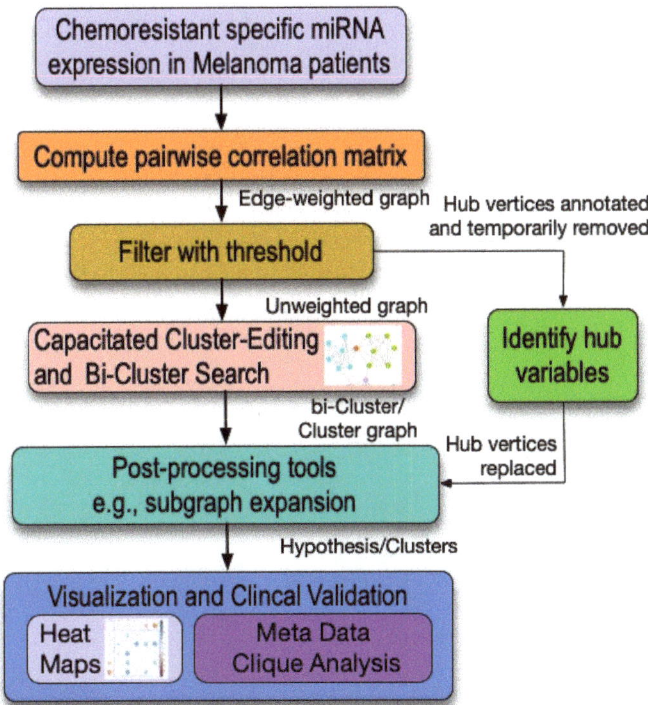

Figure 1. Network analysis toolchain. The output of the produced clustering would consist of interacting RNA or patient symptoms (also known as a generated hypothesis). To confirm and consider this hypothesis, a systematic review or meta-data search for papers that may already have considered two or more of the factors listed in these clusters should be carried out.

Random Forest Analysis

A random forest analysis provides a robust means of feature selection of miRNA expression. The results of this analysis can then be used to develop prognostic value tools, such as decision trees. By coupling a random forest analysis with the other ensemble methods, such as those provided by the R interface for 'H2O'(R H2O package), the scalable open source machine learning platform, artificial intelligence (AI) prognosis tools could also be produced [69]. Nevertheless, a random forest analysis is also an effective way of identifying robust features for predictive modelling. Alternatively, a principal component analysis (PCA) and a clustering algorithm (such as k-means) can be used. However, these techniques work, as long as the number of attributes or dimensions does not exceed five for most of the variation [70,71].

2.2. Presenting and Reporting the Review Results

This protocol was written in accordance with the PRISMA-P statement (http://www. prisma-statement.org/Extensions/Protocols; accessed on 9 September 2021) [72]. The findings will be made public in accordance with the PRISMA criteria [73]. A flowchart outlining the selection process (to be used) is shown in Figure 2. The included studies' qualitative data will be evaluated descriptively. A forest plot will be used to depict the outputs of the meta-analyses. An inverted funnel plot will be used to represent publication bias, based on Egger's graphical test for publication bias. The search strategy, PRISMA-P checklist, and the quality appraisal tool will be made available as a supplement.

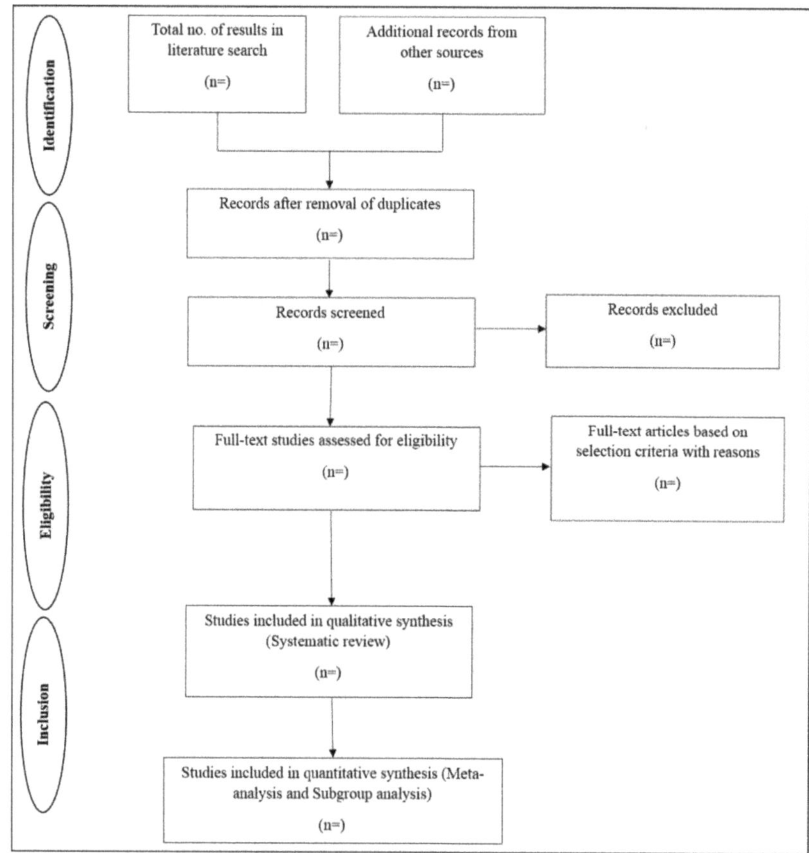

Figure 2. Flowchart for the systematic review.

2.3. Ethics and Dissemination

Because this study will not include human subjects, it will not require a formal human research ethical review or approval from a human research ethics committee. It will be carried out with publicly available anonymized data and will not need formal human study, ethical review, or permission from a human research ethics board. We want to distribute our findings by publishing them in peer-reviewed publications and discuss them in relevant conference proceedings. We also expect that the systematic review's findings will have ramifications for policy and clinical practice. We will create a policymaker-friendly summary in a validated style, which we will share via social media and email discussion groups.

2.4. Strengths and Limitations of This Study

This protocol will help researchers carry out systematic reviews and meta-analyses of the randomized data obtained from various research studies.

- PRISMA-P (Preferred Reporting Items for Systematic Reviews and Meta-Analyses Protocol) recommendations are followed in the protocol.

- It will help researchers make informed decisions, due to specific evidence obtained from organized data.
- This study will help us obtain a clear picture of the role of miRNAs on chemoresistance for melanoma patients.
- Certain forms of data obtained from various literature may be challenging to incorporate due to statistical error and, hence, may hamper the outcome.

3. Discussion

Previous research has found a relationship between miRNA expression and melanoma prognosis; however, little is known about miRNAs' prognostic value in melanoma.

There has never been a thorough investigation or meta-analysis of the function of miRNA in chemoresistance melanoma. The statistical accounts of the risk factors associated with this disease can only be uncovered by investigating more studies associated with miRNA expression in melanoma patients. The studies analyzed through this protocol can help determine the relationship between chemoresistance and patient survival. Usually, the clinical studies reported are confined to a limited population, over a short period. Hence, this protocol for systematic review and meta-analysis could provide an organized overview of the role of chemoresistance-specific miRNA expression in melanoma. The results obtained using this protocol will aid the physician's ability to make an informed decision and would result in a better quality of life for melanoma patients.

This study could provide reliable and productive results which may help in further research. The proposed protocol would build upon available studies highlighting the significance of miRNAs in effecting chemoresistance and sensitivity. Any extrapolations, unless specified in the protocol, are not recommended.

Author Contributions: Conceptualization, R.J.; methodology, R.J. and P.S. (Peter Shaw); resources, P.S. (Palanisamy Sivanandy) and S.K.; writing—original draft preparation, P.S. (Peter Shaw), G.R., K.S.T., S.B., R.R.M., S.K.G., H.C.C., P.S. (Palanisamy Sivanandy), M.R., S.S., S.K. and R.J.; writing—review and editing, P.S. (Peter Shaw), G.R., K.S.T., S.B., R.R.M., S.K.G., H.C.C., P.S. (Palanisamy Sivanandy), M.R., S.S., S.K. and R.J.; visualization, P.S. (Peter Shaw), G.R., K.S.T., S.B., R.R.M., S.K.G., H.C.C., P.S. (Palanisamy Sivanandy), M.R., S.S., S.K. and R.J.; supervision, R.J.; project administration, R.J. All authors have read and agreed to the published version of the manuscript.

Funding: Funded in part by the National Institutes of Health (NIH)/National Institute of Dental and Cra-niofacial Research (NIDCR) R01DE028105 grant to S.K.; P.S. was supported in part by the Jiangsu province, China; H.C.C. was supported by Deanship of Scientific Research, King Khalid University, Abha, Saudi Arabia; Funding #G.R.P.2/27/40.

Institutional Review Board Statement: Not applicable.

Informed Consent Statement: Not applicable.

Data Availability Statement: Not applicable.

Acknowledgments: The authors acknowledge Michael Borenstein's Meta-Analysis Concepts and Applications workshop manual for his recommendations on reporting meta-analysis, subgroup analysis, and publication bias (www.meta-analysis-workshops.com).

Conflicts of Interest: The authors declare no conflict of interest.

Disclaimer: The writers developed this protocol as part of their usual work at their individual institutions. However, the opinions stated herein are solely those of the authors and do not represent the views of their affiliated organizations.

Abbreviations

miRNA: MicroRNA; BCC: basal cell carcinoma; SCC: squamous cell carcinoma; R interface for 'H2O'(R H2O package), artificial intelligence (AI), principal component analysis (PCA), a clustering algorithm (such as k-means), HR: hazard ratio; CI: confidence interval; NOS: overall survival (OS), Disease-free survival (DFS) and other endpoint measures Newcastle-Ottawa scale; PRISMA: Preferred

Reporting Items for Systematic Review and Meta-analysis; PRISMA-P: Preferred Reporting Items for Systematic Review and Meta-analysis protocol; ICD: International Classification of Disease; CMA: Comprehensive Meta-Analysis.

References

1. Ferlay, J. *GLOBOCAN 2000. Cancer Incidence, Mortality and Prevalence Worldwide*; Version 1.0. IARC Cancerbase; IARC: Lyon, France, 2001.
2. Franceschi, S.; Levi, F.; Randimbison, L.; La Vecchia, C. Site distribution of different types of skin cancer: New aetiological clues. *Int. J. Cancer* **1996**, *67*, 24–28. [CrossRef]
3. Godbole, V.; Toprani, H.; Shah, H. Skin cancer in Saurashtra. *Indian J. Pathol. Bacteriol.* **1968**, *11*, 183.
4. Sung, H.; Ferlay, J.; Siegel, R.L.; Laversanne, M.; Soerjomataram, I.; Jemal, A.; Bray, F. Global cancer statistics 2020: GLOBOCAN estimates of incidence and mortality worldwide for 36 cancers in 185 countries. *CA A Cancer J. Clin.* **2021**, *71*, 209–249. [CrossRef]
5. Chang, D.T.; Amdur, R.J.; Morris, C.G.; Mendenhall, W.M. Adjuvant radiotherapy for cutaneous melanoma: Comparing hypofractionation to conventional fractionation. *Int. J. Radiat. Oncol. Biol. Phys.* **2006**, *66*, 1051–1055. [CrossRef] [PubMed]
6. Armstrong, B.K.; Kricker, A. The epidemiology of UV induced skin cancer. *J. Photochem. Photobiol. B Biol.* **2001**, *63*, 8–18. [CrossRef]
7. Haluska, F.G.; Hodi, F.S. Molecular genetics of familial cutaneous melanoma. *J. Clin. Oncol.* **1998**, *16*, 670–682. [CrossRef] [PubMed]
8. Koh, H.K.; Kligler, B.E.; Lew, R.A. Sunlight and cutaneous malignant melanoma: Evidence for and against causation. *Photochem. Photobiol.* **1990**, *51*, 765–779. [CrossRef]
9. Kerbel, R.S. A cancer therapy resistant to resistance. *Nature* **1997**, *390*, 335. [CrossRef]
10. Wagle, N.; Emery, C.; Berger, M.F.; Davis, M.J.; Sawyer, A.; Pochanard, P.; Kehoe, S.M.; Johannessen, C.M.; MacConaill, L.E.; Hahn, W.C.; et al. Dissecting therapeutic resistance to RAF inhibition in melanoma by tumor genomic profiling. *J. Clin. Oncol.* **2011**, *29*, 3085. [CrossRef]
11. Sharma, K.; Mohanti, B.K.; Rath, G.K. Malignant melanoma: A retrospective series from a regional cancer center in India. *J. Cancer Res. Ther.* **2009**, *5*, 173. [CrossRef]
12. Müller, D.; Bosserhoff, A. Integrin β 3 expression is regulated by let-7a miRNA in malignant melanoma. *Oncogene* **2008**, *27*, 6698. [CrossRef]
13. Dar, A.A.; Majid, S.; de Semir, D.; Nosrati, M.; Bezrookove, V.; Kashani-Sabet, M. miRNA-205 suppresses melanoma cell proliferation and induces senescence via regulation of E2F1 protein. *J. Biol. Chem.* **2011**, *286*, 16606–16614. [CrossRef]
14. Pencheva, N.; Tran, H.; Buss, C.; Huh, D.; Drobnjak, M.; Busam, K.; Tavazoie, S.F. Convergent multi-miRNA targeting of ApoE drives LRP1/LRP8-dependent melanoma metastasis and angiogenesis. *Cell* **2012**, *151*, 1068–1082. [CrossRef] [PubMed]
15. Madhav, M.R.; Nayagam, S.G.; Biyani, K.; Pandey, V.; Kamal, D.G.; Sabarimurugan, S.; Ramesh, N.; Gothandam, K.M.; Jayaraj, R. Epidemiologic analysis of breast cancer incidence, prevalence, and mortality in India: Protocol for a systematic review and meta-analyses. *Medicine (Baltimore)* **2018**, *97*, e13680. [CrossRef] [PubMed]
16. Poddar, A.; Aranha, R.R.; Muthukaliannan, G.K.; Nachimuthu, R.; Jayaraj, R. Head and neck cancer risk factors in India: Protocol for systematic review and meta-analysis. *BMJ Open* **2018**, *8*, e020014. [CrossRef]
17. Jayaraj, R.; Kumarasamy, C.; Piedrafita, D. Systematic review and meta-analysis protocol for Fasciola DNA vaccines. *Online J. Vet. Res.* **2018**, *22*, 517.
18. Kim, W.T.; Kim, W.-J. MicroRNAs in prostate cancer. *Prostate Int.* **2013**, *1*, 3–9. [CrossRef]
19. Lin Teoh, S.; Das, S. The role of MicroRNAs in diagnosis, prognosis, metastasis and resistant cases in breast cancer. *Curr. Pharm. Des.* **2017**, *23*, 1845–1859. [CrossRef]
20. Zhang, W.; Zhou, J.; Zhu, X.; Yuan, H. MiR-126 reverses drug resistance to TRAIL through inhibiting the expression of c-FLIP in cervical cancer. *Gene* **2017**, *627*, 420–427. [CrossRef]
21. Zhang, Y.; Wang, J. MicroRNAs are important regulators of drug resistance in colorectal cancer. *Biol. Chem.* **2017**, *398*, 929–938. [CrossRef]
22. Yang, W.; Ma, J.; Zhou, W.; Cao, B.; Zhou, X.; Yang, Z.; Zhang, H.; Zhao, Q.; Fan, D.; Hong, L. Molecular mechanisms and theranostic potential of miRNAs in drug resistance of gastric cancer. *Expert Opin. Ther. Targets* **2017**, *21*, 1063–1075. [CrossRef]
23. Haefliger, S.; Hudson, A.; Hayes, S.; Pavlakis, N.; Howell, V. P2. 01-012 Acquired Chemotherapy Resistance in vitro: miRNA Profiles of Chemotherapy Resistant Squamous Lung Cancer Cell Lines: Topic: Analysis of RNA. *J. Thorac. Oncol.* **2017**, *12*, S790–S791. [CrossRef]
24. Zhuang, Z.; Hu, F.; Hu, J.; Wang, C.; Hou, J.; Yu, Z.; Wang, T.T.; Liu, X.; Huang, H. MicroRNA-218 promotes cisplatin resistance in oral cancer via the PPP2R5A/Wnt signaling pathway. *Oncol. Rep.* **2017**, *38*, 2051–2061. [CrossRef]
25. Tung, S.L.; Huang, W.C.; Hsu, F.C.; Yang, Z.P.; Jang, T.H.; Chang, J.W.; Chuang, C.M.; Lai, C.R.; Wang, L.H. miRNA-34c-5p inhibits amphiregulin-induced ovarian cancer stemness and drug resistance via downregulation of the AREG-EGFR-ERK pathway. *Oncogenesis* **2017**, *6*, e326. [CrossRef]
26. Amponsah, P.S.; Fan, P.; Bauer, N.; Zhao, Z.; Gladkich, J.; Fellenberg, J.; Herr, I. microRNA-210 overexpression inhibits tumor growth and potentially reverses gemcitabine resistance in pancreatic cancer. *Cancer Lett.* **2017**, *388*, 107–117. [CrossRef]

27. Armstrong, C.M.; Liu, C.; Lou, W.; Lombard, A.P.; Evans, C.P.; Gao, A.C. MicroRNA-181a promotes docetaxel resistance in prostate cancer cells. *Prostate* **2017**, *77*, 1020–1028. [CrossRef]

28. Fattore, L.; Sacconi, A.; Mancini, R.; Ciliberto, G. MicroRNA-driven deregulation of cytokine expression helps development of drug resistance in metastatic melanoma. *Cytokine Growth Factor Rev.* **2017**, *36*, 39–48. [CrossRef]

29. Joyce, K.M. Surgical management of melanoma. *Exon Publ.* **2017**, 91–100.

30. Grossman, D.; Altieri, D.C. Drug resistance in melanoma: Mechanisms, apoptosis, and new potential therapeutic targets. *Cancer Metastasis Rev.* **2001**, *20*, 3–11. [CrossRef]

31. Royam, M.M.; Kumarasamy, C.; Baxi, S.; Gupta, A.; Ramesh, N.; Muthukaliannan, G.K.; Jayaraj, R. Current Evidence on miRNAs as Potential Theranostic Markers for Detecting Chemoresistance in Colorectal Cancer: A Systematic Review and Meta-Analysis of Preclinical and Clinical Studies. *Mol. Diagn. Ther.* **2019**, *23*, 65–82. [CrossRef]

32. Wang, W.; Li, J.; Zhu, W.; Gao, C.; Jiang, R.; Li, W.; Hu, Q.; Zhang, B. MicroRNA-21 and the clinical outcomes of various carcinomas: A systematic review and meta-analysis. *BMC Cancer* **2014**, *14*, 819. [CrossRef]

33. Nair, V.S.; Maeda, L.S.; Ioannidis, J.P. Clinical outcome prediction by microRNAs in human cancer: A systematic review. *J. Natl. Cancer Inst.* **2012**, *104*, 528–540. [CrossRef]

34. Jayawardana, K.; Schramm, S.J.; Tembe, V.; Mueller, S.; Thompson, J.F.; Scolyer, R.A.; Mann, G.J.; Yang, J. Identification, review, and systematic cross-validation of microRNA prognostic signatures in metastatic melanoma. *J. Investig. Dermatol.* **2016**, *136*, 245–254. [CrossRef]

35. Mocellin, S.; Verdi, D.; Nitti, D. DNA repair gene polymorphisms and risk of cutaneous melanoma: A systematic review and meta-analysis. *Carcinogenesis* **2009**, *30*, 1735–1743. [CrossRef]

36. Kalal, B.S.; Upadhya, D.; Pai, V.R. Chemotherapy resistance mechanisms in advanced skin cancer. *Oncol. Rev.* **2017**, *11*, 326. [CrossRef]

37. Lu, J.; Getz, G.; Miska, E.A.; Alvarez-Saavedra, E.; Lamb, J.; Peck, D.; Sweet-Cordero, A.; Ebert, B.L.; Mak, R.H.; Ferrando, A.A.; et al. MicroRNA expression profiles classify human cancers. *Nature* **2005**, *435*, 834. [CrossRef]

38. Iorio, M.V.; Visone, R.; Di Leva, G.; Donati, V.; Petrocca, F.; Casalini, P.; Taccioli, C.; Volinia, S.; Liu, C.G.; Alder, H.; et al. MicroRNA signatures in human ovarian cancer. *Cancer Res.* **2007**, *67*, 8699–8707. [CrossRef]

39. Ceylan, C.; Oztürk, G.; Alper, S. Non-Melanoma Skin Cancers between the Years of 1990 and 1999 in Izmir, Turkey: Demographic and Clinicopathological Characteristics. *J. Dermatol.* **2003**, *30*, 123–131. [CrossRef]

40. Stroup, D.F.; Berlin, J.A.; Morton, S.C.; Olkin, I.; Williamson, G.D.; Rennie, D.; Moher, D.; Becker, B.J.; Sipe, T.A.; Thacker, S.B. Meta-analysis of observational studies in epidemiology: A proposal for reporting. *JAMA* **2000**, *283*, 2008–2012. [CrossRef]

41. Wells, G.; Shea, B.; O'connell, D.; Peterson, J.; Welch, V.; Losos, M.; Tugwell, P. *The Newcastle-Ottawa Scale (NOS) for Assessing the Quality of Nonrandomised Studies in Meta-Analyses*; Ottawa Hospital Research Institute: Ottawa, ON, Canada, 2009.

42. Kumarasamy, C.; Devi, A.; Jayaraj, R. Prognostic value of microRNAs in head and neck cancers: A systematic review and meta-analysis protocol. *Syst Rev.* **2018**, *7*, 150. [CrossRef]

43. Jayaraj, R.; Kumarasamy, C. Systematic review and meta-analysis of cancer studies evaluating diagnostic test accuracy and prognostic values: Approaches to improve clinical interpretation of results. *Cancer Manag. Res.* **2018**, *10*, 4669–4670. [CrossRef]

44. Jayaraj, R.; Kumarasamy, C.; Ramalingam, S.; Devi, A. Systematic review and meta-analysis of risk-reductive dental strategies for medication related osteonecrosis of the jaw among cancer patients: Approaches and strategies. *Oral Oncol.* **2018**, *85*, 15–23. [CrossRef]

45. Sabarimurugan, S.; Royam, M.M.; Das, A.; Das, S.; Gothandam, K.M.; Jayaraj, R. Systematic Review and Meta-analysis of the Prognostic Significance of miRNAs in Melanoma Patients. *Mol. Diagn Ther.* **2018**, *22*, 653–669. [CrossRef] [PubMed]

46. Jayaraj, R.; Kumarasamy, C.; Madhav, M.R.; Pandey, V.; Sabarimurugan, S.; Ramesh, N.; Gothandam, K.M.; Baxi, S. Systematic Review and Meta-Analysis of Diagnostic Accuracy of miRNAs in Patients with Pancreatic Cancer. *Dis. Markers* **2018**, *2018*, 6904569. [CrossRef]

47. Jayaraj, R.; Kumarasamy, C. Prognostic biomarkers for oral tongue squamous cell carcinoma: A systematic review and meta-analysis. *Br. J. Cancer* **2018**, *118*, e11. [CrossRef]

48. Jayaraj, R.; Kumarasamy, C. Survival for HPV-positive oropharyngeal squamous cell carcinoma with surgical versus non-surgical treatment approach: A systematic review and meta-analysis. *J. Oral. Oncol.* **2018**, *90*, 137–138. [CrossRef]

49. Jayaraj, R.; Kumarasamy, C.; Sabarimurugan, S.; Baxi, S. Commentary: Blood-Derived microRNAs for Pancreatic Cancer Diagnosis: A Narrative Review and Meta-Analysis. *Front. Physiol.* **2018**, *9*, 1896. [CrossRef]

50. Jayaraj, R.; Kumarasamy, C. Conceptual interpretation of analysing and reporting of results on systematic review and meta-analysis of optimal extent of lateral neck dissection for well-differentiated thyroid carcinoma with metastatic lateral neck lymph nodes. *Oral Oncol.* **2019**, *89*, 153–154. [CrossRef] [PubMed]

51. Begg, C.B.; Mazumdar, M. Operating characteristics of a rank correlation test for publication bias. *Biometrics* **1994**, *50*, 1088–1101. [CrossRef]

52. Duval, S.; Tweedie, R. Trim and fill: A simple funnel-plot–based method of testing and adjusting for publication bias in meta-analysis. *Biometrics* **2000**, *56*, 455–463. [CrossRef] [PubMed]

53. Orwin, R.G. A fail-safe N for effect size in meta-analysis. *J. Educ. Stat.* **1983**, *8*, 157–159. [CrossRef]

54. Jayaraj, R.; Kumarasamy, C.; Gothandam, K.M. Letter to the editor "Prognostic value of microRNAs in colorectal cancer: A meta-analysis". *Cancer Manag Res.* **2018**, *10*, 3501–3503. [CrossRef]

55. Jayaraj, R.; Kumarasamy, C. Letter to the Editor about the Article:" Performance of different imaging techniques in the diagnosis of head and neck cancer mandibular invasion: A systematic review and meta-analysis". *J. Oncol.* **2018**, *89*, 159–160. [CrossRef]

56. Jayaraj, R.; Kumarasamy, C.; Sabarimurugan, S.; Baxi, S. Letter to the Editor in response to the article," The epidemiology of oral human papillomavirus infection in healthy populations: A systematic review and meta-analysis". *Oral Oncol.* **2018**, *84*, 121–122. [CrossRef]

57. Jayaraj, R.; Kumarasamy, C.; Samiappan, S.; Swaminathan, P. Letter to the Editor regarding, The prognostic role of PD-L1 expression for survival in head and neck squamous cell carcinoma: A systematic review and meta-analysis. *Oral Oncol.* **2018**, *77*, 92–97. [CrossRef]

58. Jayaraj, R.; Kumarasamy, C.; Madurantakam Royam, M.; Devi, A.; Baxi, S. Letter to the editor: Is HIF-1alpha a viable prognostic indicator in OSCC? A critical review of a meta-analysis study. *World J. Surg. Oncol.* **2018**, *16*, 111. [CrossRef]

59. Higgins, J.P.; Thompson, S.G.; Deeks, J.J.; Altman, D.G. Measuring inconsistency in meta-analyses. *BMJ Br. Med. J.* **2003**, *327*, 557. [CrossRef]

60. Cochran, W.G. The combination of estimates from different experiments. *Biometrics* **1954**, *10*, 101–129. [CrossRef]

61. Borenstein, M.; Hedges, L.V.; Higgins, J.P.; Rothstein, H.R. *Introduction to Meta-Analysis*; John Wiley & Sons: Hoboken, NJ, USA, 2011.

62. Renehan, A.G.; Zwahlen, M.; Minder, C.; T O'Dwyer, S.; Shalet, S.M.; Egger, M.J.T.L. Insulin-like growth factor (IGF)-I, IGF binding protein-3, and cancer risk: Systematic review and meta-regression analysis. *Lancet* **2004**, *363*, 1346–1353. [CrossRef]

63. Sterne, J.A.; Jüni, P.; Schulz, K.F.; Altman, D.G.; Bartlett, C.; Egger, M. Statistical methods for assessing the influence of study characteristics on treatment effects in 'meta-epidemiological' research. *Stat. Med.* **2002**, *21*, 1513–1524. [CrossRef] [PubMed]

64. Thompson, S.G.; Sharp, S.J. Explaining heterogeneity in meta-analysis: A comparison of methods. *Stat. Med.* **1999**, *18*, 2693–2708. [CrossRef]

65. Ben-Dor, A.; Shamir, R.; Yakhini, Z. Clustering gene expression patterns. *J. Comput. Biol.* **1999**, *6*, 281–297. [CrossRef] [PubMed]

66. Jay, J.J.; Eblen, J.D.; Zhang, Y.; Benson, M.; Perkins, A.D.; Saxton, A.M.; Voy, B.H.; Chesler, E.J.; Langston, M.A. *A Systematic Comparison of Genome-Scale Clustering Algorithms*; Springer: Berlin/Heidelberg, Germany, 2012; Volume 13.

67. Abu-Khzam, F.N. On the complexity of multi-parameterized cluster editing. *J. Discret. Algorithms* **2017**, *45*, 26–34. [CrossRef]

68. Dehne, F.; Langston, M.A.; Luo, X.; Pitre, S.; Shaw, P.; Zhang, Y. *The Cluster Editing Problem: Implementations and Experiments*; Springer: Berlin/Heidelberg, Germany, 2006; pp. 13–24.

69. Abu-Khzam, F.N.; Egan, J.; Gaspers, S.; Shaw, A.; Shaw, P. *Cluster Editing with Vertex Splitting*; Springer: Berlin/Heidelberg, Germany, 2018; pp. 1–13.

70. Barr, J.; Shaw, P. *AI Application to Data Analysis, Automatic File Processing*; IEEE: Piscataway, NJ, USA, 2018; pp. 100–105.

71. Landry, M.; Angela, B. *Machine Learning with R and H_2O*; H2O.ai, Inc.: Mountain View, CA, USA, 2018.

72. Shamseer, L.; Moher, D.; Clarke, M.; Ghersi, D.; Liberati, A.; Petticrew, M.; Shekelle, P.; Stewart, L.A. Preferred reporting items for systematic review and meta-analysis protocols (PRISMA-P) 2015: Elaboration and explanation. *BMJ* **2015**, *349*, g7647. [CrossRef]

73. Moher, D.; Liberati, A.; Tetzlaff, J.; Altman, D.G.; Prisma Group. Preferred reporting items for systematic reviews and meta-analyses: The PRISMA statement. *PLoS Med.* **2009**, *6*, e1000097. [CrossRef] [PubMed]

genes

Protocol

Clinical Theragnostic Relationship between Chemotherapeutic Resistance, and Sensitivity and miRNA Expressions in Head and Neck Cancers: A Systematic Review and Meta-Analysis Protocol

Peter Shaw [1,2], Greg Raymond [3], Raghul Senthilnathan [4], Chellan Kumarasamy [5], Siddhartha Baxi [6], Deepa Suresh [7], Sameep Shetty [8], Ravishankar Ram M [9], Harish C. Chandramoorthy [10], Palanisamy Sivanandy [11,12], Suja Samiappan [13], Mogana Rajagopal [9], Sunil Krishnan [14] and Rama Jayaraj [15,*]

[1] Department of Artificial Intelligence, Nanjing University of Information Science and Technology (NUIST), Nanjing 210044, China; 100001@nuist.edu.cn

[2] Menzies School of Health Research, Darwin, NT 0810, Australia

[3] Northern Territory Medical Program, CDU Campus, Flinders University, Ellengowan Drive, Darwin, NT 0909, Australia; greg.raymond@flinders.edu.au

[4] School of Biosciences and Technology, Vellore Institute of Technology (VIT), Vellore 632014, India; Raghulsnn@gmail.com

[5] Kumarasamy School of Health and Medical Sciences, Curtin University, Perth, WA 6102, Australia; chellank54@gmail.com

[6] MBBS, FRANZCR GAICD (SB), Genesis Care Gold Coast Radiation Oncologist, Southport, QLD 4224, Australia; Siddhartha.baxi@genesiscare.com

[7] Division of Endocrinology, Department of Internal Medicine, Mayo Clinic Florida, Jacksonville, FL 32224, USA; Deepa.Suresh@mayo.edu

[8] Department of Oral and Maxillofacial Surgery, Manipal College of Dental Sciences, Mangalore, Manipal Academy of Higher Education, A Constituent of MAHE, Manipal 576104, India; sameep.shetty@manipal.edu

[9] Department of Pharmaceutical Biology, Faculty of Pharmaceutical Sciences, UCSI University Kuala Lumpur (South Wing), No.1, Jalan Menara Gading, UCSI Heights 56000 Cheras, Kuala Lumpur 57000, Malaysia; ravishankar@ucsiuniversity.edu.my (R.R.M.); mogana@ucsiuniversity.edu.my (M.R.)

[10] Department of Microbiology and Clinical Parasitology, College of Medicine, King Khalid University, Abha 61421, Saudi Arabia; hshkonda@kku.edu.sa

[11] Department of Pharmacy Practice, School of Pharmacy, International Medical University, Bukit Jalil, Kuala Lumpur 57000, Malaysia; PalanisamySivanandy@imu.edu.my

[12] School of Postgraduate Studies, International Medical University, Bukit Jalil, Kuala Lumpur 57000, Malaysia

[13] Department of Biochemistry, Bharathiar University, Coimbatore 641046, India; sujaramalingam08@gmail.com

[14] Department of Radiation Oncology, Mayo Clinic Florida, 4500 San Pablo Road S., Jacksonville, FL 32224, USA; krishnan.sunil@mayo.edu

[15] Northern Territory Institute of Research and Training, Darwin, NT 0909, Australia

* Correspondence: jramamoorthi@gmail.com

Citation: Shaw, P.; Raymond, G.; Senthilnathan, R.; Kumarasamy, C.; Baxi, S.; Suresh, D.; Shetty, S.; Ram M, R.; Chandramoorthy, H.C.; Sivanandy, P.; et al. Clinical Theragnostic Relationship between Chemotherapeutic Resistance, and Sensitivity and miRNA Expressions in Head and Neck Cancers: A Systematic Review and Meta-Analysis Protocol. *Genes* **2021**, *12*, 2029. https://doi.org/10.3390/genes12122029

Academic Editors: Giuseppe Iacomino and Fabio Lauria

Received: 18 October 2021
Accepted: 18 December 2021
Published: 20 December 2021

Publisher's Note: MDPI stays neutral with regard to jurisdictional claims in published maps and institutional affiliations.

Abstract: Background: The microRNAs (miRNAs) are small noncoding single-stranded RNAs typically 19–25 nucleotides long and regulated by cellular and epigenetic factors. These miRNAs plays important part in several pathways necessary for cancer development, an altered miRNA expression can be oncogenic or tumor-suppressive. Recent experimental results on miRNA have illuminated a different perspective of the molecular pathogenesis of head and neck cancers. Regulation of miRNA can have a detrimental effect on the efficacy of chemotherapeutic drugs in both neoadjuvant and adjuvant settings. This miRNA-induced chemoresistance can influence the prognosis and survival rate. The focus of the study is on how regulations of various miRNA levels contribute to chemoresistance in head and neck cancer (HNC). Recent findings suggest that up or down-regulation of miRNAs may lead to resistance towards various chemotherapeutic drugs, which may influence the prognosis. **Methods:** Studies on miRNA-specific chemoresistance in HNC were collected through literary (bibliographic) databases, including SCOPUS, PubMed, Nature, Elsevier, etc., and were systematically reviewed following PRISMA-P guidelines (Preferred Reporting Items for Systematic

Review and Meta-analysis Protocol). We evaluated various miRNAs, their up and downregulation, the effect of altered regulation on the patient's prognosis, resistant cell lines, etc. The data evaluated will be represented in the form of a review and meta-analysis. **Discussion:** This meta-analysis aims to explore the miRNA-induced chemoresistance in HNC and thus to aid further researches on this topic. PROSPERO registration: **CRD42018104657.**

Keywords: head and neck cancer; miRNA; chemoresistance; protocol; systematic review; hazard ratio; patient survival; up-regulation; down-regulation

1. Introduction

Head and Neck Cancers is a collective term for cancers that usually originate in the squamous cell lining of the mucosal membrane of the inner mouth, nose, and throat [1]. Micro RNAs are a class of small, endogenous, noncoding, 18–24 nucleotide long sequences that play a crucial role in many processes essential for cancer progression, including cell death, proliferation, metastasis, and treatment resistance [2–6]. Recent studies show that chemoresistance to specific chemotherapeutic drugs such as Cisplatin, Doxorubicin, 5-fluorouracil, etc., is caused due to varying regulation of specific miRNAs [7]. These miRNAs can reduce the efficacy of drugs or even make cancer cells utterly impervious to the drug [8–24]. Such resistance can significantly worsen the prognosis of HNC patients and can also reduce survival rates. Adjuvant therapy is rendered inefficient, and patients show poor responses towards it. This study evaluates the chemoresistance due to miRNA at various stages of HNC.

HNC ranks as the sixth most prevalent type of cancer in the world. It is a heterogeneous disease with various categories based on anatomical location, etiology, and molecular characteristics [25,26]. Head and neck cancer accounts for more than 931,000 cases and 467,000 deaths annually worldwide [27]. Studies have shown that males are affected significantly more than females, with a ratio ranging from 2:1 to 4:1 [28,29]. Head and neck squamous cell carcinomas (HNSCCs) occur in the oral cavity, pharynx, or larynx and account for the bulk of HNCs [26]. Excessive use of tobacco and alcohol, as well as infection with the human papillomavirus (for oropharyngeal cancer) and Epstein-Barr virus, are all etiological factors in the development of HNSCCs (for nasopharyngeal cancer) [26]. Malignant proliferation and chemoresistance continue to be the limiting factors in HNC treatment, which leads to loco-regional relapse or distant metastasis [30,31]. Current therapy options for HNC include surgery, radiotherapy, chemotherapy, and most recently, anti-EGFR antibody treatment [32].

The focus of this study protocol is to describe the quantitative and qualitative effects of miRNA-specific chemoresistance in head and neck cancer. The primary objective of this systematic review is to explore the influence of miRNA expressions in head and neck carcinoma patients by assimilating and evaluating the evidence. The secondary objective of this proposed study is to assess the deranged regulation of miRNAs and evaluate their resistance in cell lines which may cause recurrence.

2. Rationale

Most of the publications on miRNA-specific chemoresistance of HNC are particular to evaluating the effects of specific miRNA. Published studies have been specific to the samples collected in a hospital related to a particular geographical region and were localized to that area. This systematic review and meta-analysis aims to integrate the data obtained by the publications as mentioned earlier and present it in a peer-reviewed and extensively evaluated fashion. This systematic review has been proposed after a thorough evaluation of studies and publications from various countries and hence is not restricted to the studies performed in a specific region.

This study will evaluate the data collected on various miRNAs, focusing on how their regulation can affect the sensitivity of a chemotherapeutic drug. The reviewers hope the meta-analysis will help researchers get a better perspective on how miRNA is linked to chemoresistance towards various chemotherapeutic drugs. The comprehensive meta-analysis provides quantitative synthesis of currently existing studies on miRNA specific chemoresistance in HNA patients.

This study collates risk factors from various studies and provides a broad overview of miRNA-induced chemoresistance.

Review Questions

This systematic review protocol aims to explain the methodological approaches implemented in conducting this meta-analysis on miRNA Specific Chemoresistance in HNC. The study proposes the following questions:

1. What effect does miRNA regulation have on chemotherapy?
2. What is the general prognosis of patients having miRNA-specific chemoresistance?
3. What are the miRNAs most responsible for chemoresistance in HNC patients?
4. What are the survival rates associated with each miRNA linked to chemoresistance, and how are they affected?
5. How does the prognosis vary for different miRNA causing the chemoresistance?
6. What is the future of miRNA as an active form of cancer treatment for Head and Neck Cancer?

3. Materials and Methods

This study aims to evaluate the effect of miRNA expression in HNC patients, reporting this particular kind of resistance worldwide on prognosis. The research protocol will be conducted in accordance with the Preferred Reporting Items for Systematic Reviews and Meta-Analyses (PRISMA) guidelines [33]. This study aims to detect and analyze the miRNA-specific chemoresistance in head and neck carcinoma. This study is neither confined to any datasets nor a particular region.

3.1. Search Methods

The study clearly defines the effect of miRNA expression in HNC patients and the altered up or down-regulation of miRNA in the prognosis of patients. This systematic review will also include those studies examining miRNA specific chemoresistance in HNC patients. This study will utilize a comprehensive search strategy based on the keywords listed in Table 1. The study will begin with an initial limited search of online bibliographic databases such as EMBASE, PubMed, Elsevier, Science Direct, SCOPUS, Nature, and Web of Science. This search will be then be expanded to include the words contained in the title and abstract in the papers considered in the analysis. There will be no restrictions on study participants in terms of age, gender, ethnicity, country of origin, and morbidities (for patients and the general population). The search will be limited to articles published between 2000 and 2021 to focus mainly on recent advancements in research outcomes in this field. The next level of the search will be a detailed manual search of the full-text articles to gather and retrieve all the required information for the systematic review from the bibliographic articles.

The reviewers will discard any full-text studies that do not meet the inclusion criteria. Finally, a manual search of selected articles will be conducted to extract more research from references. If there are any differences among the reviewers, they will be resolved through discussion or with the help of an unbiased third reviewer.

Table 1. A sample keyword search strategy.

S No.	Search Items
1.	"miRNA" AND "treatment" OR drug resistance" AND "HNC" OR "Head and Neck Cancer"
2.	"microRNA" AND "drug resistance" AND "HNC" OR "Head and Neck Cancer"
3.	"Up-regulation OR down-regulation in HNC" OR "Differential Expression" OR "Deregulated miRNAs "OR "Head and Neck Cancer"
4.	"miRNA" AND "chemotherapeutic resistance" OR "chemosensitivity" AND "HNC" OR "Head and Neck Cancer"
5.	"miRNA" AND "treatment resistance" OR "chemoresistance" AND "HNC" OR "Head and Neck Cancer"
6.	"microRNA" AND "chemosensitivity" AND "HNC" OR "Head and Neck Cancer"
7.	"microRNA" AND "chemoresistance" AND "HNC" OR "Head and Neck Cancer"
8.	"HNC survival outcome" OR "Hazard Ratio" AND "HNC" OR "Head and Neck Cancer"

3.2. Selection Criteria

3.2.1. Inclusion Criteria

- Studies analyzing the effect of miRNA expressions in both HNC patients and cell lines will be considered.
- Studies analyzing miRNAs and resistance/are performed in liquid biopsies (plasma, saliva....) will be included.
- Studies that discuss the clinicopathological characteristics of HNC patients along with hazard ratio or Kaplan–Meier curve will be included.
- Studies reporting resistance in HNC will be included.
- Articles that discuss the survival outcomes of almost all stages of HNC patients will be included in the meta-analysis.
- Studies reporting miRNA profiling platform and miRNA expressions analysis using in vitro assays will be included.
- Studies that differentiate between 3p and 5p in the microRNAs expressions in HNC
- Genes and/or pathways involved in chemoresistance or sensitivity in HNC patients will also be considered.
- miRNA expression analysis, HR, and associated 95% CI or Kaplan–Meier (KM) curve is required for the eligible studies.
- Studies appropriate to PRISMA guidelines for systematic review and meta-analysis will be included.

3.2.2. Exclusion Criteria

- Studies published in languages other than English.
- Any information or results from letters to the editors, case studies, conference abstracts, case reports, and review articles of HNC will be removed.
- Studies performed only in vitro will be excluded and will not be considered for the systematic review.
- Studies in which proper discussion about miRNA profiling and pathways related to that are not available will be excluded.
- Studies with no accessibility to survival outcomes, HR values, or Kaplan–Meier (KM) curve will not be considered for the meta-analysis.
- Studies using patients' information from datasets or cancer registries will be removed.
- Studies whose full texts are not accessible will also be excluded.
- Duplicates will be removed, and the study will be excluded if it falls within the exclusion criteria.

3.2.3. Participants

The systematic review and meta-analysis will add studies involving patients suffering from all types of HNC. Participants with evidently established diagnoses of HNC will be included. There will be no restrictions on study participants in terms of age, gender, ethnicity, country of origin, and morbidities (for patients and the general population).

3.3. Data Collection and Management

The data collected by the reviewers will be saved in a Microsoft Excel spreadsheet. The data will be integrated with pertinent material, with repeats removed and reviewed later. Only full-text articles that meet the inclusion and exclusion criteria will be retrieved. The following data will be collected from the selected papers:

- Author name and information
- Type of head or neck cancer
- Date and journal of publication
- Type of micro-RNA studied
- Hazard Ratio and 95% Confidence Interval
- Patients origin (by country)
- Type/origin of Samples collected for analysis
- The total number of samples used
- Cell lines and the pathways affected
- Clinical stage of the affected HNC patients and their details
- Type of chemoresistant cells and chemoresistant drug
- Micro-RNA profiling platform

Data Items Included in this Study

- Characteristics of study material (author names, geographical area of study, type of study, year of publication)
- Characteristics of study participants (country of origin, clinical stage of cancer, type of expressed miRNA, drug to which resistance is expressed)
- Characteristics of study methods and results (miRNA profiling platform, number of samples, statistical analysis)

3.4. Study Selection Process

The reviewers will initially analyze abstracts and titles retrieved through the primary search approach against the selection criteria individually. Full re-reports will be obtained for any titles that appear to fulfill the study selection criteria. Second, the reviewers will examine the full-text articles to acquire research-related information and address any issues about the eligibility of the selected articles. Any study that fails to fulfill the required selection criteria will be excluded. The writers will additionally look for extra material in the references of the chosen publications to ensure that all of the necessary information for the study is gathered and nothing vital is overlooked.

Any divisive viewpoint will be settled through discussion. The reviewers will record the reasons for excluding studies. Neither of the review authors will be blind to the journal titles or the study authors and their respective institutions. The quality appraisal factors listed below will be assessed, graded, and documented: the author's name, sample size, patient age, and gender, country of origin, disease stage, miRNA expression, study period, survival outcome, miRNA profiling platform, gene/pathways related, statistical data, drugs or chemotherapy, resistant cell lines, outcome variables, and other factors. The search results will be published following PRISMA criteria. A PRISMA flow chart will be utilized to summarize the selection procedure used to filter through the studies initially gathered (Figure 1).

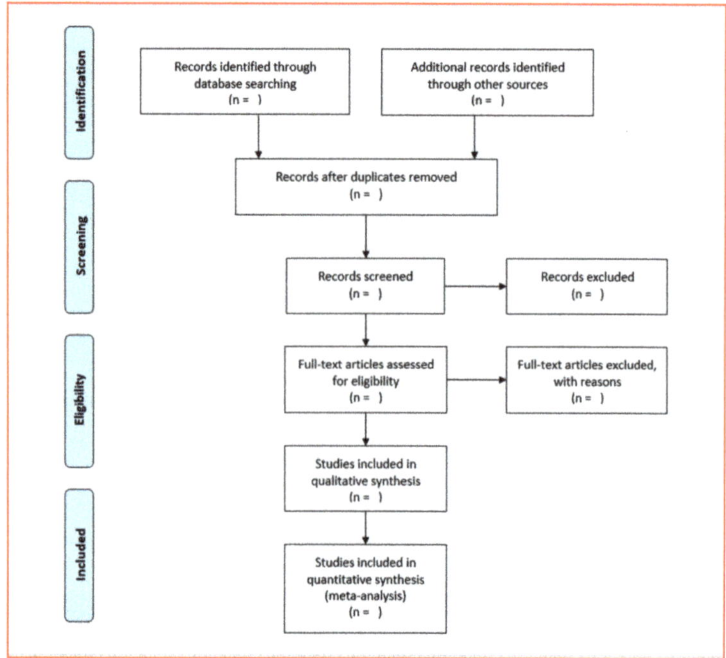

Figure 1. Schematic representation of the selection process.

3.5. Study Outcomes

Primary outcome: The study's primary outcome is to evaluate the miRNA expression in the head and neck cancer patients, gender standardized prevalence of the disease, and the survival outcome rate linked with HNC patients.

Secondary outcome: The study's secondary outcome is to analyze the effect of miRNA-specific chemoresistance in HNC patients. It also aims to investigate the survival rates associated with each miRNA linked to chemoresistance and how they are affected.

3.6. Mitigating Risk of Bias in Individual Studies

The National Heart, Lung, and Blood Institute's (NHLBI) quality evaluation tool for observational and cross-sectional studies will be utilized to assess the quality of selected studies [34]. This tool will be used to rate all full-text articles classified as good, fair, or bad. The risk of bias in the studies will be assessed by reviewers. Each reviewer will work independently and should have the needed expertise to determine the validity of the studies collected based on factors such as the number of samples collected, the number of patients, the miRNA profiling platforms used, the year of publication, the duration of the study, and geographical area analysis. Other considerations will be evaluated if necessary.

3.7. Data Synthesis

Each author will gather information from several databases. The data collected will be collated and sent to the reviewers following data collection. Then, data will be extracted and tabulated to determine the type of survival (OS, DFS, DSS, PS, and LCR), type of miRNA, HR, and CI. The collected data will be utilized to create statistical results for all of the studies included in the meta-analysis, such as the pooled HR and CI. The statistical data for the study will be generated using the 'Comprehensive Meta-Analysis Soft-ware.' The miRNA will be analyzed and classified based on expression (up-regulation and down-regulation).

3.8. Meta-Analysis and Subgroup Analysis

Meta-analyses for this particular study will be conducted using "Comprehensive Meta-Analysis V.3.0" software for the HR and 95% CI values obtained from the relevant materials collected for this study. The pooled HR value as an estimated effect size provides more clinical utility as it examines the survival pattern of the patients of the included studies. Heterogeneity will be calculated using Tau Square, Cochran's Q test [35] and I^2 statistic [36]. These parameters increased robustness in analysis of between study heterogenicity. Additionally, Z statistics will be used to calculate the heterogeneity.

The I^2 statistic will be used to assess the degree of heterogeneity between studies, with an I^2 value more than 50% indicating considerable heterogeneity. The meta-analysis will use a random or fixed effects model based on the heterogeneity. A P value of less than 0.01 will be considered statistically significant for the Q test. In the meta-analysis, the z-test will also be used to estimate how many standard deviations each research can deviate from the study mean. To detect heterogeneity, the Eggers bias indicator test will be performed [37]. Quality assessment and statistical analysis would be performed [34]. CMA will be used to compute the pooled HR and 95% CI.

Publications bias will be evaluated using Orwin and classic fail-safe N test (demonstrates the likelihood that studies are absent from the current meta-analysis and these studies if included in the analysis, would shift the Hazard Ratio of the included studies toward the null) [38], Egger's bias indicator test (gold standard regression test), Begg and Mazumdar Rank collection test (defines the estimated or computed Tau between Hazard Ratio and standard error), Duval and Tweedie's trim and fill calculation (explores the missing studies that likely to fall, adds them to the analysis and then recommutes the pooled HR) [39], and inverted funnel plot. The inclusion of these publication bias indicators will explain the possible publication bias from small or missing studies.

Subgroup analysis will be performed considering location or origin of the reported incidence, anatomical sites of cancer growth, variation and resistance to HNC among different age groups and different genders. Subgroup analysis will be performed on all miRNAs found to be differentially expressed in the studies. Different miRNA expression patterns in HNC will be investigated, as will their influence on the development of chemoresistance. Subgroup analysis will be performed based on the studies that differentiate between 3p and 5p in the microRNAs expressions in HNC.

3.9. Reporting of the Review

The PRISMA 2015 guidelines will be followed when reporting and communicating the results of this study. A flow diagram indicating the selection process of selected studies will be made. These results also include detailed inclusion and exclusion criteria to aid in the selection process of the appropriate material. A search strategy will also be included to identify the required data. The collected data such as HR and 95% CI values will be used to create Forest plots. Risk of bias, Publication bias, and Heterogeneity tests will be calculated to evaluate the validity and accuracy of the data collected.

4. Discussion

The pathophysiology underlying chemotherapy and polydrug resistance in humans involving polymorphic miRNAs play a vital role in regulating apoptosis, DNA patch-up, and epithelial-mesenchymal transition cell cycle regulation.

This study focuses on evaluating the effect of miRNA on chemoresistance in HNC. It aims to determine the effect of the regulation of miRNA on chemoresistance. The collected HR and 95% CI will be used to create Forest plots. The outcome will be used to determine the effect of deregulation of miRNA on the prognosis of patients diagnosed with HNC. This systematic review and meta-analysis will provide a better understanding of the effects of miRNA on chemoresistance and prognosis.

Resistance to chemotherapeutic drugs is known to drastically deteriorate the prognosis of patients who show miRNA-specific chemoresistance. Many miRNAs can symbolize the growth of cancer by acting like biomarkers. For example, some miRNA from the 'let 7-g' family are thought to act as tumor suppressors, and an up-regulation of this miRNA can be used as a prognostic biomarker for detecting cancers [40].

The miRNA creates acquired chemoresistance by silencing genes/pathways that are directly or indirectly linked with the action of the chemotherapeutic drug. For example, in a study by Martz et al., the authors demonstrated that activation of the mitogen-activated protein kinase (RAS-MAPK), Notch-1, phosphoinositide 3-kinase (PI3K), and mammalian target of rapamycin (mTOR), PI3K/AKT, and estrogen receptor (ER) signaling pathways induced resistance in a selection of different drugs [41]. MiRNA dysregulation has an important role in modulating the principal mechanisms that induce HNC drug resistance that are currently known. Abnormal miRNA expression can disrupt the expression levels of multiple genes or important cellular pathways, which has a direct impact on the creation of chemotherapy resistance in HNC [42]. A study by Qin et al. clearly validates how the over expression of a miRNA interrupts the prevailing pathways and induce chemoresistance by directing the tumor suppressor genes [17]. As a result, it is critical to investigate the role of miRNA expression in controlling the prevalent pathways developing chemo resistance. It remains elusive to determine if the miRNA signature established for HNC can be replicated in future studies or for different tumor entities.

5. Ethics and Dissemination

This protocol was designed in accordance with the PRISMA-P guidelines. This study will be carried out using publically accessible data and will not include any human volunteers. As a result, no formal human research ethics committee assessment is required. Our findings will be published in peer-reviewed publications and conference proceedings. Furthermore, this study aims to provide a publicly reviewed standard for the systematic review with the expectation of maintaining standards.

Author Contributions: R.J. predominantly conceived this review and P.S. (Peter Shaw) and P.S. (Palanisamy Sivanandy) led the development of the letter to the editor. R.J., G.R., R.S., C.K., S.B., D.S., S.S. (Sameep Shetty), R.R.M., H.C.C., P.S. (Peter Shaw) and P.S. (Palanisamy Sivanandy), S.S. (Suja Samiappan), M.R. and S.K. wrote the first draft of the letter, and critically revised and edited successive drafts of the manuscript. All authors have read and agreed to the published version of the manuscript.

Funding: Funded in part by the National Institutes of Health (NIH)/National Institute of Dental and Craniofacial Research (NIDCR) R01DE028105 grant to S.K. Peter Shaw was supported in part by the Jiangsu province, China, 100 Talent project fund (BX2020100) and Double Innovation grant, Jiangsu (JSSCR2021520). Deanship of Scientific Research, King Khalid University, Abha, Saudi Arabia. Funding # G.R.P. 2/27/40.

Institutional Review Board Statement: Not applicable.

Informed Consent Statement: Not applicable.

Data Availability Statement: Not applicable.

Conflicts of Interest: The authors declare no conflict of interest.

References

1. Head and Neck Cancers-National Cancer Institute. Available online: https://www.cancer.gov/types/head-and-neck/head-neck-fact-sheet (accessed on 29 October 2021).
2. Nakashima, H.; Yoshida, R.; Hirosue, A.; Kawahara, K.; Sakata, J.; Arita, H.; Yamamoto, T.; Toya, R.; Murakami, R.; Hiraki, A. Circulating miRNA-1290 as a potential biomarker for response to chemoradiotherapy and prognosis of patients with advanced oral squamous cell carcinoma: A single-center retrospective study. *Tumor Biol.* **2019**, *41*, 1010428319826853. [CrossRef]
3. Yang, F.; Li, Y.; Xu, L.; Zhu, Y.; Gao, H.; Zhen, L.; Fang, L. miR-17 as a diagnostic biomarker regulates cell proliferation in breast cancer. *Onco Targets Ther.* **2017**, *10*, 543. [CrossRef]
4. Wu, S.-G.; Chang, T.-H.; Liu, Y.-N.; Shih, J.-Y. MicroRNA in lung cancer metastasis. *Cancers* **2019**, *11*, 265. [CrossRef] [PubMed]
5. Chan, S.-H.; Wang, L.-H. Regulation of cancer metastasis by microRNAs. *J. Biomed. Sci.* **2015**, *22*, 9. [CrossRef]

6. Shirjang, S.; Mansoori, B.; Asghari, S.; Duijf, P.H.; Mohammadi, A.; Gjerstorff, M.; Baradaran, B. MicroRNAs in cancer cell death pathways: Apoptosis and necroptosis. *Free Radic. Biol. Med.* **2019**, *139*, 1–15. [CrossRef] [PubMed]

7. Rolfo, C.; Fanale, D.; Hong, D.S.; Tsimberidou, A.M.; Piha-Paul, S.A.; Pauwels, P.; Van Meerbeeck, J.P.; Caruso, S.; Bazan, V.; Cicero, G. Impact of microRNAs in resistance to chemotherapy and novel targeted agents in non-small cell lung cancer. *Curr. Pharm. Biotechnol.* **2014**, *15*, 475–485. [CrossRef] [PubMed]

8. Berania, I.; Cardin, G.B.; Clément, I.; Guertin, L.; Ayad, T.; Bissada, E.; Nguyen-Tan, P.F.; Filion, E.; Guilmette, J.; Gologan, O. Four PTEN-targeting co-expressed mi RNA s and ACTN 4-targeting mi R-548b are independent prognostic biomarkers in human squamous cell carcinoma of the oral tongue. *Int. J. Cancer* **2017**, *141*, 2318–2328. [CrossRef]

9. Jin, Y.; Li, Y.; Wang, X.; Yang, Y. Dysregulation of MiR-519d affects oral squamous cell carcinoma invasion and metastasis by targeting MMP3. *J. Cancer* **2019**, *10*, 2720. [CrossRef]

10. Zhao, Y.; Wang, P.; Wu, Q. miR-1278 sensitizes nasopharyngeal carcinoma cells to cisplatin and suppresses autophagy via targeting ATG2B. *Mol. Cell. Probes* **2020**, *53*, 101597. [CrossRef] [PubMed]

11. Hamano, R.; Miyata, H.; Yamasaki, M.; Kurokawa, Y.; Hara, J.; ho Moon, J.; Nakajima, K.; Takiguchi, S.; Fujiwara, Y.; Mori, M. Overexpression of miR-200c induces chemoresistance in esophageal cancers mediated through activation of the Akt signaling pathway. *Clin. Cancer Res.* **2011**, *17*, 3029–3038. [CrossRef]

12. Li, B.; Hong, P.; Zheng, C.-C.; Dai, W.; Chen, W.-Y.; Yang, Q.-S.; Han, L.; Tsao, S.W.; Chan, K.T.; Lee, N.P.Y. Identification of miR-29c and its target FBXO31 as a key regulatory mechanism in esophageal cancer chemoresistance: Functional validation and clinical significance. *Theranostics* **2019**, *9*, 1599. [CrossRef] [PubMed]

13. Shi, J.; Bao, X.; Liu, Z.; Zhang, Z.; Chen, W.; Xu, Q. Serum miR-626 and miR-5100 are promising prognosis predictors for oral squamous cell carcinoma. *Theranostics* **2019**, *9*, 920. [CrossRef] [PubMed]

14. Sannigrahi, M.K.; Sharma, R.; Singh, V.; Panda, N.K.; Rattan, V.; Khullar, M. Role of host miRNA hsa-mir-139-3p in hpv-16–induced carcinomas. *Clin. Cancer Res.* **2017**, *23*, 3884–3895. [CrossRef] [PubMed]

15. Hess, A.-K.; Müer, A.; Mairinger, F.D.; Weichert, W.; Stenzinger, A.; Hummel, M.; Budach, V.; Tinhofer, I. MiR-200b and miR-155 as predictive biomarkers for the efficacy of chemoradiation in locally advanced head and neck squamous cell carcinoma. *Eur. J. Cancer* **2017**, *77*, 3–12. [CrossRef] [PubMed]

16. Just, C.; Knief, J.; Lazar-Karsten, P.; Petrova, E.; Hummel, R.; Röcken, C.; Wellner, U.; Thorns, C. MicroRNAs as potential biomarkers for chemoresistance in adenocarcinomas of the esophagogastric junction. *J. Oncol.* **2019**, *2019*, 4903152. [CrossRef]

17. Qin, X.; Guo, H.; Wang, X.; Zhu, X.; Yan, M.; Wang, X.; Xu, Q.; Shi, J.; Lu, E.; Chen, W. Exosomal miR-196a derived from cancer-associated fibroblasts confers cisplatin resistance in head and neck cancer through targeting CDKN1B and ING5. *Genome Biol.* **2019**, *20*, 12. [CrossRef] [PubMed]

18. Guan, G.-F.; Zhang, D.-J.; Wen, L.-J.; Xin, D.; Liu, Y.; Yu, D.-J.; Su, K.; Zhu, L.; Guo, Y.-Y.; Wang, K. Overexpression of lncRNA H19/miR-675 promotes tumorigenesis in head and neck squamous cell carcinoma. *Int. J. Med Sci.* **2016**, *13*, 914. [CrossRef]

19. Song, J.; Zhang, N.; Cao, L.; Xiao, D.; Ye, X.; Luo, E.; Zhang, Z. Down-regulation of miR-200c associates with poor prognosis of oral squamous cell carcinoma. *Int. J. Clin. Oncol.* **2020**, *25*, 1072–1078. [CrossRef]

20. Yu, E.-H.; Tu, H.-F.; Wu, C.-H.; Yang, C.-C.; Chang, K.-W. MicroRNA-21 promotes perineural invasion and impacts survival in patients with oral carcinoma. *J. Chin. Med Assoc.* **2017**, *80*, 383–388. [CrossRef]

21. Tanaka, K.; Miyata, H.; Sugimura, K.; Fukuda, S.; Kanemura, T.; Yamashita, K.; Miyazaki, Y.; Takahashi, T.; Kurokawa, Y.; Yamasaki, M. miR-27 is associated with chemoresistance in esophageal cancer through transformation of normal fibroblasts to cancer-associated fibroblasts. *Carcinogenesis* **2015**, *36*, 894–903. [CrossRef] [PubMed]

22. Arantes, L.M.R.B.; Laus, A.C.; Melendez, M.E.; de Carvalho, A.C.; Sorroche, B.P.; De Marchi, P.R.M.; Evangelista, A.F.; Scapulatempo-Neto, C.; de Souza Viana, L.; Carvalho, A.L. MiR-21 as prognostic biomarker in head and neck squamous cell carcinoma patients undergoing an organ preservation protocol. *Oncotarget* **2017**, *8*, 9911. [CrossRef] [PubMed]

23. Chang, Y.-C.; Jan, C.-I.; Peng, C.-Y.; Lai, Y.-C.; Hu, F.-W.; Yu, C.-C. Activation of microRNA-494-targeting Bmi1 and ADAM10 by silibinin ablates cancer stemness and predicts favourable prognostic value in head and neck squamous cell carcinomas. *Oncotarget* **2015**, *6*, 24002. [CrossRef] [PubMed]

24. Tu, H.-F.; Liu, C.-J.; Chang, C.-L.; Wang, P.-W.; Kao, S.-Y.; Yang, C.-C.; Yu, E.-H.; Lin, S.-C.; Chang, K.-W. The association between genetic polymorphism and the processing efficiency of miR-149 affects the prognosis of patients with head and neck squamous cell carcinoma. *PLoS ONE* **2012**, *7*, e51606. [CrossRef]

25. Klussmann, J.P. Head and neck cancer-new insights into a heterogeneous disease. *Oncol. Res. Treat.* **2017**, *40*, 318–319. [CrossRef] [PubMed]

26. Ahmad, P.; Sana, J.; Slávik, M.; Gurín, D.; Radová, L.; Gablo, N.A.; Kazda, T.; Smilek, P.; Horáková, Z.; Gal, B. MicroRNA-15b-5p predicts locoregional relapse in head and neck carcinoma patients treated with intensity-modulated radiotherapy. *Cancer Genom. Proteom.* **2019**, *16*, 139–146. [CrossRef]

27. Sung, H.; Ferlay, J.; Siegel, R.L.; Laversanne, M.; Soerjomataram, I.; Jemal, A.; Bray, F. Global cancer statistics 2020: GLOBOCAN estimates of incidence and mortality worldwide for 36 cancers in 185 countries. *CA Cancer J. Clin.* **2021**, *71*, 209–249. [CrossRef]

28. Iarc, W. GLOBOCAN 2012: Estimated Cancer Incidence, Mortality and Prevalence Worldwide in 2012. 2012. Available online: https://gco.iarc.fr/ (accessed on 15 September 2021).

29. Lambert, R.; Sauvaget, C.; de Camargo Cancela, M.; Sankaranarayanan, R. Epidemiology of cancer from the oral cavity and oropharynx. *Eur. J. Gastroenterol. Hepatol.* **2011**, *23*, 633–641. [CrossRef]

30. Shen, D.-W.; Pouliot, L.M.; Hall, M.D.; Gottesman, M.M. Cisplatin resistance: A cellular self-defense mechanism resulting from multiple epigenetic and genetic changes. *Pharmacol. Rev.* **2012**, *64*, 706–721. [CrossRef]

31. Forastiere, A.A.; Ang, K.; Brizel, D.; Brockstein, B.E.; Dunphy, F.; Eisele, D.W.; Goepfert, H.; Hicks, W.L.; Kies, M.S.; Lydiatt, W.M. Head and neck cancers: Clinical practice guidelines. *JNCCN J. Natl. Compr. Cancer Netw.* **2005**, *3*, 316–391.

32. Zibelman, M.; Mehra, R. Overview of current treatment options and investigational targeted therapies for locally advanced squamous cell carcinoma of the head and neck. *Am. J. Clin. Oncol.* **2016**, *39*, 396–406. [CrossRef]

33. Moher, D.; Shamseer, L.; Clarke, M.; Ghersi, D.; Liberati, A.; Petticrew, M.; Shekelle, P.; Stewart, L.A. Preferred reporting items for systematic review and meta-analysis protocols (PRISMA-P) 2015 statement. *Syst. Rev.* **2015**, *4*, 1. [CrossRef]

34. Health NIo. Quality Assessment Tool for Observational Cohort and Cross-Sectional Studies: National Heart, Lung, and Blood Institute. Available online: https://www.nhlbi.nih.gov/health-topics/study-quality-assessment-tools (accessed on 10 December 2021).

35. Cochran, W.G. The combination of estimates from different experiments. *Biometrics* **1954**, *10*, 101–129. [CrossRef]

36. Higgins, J.P.; Thompson, S.G.; Deeks, J.J.; Altman, D.G. Measuring inconsistency in meta-analyses. *BMJ* **2003**, *327*, 557–560. [CrossRef]

37. Egger, M.; Smith, G.D.; Schneider, M.; Minder, C. Bias in meta-analysis detected by a simple, graphical test. *BMJ* **1997**, *315*, 629–634. [CrossRef]

38. Orwin, R.G. A fail-safe N for effect size in meta-analysis. *J. Educ. Stat.* **1983**, *8*, 157–159. [CrossRef]

39. Duval, S.; Tweedie, R. A nonparametric "trim and fill" method of accounting for publication bias in meta-analysis. *J. Am. Stat. Assoc.* **2000**, *95*, 89–98.

40. Biamonte, F.; Santamaria, G.; Sacco, A.; Perrone, F.M.; Di Cello, A.; Battaglia, A.M.; Salatino, A.; Di Vito, A.; Aversa, I.; Venturella, R. MicroRNA let-7g acts as tumor suppressor and predictive biomarker for chemoresistance in human epithelial ovarian cancer. *Sci. Rep.* **2019**, *9*, 5668. [CrossRef] [PubMed]

41. Magee, P.; Shi, L.; Garofalo, M. Role of microRNAs in chemoresistance. *Ann. Transl. Med.* **2015**, *3*, 332. [PubMed]

42. Dai, F.; Dai, L.; Zheng, X.; Guo, Y.; Zhang, Y.; Niu, M.; Lu, Y.; Li, H.; Hou, R.; Zhang, Y. Non-coding RNAs in drug resistance of head and neck cancers: A review. *Biomed. Pharmacother.* **2020**, *127*, 110231. [CrossRef] [PubMed]

MDPI
St. Alban-Anlage 66
4052 Basel
Switzerland
Tel. +41 61 683 77 34
Fax +41 61 302 89 18
www.mdpi.com

Genes Editorial Office
E-mail: genes@mdpi.com
www.mdpi.com/journal/genes